PARTICLE INTERACTIONS AT HIGH ENERGIES

Earlier Volumes in this Series

1. *Dispersion Relations*, edited by G. R. Screaton, M.A., PH.D.
(Scottish Universities' Summer School, 1960)

2. *Fluctuation, Relaxation and Resonance in Magnetic Systems*, edited by D. ter Haar, M.A., DR.SC.
(Scottish Universities' Summer School, 1961)

3. *Polarons and Excitons*, edited by C. G. Kuper, B.SC., M.A., PH.D. and G. D. Whitfield, M.A., PH.D.
(Scottish Universities' Summer School, 1962)

4. *Strong Interactions and High Energy Physics*, edited by R. G. Moorhouse
(Scottish Universities' Summer School, 1963)

5. *Nuclear Structure and Electromagnetic Interaction*, edited by N. MacDonald, B.SC., PH.D.
(Scottish Universities' Summer School, 1964)

6. *Phonons in Perfect Lattices and in Lattices with Point Imperfections*, edited by R. W. H. Stevenson, B.SC., PH.D.
(Scottish Universities' Summer School, 1965)

PARTICLE INTERACTIONS AT HIGH ENERGIES

Scottish Universities' Summer School
1966

Edited by

T. W. PREIST, B.Sc., Ph.D.
Department of Mathematical Physics
University of Edinburgh, Scotland

and

L. L. J. VICK, B.Sc., Ph.D.
Department of Mathematical Physics
University of Edinburgh, Scotland

PLENUM PRESS
NEW YORK
1967

Published in the U.S.A. by
PLENUM PRESS
a division of PLENUM PUBLISHING CORPORATION
227 West 17th Street, New York, N. Y. 10011

Library of Congress Catalog Number: 67-26706

Printed in Great Britain by
Robert Cunningham and Sons Ltd
Alva, Scotland

PREFACE

The seventh Scottish Universities' Summer School in Physics was held at Newbattle Abbey, near Edinburgh, from 1st to 20th August. Like the first and fourth annual Schools this School was devoted to topics in High Energy Physics. In this field the School is now in competition with many new successful Schools established since 1960, but nevertheless we had strong and early evidence that there was a widespread demand for a continuation of the triennial series of High Energy Schools. This early impression was amply confirmed by the number of applications eventually received—96, of which we were able to accept 55, even though detailed arrangements for the School were exceptionally delayed this year.

The delays were related to a serious problem that faced the Governing Committee of the School for the first time. In the past the generous support of the N.A.T.O. programme for Advanced Study Institutes was offered early enough for the organisers to be assured of adequate financial backing at the start of planning. This year, however, some understandable hesitation in view of the number of competing Schools put N.A.T.O. support in question at a critical state in preliminary organization. Somewhat later the N.A.T.O. authorities did proffer support, still at a generous level, but in the intervening period it had been necessary to contemplate organizing a School on a reduced scale, financed entirely by the Scottish Universities.

As Director of the School it is my pleasant duty to acknowledge that the four ancient Universities of Aberdeen, Edinburgh, Glasgow and St. Andrews, in spite of grave financial stringency, readily sprang to the aid of the Summer Schools Committee, by offering sufficient emergency support to allow continued planning of the School at essentially the normal level. When subsequently assurance of considerable N.A.T.O. support was restored, it became possible to reduce the call on University funds.

While it is clear that present financial conditions will not allow the Scottish Universities to bear entirely the expense of future Schools, it may be hoped that the general pattern eventually established this year—with N.A.T.O. and the Scottish Universities spending comparable sums—may be permanently retained. This (as well as the task of running the School) may be made easier by the fact that the number of Scottish Universities is about to be doubled and it is already understood that the newer Universities are interested both in participation and in future planning.

The scientific programme of the School was also only finalized after some delays—partly caused by the events outlined above and partly by a genuine uncertainty in 1965 as to what topics would be ripe for discussion in August

1966. In the event the choice made—high energy scattering, electromagnetic interactions and current algebras, and weak interactions, proved a satisfactory one; many participants have commented on the particular aptness for their special interests of the choice made.

It is usual for the Director of the School to express his indebtedness to the other officials of the School. This year other unexpected duties made my own share in organization far less than a Director's should be and I have a particularly strong reason to extend my thanks to Drs D. J. Candlin (Secretary), J. Cumming (Treasurer) and T. W. Preist and L. L. J. Vick (Joint Editors and Stewards) for making the school the organizational success it certainly was. They have been most ably assisted in their clerical work before, during and after the School by Mrs R. W. Chester, who was also responsible for a large part of the work involved in preparing lecture notes, and by many other members of the Tait Institute. During the School Misses Jane McGhee and Molly Keys ably assisted in the Conference office. I am sure that all participants in the School will join me in expressing their sincere thanks to all these willing helpers, as well as to all the members of the Staff of Newbattle Abbey, who contributed so essentially to the smooth and pleasant running of the School.

N. Kemmer
Tait Professor of Mathematical Physics,
University of Edinburgh

EDITORS' NOTE

The lecture notes were prepared by the lecturers themselves before the Summer School and revised, in some cases, during and shortly after the course of the School in order to incorporate changes produced by new experimental information. The galley proofs, but not the page proofs, were checked by the lecturers and so the ultimate responsibility for errors and misprints lies with the Editors.

Executive Committee of the 1966 Summer School

Professor N. Kemmer, Edinburgh, *Chairman.*
Dr J. Cumming, Glasgow, *Treasurer.*
Dr D. J. Candlin, Edinburgh, *Secretary.*
Professor J. C. Gunn, Glasgow.
Professor G. Bishop, Glasgow.
Dr T. W. Preist, Edinburgh, *Joint Editor.*
Dr L. L. J. Vick, Edinburgh, *Joint Editor.*

Lecturers

Dr A. Donnachie, C.E.R.N., Geneva.
Professor S. D. Drell, Stanford, California.
Dr R. J. Eden, Cambridge.
Dr P. K. Kabir, Rutherford Laboratory.
Professor L. A. Radicati, Pisa.
Professor E. J. Squires, Durham.
Professor L. Van Hove, C.E.R.N., Geneva.
Professor R. Wilson, Harvard.

Participants

Dr George R. Allcock, Liverpool.
Mr Ananda Man Singh Amatya, Edinburgh.
Mr Nicolas Antoniou, Birmingham.
Mr Jean-Eudes Augustin, Orsay.
Mr James E. Bateman, Glasgow.
Dr Grzegorz Białkowski, Warsaw.
Mr Klaus Buchner, Munich.
Mr Nigel H. Buttermore, Cambridge.
Mr Peter A. Cook, Manchester.
Mr Ronald L. Crawford, Glasgow.
Dr Arnon Dar, Rehovoth.
Dr Riccardo D'Auria, Turin.
Dr Francesco Felicetti, Frascati.
Dr Luciano Girardello, Milan.
Mr George S. Gordon, Glasgow.
Dr Peter Grassberger, Bonn.
Mr Dietrich Haase, Marburg.
Mr Anthony G. Hall, Newcastle.
Mr Bodo Hamprecht, Cambridge.
Mr Geoffrey H. Henderson, Edinburgh.
Dr Archibald W. Hendry, Rutherford Laboratory.
Dr Herbert S. Hoffman, Tel Aviv.
Mr Stephen Humble, Durham.
Mr Bruce H. Kellet, London.
Dr John M. Kennedy, Glasgow.
Mr Robert G. Kirsopp, London.
Mrs Naomi Levy, Haifa.
Mr Christopher H. Llewellyn-Smith, Oxford.
Dr David Lyth, Birmingham.
Dr Giuseppe Marchesini, Milan.
Dr Mladen Martinis, Edinburgh.
Mr Christopher Michael, Oxford.
Dr Marvin K. Moss, North Carolina.
Mr Ernesto Napolitano, Turin.
Dr Jean Pestieau, Louvain.
Dr Anthony C. Phillips, Sussex.
Mr Frank Poole, Durham.
Mr Nuthakki V. P. Rao, Manchester.
Dr Muhammad S. K. Razmi, Edinburgh.
Mr George F. Reid, Edinburgh.
Mr Gordon Ringland, Durham.
Mr Alan Ritchie, Edinburgh.
Mr Hans Rosdolsky, Michigan.
Dr Morton H. Rubin, Wisconsin.
Mr Juan J. Salazar, Paris.
Mr Vanniarachchige K. Samaranayake, London.
Dr Dieter Schildknecht, D.E.S.Y., Hamburg.

Dr Klaus Schlüpmann, C.E.R.N., Geneva.
Mr Graham Shaw, London.
Dr Robert L. Sugar, Columbia.
Mr Roger F. Templeman, Manchester.
Mr Robin W. Tucker, Cambridge.
Dr Henry D. D. Watson, Edinburgh.
Mr Trevor J. Weare, London.
Mr James N. J. White, Edinburgh.
Mr Edwin G. H. Williams, Liverpool.

CONTENTS

PAGE

PREFACE v

Executive Committee of the 1966 Summer School vii
Lecturers vii
Participants vii

1. PROPERTIES OF SCATTERING AMPLITUDES AT HIGH ENERGY *R. J. EDEN*

1. Introduction 1

2. Experimental Survey 2
 2.1. Total cross sections 3
 2.2. Real part of the forward scattering amplitude 4
 2.3. The forward peak for elastic scattering 4
 2.4. Large angle scattering 6
 2.5. High-energy reactions 7

3. Theoretical Background and Assumptions 8
 3.1. Kinematics, the optical theorem and partial wave expansions 8
 3.2. Dispersion relations and the Mandelstam representation 15
 3.3. Complex angular momenta and Regge theory 19
 3.4. Review of assumptions 23

4. Upper Bounds on Cross Sections and Other Inequalities 24
 4.1. Partial wave amplitudes 24
 4.2. Total cross sections; subtractions in dispersion relations 26
 4.3. The forward peak, its slope and higher derivatives 27
 4.4. Elastic cross sections 29
 4.5. Non-forward scattering 31

5. Lower Bounds 32
 5.1. Lower bounds for large angle scattering 32
 5.2. Lower bounds for forward scattering and total cross sections 37
 5.3. The problem of oscillations 39

6. Crossing Symmetry and Analyticity in the Energy 39
 6.1. Relations between different processes; the Phragmén-Lindelöff theorem 39
 6.2. The real part of the forward scattering amplitude 42
 6.3. Fixed transfer dispersion relations 44
 6.4. Martin's enlargement of the Lehmann ellipse 45

7. The Pomeranchuk Theorems 48
 7.1. Crossing symmetry and analyticity 48
 7.2. Asymptotic equality of total cross sections 50

	PAGE
8. Partial Wave Amplitudes	52
8.1. Bounds for large *s* or for large *l*	52
8.2. Partial wave phase shifts at high energy	55
9. Summary of Bounds	59
9.1. Upper bounds (reference section 4)	60
9.2. Lower bounds (reference section 5)	60

2\. VERY HIGH-ENERGY SCATTERING OF STRONGLY INTERACTING PARTICLES — *L. VAN HOVE*

1. Introduction and Survey of Experimental Data	63
1.1. Kinematical variables and cross-section formulae	64
1.2. Summary of experimental information	66
1.3. General remarks on theoretical methods	68
2. High-Energy Expansion of Two Particle Collision Amplitude at Finite Momentum Transfers. Regge Poles	70
2.1. Form of the expansion	70
2.2. Conservation of internal quantum numbers	72
2.3. Analyticity and crossing symmetry between *s* and *u* channels	74
2.4. Interpretation of the high-energy expansion in terms of the *t* channel	78
2.5. The Regge pole model	82
2.6. Comparison with experiment	86
3. Diffraction as Shadow Scattering	88
3.1. Unitarity condition and overlap function	88
3.2. Inequalities between cross-sections	90
3.3. The *t* dependence of the overlap function	93
4. Quark Model of High-Energy Processes	94
4.1. SU_6 symmetry, current algebras and quark model	94
4.2. Additivity assumption for scattering amplitude	96
4.3. Relations between total cross-sections	99
4.4. Differential cross-sections and consistency of the additivity assumption	101
4.5. Relation to Regge pole model and inelastic collisions in quark model	106

3\. SOME FEATURES OF ELECTROMAGNETIC INTERACTIONS — *RICHARD WILSON*

1. Introduction	112
2. Elastic Electron Scattering	112
2.1. Kinematics and definitions of form factors	112
2.2. Experiments on elastic *ep* scattering	117
2.2a. General	117
2.2b. Experimental methods	117
2.2c. Beam energy	118
2.2d. Beam direction	118
2.2e. Intensity monitor	119
2.2f. Target	119
2.2g. Solid angle	119
2.2h. Spectrometers	120
2.2i. Detectors	120
2.2j. Radiative corrections	124

PAGE

2.3. Experimental results 124
2.4. Annihilation processes 125
2.5. Threshold constraint 128

3. Neutron Form Factors 128
3.1. Neutron-electron interactions 129
3.2. Elastic *ed* scattering 130
3.3. Electron-deuteron inelastic scattering 135

4. Theory of Form Factors 140
4.1. Pion form factor 140
4.2. Nucleon isovector form factors 141
4.3. Nucleon isoscalar form factors 142
4.4. Relation to *pp* scattering 145

5. Further Experiments on Form Factors 148
5.1. Leptonic decays of vector mesons 148

6. Colliding Beam Devices 152
6.1. Colliding beams for form factors 152
6.2. Technical status of colliding beams 152
6.3. List of experiments for colliding beams 155

7. Inelastic Electron-Proton Scattering 156
7.1. Kinematics and form factors 156
7.2. Detection of electron only—multipole terms 160
7.3. Radiative corrections 162
7.3a. General 162
7.3b. Minimizing radiative background 166
7.4. Neutron-proton mass difference 167
7.5. Coincidence experiments—pion form factors 168
7.6a. Photon tagging and virtual photon tagging 170
7.6b. Illustrative numbers 173
7.6c. Electron tagging with low momentum-transfer cut-off 174
7.7. Time-reversal experiments 175

8. Two Photon Exchange Effects 176
8.1. Positron scattering 176
8.2. Hydrogen hyperfine structure 181
8.3. Unitarity 184

9. Quantum Electrodynamics Experiments 185
9.1. General 185
9.2. Lamb shift 185
9.3. *g*-factor of electron 186
9.4. Positronium triplet-singlet separation 186
9.5. Modification of propagators 186
9.6. *g*-factor of free muon 187
9.7. Electron-electron scattering 187
9.8. Electron-positron scattering 187
9.9. Electron-proton scattering 188
9.10. Electron-positron pair production 188
9.11. Excited lepton states 194

4. PRECISION TESTS OF QUANTUM ELECTRODYNAMICS AT LOW ENERGIES — *S. D. DRELL*

	PAGE
Preface	199
1. The Hyperfine Structure in Hydrogen	201
1.1. Introduction	201
1.2. Corrections to the hyperfine splitting	202
1.3. Non-relativistic model of proton structure	205
1.4. Relativistic calculation	220
1.5. Conclusions	231
2. Muon $g-2$ Value	235
2.1. Introduction	235
2.2. QED contributions	238
2.3. Non QED contributions	242

5. THE SYMMETRIES OF THE WEAK INTERACTIONS — *P. K. KABIR*

1. Introduction	248
2. The TCP Theorem	249
2.1. Consequences of TCP invariance	253
3. Superselection Rules	257
4. Space-Time Symmetries	258
4.1. Parity	258
4.2. Charge-conjugation invariance	260
4.3. Time-reversal invariance	261
5. K^0 and η Decays	264
5.1. K^0 decay	264
5.2. η decay	267
6. Internal Symmetries	269
7. $SU(3)$ Transformation Properties	275
8. Non-leptonic Decays	281
9. Algebra of Vector and Axial-Vector Charges	285

6. THE ALGEBRA ASSOCIATED WITH THE VECTOR AND AXIAL-VECTOR CURRENT — *L. A. RADICATI*

Summary	294
1. Historical Prelude	294
2. The Algebra Associated with the Hadron Vector Current	297
2.1. The $SU(3)$ algebra	297
2.2. Universality of the weak coupling	300
2.3. The isospinor vector current	302
3. The Algebra generated by the Vector and Axial-Vector Currents	305
3.1. The quark model	305

PAGE

3.2. The chiral $SU(2) \otimes SU(2)$ generated by the hadron current $h_\lambda^\pm$ 307
3.3. The Adler-Weisberger sum rule 308
3.4. Approximate saturation of the sum rule 312

4. Current Density Commutators 314
4.1. Low-energy theorems 314
4.2. The axial-vector form factor 316
4.3. Sum rule for the isovector charge radii 318

5. The Infinite-Parameter Algebra and the Dispersion Method 322
5.1. The infinite-parameter algebra associated with the current densities 322
5.2. The dispersion method of Fubini 326
5.3. Final remarks 328

7. PION-NUCLEON PHASE SHIFT ANALYSIS *A. DONNACHIE*

1. Introduction 330
2. Kinematics and Formulae 331
3. Phase Shift Analysis I 336
4. Partial Wave Dispersion Relations 341
5. Phase Shift Analysis II 349
6. Resonances 352
7. Results 358

Appendix 1. The Evaluation of Partial Wave Dispersion Relations 369
Appendix 2. Conformal Mapping Methods in Evaluating Dispersion Integrals 376

8. QUARK MODELS *EUAN J. SQUIRES*

1. Introduction 380
2. Unitary Triplets 380
3. Do Quarks Exist? 382
4. Are Quarks Fundamental? 383
5. Are Quark Models Reasonable? 384
6. Mass Shifts Due to Lower Threshold Channels 386
7. Effect of Lower Threshold Channels on Couplings 392
8. The "Long Range" Quark Model 395
9. The "Short Range" Quark Model 397
10. Further Possible Predictions of Quark Models 399

1

PROPERTIES OF SCATTERING AMPLITUDES AT HIGH ENERGY

R. J. EDEN
Cavendish Laboratory, Cambridge

1

INTRODUCTION

THE properties of scattering amplitudes at high energy can be discussed at various levels of sophistication ranging from rigorous consequences of the axioms of quantum field theory to the use of models based on less general assumptions. The aim of these lectures is to describe the principal methods that are of general validity in studying high energy behaviour. For illustration these methods will be used to derive results from properties that have been rigorously established in quantum field theory, but they will also be used to obtain results when it is assumed that the scattering amplitudes have specially simple form. I shall not however be concerned with the special features of models or their use in phenomenological calculations; both of these topics are discussed in the lecture notes of L. Van Hove in this book. (See also *High Energy Collisions of Elementary Particles* by R. J. Eden (1967)).

Much of the material for these lectures is based on the work of Andre Martin and collaborators. I shall give references in context to some of the papers by Martin and collaborators and to other authors whose work I describe, but I have not attempted to produce a comprehensive list of relevant papers so I apologise to those authors whose work is omitted from the references.

The types of experiment that are important to the study of high energy scattering are described briefly in section 2 and their main features are noted. More detailed results are given in the Proceedings of the Stony Brook Conference on High Energy Scattering 1966. In section 3 the theoretical background and principal assumptions are described.

In sections 4, 5 and 6 the general methods are described that form the main content of these lectures. They are used to discuss upper bounds on cross sections and related quantities (§ 4), lower bounds (§ 5) and consequences of crossing symmetry (§ 6). The special aspects of crossing symmetry related to the Pomeranchuk theorem are described in section 7. For partial wave amplitudes it is at present necessary to add special assumptions to obtain useful properties; these are described in section 8. Finally in section 9 I summarize the main methods and the consequent results on general properties of scattering amplitudes at high energy.

2

Experimental Survey

We shall be concerned in these lectures with theoretical properties of scattering amplitudes at high energy. These properties are directly or indirectly related to the following experiments:

1. Total cross sections $\sigma(E)$ as functions of the laboratory energy E. From the optical theorem (§ 3.1) the total cross section determines the imaginary part of the forward scattering amplitude Im $F(E, 0)$.
2. The total elastic cross section.
3. The real part of the forward scattering amplitude; this can be obtained from Im $F(E, 0)$ and extrapolation to zero t of the differential cross section

$$\frac{d\sigma}{dt} = \frac{C}{E^2} \mid F(E, t) \mid^2$$

   where t is minus the momentum transfer squared (§ 3.1). More precisely, Re $F(E, 0)$ can be obtained from Coulomb interference near $t = 0$, (§ 2.2).
4. The forward peak for elastic scattering; at 10 GeV$-$30 GeV this lies in the range $0 > t > -1{\cdot}0(\text{GeV}/c)^2$.
5. Large angle scattering; at high enough energies this is the region of fixed angle scattering, $t = cs$ where c is some constant less than one; at energies of 10-30 GeV large angle means 30° to 150°.
6. Cross sections for high-energy reactions (§ 2.5)

$$A + B \rightarrow C + D$$

   where A, B are elementary particles, and C, D are elementary particles or excited states (resonances). For each such two body reaction the main features are analogous to (2) to (5) listed above for elastic scattering. In analogy to (4) for example we include forward scattering for

$$A + B \rightarrow B + A$$

   which is backward scattering for the process

$$A + B \rightarrow A + B.$$

   Similarly charge exchange scattering is a special example of a two body reaction.

In this survey I shall not list detailed results which will rapidly become out of date but will note for each type of measurement the main features which will guide our theoretical studies. Further details are given in the Proceedings of the Stony Brook Conference 1966 and the Berkeley Conference 1966.

2.1. *Total cross sections*

The high energy behaviour of total cross sections is indicated in Fig. 2.1 taken from the report by Lindenbaum (1965).

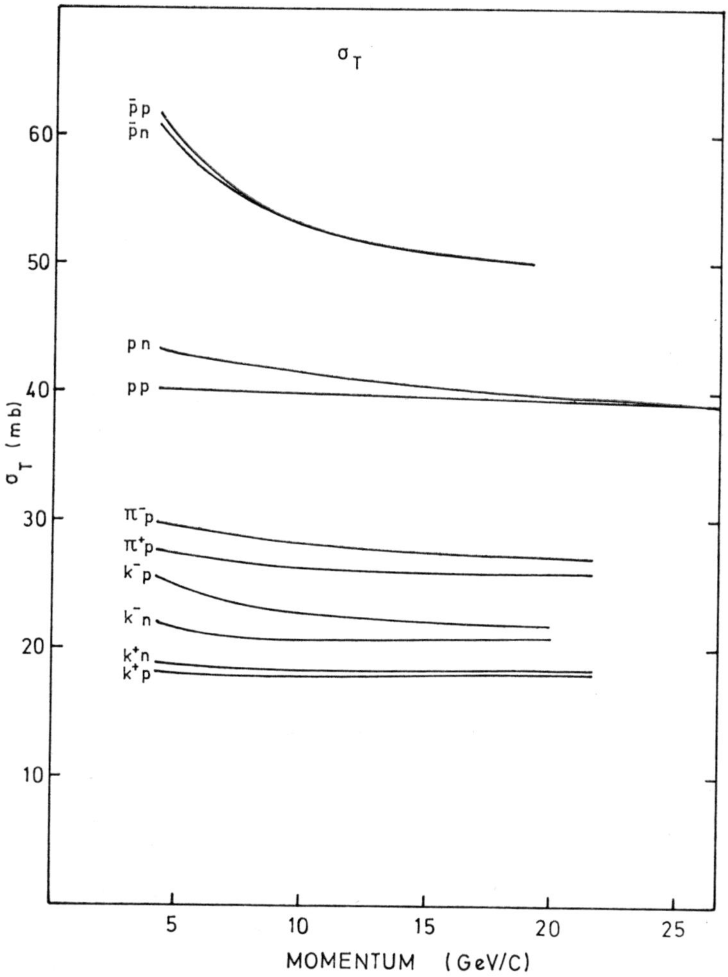

FIG. 2.1. Total cross sections at high energies, taken from the report by S. J. Lindenbaum, International Conference on High Energy Physics, Dubna (1964).

The particular points to notice are:

(i) Each total cross section decreases towards a constant value.

(ii) The total cross sections for particle-particle and for particle-antiparticle scattering are tending towards equality at high energy. Their equality is predicted by the Pomeranchuk theorem (§ 7).

(iii) There is an indication of near-equality for large E, of all baryon-baryon total cross sections at about 36mb, and of meson-baryon total cross sections at 18-24mb.

2.2. *The real part of the forward scattering amplitude*

This is measured by Coulomb interference in the near forward direction. For 10 GeV π^+p scattering this interference is strong for $0 > t > -0{\cdot}01(\text{GeV/c})^2$. For πp scattering, invariance under rotation, parity, and time reversal, shows that the scattering amplitude has the form

$$F(s, t) = F_{\text{ord}} + \boldsymbol{\sigma}\cdot(\boldsymbol{k}_{\text{inc}} \times \boldsymbol{k}_{\text{fin}}) F_{\text{sp. flip.}} \qquad \ldots[2.2.1]$$

In the forward direction the spin flip term is zero so that only F_{ord} is involved in Coulomb interference. (The situation is more complicated for pp scattering where there is a residual spin flip term at $t = 0$.) The Coulomb phase is known and the effects of $\gamma\pi$ and γN form factors can be included. This leads to the phase of the nuclear part of the scattering amplitude (defined as the total amplitude minus the amplitude for single photon exchange).

It is expected that full experimental details on Re F will be given by Lindenbaum *et al.* early in 1967. I also expect (for πp but not for pp and $p\bar{p}$) that the characteristic form of these results will be as indicated in Fig. 2.2 which is partly based on earlier work by Lindenbaum and collaborators (Foley *et al.* 1965a).

2.3. *The forward peak for elastic scattering*

The measurement of the forward peak for energies above 7 GeV and for $0 > t > -1{\cdot}0(\text{GeV/c})^2$ was carried out by Lindenbaum and collaborators (Foley *et al.* 1965b). Their results are summarized by the formula (see also Fig. 2.3.)

$$\frac{d\sigma}{dt} = \left(\frac{d\sigma}{dt}\right)_{(t=0)} \exp(at). \qquad \ldots[2.3.1]$$

(i) The coefficient a is of the order $10(\text{GeV/c})^{-2}$ but its precise value depends on the process being considered.

(ii) There is very little variation of a with energy for πp scattering, but for pp scattering it increases like $\log E$. The simple one pole approximation (Pomeron exchange) of Regge theory predicts $a \propto \log E$ for small t. It is evident that this is not a good approximation at energies up to 30 GeV for elastic scattering.

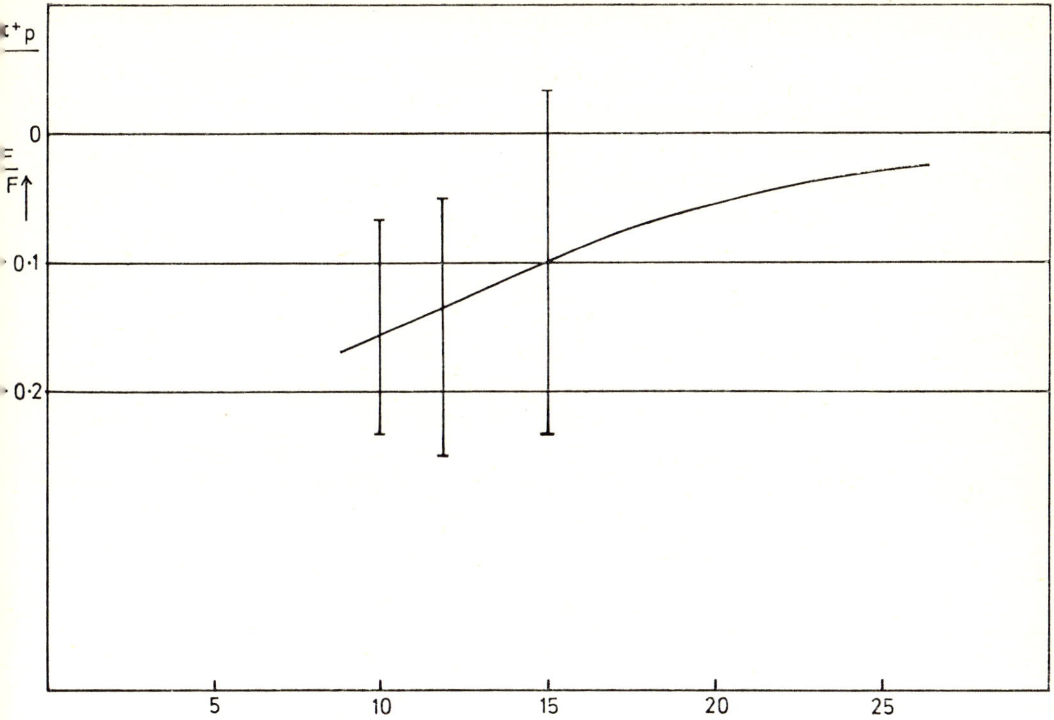

FIG. 2.2*a*

Re $F(s, 0)$ / Im $F(s, 0)$ taken from Foley *et al.*, *Phys. Rev. Lett.* (1965) **14**, 863. The continuous curve is estimated from the theoretical discussion in section 6.

FIG. 2.2*b*

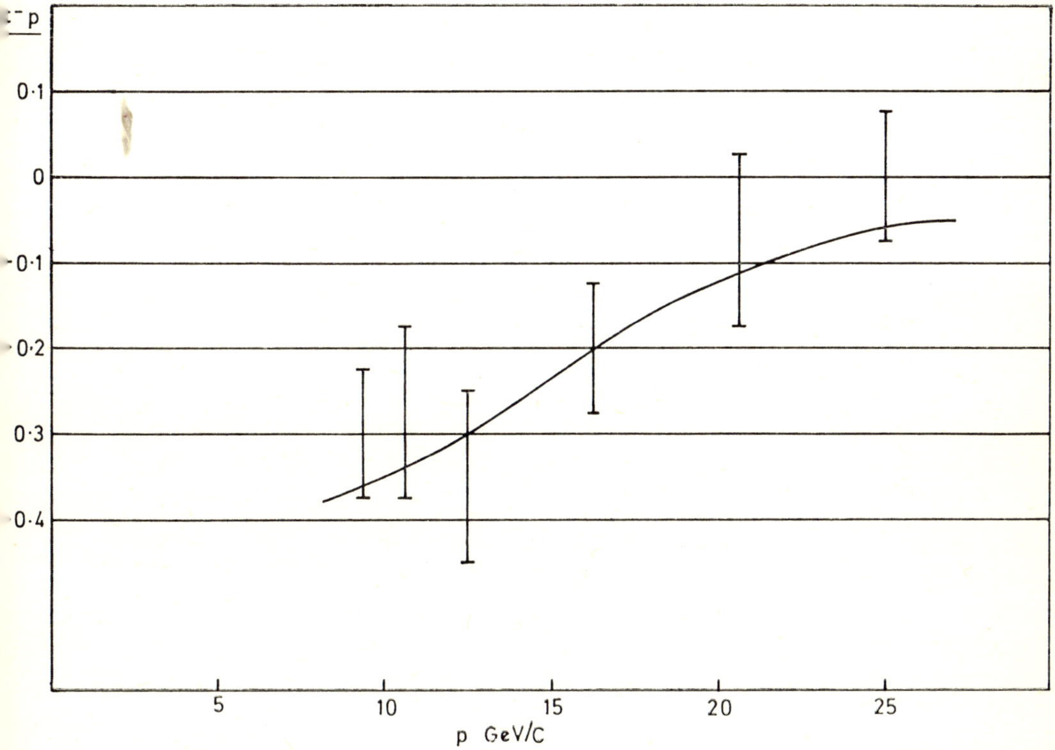

2.4. *Large angle scattering*

(Reference Orear *et al.* (1965) see Fig. 2.3.)

(i) For scattering outside the forward peak there is approximate agreement with the formula

$$\frac{d\sigma}{dt} = C \exp(-bp \sin\theta), \qquad \ldots[2.4.1]$$

where p is the momentum in the centre of mass system. Thus for fixed θ the energy dependence is $\exp(-b'E^{\frac{1}{2}})$ rather than $\exp(-b'E)$, which is the behaviour in the forward peak. The angular dependence for $60° < \theta < 120°$ is observed to be very slight at fixed energy.

(ii) For energy $E \sim 10$ GeV, the change in differential cross section from 0° to 90° is typically in the region of 10^5. This rapid decrease with angle is one of the outstanding theoretical problems to which rigorous

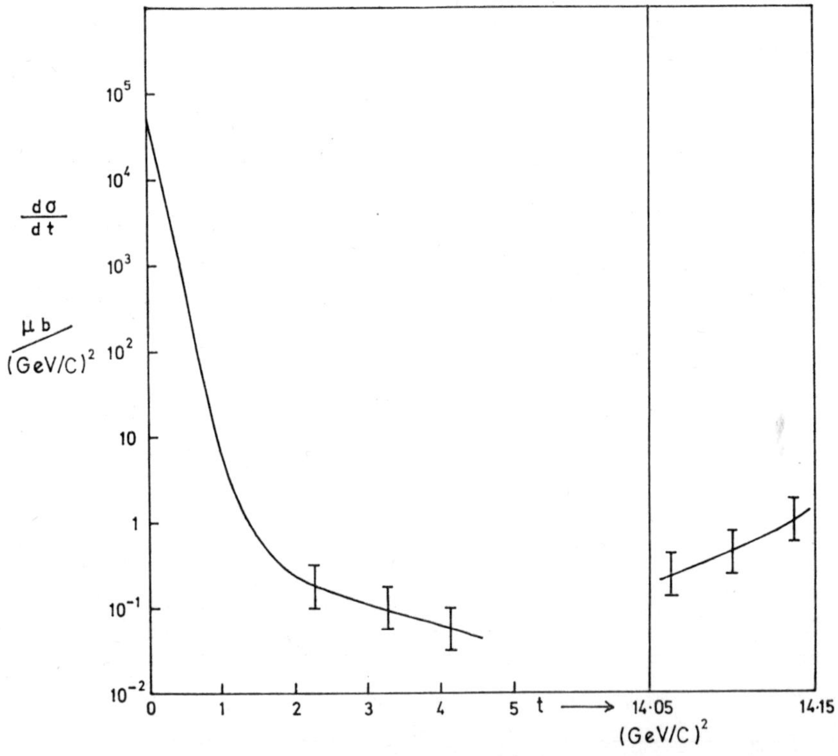

FIG. 2.3. Elastic pion-proton differential cross sections for 8 GeV/c incident negative pions. The curve is interpolated from the data reported by Orear *et al.*, *Phys. Rev. Lett.* (1965) **15**, 309, and includes also the small angle data discussed in section 2.3. The errors indicated are only to indicate their order of magnitude, for small angle scattering they are very small.

theory has yet made no contribution, although various models have been proposed (for example by Wu and Yang (1965)). These are discussed in detail in L. Van Hove's lectures.

2.5. *High-energy reactions*

(i) *Backward scattering*

In the introduction to this section it was noted that backward elastic scattering corresponds to forward scattering for the exchange process. The basic theoretical explanation is therefore likely to be similar. Experimentally, as indicated in Fig. 2.3 for $\pi^- p$ and $\pi^+ p$ scattering, there is a backward peak. For $\pi^+ p$ there is also a diffraction minimum (Selove 1966).

(ii) *Charge-exchange scattering*

From isospin invariance, this is related to the difference of particle-particle and particle-antiparticle scattering. The total charge-exchange cross section tends to zero at high energies. With certain charge-exchange processes Regge theory may take a simple form due to limitations on the type of single particle that can be exchanged (for example ρ in $\pi^- p$ charge-exchange scattering).

(iii) *Quasi-elastic scattering*

I call the process

$$A+B \rightarrow A^* + B^*$$

quasi-elastic if A^* and A have the same quantum numbers except for spin. It is indicated (Collins 1966) that the total quasi-elastic cross section may remain constant with energy at large energy.

(iv) *Exchange scattering*

$$A+B \rightarrow C+D$$

where there is an exchange of quantum numbers. These processes have total cross sections that increase above threshold and then decrease for large E. Their differential cross sections show angular dependence similar to that of elastic scattering including a forward peak exponential in t, and a lower rate of decrease $\exp(-|t|^{\frac{1}{2}}.c)$ for large angles. Experimental details are given in the Stony Brook Conference Proceedings, particularly by Collins (1966) and Morrison (1966).

(v) *Production processes*

$$A+B \rightarrow C+D+X.$$

Multiparticle final states that cannot be approximated by resonances are of considerable importance and probably form an increasing proportion of the total cross section. They are difficult to analyse theoretically but undoubtedly there should be some effort made to extend to multiparticle processes the work on two body processes at high energy.

3

Theoretical Background and Assumptions

This is not primarily a course on dispersion theory, or Regge theory, or the Mandelstam representation, or even on elementary quantum theory. However, all of these topics will be required and used in a limited fashion during our discussion, so I will briefly survey those aspects that seem most essential for later lectures. For more detailed questions there are several books available, for example, Omnes and Froissart, *Mandelstam Theory and Regge Poles*; Chew, *The Analytic S-Matrix*; Eden, *High Energy Collisions of Elementary Particles*; Frautschi, *Regge Poles and S-Matrix Theory*; Barton, *Dispersion Techniques in Field Theory*; Chew and Jacob, *Strong Interactions and S-matrix theory*. A more general account of analytic properties is given by Eden, Landshoff, Olive, Polkinghorne, *The Analytic S-Matrix* (Cambridge U.P., 1966) and by Goldberger (1960) and Wightman (1960).

In this section we shall consider the following:

1. Kinematics and partial wave expansions
2. Dispersion relations and the Mandelstam representation
3. Complex angular momenta and Regge theory
4. Review of assumptions in the study of high-energy scattering.

3.1. *Kinematics, the optical theorem and partial wave expansions*

We will introduce special mass values when required, but for our general discussion we consider collisions between equal mass particles. Also for simplicity we assume them to be spinless bosons except where spin properties are essential. For the reaction I,

$$1+2 \rightarrow 3+4, \qquad \text{I}, \qquad \ldots[3.1.1]$$

the four-momenta will be labelled p_i, having directions as indicated in Fig. 3.1a. The square of the energy in the centre of mass system for reaction [*3.1.1*] is

$$s = (p_1+p_2)^2 = (p_3+p_4)^2. \qquad \ldots[3.1.2]$$

The square of the momentum transfer is

$$t = (p_1+p_4)^2 = (p_2+p_3)^2 \qquad \ldots[3.1.3]$$

and the square of exchange-momentum transfer is

$$u = (p_1+p_3)^2 = (p_2+p_4)^2. \qquad \ldots[3.1.4]$$

Since each particle is on its mass shell

$$p_i^2 = m_i^2, \quad (m_i = m, i = 1, 2, 3, 4), \qquad \ldots[3.1.5]$$

$$s+t+u = \sum m_i^2 = 4m^2. \qquad \ldots[3.1.6]$$

The diagram in Fig. 3.1a can be used also to represent the scattering process II,

$$1+\bar{4} \rightarrow \bar{2}+3, \qquad \text{II}, \qquad \ldots[3.1.7]$$

or III,

$$1+\bar{3} \rightarrow \bar{2}+4, \qquad \text{III}. \qquad \ldots[3.1.8]$$

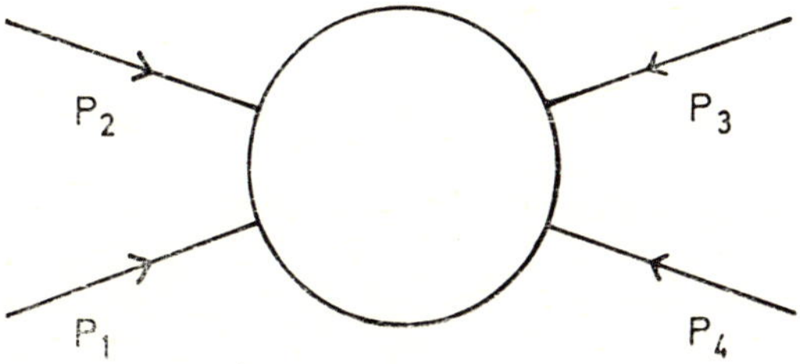

FIG. 3.1*a*

Each of the processes I, II, III becomes physically allowed for appropriate values of the variables. For process I, in the centre-of-mass system, let k denote three-momentum and θ the scattering angle of each particle. Then

$$s = 4(m^2+k^2)$$
$$t = -2k^2(1-\cos\theta)$$
$$u = -2k^2(1+\cos\theta). \qquad \ldots[3.1.9]$$

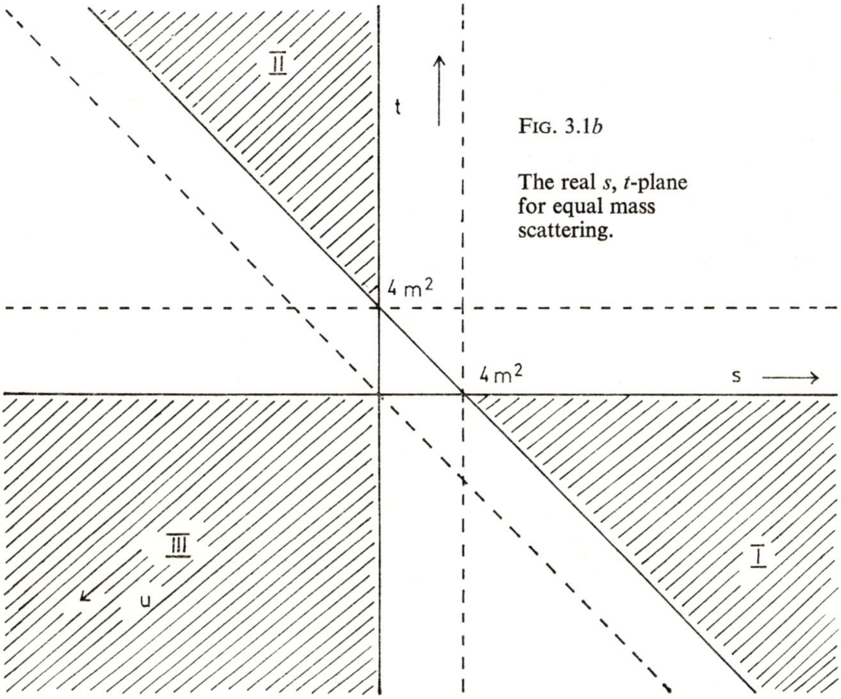

FIG. 3.1*b*

The real s, t-plane for equal mass scattering.

Process I is physical when k and θ are real. This gives the area shaded and labelled in Fig. 3.1b, for which

$$s > 4m^2, \quad t < 0, \quad u < 0, \qquad \text{I.} \qquad \ldots[3.1.10]$$

Similarly II and III become physical processes as shown shaded in Fig. 3.1b,

$$t > 4m^2, \quad u < 0, \quad s < 0, \qquad \text{II} \qquad \ldots[3.1.11]$$

$$u > 4m^2, \quad s < 0, \quad t < 0, \qquad \text{III.} \qquad \ldots[3.1.12]$$

The scattering amplitude F is a function of any two of the invariants s, t, u $(s+t+u = 4m^2)$. Thus $F(s, t)$ is the scattering amplitude for process I in the region [3.1.10] which we may call the "s region" since s is the square of the energy. The assumptions of dispersion theory show that $F(s, t)$ gives also the scattering amplitude for processes II and III when it is continued analytically in an appropriate way to regions II and III of the (s, t) plane.

The differential cross section is

$$\frac{d\sigma}{d\Omega} = \left| \frac{F(s, t)}{8\pi W} \right|^2 \qquad \ldots[3.1.13]$$

where $W = s^{\frac{1}{2}}$ is the centre-of-mass energy. The total cross section is (in the elastic region),

$$\sigma = \int \frac{d\sigma}{d\Omega} d\Omega = \frac{1}{32\pi s} \int_{-1}^{1} | F(s, t) |^2 d(\cos \theta) = \frac{1}{64\pi^2 W^2} \int | F(s, t) |^2 d\Omega. \qquad \ldots[3.1.14]$$

The scattering amplitude F is related to the S-matrix for scattering (or reaction) $a \to b$,

$$S_{ba} = \delta_{ba} + i(2\pi)^4 \delta^{(4)}(p-p') F_{ba}. \qquad \ldots[3.1.15]$$

Unitarity gives

$$\sum_c S_{bc} S^*_{ac} = \delta_{ba} \qquad \ldots[3.1.16]$$

where $\sum_c$ denotes appropriate sums and integrations over intermediate states (which must all be physical states of free particles conserving total energy and momentum). Hence if we take the simple case of an energy that permits only two-particle states,

$$i(F_{ba} - F^*_{ab}) = -(2\pi)^4 \int F_{bc} F^*_{ac} \delta^{(4)}(p_a - p_b) \frac{d^3k_1^c d^3k_2^c}{(2\pi)^6} \qquad \ldots[3.1.17]$$

$$= -\frac{1}{(2\pi)^2} \frac{k}{4W} \int F_{bc} F^*_{ac} d\Omega_c. \qquad \ldots[3.1.18]$$

For elastic scattering in a forward direction $b = a$. Then c denotes scattering through other angles,

$$2 \operatorname{Im} F(s, t=0) = \frac{1}{(2\pi)^2} \frac{k}{4W} \int | F_{ac} |^2 d\Omega_c \qquad \text{...[3.1.19]}$$

$$\operatorname{Im} F(s, t=0) = 2kW\sigma. \qquad \text{...[3.1.20]}$$

This is the optical theorem.

More generally this gives the optical theorem with $\sigma = \sigma$ (total) since the states c in [*3.1.17*] include all allowed inelastic states.

Rewriting [*3.1.18*] we obtain

$$\frac{ik}{8\pi W}(F_{ba} - F_{ab}^*) = -\int_{-1}^{1} d(\cos\theta)\left(\frac{k}{8\pi W}\right)^2 F_{bc}F_{ac}^* - \frac{1}{2\pi}\int\left(\frac{k}{8\pi W}\right)^2 F_{bx}F_{ax}^* d\Omega_x$$

...[*3.1.21*]

where the first term on the right includes only elastic processes c, and the second includes all inelastic processes x.

Expanding in Legendre polynomials below the inelastic threshold (when no states x are allowed),

$$F_{ab} = F(s, t) = F\{s, -2k^2(1-\cos\theta)\} = \frac{8\pi W}{k}\sum_{0}^{\infty}(2l+1)f_l(s)P_l(\cos\theta).$$

...[*3.1.22*]

Substitute in [*3.1.21*], and use orthogonality of the Legendre polynomials

$$i\{f_l^*(s) - f_l(s)\} = 2\,|f_l(s)|^2. \qquad \text{...[3.1.23]}$$

Hence

$$f_l(s) = e^{i\delta_l}\sin\delta_l = (e^{2i\delta_l}-1)/2i \qquad \text{...[3.1.24]}$$

where δ_l is real in the elastic region.

More generally, if inelastic scattering can occur (e.g. $s > 9m^2$), other channels x in [*3.1.21*] must also be considered. The partial wave S-matrix must be written so as to include all allowed channels,

$$S_l = 1 + 2i\begin{bmatrix} f_l(s) & , g_{12}^l(k_1, k_2), & \ldots, & g_{1n}^l(k_1, \ldots) \\ g_{21}^l(k_1, k_2), & — & ,\ldots, & — \\ — \quad , & — & ,\ldots, & — \\ — \quad , & — & ,\ldots, & — \\ — \quad , & — & ,\ldots, & — \\ g_{n1}^l(k_1 \ldots k_n), & — & ,\ldots, & — \end{bmatrix}. \qquad \text{...[3.1.25]}$$

Writing unitarity including the inelastic channel we get instead of [*3.1.23*],

$$i(f_l^* - f_l) = 2f_l^* f_l + 2\sum_{r=2}^{n}\int (g_{1r}^l)^* g_{1r}^l dx. \qquad \text{...[3.1.26]}$$

The integrand is subject to overall energy-momentum conservation. Write $2\mathscr{P}_l$ for the last term in [3.1.26], then

$$i(f_l^* - f_l) = 2f_l^* f_l + 2\mathscr{P}_l. \qquad ...[3.1.27]$$

Whether or not inelastic scattering is possible, the probability for elastic scattering must be less than or equal to one,

$$|S_l(\text{elastic})| = |1 + 2if_l| \leqq 1. \qquad ...[3.1.28]$$

Writing $if_l = ia - b$, this gives

$$0 \leqq 1 - 4b + 4|f_l|^2 \leqq 1$$

giving

$$|f_l|^2 \leqq b \leqq \tfrac{1}{4} + |f_l|^2. \qquad ...[3.1.29]$$

But

$$|f_l|^2 = \tfrac{1}{2}i(f_l^* - f_l) - \mathscr{P}_l = b - \mathscr{P}_l.$$

Hence from [3.1.29]

$$0 \leqq b - |f_l|^2 \leqq \tfrac{1}{4},$$

or

$$0 \leqq \mathscr{P}_l \leqq \tfrac{1}{4}. \qquad ...[3.1.30]$$

If with inelasticity we now write, instead of [3.1.24],

$$f_l(s) = \frac{\eta_l \exp(2i\delta_l) - 1}{2i}. \qquad ...[3.1.31]$$

Then

$$\mathscr{P}_l = b - |f_l|^2 = \tfrac{1}{4}(1 - \eta_l^2). \qquad ...[3.1.32]$$

From [3.1.28] $\eta_l = |S_l(\text{elastic})|$, and

$$0 \leqq \eta_l \leqq 1. \qquad ...[3.1.33]$$

The real quantity η_l is unity when there is no inelasticity. With inelastic effects there are three associated partial wave cross sections $\sigma^l_{\text{elastic}}$, $\sigma^l_{\text{inelastic}}$ and σ^l_{total}. The latter is related to the total cross section by

$$\sigma\,(\text{total}) = \sum_{l=0}^{\infty} \sigma^l_{\text{total}} \qquad ...[3.1.34]$$

and there are similar relations for σ_{elastic} and $\sigma_{\text{inelastic}}$. From the optical theorem

$$\sigma\,(\text{total}) = \frac{1}{2kW}\,\text{Im}\,F(s, 0)$$

$$= \frac{1}{2kW}\,\frac{8\pi W}{k} \sum_{l=0}^{\infty} (2l+1)\,\text{Im}\,f_l(s) P_l(1). \qquad ...[3.1.35]$$

We therefore have

$$\sigma^l_{\text{total}} = \frac{(2l+1)\pi}{k^2}\{2-2\eta_l \cos 2\delta_l\} \qquad \text{...[3.1.36]}$$

$$\sigma^l_{\text{elastic}} = \frac{(2l+1)\pi}{k^2}\{1-2\eta_l \cos 2\delta_l+\eta_l^2\} \qquad \text{...[3.1.37]}$$

$$\sigma^l_{\text{inelastic}} = \frac{(2l+1)\pi}{k^2}\{1-\eta_l^2\}. \qquad \text{...[3.1.38]}$$

From [*3.1.30*] and [*3.1.32*] we see that

$$\sigma^l_{\text{inelastic}} = \frac{(2l+1)}{k^2} 4\mathscr{P}_l$$

has a maximum value when $\mathscr{P}_l = \frac{1}{4}$, which gives

$$\eta_l = 0, \qquad \text{...[3.1.39]}$$

and

$$\sigma^l_{\text{elastic}} = \sigma^l_{\text{inelastic}} = \frac{(2l+1)}{k^2}. \qquad \text{...[3.1.40]}$$

However this does not give the maximum *proportion* of inelastic scattering. To obtain this we must maximize

$$R_l = \frac{\sigma^l_{\text{total}}}{\sigma^l_{\text{elastic}}} = \frac{2-2\eta_l \cos 2\delta_l}{1-2\eta_l \cos 2\delta_l+\eta_l^2}. \qquad \text{...[3.1.41]}$$

For no inelasticity $\eta_l = 1$ and $R_l = 1$. If f_l is pure imaginary $\eta_l = 0$ and $R_l = 2$, then $\sigma^l_{\text{inelastic}}$ has its largest value, but not its largest share of σ^l_{total}. For the general case, write in [*3.1.41*]

$$x = 1-\eta_l;\; y = 1-\cos 2\delta_l. \qquad \text{...[3.1.42]}$$

Then

$$R_l = \frac{2(x+y-xy)}{2y-2xy+x^2}. \qquad \text{...[3.1.43]}$$

This can take any value from 1 to ∞. If $x \to 0$, $(\eta_l \to 1)$ for $y \neq 0$, $(\delta_l \neq 0)$

$$R_l = 1; \qquad \text{...[3.1.44]}$$

this is the elastic scattering situation.

If $y = 0$, $(\delta_l = 0)$,

$$R_l = \frac{2}{x} \to \begin{Bmatrix} \infty & \text{as} & x \to 0, \\ 2 & \text{as} & x \to 1. \end{Bmatrix} \qquad \text{...[3.1.45]}$$

Remembering that these variables are functions of the energy, it is evident that for high-energy behaviour of R_l we will have to ascertain the ratio at which η_l and δ_l approach their asymptotic values if these involve the singular point $\eta_l = 1$, $\delta_l = 0$.

Levinson's theorem (non-relativistic)

(Reference N. Levinson (1949). See also Vaughn, Aaron, Amado (1961), who give the following derivation.) This theorem states that the phase shift $\delta_l(E)$ changes by an integer multiple of π as the energy E varies from 0 to ∞,

$$\delta_l(0) - \delta_l(\infty) = (n_b - n_p)\pi \qquad \text{...[3.1.46]}$$

where n_b is the number of bound states of angular momentum l and n_p is the number of CDD poles (see Castillejo, Dalitz, Dyson (1956)). Write the amplitude

$$f_l(E) = \exp(i\delta_l) \sin \delta_l = N_l(E)/D_l(E) \qquad \text{...[3.1.47]}$$

where $D_l(E)$ is the boundary value $(E+i0)$ of a function analytic in E, and regular in the E-plane cut along the real axis from 0 to ∞, and satisfying $D_l(E) \to 1$, as $E \to \infty$. N_l is defined by the equation,

$$N_l(E) = \frac{1}{2ik} \lim_{\varepsilon \to 0} [D_l(E+i\varepsilon) - D_l(E-i\varepsilon)]. \qquad \text{...[3.1.48]}$$

Since $N_l(E)$ is real in $0 \leqq E \leqq \infty$, the function $D_l(E)$ has the phase $(-\delta_l)$ in this range. The S-matrix element is

$$S_l(E) = \exp(2i\delta_l) = \frac{D_l^*(E)}{D_l(E)} = \frac{D_l(E-i0)}{D_l(E+i0)}. \qquad \text{...[3.1.49]}$$

The phase shift is given by

$$\delta_l(E) = \frac{1}{2i} \log S_l = \frac{1}{2i} [\log D_l(E-i0) - \log D_l(E+i0)]. \qquad \text{...[3.1.50]}$$

Consider the integral [3.1.51] round the contour C shown in Fig. 3.1c, and assume the curved part of C gives zero contribution as its radius tends to infinity. Then

$$\frac{1}{2\pi i} \int_C \frac{dE}{D_l} \cdot \frac{dD_l}{dE} = \frac{1}{2\pi i} \int_0^\infty dE \frac{d}{dE} [\log D_l(E+i0) - \log D_l(E-i0)] \qquad \text{...[3.1.51]}$$

$$= \frac{1}{\pi} [\delta_l(0) - \delta_l(\infty)]. \qquad \text{...[3.1.52]}$$

It can be arranged that zeros of $D_l(E)$ in the E-plane determine all the bound state poles of the amplitude $f_l(E)$ (on the real axis for $E<0$). The poles of $D_l(E)$ are CDD poles and do not correspond to singularities of the amplitude. The left hand side of [*3.1.51*] gives the number of zeros n_b, minus the number of poles n_p, of $D_l(E)$. Levinson's result follows,

$$\delta_l(0)-\delta_l(\infty) = (n_b-n_p)\pi. \qquad ...[3.1.53]$$

It is usual to adopt the convention that $\delta_l(0) = 0$.

This theorem has not been proved in a relativistic theory but it is often assumed to hold either for the phase or for the eigenphase.

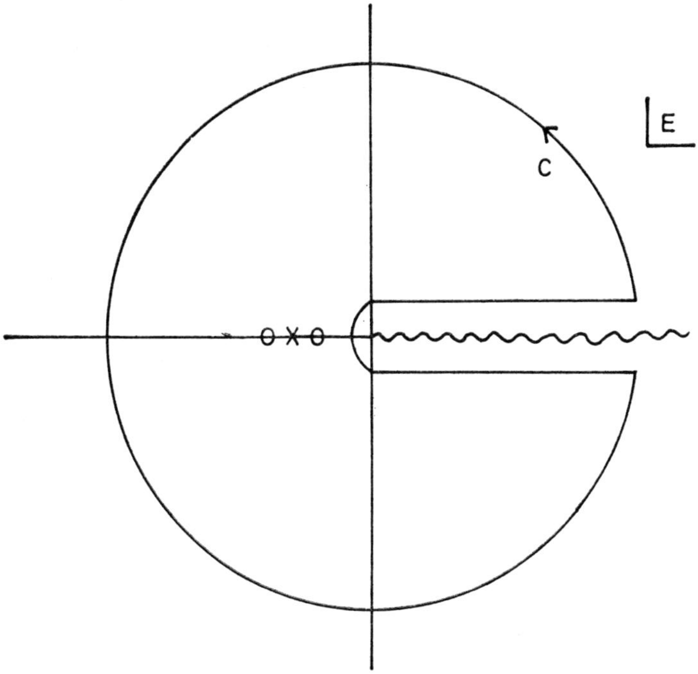

FIG. 3.1*c*. The contour C in [3.1.51]. Poles of *D* are shown by X and zeros by O.

3.2. *Dispersion relations and the Mandelstam representation*

If a function $f(z)$ of the complex variable z is regular in the entire complex z-plane except for a cut along the real axis from x_0 to ∞, and if

$$|f(z)| \to 0 \text{ as } |z| \to \infty \qquad ...[3.2.1]$$

and

$$f^*(z) = f(z^*) \qquad ...[3.2.2]$$

then $f(z)$ satisfies the dispersion relation

$$f(z) = \frac{1}{\pi}\int_{x_0}^{\infty}\frac{f_1(z')\,dz'}{z'-z}. \qquad ...[3.2.3]$$

This result can be proved from Cauchy's theorem for any contour C that encloses the point z. The contour C can then be distorted to yield [*3.2.3*] where (with x real)

$$2f_1(x) = \lim_{\varepsilon\to 0}\{f(x+i\varepsilon)-f(x-i\varepsilon)\}. \qquad \ldots[3.2.4]$$

If, instead of [*3.2.1*], $f(z) \ll O(|z|)$ as $|z| \to \infty$, a dispersion relation can be written for the function

$$\frac{f(z)-f(x_1)}{(z-x_1)}$$

where $x_1 < x_0$. This gives the dispersion relation

$$f(z) = f(x_1)+\frac{(z-x_1)}{\pi}\int_{x_0}^{\infty}\frac{f_1(x)\,dx}{(x-x_1)(x-z)}. \qquad \ldots[3.2.5]$$

This is described as a dispersion relation with one subtraction. More generally if $f(z)$ is bounded by a polynomial of order N, then N subtractions will be needed (or $(N+1)$ if $f \sim z^N$).

The kinematics for scattering of equal mass particles have been given in section 3.1, for the scattering amplitude $F(s, t)$. This function is essentially the Feynman amplitude that is given formally by a perturbation series corresponding to a series of Feynman diagrams. It has been proved to all orders in perturbation theory that for $s = z$ complex and for

$$-4m^2 < t < 4m^2 \qquad \ldots[3.2.6]$$

$F(z, t)$ satisfies a dispersion relation (see Eden (1961))

$$F(z, t) = \frac{1}{\pi}\int_{4m^2}^{\infty}\frac{F_s(s, t)\,ds}{s-z}+\frac{1}{\pi}\int_{-\infty}^{-t}\frac{F_s'(s, t)\,ds}{s-z}. \qquad \ldots[3.2.7]$$

The proof of a dispersion relation in perturbation theory does not give any information about the number of subtractions required. The formula [*3.2.7*] assumes that $F(z, t) \to 0$ as $|z| \to \infty$. More generally it is usual to assume that $F(z)$ is bounded by a polynomial of order N, so that a finite number of subtractions is sufficient. It can then be shown (as we shall see later) that $N \leqq 2$ for the range [*3.2.6*] (Martin (1965b) has established this result from quantum field theory).

The complex s, (z)-plane corresponding to [*3.2.7*] is shown in Fig. 3.2a, and its real axis is the line shown in Fig. 3.2b. Since the latter intersects the physical region where u is the energy it is clearly more symmetric to write [*3.2.7*] in the form

$$F(z, t) = \frac{1}{\pi}\int_{4m^2}^{\infty}\frac{F_s(s, t)\,ds}{s-z}+\frac{1}{\pi}\int_{4m^2}^{\infty}\frac{F_u(u, t)\,du}{u-(4m^2-t-z)}. \qquad \ldots[3.2.8]$$

The symmetric notation implies the assumption that F_u is the imaginary part of F when u is the energy just as F_s is the imaginary part when s is the physical energy. The assumption can be proved, and is called crossing symmetry. (See Bros, J., Epstein, H., Glaser, V. (1965) and Hepp, K. (1964).) For equal mass identical particles, the scattering in the u physical region is the same as in the s physical region. Then crossing symmetry gives

$$F(s+i0, t) = F(4m^2-t-s-i0, t) \qquad \text{...[3.2.9]}$$

$$= F(u+i0, t) \quad \text{with} \quad u = s. \qquad \text{...[3.2.10]}$$

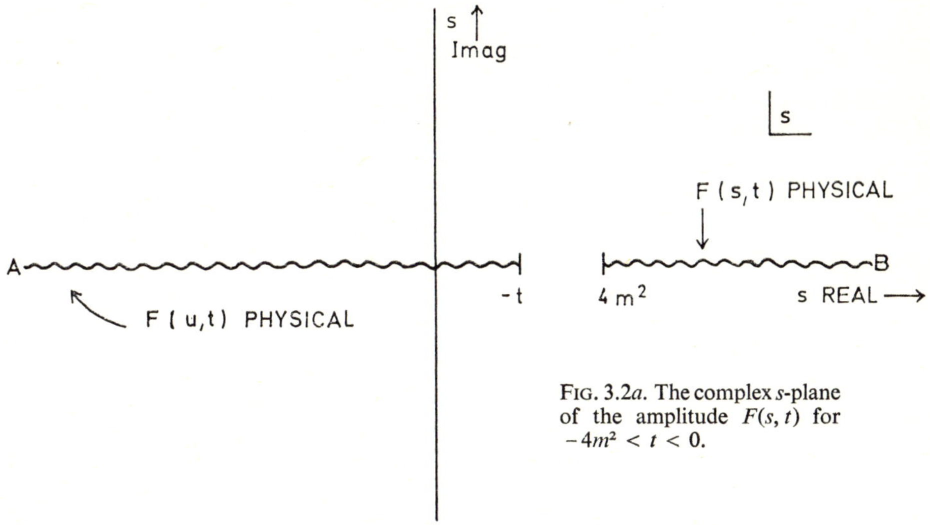

FIG. 3.2*a*. The complex s-plane of the amplitude $F(s, t)$ for $-4m^2 < t < 0$.

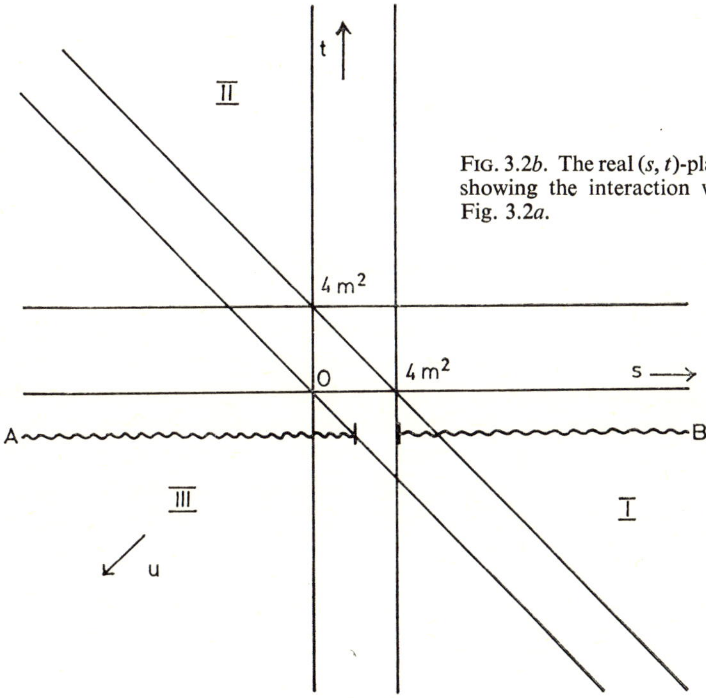

FIG. 3.2*b*. The real (s, t)-plane, showing the interaction with Fig. 3.2*a*.

For forward scattering, this result becomes very simple in terms of the laboratory energy E. We have for $t = 0$,

$$(E+m)2m = s. \qquad ...[3.2.11]$$

Writing $f(E)$ for $F(s, 0)$, crossing symmetry gives for E real

$$f(E+i0) = f(-E-i0). \qquad ...[3.2.12]$$

In the complex E plane $f(E)$ is analytic except for the branch cuts along $(-\infty, -m^2)$ and (m^2, ∞). Hermitean analyticity [3.2.2] gives for E real

$$f(E+i0) = f^*(E-i0). \qquad ...[3.2.13]$$

These relations for $f(E)$ will be used later.

Mandelstam has generalized the dispersion relation [3.2.8] by assuming that the imaginary part of F, say $F_s(s, t)$ is itself an analytic function of complex t, regular in a plane cut along the real axis. Thus

$$F_s(s, z_2) = \frac{1}{\pi}\int_{4m^2}^{\infty} \frac{F_{st}(s, t)\,dt}{t-z_2} + \frac{1}{\pi}\int_{4m^2}^{\infty} \frac{F_{su}(s, u)\,du}{u-u_1} \qquad ...[3.2.14]$$

where $u_1 = 4m^2 - s - z_2$.

Applying this to both terms in [3.2.8] and making $(s, t) \to (z_1, z_2)$ both complex variables, Mandelstam (1958) obtained, after combining two of the 4 terms,

$$\begin{aligned} F(z_1, z_2) = {} & \frac{1}{\pi^2}\iint_{4m^2}^{\infty} ds\,dt\, \frac{\rho(s, t)}{(s-z_1)(t-z_2)} \\ & + \frac{1}{\pi^2}\iint_{4m^2}^{\infty} dt\,du\, \frac{\rho(t, u)}{(t-z_2)(u-z_3)} \qquad ...[3.2.15] \\ & + \frac{1}{\pi^2}\iint_{4m^2}^{\infty} du\,ds\, \frac{\rho(u, s)}{(u-z_3)(s-z_1)} \end{aligned}$$

where $z_3 = 4m^2 - z_1 - z_2$.

The Mandelstam representation implies that:

1. $F(z_1, z_2)$ is analytic in the product of the z_1, z_2, z_3 planes cut along their real axes from $4m^2$ to ∞ and with $\sum z_i = 4m^2$.
2. Including single particle exchange there may also be pole terms like $g^2/(z_1 - m^2)$, but there is no singularity closer than a distance m^2 to the physical region.
3. There may also be subtraction terms, but the representation implies that their number is finite; it seems probable that this is satisfied for $F(z_1, t)$ when t is finite, but not when $t \to \infty$. This difficulty does not affect single variable dispersion relations for fixed t but does cause trouble in the double dispersion relation.

Partial wave dispersion relations can be derived either directly from perturbation theory to all orders, or from the Mandelstam representation. For equal mass scattering, neglecting possible subtractions, with $t = -2k^2(1-x)$,

$$g_l(s) = \frac{1}{k} f_l(s) = \tfrac{1}{2} \int_{-1}^{1} \frac{dx}{8\pi W} F(s, t) P_l(x) \qquad \ldots[3.2.16]$$

satisfies a dispersion relation

$$g_l(z) = \frac{1}{\pi} \int_{4m^2}^{\infty} \frac{g_1^l(s)\,ds}{s-z} + \frac{1}{\pi} \int_{-\infty}^{0} \frac{g_2^l(s)\,ds}{s-z}. \qquad \ldots[3.2.17]$$

If there is a single particle pole the second term is integrated up to $3m^2$. The spectral function $g_1^l(s)$ is related to the phase by

$$g_1^l(s) = \frac{1}{k} \operatorname{Im} f_l(s)$$

$$= (1-\eta_l \cos 2\delta_l)/2k. \qquad \ldots[3.2.18]$$

There is no such simple relation for $g_2^l(s)$ on the left hand cut, and it is an important unsolved problem how to obtain $g_2^l(s)$. Although many approximations have been suggested none is fully satisfactory.

3.3. *Complex angular momenta and Regge theory*

The basis of Regge theory is the interpolation of the partial wave amplitudes $g_l(s)$, which are defined only for integer values of l, by functions $g(l, s)$ which are defined for real or complex values of l such that

$$g(l, s) = g_l(s), \quad (l \text{ integer}). \qquad \ldots[3.3.1]$$

Define

$$g_l(s) = \tfrac{1}{2} \int_{-1}^{1} dx G(s, t(x)) P_l(x) \qquad \ldots[3.3.2]$$

where

$$G(s, t) = \frac{1}{8\pi W} F(s, t). \qquad \ldots[3.3.3]$$

The single variable dispersion relation in t for F obtained from the Mandelstam representation gives a similar one for G, namely

$$G(s, t) = \sum_{n=0}^{N-1} c_n(s) t^n + \frac{t^N}{\pi} \int_{t_0}^{\infty} \frac{dt' G_1(s, t')}{(t'-t)(t')^N} + \frac{u^N}{\pi} \int_{u_0}^{\infty} \frac{du' G_2(s, u')}{(u'-u)(u')^N}. \qquad \ldots[3.3.4]$$

We have assumed that N is large enough to ensure convergence of the integrals; that is to say G is bounded by a polynomial in t of order N for given s. In general $N = N(s)$.

Substituting [3.3.4] into [3.3.2], we find for $l > N$ that the polynomial does not contribute to the integral, nor do the subtractions affect it. We obtain after an integration following change of variables,

$$g_l(s) = \frac{1}{\pi}\int_{t_0}^{\infty}\frac{dtQ_l(1+t/2k^2)}{2k^2}\{G_1(s,t)+(-1)^l G_2(s,t)\} \qquad \text{...[3.3.5]}$$

for $l > N$. We have used

$$Q_l(z) = \tfrac{1}{2}\int_{-1}^{1} dx\,\frac{P_l(x)}{x-z} \qquad \text{...[3.3.6]}$$

$$Q_l(-z) = -(-1)^l Q_l(z). \qquad \text{...[3.3.7]}$$

The formula [3.3.5] for the partial wave amplitude $g_l(s)$ has been derived for l = integer $> N$. However it also provides a definition for $g(l, s)$ simply by making l continuous or complex, since $Q_l(x)$ is well defined for l complex. This definition can also be made unique by an application of Carlson's theorem which states that if

$$g(l, s) = g_l(s) \quad \text{for} \quad l = N, N+1, \ldots \infty, \qquad \text{...[3.3.8]}$$

and if $g(l, s)$ is analytic for Re $(l) > N$, and has a bound

$$|\,g(l, s)\,| \leqq O\{\exp a\,|\,l\,|\} \qquad \text{...[3.3.9]}$$

where $0 \leqq a < \pi-\varepsilon$, $\varepsilon > 0$, then $g(l, s)$ is unique.

These uniqueness conditions can be established from [3.3.5] using the assumption of convergence of the integral, and the properties of $Q_l(x)$ for complex l with Re $(l) > N$ and x (real) > 1, subject to one important modification. The factor $(-1)^l$ behaves like exp $(i\pi l)$ for complex l and does not satisfy the Carlson conditions. It is therefore necessary to interpolate two functions

$$g_+(l, s) = \frac{1}{\pi}\int_{t_0}^{\infty}\frac{dt}{2k^2}\,Q_l\{G_1+G_2\} \qquad \text{...[3.3.10]}$$

$$g_-(l, s) = \frac{1}{\pi}\int_{t_0}^{\infty}\frac{dt}{2k^2}\,Q_l\{G_1-G_2\}. \qquad \text{...[3.3.11]}$$

Both g_+ and g_- satisfy the Carlson conditions. The former interpolates the even l amplitudes and the latter does the odd l amplitudes. In the special case of equal mass scattering of identical particles $G_1 = G_2$ and g_- vanishes identically, as do the odd partial waves.

Using $g(l, s)$ the partial wave series can be modified to give instead an integral representation of the amplitude.

$$G(s, t) = \sum_{0}^{\infty}(2l+1)g_l(s)P_l(x) \qquad \text{...[3.3.12]}$$

$$= \frac{1}{2i}\int_C \frac{(l+\frac{1}{2})dl}{\sin(\pi l)}\,P_l(x)(-1)^l g(l, s) \qquad \text{...[3.3.13]}$$

where the contour C surrounds the positive integers and zero as shown in Fig. 3.3.

Provided that the partial wave amplitude $g(l, s)$ satisfies suitable boundary conditions for large $|l|$ with Re $(l) > -\frac{1}{2}$, the contour of integration C in [3.3.13] can be deformed to give the contour C′ along Re $(l) = -\frac{1}{2}$, together with various contours C″ surrounding the singularities of $g(l, s)$ in the right half l-plane.

The simplest assumption about singularities of $g(l, s)$ for complex l is that it is meromorphic (the only singularities are poles) in

$$\text{Re}\,(l) > -\tfrac{1}{2}. \qquad ...[3.3.14]$$

This assumption was proved in potential scattering by Regge for Yukawa potentials. Then he also showed (1958), (1960), that for large t, G is dominated by the contribution from the contour C″. We have from [*3.3.13*] and Fig. 3.3,

$$G(s, t) = \frac{1}{2i}\int_{C'} \frac{(l+\frac{1}{2})\,dl}{\sin(\pi l)} P_l(x)(-1)^l g(l, s) + \pi \sum_n b_n P_{\alpha_n}(-x) \qquad ...[3.3.15]$$

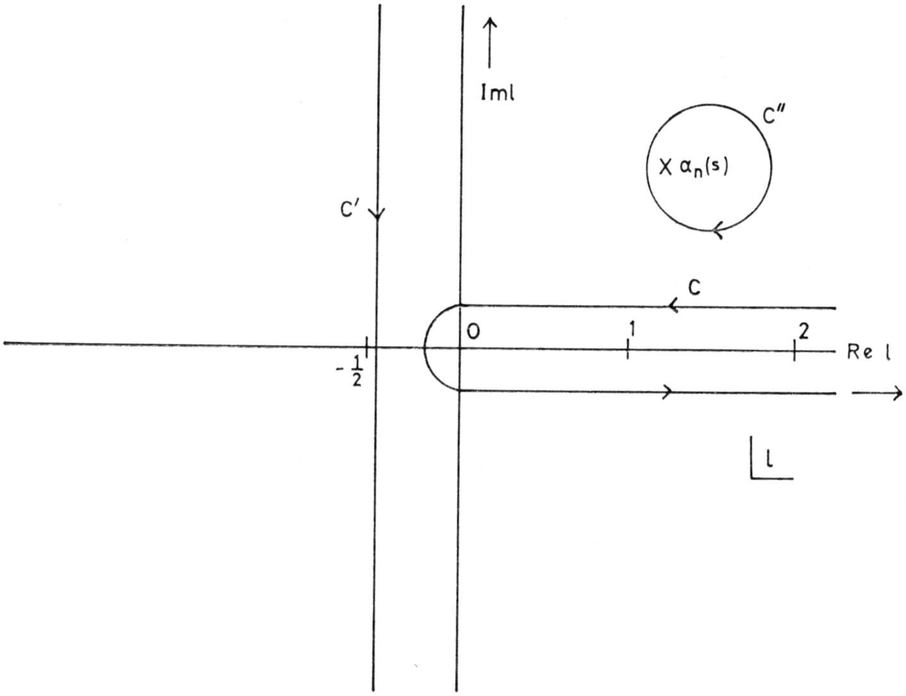

FIG. 3.3. The contour C, C′, and C″ used in equations [3.3.13] and [3.3.15] in the complex l-plane; $\alpha_n(s)$ denotes the position of a pole of the partial wave amplitude $g(l, s)$.

where $\sum_n$ is over all poles of $g(l, s)$ in the region [3.3.14]. For large values of t,

$$x = 1+2t/(s-4m^2) \qquad ...[3.3.16]$$

is also large for fixed values of s.

Also for large x,

$$P_l(-x) \sim O(|x|^{\mathrm{Re}(l)}). \qquad ...[3.3.17]$$

It is therefore intuitively clear that the largest term in [3.3.15] for large t (i.e. large x) is given by the pole $\alpha(s)$ that has the largest real part amongst the Regge poles $\alpha_n(s)$. Thus for fixed s,

$$G(s, t) \sim t^{\alpha(s)} \quad \text{as} \quad t \to \infty \qquad ...[3.3.18]$$

where $\alpha(s)$ is the right-most pole of $g(l, s)$ the partial wave amplitude in the s channel.

The above result [3.3.18] has been proved by Regge for potential scattering with Yukawa potentials. Its significance in relativistic theory was noted by Blankenbecler and Goldberger (1961 La Jolla Conference unpublished, and 1962) and by Chew and Frautschi (1961), and by Gribov (1962).

If the assumption of meromorphy in l of $g(l, s)$ is correct in a relativistic theory, then the result [3.3.18] can be combined with crossing symmetry to give for fixed t,

$$G(s, t) \sim cs^{\beta(t)} \quad \text{as} \quad s \to \infty \qquad ...[3.3.19]$$

for fixed t. In this case $\beta(t)$ is the pole of the partial wave amplitude $g'(l, t)$ evaluated in the t channel.

We will refer to [3.3.19] as the Regge form for high-energy behaviour and will discuss it in more detail later. However it should be noted that Mandelstam has shown (1963) (see also Polkinghorne (1963)), that for large values of t, there are branch cuts in the l-plane of $g'(l, t)$ that lie to the right of the Regge poles, so the formula [3.3.19] has to be modified to include contributions from the contour C″ of Fig. 3.3 that surrounds these branch cuts. For small values of t, particularly near $t = 0$, there is no reason to disbelieve the result [3.3.19] on theoretical grounds although it has not yet been proved in a relativistic theory. We will see later that experimental results do not confirm the dominance of a single Regge pole term at present-day energies (up to 30 BeV).

One profound difficulty arising from Mandelstam's result on branch cuts is that it provides considerable evidence that if

$$F(s, t) \sim s^{\beta(t)} \quad \text{as} \quad s \to \infty, \qquad ...[3.3.20]$$

then as $t \to \infty$, so also does $\beta(t) \to \infty$. This implies a need for an infinite number of subtractions in the Mandelstam representation, so that one must be rather cautious in using it to study high-energy behaviour.

3.4. *Review of assumptions*

I will take as basic assumptions the analyticity properties that have been established rigorously from quantum field theory. These are:

1. $F(s, t)$ is analytic in the s-plane cut along the real axis from $-\infty$ to $4m^2-t$, and $4m^2$ to ∞, when $0 \geqq t \geqq -4m^2$ (see Hepp 1964).
2. For any fixed physical energy s, the amplitude $F(s, t)$ and Im $F(s, t)$ are regular in the Lehmann ellipse (Lehmann 1958). This is an ellipse in the $\cos\theta$ plane with foci at ± 1 and semi-major axis $1+c/s^2$ where c is a constant.
3. In the neighbourhood of any point (s_0, t_0), with $0 \geqq t_0 \geqq -4m^2$ and s_0 outside the cuts, there is simultaneous analyticity in s and t in

$$|s-s_0| < \eta(s_0, t_0),$$

$$|t-t_0| < \eta(s_0, t_0).$$

This result has been established by Bros, Epstein and Glaser (1964).

The domain of analyticity has been rigorously extended by Martin (1965b). The extension that is of most crucial importance is the enlargement of the Lehmann ellipse giving

4. $F(s, t)$ is analytic in the topological product

$$\{|t| < R\} \times \{s \text{ in a cut plane}\}$$

where R is a constant (independent of s; contrast this with (2) above where R shrinks like c/s).

This domain has been further enlarged by Martin (1966a), and it is possible (with restrictions on the masses of particles) that full Mandelstam analyticity in two variables may be established eventually. But it is more doubtful whether the number of subtractions is finite as is required for double dispersion relations.

In addition to the above basic assumptions we have the requirements of unitarity. For physical values of s, two-body unitarity implies that

5. $$0 \leqq |f_l(s)|^2 \leqq \operatorname{Im} f_l(s) \leqq 1.$$

This leads to conditions on the amplitude $F(s, t)$ and its derivatives with respect to t at $t = 0$, when combined with our knowledge of analyticity.

The final basic assumption that we require is that the amplitude is polynomial bounded. This is often called the temperedness (or temperateness) condition as it is related to the assumptions about tempered distributions used in quantum field theory.

6. $$|F(s, t)| < |s|^N, \quad \text{as} \quad |s| \to \infty,$$

for t in a bounded domain. This has been proved from quantum field theory for $0 \geqq t \geqq$ negative constant, by Hepp (1964). We will use it for $|t| < t_0(s)$ where t_0 is (c/s) with assumption (2) above and is constant with assumption (4).

In the following sections we will apply the above basic assumptions to determine bounds on the behaviour of $F(s, t)$ for large values of s. From these bounds one can obtain limits on the cross sections and other physically measured quantities. In the first instance we will assume Martin's result (4) above, but in section 6 we will indicate how it is derived.

4

Upper Bounds on Cross Sections and Other Inequalities

4.1. *Partial wave amplitudes*

Write

$$8\pi A(s, \cos\theta) = \text{Im}\, F(s, t) \qquad \ldots[4.1.1]$$

where Im F is taken for physical s.
Then

$$A(s, \cos\theta) = \frac{s^{\frac{1}{2}}}{k} \sum_{0}^{\infty} (2l+1)\, a_l(s) P_l(\cos\theta) \qquad \ldots[4.1.2]$$

where $a_l(s) = \text{Im}\, f_l(s)$.
For equal masses (taken to be unity),

$$\cos\theta = 1 + \frac{t}{2k^2} = 1 + \frac{2t}{s-4}. \qquad \ldots[4.1.3]$$

The series expansion [4.1.2] converges in the complex $\cos\theta$-plane inside the Lehmann ellipse from assumption (2) of section 3.4, and inside the larger Martin domain from assumption (4) of section 3.4. Note that both these assumptions are derivable from the axioms of quantum field theory. We can include what we need from both by the statement that A is regular in

$$|t| \leqq t_0(s) \qquad \ldots[4.1.4]$$

where t_0 is c/s for the Lehmann ellipse and is constant for the Martin domain (where the constant is $4m^2$), as we will see in section 6.4.

Write $t = t_0$ (which is positive) in the partial wave series [4.1.2]

$$A(s, \cos\theta) = \frac{s^{\frac{1}{2}}}{k} \sum_{l=0}^{\infty} (2l+1)\, a_l(s) P_l(1 + t_0/2k^2). \qquad \ldots[4.1.5]$$

From unitarity

$$0 \leqq |f_l(s)|^2 \leqq a_l(s) \leqq 1. \qquad ...[4.1.6]$$

From polynomial boundedness (assumption (6) of section 3.4)

$$A(s, 1+t_0/2k^2) < s^N \qquad ...[4.1.7]$$

where N does not depend on s.

Since each term in [4.1.5] is positive for $t_0 > 0$, we deduce that as $s \to \infty$

$$(2l+1)a_l(s)P_l(1+t_0/2k^2) < s^N(k/s^{\frac{1}{2}}).$$

In the following, C will denote a constant that may be a different constant in different equations.

For large l, and $x > 1$,

$$P_l(x) > \frac{C}{(2l+1)^{\frac{1}{2}}}[1+(2x-2)^{\frac{1}{2}}]^l. \qquad ...[4.1.8]$$

Hence as $l \to \infty$,

$$a_l(s) < Cs^N[1+(t_0/k^2)^{\frac{1}{2}}]^{-l}. \qquad ...[4.1.9]$$

giving

$$a_l(s) < C\exp[-l(t_0/s)^{\frac{1}{2}}+\varepsilon l+N\log s] \qquad ...[4.1.10]$$

where $\varepsilon \to 0$ as $s \to \infty$.

We see from [4.1.6] and [4.1.10] that for

$$l > L \sim C(s/t_0)^{\frac{1}{2}}\log s \qquad ...[4.1.11]$$

both real and imaginary parts of the partial wave amplitudes f_l decrease exponentially with l.

With the Lehmann ellipse

$$L \sim L_0 \sim Cs(\log s). \qquad ...[4.1.12]$$

With the Martin domain

$$L \sim L_1 \sim Cs^{\frac{1}{2}}(\log s). \qquad ...[4.1.13]$$

For $|t| < xt_0$, where $0 < x < 1$, we can write

$$F(s, t) = 8\pi\frac{s}{k}\sum_0^{\frac{1}{2}L}(2l+1)f_l(s)P_l(\cos\theta)+R_L(s, t) \qquad ...[4.1.14]$$

where

$$|R_L(s, t)| < s^{-M} \qquad ...[4.1.15]$$

and by choice of C in [4.1.12] or [4.1.13] we can make M arbitrarily large.

In particular, these relations can be used to bound cross sections, as we will see in the next section. For a more detailed discussion of the above method one should refer to the lectures by Martin at the Scottish Universities Summer School in 1963 (reference Moorhouse 1963).

4.2. *Total cross sections; subtractions in dispersion relations*

From the optical theorem

$$\sigma_{\text{total}}(s) = \frac{4\pi}{ks^{\frac{1}{2}}} A(s, 1) = \frac{1}{2ks^{\frac{1}{2}}} \operatorname{Im} F(s, 0). \qquad \ldots[4.2.1]$$

For $t = 0$, $(\cos\theta = 1)$ from [*4.1.5*], using [*4.1.6*], as $s \to \infty$

$$A(s, 1) < \frac{s^{\frac{1}{2}}}{k} \sum_{l=0}^{L} (2l+1) < L^2. \qquad \ldots[4.2.2]$$

This gives from the Lehmann ellipse condition [*4.1.12*],

$$\sigma_{\text{total}} < Cs \log^2 s \qquad \ldots[4.2.3]$$

and from the Martin condition [*4.1.13*]

$$\sigma_{\text{total}} < C \log^2 s. \qquad \ldots[4.2.4]$$

The bound [*4.2.3*] was first obtained by Greenberg and Low (1961) using essentially the method given here. The bound [*4.2.4*] is the Froissart bound first obtained by him (1961), assuming the Mandelstam representation but later shown by Martin (1965b) to be derivable from the properties of quantum field theory, as stated in section 3.4.

In the following discussion we will use the Martin result (4) from section 3.4 and the corresponding condition [*4.1.13*] yielding the Froissart bound, except in section 6.4 where we discuss Martin's derivation of the result (4).

Analyticity in the cut s-plane for t in $(-4m^2, 4m^2)$ gives a dispersion relation for $F(s, t)$. The number of subtractions required for convergence is determined from [*4.1.14*] using crossing symmetry. The result is

$$F(s, t) = C_0 + C_1 s + \frac{s^2}{\pi} \int_4^{\infty} \frac{\operatorname{Im} F(s', t)\, ds'}{s'^2(s'-s)} + \frac{s^2}{\pi} \int_{-\infty}^{-t} \frac{\operatorname{Im} F(s', t)\, ds'}{s'^2(s'-s)}. \qquad \ldots[4.2.5]$$

Crossing symmetry for scattering of uncharged pions gives

$$\operatorname{Im} F(4-s', 0) = -\operatorname{Im} F(s', 0). \qquad \ldots[4.2.6]$$

Similarly for $\pi^+ p$ scattering the contribution from the left hand cut is related to $\pi^- p$ scattering for physical values of s' for which [*4.1.14*] holds. Provided $|t| < 4m^2$, the growth of $P_l(1+2t/s)$ as $l \to \infty$ is always dominated by the decrease of $f_l(s)$ if we choose C appropriately in [*4.1.13*]. Thus in this range of t only two subtractions are required. This is true also for any fixed negative value of t, but *not* for positive $t > 4m^2$ (above the first normal threshold).

4.3. *The forward peak, its slope and higher derivatives*

If special functional forms are assumed for the non-forward amplitude it is possible to obtain improvements on the Froissart bound. These improvements were noted by Martin (1963a). The following method of derivation was suggested by Leader (1963). The elastic cross section is

$$\sigma(\text{elastic}) = \frac{1}{s(s-4)} \int_{-(s-4)}^{0} \frac{|F(s,t)|^2 \, dt}{64\pi^2}. \qquad \ldots[4.3.1]$$

Assume a Regge term dominates in F near the forward direction so that the integral can be approximated by taking

$$F(s,t) = F_0(s,0) s^{\alpha(t)-1}. \qquad \ldots[4.3.2]$$

For large s we can approximate

$$s^{\alpha(t)-1} \sim \exp(\alpha' t \log s). \qquad \ldots[4.3.3]$$

From the optical theorem, for large s,

$$\sigma(\text{total}) \sim \frac{\operatorname{Im} F(s,0)}{s}. \qquad \ldots[4.3.4]$$

Hence, using unitarity in the form

$$\sigma(\text{elastic}) \leqq \sigma(\text{total}) \equiv \sigma_t, \qquad \ldots[4.3.5]$$

we obtain

$$\sigma_t \geqq \frac{2\pi}{(8\pi)^2} \int_{-(s-4)}^{0} \left| \frac{\operatorname{Im} F(s,t)}{s} \right|^2 dt$$

$$\geqq \frac{2\pi}{(8\pi)^2} \int_{-(s-4)}^{0} (\sigma_t)^2 \exp(2\alpha' t \log s) \, dt. \qquad \ldots[4.3.6]$$

Rearranging, as $s \to \infty$, we obtain

$$\sigma_t \leqq C \log s, \quad C = 64\pi\alpha'. \qquad \ldots[4.3.7]$$

This is a special case of the fact that if the logarithmic derivative of Im $F(s,t)$ is given at $t = 0$, one can immediately obtain an upper bound on σ_t. We will consider this more generally in terms of the partial wave expansion for $A = 8\pi \operatorname{Im} F$.

For large s,

$$A(s,t) \sim 2 \sum_{0}^{L} (2l+1) a_l(s) P_l(1+2t/s) \qquad \ldots[4.3.8]$$

where $a_l = \operatorname{Im} f_l > 0$.
Rearrange this series to give

$$A(s,t) \sim 4 \sum_{0}^{L} b_n(s) \frac{t^n}{(n!)^2}. \qquad \ldots[4.3.9]$$

Then

$$b_0(s) = \sum_0^L l a_l(s) \qquad \text{...[4.3.10]}$$

$$b_1(s) = \frac{1}{s}\sum_1^L l^2(l-1)a_l \sim \frac{1}{s}\sum_0^L l^3 a_l \qquad \text{...[4.3.11]}$$

$$b_n(s) = \frac{n!}{4}\frac{d^n}{dt^n}A(s,t)\Big|_{t=0} \qquad \text{...[4.3.12]}$$

$$b_n(s) \sim \frac{1}{s^n}\sum_0^L l^{2n+1}a_l(s) \qquad \text{...[4.3.13]}$$

provided $n \ll L \sim s^{\frac{1}{2}}(\log s)$.

The optical theorem gives

$$b_0 \sim s\sigma_t/32\pi. \qquad \text{...[4.3.14]}$$

Define $g(s)$ by differentiating with respect to t at $t = 0$,

$$g(s) = \frac{d}{dt}\log A(s,0) = \frac{d}{dt}\log A(s,t)\Big|_{t=0} \qquad \text{...[4.3.15]}$$

Then

$$b_1 = gb_0 = gs\sigma_t/32\pi. \qquad \text{...[4.3.16]}$$

Now since $0 \leqq a_l \leqq 1$,

$$\left[2\sum_0^L la_l \sum_0^{l-1} ma_m - \sum_0^L l^3 a_l\right] \leqq 0, \qquad \text{...[4.3.17]}$$

this gives for large s, taking only leading terms,

$$b_0^2 - sb_1 \leqq 0, \qquad \text{...[4.3.18]}$$

or

$$b_0^2 - sgb_0 \leqq 0,$$

or

$$\sigma(\text{total}) \leqq 32\pi g(s). \qquad \text{...[4.3.19]}$$

More general relations involving higher derivatives can be obtained using Cauchy's inequality for two series of positive terms,

$$\sum_0^N x_n^2 \sum_0^N y_n^2 \geqq \left[\sum_0^N x_n y_n\right]^2. \qquad \text{...[4.3.20]}$$

This gives

$$b_2 b_0 = \sum_2^L l^5 a_l \sum_0^L la_l \geqq \left[\sum_2^L l^3 a_l\right]^2 \sim (b_1)^2 \sim g^2 b_0^2. \qquad \text{...[4.3.21]}$$

More generally, using $n \ll L$,

$$b_n \geqq g^n b_0 \geqq g^n s\sigma_t/32\pi. \qquad \text{...[4.3.22]}$$

We have also from [*4.3.20*],

$$b_{n+1}\, b_{n-1} \geqq b_n^2. \qquad \text{...[4.3.23]}$$

Writing

$$b_n = gC_n, \qquad \text{...[4.3.24]}$$

$$\frac{C_{n+1}}{C_n} \geqq \frac{C_n}{C_{n-1}}, \qquad \text{...[4.3.25]}$$

and for $t \geqq 0$,

$$A(s,t) \sim \frac{s\sigma_t}{8\pi} \sum_0^L \frac{C_n(s)}{(n!)^2} [tg(s)]^n. \qquad \text{...[4.3.26]}$$

By taking $C_n = 1/n$, we obtain with $g(s) \sim \log^2 s$ a series for $A(s, t)$ that saturates the Froissart bound.

For positive values of t, using [*4.2.25*] and [*4.2.26*],

$$A(s,t) > \frac{s\sigma_t}{8\pi} \sum_0^L \frac{(tg)^n}{(n!)^2}. \qquad \text{...[4.3.27]}$$

If $g(s) \to \infty$ as $s \to \infty$, for fixed t and large s

$$A(s,t) > s\sigma_t \exp [(tg)^{\frac{1}{2}}]. \qquad \text{...[4.3.28]}$$

However A must be polynomial bounded, hence

$$g(s) \leqq \log^2 s. \qquad \text{...[4.3.29]}$$

One can generalize this use of entire functions and see from [*4.3.26*] that as $s \to \infty$, if $g(s) \to \infty$, then $A(s, t)$ is bounded below for positive t by an entire function of $tg(s)$ of order ρ, where $\rho = \frac{1}{2}$ from [*4.3.25*] and [*4.3.26*]. This is discussed further by Eden (1966c).

4.4. *Elastic cross sections*

In terms of partial wave amplitudes f_l the elastic cross section is given by,

$$\sigma(\text{elastic}) = \frac{16\pi}{s-4} \sum_0^\infty (2l+1)\,|f_l(s)|^2. \qquad \text{...[4.4.1]}$$

As in section 4.2 we write a_l for $\text{Im} f_l$, hence

$$f_l(s) = C_l + ia_l. \qquad \text{...[4.4.2]}$$

Then σ(elastic) is greater than σ(el. im.), which is given by

$$\sigma(\text{el. im.}) \sim \frac{32\pi}{s} \sum_0^L la_l^2. \qquad \text{...[4.4.3]}$$

Using the fact that only the first L partial waves contribute significantly to the total or elastic cross sections, MacDowell and Martin (1964) have obtained a lower bound on σ(elastic) in terms of σ(total) and the logarithmic derivative $g(s)$ defined in [*4.3.15*]. For large s,

$$\sigma_t = \sigma(\text{total}) \sim \frac{32\pi}{s} \sum_0^L l a_l, \qquad \ldots[4.4.4]$$

$$b_1(s) \sim \frac{1}{s} \sum_1^L l^3 a_l \sim g s \sigma_t / 32\pi. \qquad \ldots[4.4.5]$$

Minimise $b_1(s)$ assuming that σ(el. im.) and σ(total) are given constants, so [*4.4.3*] and [*4.4.4*] are constraints on the variables a_l. This gives, using Lagrange multipliers,

$$a_l = \lambda - \mu l^2, \qquad \ldots[4.4.6]$$

with the additional unitarity constraint $0 \leqq a_l \leqq 1$. There are now two possibilities: (a) $\lambda > 1$ and (b) $\lambda < 1$. The solution (a) leads to the result $\sigma(\text{el. im.}) > \frac{2}{3}\sigma(\text{total})$ which is unlikely on physical grounds since it does not agree with existing results at high energies. The solution (b) leads to

$$\begin{aligned} a_l &= \lambda - \mu l^2, & &\text{for } l^2 < \lambda/\mu \\ a_l &= 0, & &\text{for } l^2 > \lambda/\mu. \end{aligned}$$

This gives from [*4.4.3, 4* and *5*], replacing sums by integrals,

$$\sigma_t = \frac{32\pi}{s}\left(\frac{\lambda^2}{4\mu}\right) \qquad \ldots[4.4.7]$$

$$\sigma(\text{el. im.}) = \frac{32\pi}{s}\left(\frac{\lambda^3}{6\mu}\right) \qquad \ldots[4.4.8]$$

and, bounding b_1 by its minimum value,

$$g(s)\sigma_t > \frac{32\pi}{s}\left(\frac{\lambda^3}{12\mu^2}\right) \qquad \ldots[4.4.9]$$

$$g(s) > \frac{\lambda}{3\mu} \geqq \frac{\sigma_t}{9 \times 4\pi}\left(\frac{\sigma_t}{\sigma(\text{el. im.})}\right). \qquad \ldots[4.4.10]$$

Hence as $s \to \infty$,

$$\sigma(\text{elastic}) > \left\{\frac{\sigma_t(s)}{36\pi g(s)}\right\} \sigma_t(s) \qquad \ldots[4.4.11]$$

where

$$g(s) = \frac{d}{dt} \log \operatorname{Im} F(s, t)\bigg|_{t=0}. \qquad \ldots[4.4.12]$$

In deriving [*4.4.11*] I have omitted terms that are of order $1/s$ compared with σ_t. If these are included one obtains (MacDowell and Martin 1964)

$$g(s) > \frac{\sigma_t^2}{36\pi\sigma(\text{el. im.})} - \frac{4}{9(s-4m^2)}. \qquad \ldots[4.4.13]$$

If we use also the bound [*4.3.29*], $g(s) < \log^2 s$, we obtain from [*4.4.11*]

$$\sigma(\text{elastic}) > \text{C}\,\frac{\sigma_t^2}{\log^2 s}. \qquad \ldots[4.4.14]$$

This is, of course, a much weaker statement than is contained in [*4.4.11*]. The experimental implications of [*4.4.13*] were investigated by Martin and MacDowell who found that experimental values came near to saturating this inequality at high energy.

4.5. Non-forward scattering

Bounds for the forward amplitude (total cross section) are approached to within $\log^2 s$ by experimental results. This is not the situation for non-forward scattering either at fixed t or fixed angle. It is evident that non-forward scattering involves complicated cancellations that are not revealed by present rigorous methods. In this section I will outline the upper bounds obtained by Martin (1963a) and by Kinoshita, Loeffel and Martin (1964). In the next section we will consider lower bounds.

For $|\cos\theta| < 1$,

$$|P_l(\cos\theta)| \leqq 1 \quad \text{and} \quad |P_l(\cos\theta)| \leqq \left(\frac{2}{\pi l \sin\theta}\right)^{\frac{1}{2}}. \qquad \ldots[4.5.1]$$

This gives from

$$A(s, \cos\theta) \sim \frac{s^{\frac{1}{2}}}{k} \sum_0^L (2l+1)\,a_l(s)\,P_l(\cos\theta) \qquad \ldots[4.5.2]$$

that

$$|A(s,\cos\theta)| \leqq \sum_0^L (2l+1)\left(\frac{2}{\pi l \sin\theta}\right)^{\frac{1}{2}} \leqq 4\int_0^L \frac{l^{\frac{1}{2}}\,dl}{(\sin\theta)^{\frac{1}{2}}} \leqq \frac{CL^{\frac{3}{2}}}{(\sin\theta)^{\frac{1}{2}}}. \qquad \ldots[4.5.3]$$

Hence, for $\theta \neq 0$,

$$|A(s,\cos\theta)| \leqq C\,\frac{s^{\frac{3}{4}}\,(\log s)^{\frac{3}{2}}}{(\sin\theta)^{\frac{1}{2}}}. \qquad \ldots[4.5.4]$$

Using more refined methods Kinoshita *et al.* (1964) obtained

$$|F(s,\cos\theta)| \leqq C\,\frac{(\log s)^{\frac{3}{2}}}{\sin^2\theta}. \qquad \ldots[4.5.5]$$

Note that for large s,

$$\sin^2\theta \sim \frac{2t}{s}+\frac{t^2}{s^2} \qquad \text{...[4.5.6]}$$

so [*4.5.5*] and [*4.5.4*] are compatible with each other. Although [*4.5.5*] improves over [*4.5.4*] by a factor $s^{\frac{3}{4}}$ it does not approach the experimental situation where

$$|F(s, \cos\theta)|_{(\text{experimental})} \sim C_1 \exp\left[-s^{\frac{1}{2}}\, C_2 \sin\,\theta\right]. \qquad \text{...[4.5.7]}$$

5

Lower Bounds

5.1. *Lower bounds for large angle scattering*

An interesting study of large angle scattering has been made by Cerulus and Martin (1964). Their method provides a nice illustration of the power of analytic function theory. Unfortunately, as the authors note, their results depend on the assumption of a polynomial bound for the amplitude which extends into the large s large t region; this is the region where the Regge branch cuts probably dominate the asymptotic behaviour.

This work is concerned primarily with a bound for the amplitude

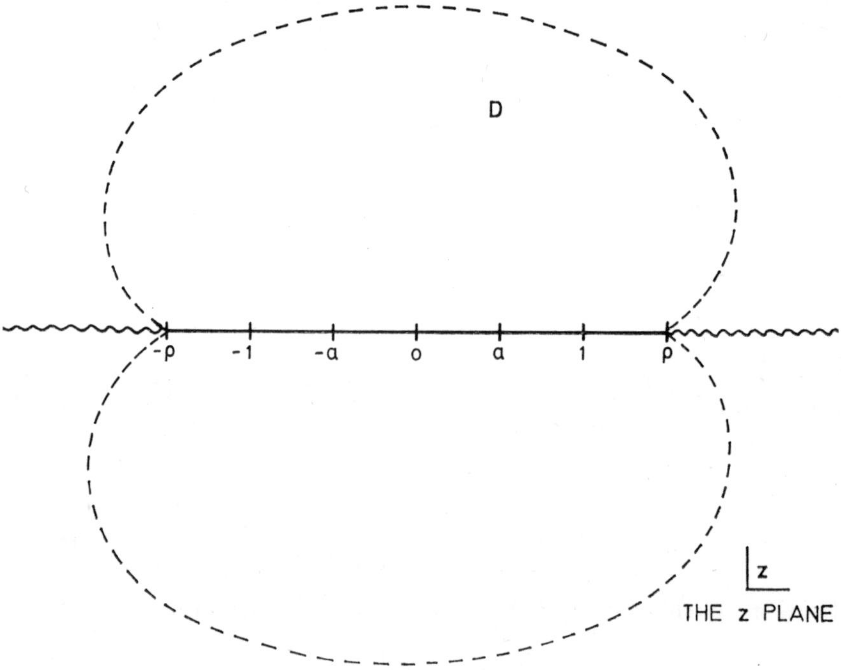

FIG. 5.1*a*

$f(s, \cos\theta)$ near to $z = \cos\theta = 0$. Incidentally they obtain also a new bound on the amplitude at fixed (physical) momentum transfer, provided certain conditions are satisfied. As with previously discussed work, a bound is obtained for large s, using analyticity in z.

The assumptions are:

1. Analyticity in z in a domain D whose approximate form is shown in the z-plane in Fig. 5.1a and whose detailed form is given later. D meets the real z-axis at $z = \rho$, which corresponds to a fixed point in the variable t (e.g. $t = m^2$, or $4m^2$), call this point

$$t = 2b. \qquad ...[5.1.1]$$

Then the z-plane is assumed to have cuts from $-\infty$ to $-\rho$ and ∞ to ρ, where

$$\rho = 1 + \frac{b}{k^2}. \qquad ...[5.1.2]$$

2. Temperedness in the form of a polynomial bound

$$|f(s, z)| < s^N \qquad ...[5.1.3]$$

where N is independent of s and z, within D.

3. On a segment of the real z axis, $(-a, a)$, the modulus of f is bounded by a given function,

$$|f(s, z)| \leqq \exp[-\phi(s)], \quad -a < z < a, \qquad ...[5.1.4]$$

where $0 < a < 1$. Thus for z in the range $(-a, a)$ the momentum transfer is large, $t \sim -s$, but is physical.

From these hypotheses a bound can be obtained on $f(s, 1)$, the forward scattering amplitude. The basic idea is that an analytic function that has the bound [5.1.4] near $z = 0$, and [5.1.3] for $|z|$ at a specified value (greater than one) within D, must increase reasonably steadily through intermediate values. This can be put in a precise form through Hadamard's three circles theorem (ref. Titchmarsh, *Theory of Functions*, and note that the statement of this theorem has been misprinted in the Cerulus-Martin letter).

In order to apply Hadamard's theorem the domain D and the line segment [5.1.4] must be made circular by two conformal transformations:

1. $$\omega = \frac{\rho}{z}[\rho - (\rho^2 - z^2)^{\frac{1}{2}}]. \qquad ...[5.1.5]$$

This transforms the cut z-plane of Fig. 5.1a into the interior of a circle of radius ρ in the ω-plane of Fig. 5.1b. Also

$$z = \pm 1 \rightarrow \omega = \pm e = \pm\rho[\rho - (\rho^2 - 1)^{\frac{1}{2}}], \qquad ...[5.1.6]$$

$$z = \pm a \rightarrow \omega = \pm a' = \pm\rho[\rho - (\rho^2 - a^2)^{\frac{1}{2}}]/a. \qquad ...[5.1.7]$$

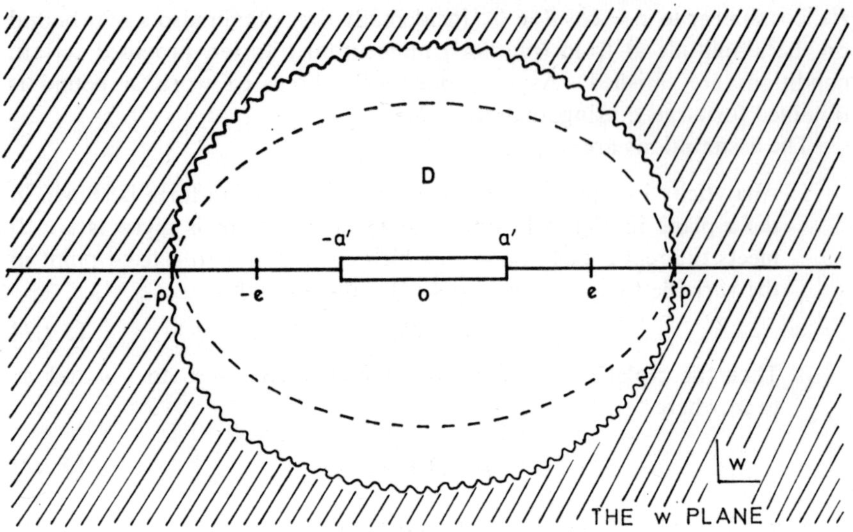

FIG. 5.1*b*

2. $$\tau = \frac{1}{a'}\,[\omega+(\omega^2-a'^2)^{\frac{1}{2}}].\qquad ...[5.1.8]$$

This transforms the cut in ω, $(-a', a')$, into the unit circle in τ. Also

$$\omega = \rho \rightarrow \tau = R = \frac{1}{a'}\,[\rho+(\rho^2-a^2)^{\frac{1}{2}}],\qquad ...[5.1.9]$$

$$\omega = e \rightarrow \tau = E = \frac{1}{a'}\,[e+(e^2-a^2)^{\frac{1}{2}}].\qquad ...[5.1.10]$$

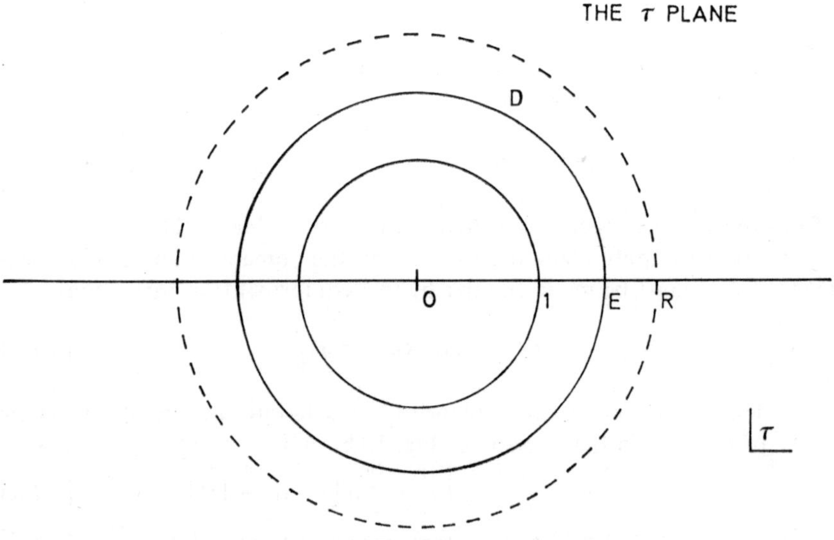

FIG. 5.1*c*

The branch cut in the z-plane, which in the ω-plane is the circle radius ρ, lies outside the circle radius R in the τ-plane except for the end points at $+R$ and $-R$.

The domain D is defined by the requirement that its transform is the circle radius R in the τ-plane. Its precise form in the z-plane need not concern us except to note that D includes momentum transfers with $|t| \sim |s|$, which is the cause of the cautionary doubt about the temperedness assumption.

Hadamard's three circles theorem states (ref. Titchmarsh, *Theory of Functions*, page 172):

Let $f(z)$ be an analytic function, regular for $r_1 \leqq |z| \leqq r_3$. Let $r_1 < r_2 < r_3$, and let M_1, M_2, M_3 be the maxima of $|f(z)|$ on the three circles $|z| = r_1, r_2, r_3$ respectively. Then

$$M_2^{\log(r_3/r_1)} \leqq M_1^{\log(r_3/r_2)} \,.\, M_3^{\log(r_2/r_1)}. \qquad \ldots[5.1.11]$$

Applying this theorem to the three circles in the τ-plane of Fig. 5.1c, one gets

$$\log M \leqq \left\{\frac{\log R - \log E}{\log R}\right\} \log M_1 + \left\{\frac{\log E}{\log R}\right\} \log M_2. \qquad \ldots[5.1.12]$$

Now

$$\frac{\log E}{\log R} = 1 - \frac{C(2b)^{\frac{1}{2}}}{k} + O\left(\frac{1}{k^2}\right) \qquad \ldots[5.1.13]$$

where

$$C = a[a^2 - \{1-(1-a^2)^{\frac{1}{2}}\}^2]^{-\frac{1}{2}}.$$

Substituting [*5.1.13*] into [*5.1.12*],

$$|f(s,1)| \leqq M \leqq \exp\left[-\phi(s)\, C\frac{(2b)^{\frac{1}{2}}}{k}\right] \exp\left[N \log s \left(1 - C\frac{(2b)^{\frac{1}{2}}}{k}\right)\right]. \qquad \ldots[5.1.14]$$

By taking a circle of radius less than E, a bound in $f(s, t)$ can be obtained. With

$$t = -2k^2(1-z)$$

$$0 > t, \quad \text{and} \quad |t| = O(s),$$

then

$$|f(s,t)| < \exp\left[-\phi(s)\frac{C(2b-t)^{\frac{1}{2}}}{k}\right] \exp\left[N \log s \left(1 - \frac{C(2b-t)^{\frac{1}{2}}}{k}\right)\right]. \qquad \ldots[5.1.15]$$

Cerulus and Martin discuss the consequences of taking the bound [*5.1.4*] near $\cos\theta = 0$ to have the form

$$\phi(s) = C's^{\alpha}(\log s)^{\beta}. \qquad \ldots[5.1.16]$$

If $\alpha > \frac{1}{2}$, then [*5.1.14*] gives

$$|f(s, 1)| < s^{-M} \quad \text{as} \quad s \to \infty$$

where M may be as large as we please. Presumably this is ruled out by experiment since $|f| \sim s$ in order to give a constant cross section.

If $\alpha < \frac{1}{2}$ then

$$|f(s, 1)| < s^{N} \quad \text{as} \quad s \to \infty. \qquad \text{...[5.1.17]}$$

Nothing has been gained over the initial assumptions in this case.

If $\alpha = \frac{1}{2}$, $\beta > 1$,

$$|f(s, 1)| < s^{-(\log s)^{\beta - 1}} \quad \text{as} \quad s \to \infty. \qquad \text{...[5.1.18]}$$

This rapid decrease is also inconsistent with experiment unless the relevant constants delay the onset of this asymptotic behaviour to larger s than is met in existing experiments. If $\alpha = \frac{1}{2}$ and $\beta < 1$, the result [*5.1.17*] holds, so nothing new is obtained.

If $\alpha = \frac{1}{2}$, $\beta = 1$,

$$|f(s, 1)| \leqq \exp\left[\{N - CC'(2b)^{\frac{1}{2}}\} \log s\right]. \qquad \text{...[5.1.19]}$$

In this case if, on experimental grounds, we require the cross section to be constant (to within a power of a logarithm), we obtain

$$N \leqq CC'(2b)^{\frac{1}{2}}. \qquad \text{...[5.1.20]}$$

Cerulus and Martin give an example within this class for which they verify that unitarity conditions are satisfied. Thus one cannot obtain more restrictive information by using s channel unitarity.

I conclude by discussing the possible replacement of the temperedness condition by a weaker one. Replace [*5.1.3*] by a condition that could reasonably apply if Regge branch cuts extend to $\alpha(t) = t^{\frac{1}{2}}$. Then instead of [*5.1.3*] we assume in D,

$$|f(s, z)| < \exp\left[C' s^{\frac{1}{2}} \log s\right]. \qquad \text{...[5.1.21]}$$

Then [*5.1.14*] is replaced by

$$|f(s, 1)| \leqq M \leqq \exp\left[-\phi(s)\, C\frac{(2b)^{\frac{1}{2}}}{k}\right] \exp\left[C' s^{\frac{1}{2}} (\log s)\right] \qquad \text{...[5.1.22]}$$

$$\leqq \exp\left[\frac{C'}{k}\left\{s - \frac{C(2b)^{\frac{1}{2}}}{C'}\right\}\right]. \qquad \text{...[5.1.23]}$$

It is possible now to have a constant cross section when $\phi(s)$ takes the form

$$\phi(s) = C' s^{\alpha} (\log s)^{\beta} \qquad \text{...[5.1.24]}$$

for $\alpha < 1$, and any value of β; or for $\alpha = 1$ and β negative or zero.

5.2. *Lower bounds for forward scattering and total cross sections*

Although the results for lower bounds on the asymptotic form for the forward amplitude are rather weak when derived from general assumptions, it is of considerable theoretical importance that they exist. The most convenient theoretical technique is based on Herglotz functions and was used in this context by Jin and Martin (1964), whose method is outlined here.

A Herglotz function $H(z)$ satisfies the conditions:

(a) it is regular in Im $z > 0$, such that Im $H(z) > 0$, when Im $z > 0$,
(b) it has a representation

$$H(z) = A+Bz+\frac{1}{\pi}\int_{-\infty}^{\infty}\frac{\operatorname{Im} H(x)(1+zx)\,dx}{(1+x^2)(x-z)} \qquad \ldots[5.2.1]$$

with $B \geqq 0$, Im $H(x) \geqq 0$, and
(c) the integral

$$\int_{-\infty}^{\infty}\frac{\operatorname{Im} H(x)\,dx}{1+x^2} \quad \text{is convergent.} \qquad \ldots[5.2.2]$$

In addition $-[H(z)]^{-1}$ is a Herglotz function which means that

$$\int_{-\infty}^{\infty}\frac{\operatorname{Im} H(x)\,dx}{|H(x)|^2(1+x^2)} \quad \text{is convergent.} \qquad \ldots[5.2.3]$$

For z complex, $\varepsilon < \arg z < \pi-\varepsilon$,

$$\frac{C}{|z|} < |H(z)| < C|z|. \qquad \ldots[5.2.4]$$

An analogous condition is required for z real. This can be obtained from the function

$$G(z) = \frac{1}{z}\int_0^z H(z')\,dz'. \qquad \ldots[5.2.5]$$

Using a straight line for the integration path it can be seen that $G(z)$ is also a Herglotz function. In particular putting G into [5.2.3]

$$\int_A^{\infty}\frac{dx}{x}\left[\int_0^x \operatorname{Im} H(x')\,dx'\right]\left|\int_0^x H(x')\,dx'\right|^{-2} \text{converges.} \qquad \ldots[5.2.6]$$

Since $\int_0^x \operatorname{Im} H(x')\,dx'$ is non-decreasing, this means that

$$\int_0^x |H(x')|\,dx' > C(\log x)^{\frac{1}{2}}. \qquad \ldots[5.2.7]$$

Hence

$$\lim_{x\to\infty} x\,|H(x)|\,(\log x)^{\frac{1}{2}} \to +\infty. \qquad \ldots[5.2.8]$$

These results are extended by Jin and Martin (1964) to functions that are analytic in a cut plane, with Im $F(x+i0)$ positive on the right-hand cut and negative on the left-hand cut.

The symmetric scattering amplitude $F(z)$ may have complex zeros at z_r, z_r^* and real zeros at $x_1 \ldots x_{2n+1}$. Write it as

$$F(z) = \prod (z-z_r)(z-z_r^*) \prod_{1}^{2n} (z-x_p) H(z). \qquad \ldots[5.2.9]$$

Then it can be shown that $H(z)$ is Herglotz (ref. Symanzik (1960)). By leaving one zero x_{2n+1} on the real axis between the cuts, Im $H(x+i0)$ becomes positive (or zero) along the entire x axis.

The conditions on $F(x)$ analogous to [*5.2.7*] or [*5.2.8*] can be shown to be

$$\int_{x_0}^{x} dx'x' \,|\, F(x') \,|\, > (\log x)^{\frac{1}{2}} \qquad \ldots[5.2.10]$$

$$\lim_{x \to \infty} x^2 (\log x)^2 \,|\, F(x) \,|\, = +\infty. \qquad \ldots[5.2.11]$$

If $F(x)$ oscillates at infinity one uses the least upper bound in [*5.2.11*].

From these conditions it follows that $F(x)$ cannot decrease arbitrarily fast. In particular using [*5.2.11*],

$$\text{if} \quad |\, F(s, 0) \,| \sim s^{\alpha}, \quad \text{then} \quad \alpha \geqq -2. \qquad \ldots[5.2.12]$$

For the non-forward amplitude, when $t > 0$, we have Im $F(s, t) > 0$ and one can proceed as before. However for physical values of t, ($t < 0$) the situation is much more difficult unless special assumptions are made. For any fixed t as $s \to \infty$, it is possible that Im $F(s, t)$ oscillates an infinite number of times, having an infinity of zeros (see section 5.3). Jin and Martin (1964) make the special assumption that

$$F(s, t) \sim s^{\alpha(t)} f(t) \quad \text{for} \quad s \to \infty. \qquad \ldots[5.2.13]$$

Then there can be only a finite number $v(t)$ of oscillations of Im $F(s, t)$ and [*5.2.10*] is replaced by

$$\int_{x_0}^{x} dy y^{v+1} \,|\, F(y) \,|\, > (\log x)^{\frac{1}{2}} \qquad \ldots[5.2.14]$$

giving

$$\alpha \geqq -2 - v(t). \qquad \ldots[5.2.15]$$

Under certain assumptions about low energy behaviour Jin and Martin (1964) show that these lower bounds can be improved. Their work is also extended to partial wave amplitudes (see also Jin (1966) and Kinoshita (1966)) and it is undoubtedly of importance for models and dynamics of strong interactions. However it goes outside the general theory of these lectures and will not be considered here.

5.3. *The problem of oscillations*

If it is assumed that the amplitude $F(s, t)$ does not have oscillations as $s \to \infty$ for fixed t one obtains considerable simplification in the theory. We have noted these simplifications in section 4 and will use them again in sections 6 and 7. In general, the assumption of no oscillations leads to stronger results though this is not so for the Froissart bound or for Martin's (1965b) enlargement of analyticity domains (§ 6.4).

The nature of the general problem of oscillations has been considered by Eden and Łukaszuk (1966). Theorems on the oscillations of Im $F(s)$ for s real that are compatible with cut plane analyticity and polynomial boundedness of $F(s)$ do not seem to yield useful results. By considering entire functions of $\log s$ that are polynomial bounded in the cut s-plane one can obtain an indication that oscillations become slower as $s \to \infty$ in the sense that

$$F(s+a)-F(s) \sim \frac{1}{s} \max\left[F(s), F(s+a)\right]. \qquad \ldots[5.3.1]$$

An example of a function of this type is

$$F(s, 0) \sim is[C-\sin\{\log^{\beta} s\}]. \qquad \ldots[5.3.2]$$

From unitarity, $C \geqq 1$, and from polynomial boundedness, $\beta \leqq 2$. For this example the distance between successive oscillations is

$$\frac{\pi s}{\beta(\log s)^{\beta-1}}. \qquad \ldots[5.3.3]$$

This can be compared with the distance $s^{\frac{1}{2}}$ between successive thresholds. The simplifying assumption that F is an entire function of $(\log s)$ may be crucial to this result. The problem of oscillations is vital to our understanding of high-energy behaviour but it has so far proved rather baffling except under special assumptions.

6

Crossing Symmetry and Analyticity in the Energy

6.1. *Relations between different processes; the Phragmén-Lindelöff theorem*

Up to now our discussion has utilised analyticity in the momentum transfer, t, but has not made significant use of analyticity in the energy variable s, although polynomial boundedness for large s has been utilised.

For real t in $-4m^2 < t < 4m^2$, the scattering amplitude $f(s, t)$ is analytic in the complex s-plane cut along $(-\infty, -t)$ and $(4m^2, \infty)$. If we consider only large $|s|$, this analyticity does not seem to give useful restrictions by itself, but when it is combined with crossing symmetry and hermiticity there are found to be restrictive relations on $F(s, t)$. These have been studied under quite general conditions in papers by Khuri and Kinoshita (1965a, b, c). A rigorous discussion of bounds using crossing and analyticity in s has also been made by Jin and MacDowell (1965) using the phase representation for the amplitude (ref. Sugawara and Tubis (1963)). I shall discuss the general results later (§ 6.2) but first I shall use a more simple approach which obtains the main features of the dominant terms in the scattering amplitude. This method has been used by Meiman (1962), Logunov *et al.* (1963) and Van Hove (1964), following an earlier paper by Sugawara and Kanezawa (1961).

We shall consider equal mass particles

$$\begin{aligned} s &= 4(m^2+k^2) \\ t &= -2k^2(1-\cos\theta) \\ u &= -2k^2(1+\cos\theta). \end{aligned} \qquad \text{...[6.1.1]}$$

It is convenient to work in terms of the laboratory energy E instead of s. The incident particle has four vector (E, p) with $E^2 = m^2+p^2$. Then the invariant energy squared is

$$\begin{aligned} s &= (E+m)^2-p^2 \\ s &= 2m(E+m). \end{aligned} \qquad \text{...[6.1.2]}$$

For forward scattering $\theta = 0$, and the scattering amplitude

$$F(s, t) = F(s, 0) = F(E) \qquad \text{...[6.1.3]}$$

has branch cuts along

$$-\infty < s < 0, \quad 4m^2 < s < \infty. \qquad \text{...[6.1.4]}$$

From our discussion in section 3, we know that, with identical bosons, crossing symmetry gives the same scattering amplitude under the substitution

$$s+i\varepsilon \rightarrow u+i\varepsilon = 4m^2-s-i\varepsilon \qquad \text{...[6.1.5]}$$

when $t = 0$, which corresponds to $E \rightarrow -E$ (real or complex E).

Hence the forward scattering amplitude F has the following properties:

1. $F(E)$ is analytic in E, regular in a plane cut along $(-\infty < E < -m)$ and $(m < E < \infty)$.
2. $F(E)$ is hermitean, in particular for E real,

$$F(E-i\varepsilon) = F^*(E+i\varepsilon) \qquad \text{...[6.1.6]}$$

   where ε is small.
3. $F(E)$ is crossing symmetric,

$$F(E+i\varepsilon) = F(-E-i\varepsilon). \qquad \text{...[6.1.7]}$$

Hence from [*6.1.6*] and [*6.1.7*]

$$F(E+i\varepsilon) = F^*(-E+i\varepsilon) \qquad \ldots[6.1.8]$$

or, writing E for $E+i0$,

$$F(E \exp i\pi) = F^*(E). \qquad \ldots[6.1.9]$$

If we now assume any given functional form for $|F|$ as $E \to \infty$ along the real axis, the phase of F can be found from [*6.1.9*] and the Phragmén-Lindelöff theorems (ref. Titchmarsh (1939) and Boas (1954)). We state the most relevant of these:

A Phragmén-Lindelöff Theorem

Let $f(z)$ be an analytic function of $z = r \exp i\theta$, regular in the region D between two straight lines making an angle π/α at the origin and continuous on the lines themselves. Suppose that as $r \to \infty$,

$$f(z) = O\,[\exp(r^\beta)],\ \beta < \alpha, \quad \text{(A)}$$

in the region D, and that

$$|f(z)| \leqq M \qquad \text{(B)}$$

on the lines, then the inequality (B) holds throughout the region D.

We shall use this theorem and its variations as given for example in Boas (1954). In particular, when D is the half-plane, $\text{Im}\, z > 0$, we will use the consequence that:

Corollary, If $f(z) \to a$ along two rays then $f(z) \to a$ along any ray in D. This corollary includes the rays $z = x \to \pm\infty$ (along the real axis).

Now assume a given functional form for the symmetric scattering amplitude $|F(E)|$ as $E \to \infty$ along the real axis, namely for some α and constant M

$$\left[\frac{F(E)}{E^\alpha}\right] \to M \exp i\theta \quad \text{as} \quad E \to \infty, \qquad \ldots[6.1.10]$$

then to determine θ we use the above Corollary and [*6.1.9*]

$$\frac{F(E \exp i\pi)}{E^\alpha \exp i\pi\alpha} \to M \exp i\theta \qquad \ldots[6.1.11]$$

and

$$F(E \exp i\pi) = F^*(E). \qquad \ldots[6.1.12]$$

Hence, dividing [*6.1.10*] and [*6.1.11*], and using [*6.1.12*],

$$\exp(2i\theta) = \exp(-i\pi\alpha), \qquad \ldots[6.1.13]$$

$$\theta = n\pi - \tfrac{1}{2}\alpha\pi. \qquad \ldots[6.1.14]$$

For large E

$$F(E) \sim \pm E^\alpha M \exp(-\pi\alpha i/2). \qquad \ldots[6.1.15]$$

The $\pm$ sign can be resolved by the unitarity requirement

$$\text{Im}\, F(E) > 0, \qquad \text{...[6.1.16]}$$

for E real above the right-hand branch cut.

Thus if we are given a polynomial (high E) form for the total cross section for a symmetric scattering process, such that

$$\sigma(\text{total}) \sim CE^{\alpha-1} \qquad \text{...[6.1.17]}$$

then the optical theorem gives

$$\text{Im}\, F(E) \sim CE^{\alpha}, \qquad \text{...[6.1.18]}$$

and crossing symmetry as stated above gives

$$F(E) \sim CE^{\alpha}\left[-\cot\frac{\pi\alpha}{2}+i\right]. \qquad \text{...[6.1.19]}$$

The central result is that if $F(E)$ is a function of $(-iE) = E \exp(-i\pi/2)$ with real coefficients then it is explicitly crossing symmetric. Another example is given by

$$F(E) \sim \pm(-iE)^{\alpha}\,[\log E - i\pi/2]^{\beta} \qquad \text{...[6.1.20]}$$

where α and β are constants.

6.2. *The real part of the forward scattering amplitude*

We have noted in section 2.2 that the real part of a forward scattering amplitude can be measured by Coulomb interference. In an independent experiment the rate of change with energy of the total cross section can be measured. In this section we discuss the relation between these two measurements for a crossing symmetric amplitude—for example the sum of π^-p and π^+p scattering in the forward direction ($t = 0$)

$$F(E) = F_{\pi^+p}(E, 0)+F_{\pi^-p}(E, 0). \qquad \text{...[6.2.1]}$$

Then $F(E)$ satisfies the relation [*6.1.8*]

$$F(E+i\varepsilon) = F^*(-E+i\varepsilon). \qquad \text{...[6.2.2]}$$

Also

$$\frac{\text{Im}\, F(E)}{E} \sim \sigma(\pi^+p)+\sigma(\pi^-p). \qquad \text{...[6.2.3]}$$

Let us assume an explicit functional form for these total cross sections that corresponds to the experimental results that they decrease towards a constant value. This can be represented by

$$\text{Im}\, F(E) = E(C+C'E^{-\alpha}) \qquad \text{...[6.2.4]}$$

where C, C' are real and positive, and $\alpha \sim 0{\cdot}5$.

Applying the rules of section 6.1,

$$F(E) = iEC - \frac{C'E^{1-\alpha}\exp[-i\pi(1-\alpha)/2]}{\sin[\pi(1-\alpha)/2]}. \qquad \ldots[6.2.5]$$

This gives

$$\frac{\operatorname{Re} F}{\operatorname{Im} F} \sim -(E^{-\alpha})\frac{C'\cot[\pi(1-\alpha)/2]}{C+C'E^{-\alpha}}. \qquad \ldots[6.2.6]$$

Thus the ratio tends to zero from negative values, as E tends to infinity. The above result is a feature of any symmetric amplitude that corresponds to cross sections that decrease monotonically towards a constant. If more general behaviour of cross sections is allowed, including possible oscillations, it is possible to obtain bounds on the rate of change of $|F(E)|$ from given bounds on the ratio of real to imaginary parts. These have been investigated by Khuri and Kinoshita (1965c) using the univalent function $g(E)$ defined by

$$g(E) = \int_0^E dE' \frac{F(E')-F(0)}{(E')^2} \qquad \ldots[6.2.7]$$

where the integration is to be taken entirely in $\operatorname{Im} E > 0$. It is assumed that $|F(E)|$ does not increase faster than $|E|^2$ as $|E| \to \infty$ (see the next section), so

$$\left|\frac{g(E)}{E}\right| \leqq \text{constant}. \qquad \ldots[6.2.8]$$

Khuri and Kinoshita (1965c) show that if, for E real $> E_0$,

$$\tan \pi\beta \leqq \frac{\operatorname{Re} g(E)}{\operatorname{Im} g(E)} \leqq \tan \pi\gamma \qquad \ldots[6.2.9]$$

with $0 < \beta < \gamma < \frac{1}{2}$, then for $E > E_0$,

$$C\left(\frac{E}{E_0}\right)^{2\beta} \leqq |g(E)| \leqq C\left(\frac{E}{E_0}\right)^{2\gamma}. \qquad \ldots[6.2.10]$$

In the special case where $\beta = \gamma$ this result corresponds to the example noted above in equation [6.2.6]. They also obtain the following theorem: if, for real $E > E_0$,

$$\frac{\operatorname{Re} g(E)}{\operatorname{Im} g(E)} < \frac{C}{(\log E)^\alpha}, \quad \alpha \geqq 1 \qquad \ldots[6.2.11]$$

then, for $E > E_0$, $\alpha > 1$,

$$|g(E)| \leqq \text{constant}. \qquad \ldots[6.2.12]$$

If $\alpha = 1$, then

$$|g(E)| \leqq C''(\log E)^{C'} \qquad \ldots[6.2.13]$$

where $C' \sim 2^{\frac{3}{2}}C$ as $E_0 \to \infty$.

Many other theorems of this type can be obtained, some are listed by Khuri and Kinoshita (1965c). In general for heuristic purposes it is simpler to assume specific smooth functional forms for $F(E)$ as discussed in section 6.1.

An alternative approach by Jin and MacDowell (1965) makes use of the phase representation of Sugawara and Tubis (1963). They obtain results that are similar to those given above. Crossing symmetry and phase relations are also discussed by Eden (1965) and (1967).

Note on crossing symmetry for non-forward amplitudes

If t is non-zero, crossing symmetry involves a more inconvenient variable than E. We must still take the symmetric amplitude

$$F(E, t) = F(\pi^+ p) + F(\pi^- p) \qquad \ldots[6.2.14]$$

but must use instead of E, the variable W. For equal mass particles this is

$$W = s + \tfrac{1}{2}t - 2m^2 \qquad \ldots[6.2.15]$$

so the substitution, for fixed t, of $s = 4m^2 - u - t$ gives

$$-W = u + \tfrac{1}{2}t - 2m^2. \qquad \ldots[6.2.16]$$

Thus crossing symmetry between the physical regions where s and u respectively are the energies gives for the symmetric amplitude F, written as a function of W and t,

$$F(W + i\varepsilon, t) = F^*(-W + i\varepsilon, t) \qquad \ldots[6.2.17]$$

6.3. *Fixed transfer dispersion relations*

The importance of dispersion relations for strongly interacting particles was first noted by Goldberger (1955).
The Froissart bound

$$\sigma(E) \leqq C \log^2 E \qquad \ldots[6.3.1]$$

follows from the result of Martin (§ 6.4). This establishes the convergence of the forward scattering dispersion relation with two subtractions,

$$F(E) = F(0) + EF'(0) + \frac{E^2}{\pi}\int_m^\infty \frac{dE' \operatorname{Im} F(E')}{E'^2(E'-E)} + \frac{E^2}{\pi}\int_{-\infty}^{-m} \frac{dE' \operatorname{Im} F(E')}{E'^2(E'-E)}. \qquad \ldots[6.3.2]$$

If $F(E)$ is the symmetric amplitude used in section 6.2 then the subtraction, constant $F'(0)$ is zero, and [6.3.2] becomes

$$F(E) - F(0) = \frac{2E^2}{\pi}\int_m^\infty \frac{dE'\sigma(E')}{(E'^2 - E^2)} \qquad \ldots[6.3.3]$$

where σ is the sum of the $\pi^+ p$ and $\pi^- p$ total cross sections. The real part of this equation gives a direct check against experiment of possible forms of $\sigma(E)$ as $E \to \infty$,

$$\text{Re}\,[F(E)-F(0)] = \frac{2E^2}{\pi}\,P\int_m^\infty \frac{dE'\sigma(E')}{(E'^2-E^2)}. \qquad \ldots[6.3.4]$$

It is generally more convenient (Lindenbaum 1965) to make the subtraction not at $E = 0$ but at a physical value of E in the high-energy range.

For fixed transfer $t < 0$, we have for physical E and t (§ 6.4)

$$|\text{Im}\,F(E,t)| < \text{Im}\,F(E,0). \qquad \ldots[6.3.5]$$

It follows that fixed transfer dispersion relations for $t < 0$ require only two subtractions; this can also be shown to hold for $t < 4m^2$.

6.4. *Martin's enlargement of the Lehmann ellipse*

In this section we do not assume analyticity in t outside the Lehmann ellipse. This means that, in the first instance, we have only the Greenberg-Low bound

$$\text{Im}\,F(E,0) < E^2 \log^2 E \qquad \ldots[6.4.1]$$

as $E \to \infty$. However we will assume the improvement on this bound that was derived by Martin (1965a) and by Eden (1966a), namely

$$\text{Im}\,F(E,0) < E^2(\log E)^{-2}. \qquad \ldots[6.4.2]$$

This ensures convergence of the forward dispersion relation, with two subtractions,

$$F(E,0) = A+BE+\frac{E^2}{\pi}\int_m^\infty dE'\,\frac{\text{Im}\,F(E',0)}{E'^2(E'-E)}+\frac{E^2}{\pi}\int_{-\infty}^{-m} dE'\,\frac{\text{Im}\,F(E',0)}{E'^2(E'-E)}. \qquad \ldots[6.4.3]$$

We now describe a simplified version of Martin's (1965b) enlargement of the Lehmann ellipse (see also Eden 1966c). The simplifications to be used are the neglect of the left-hand cut and the subtraction constants in [*6.4.3*] and in other dispersion relations that we use. The left-hand cut causes considerable complications but they are purely of a technical or algebraic nature and do not concern the main principles underlying the method. We therefore omit the second integral in [*6.4.3*] in the following.

We have (with equal mass particles $s = 2mE$)

$$\text{Im}\,F(E,t) = \frac{(2mE)^{\frac{1}{2}}}{8\pi k}\sum_0^\infty (2l+1)\,a_l(E)P_l(\cos\theta). \qquad \ldots[6.4.4]$$

From unitarity, for real positive E,

$$0 \leqq a_l(E) \leqq 1 \qquad \ldots[6.4.5]$$

and for $t < 0$, $|\cos\theta| < 1$,

$$|P_l(\cos\theta)| < 1. \qquad ...[6.4.6]$$

Hence for physical values of E and t, $(t < 0)$,

$$|\operatorname{Im} F(E, t)| < \operatorname{Im} F(E, 0). \qquad ...[6.4.7]$$

Also, for $|\cos\theta| < 1$,

$$\left|\left(\frac{d}{d\cos\theta}\right)^n P_l(\cos\theta)\right| < \left(\frac{d}{d\cos\theta}\right)^n P_l(1). \qquad ...[6.4.8]$$

From the inequality [6.4.7] it follows that our (simplified) dispersion relation converges for $t < 0$,

$$F(E_0, t) = \frac{E_0^2}{\pi}\int_m^\infty dE' \frac{\operatorname{Im} F(E', t)}{E'^2(E'-E_0)}. \qquad ...[6.4.9]$$

It is convenient to choose E_0 in the range $(-m < E_0 < m)$. In this range $F(E_0, t)$ is regular in some finite domain (at least) in t,

$$|t| \leqq R. \qquad ...[6.4.10]$$

Hence, using Cauchy's inequalities (Martin 1965b) we obtain, at $t = 0$,

$$\left|\left(\frac{d}{dt}\right)^n F(E_0, 0)\right| \leqq \frac{Mn!}{R^n} \qquad ...[6.4.11]$$

where $M(E_0)$ is the maximum of $|F(E_0, t)|$ on the circle $|t| = R$.

For $|E_0| < m$, the denominator in [6.4.9] is positive. The inequality [6.4.8], with $n = 1$, can be used with [6.4.4] and [6.4.9] to prove that

$$\frac{d}{dt}F(E_0, 0) = \frac{E_0^2}{\pi}\int_m^\infty dE' \frac{\frac{d}{dt}\operatorname{Im} F(E', 0)}{E'^2(E'-E_0)}. \qquad ...[6.4.12]$$

The inequality [6.4.8] is of particular importance in that it shows that the integral [6.4.12] converges. Similarly, Martin shows that, for any n,

$$\left(\frac{d}{dt}\right)^n F(E_0, 0) = \frac{E_0^2}{\pi}\int_m^\infty dE' \frac{\left(\frac{d}{dt}\right)^n \operatorname{Im} F(E', 0)}{E'^2(E'-E_0)}. \qquad ...[6.4.13]$$

Using [6.4.11] we have

$$\frac{Mn!}{R^n} > \frac{E_0^2}{\pi}\int_m^\infty dE' \frac{\left(\frac{d}{dt}\right)^n \operatorname{Im} F(E', 0)}{E'^2(E'-E_0)}. \qquad ...[6.4.14]$$

From the inequalities obtained in section 4.3 we have (Eden 1966c), for large E',

$$\frac{d^n}{dt^n}\,\text{Im}\,F(E',0) > E'\sigma^{n+1} \qquad \ldots[6.4.15]$$

where $\sigma = \sigma(E')$ is the total cross section.
If, for large E',

$$\sigma \sim (E')^{\alpha} \qquad \ldots[6.4.16]$$

then, from [*6.4.15*], we see that if $\alpha > 0$ the integral in [*6.4.14*] does not converge if n is chosen sufficiently large. Hence $\alpha \leqq 0$.
If, for large E',

$$\sigma \sim (\log E')^{\beta} \qquad \ldots[6.4.17]$$

then

$$\frac{Mn!}{R^n} > \frac{E_0^2}{\pi}\int_m^{\infty}\frac{(\log E)^{(n+1)\beta}dE}{E^2(E-E_0)}. \qquad \ldots[6.4.18]$$

The integral is dominated by its value near

$$\log E \sim n \qquad \ldots[6.4.19]$$

when n is large. It follows that in order not to contradict this inequality

$$\beta \leqq 2, \qquad \ldots[6.4.20]$$

so that for large E we have the Froissart bound

$$\sigma(E) \leqq (\log E)^2. \qquad \ldots[6.4.21]$$

We return to Martin's method. The relation [*6.4.13*] holds for any E_0 in $(-m, m)$. This relation can therefore be continued analytically from E_0 to complex E_1 as long as the integral converges. Now the integral converges for arbitrary complex E_1 because

$$\left|\frac{\left(\frac{d}{dt}\right)^n \text{Im}\,F(E',0)}{E'-E_1}\right| \leqq \frac{\left(\frac{d}{dt}\right)^n \text{Im}\,F(E',0)}{E'-E_0}\,B(E_1,E_0) \qquad \ldots[6.4.22]$$

where

$$B(E_1,E_0) = \underset{E'>m}{\text{Sup}}\left|\frac{E'-E_0}{E'-E_1}\right|. \qquad \ldots[6.4.23]$$

Hence

$$\left|\left(\frac{d}{dt}\right)^n F(E_1,0)\right| < B(E_1,E_0)\,\frac{Mn!}{R^n}. \qquad \ldots[6.4.24]$$

This establishes that the series

$$F(E_1, t) = \sum_0^\infty C_n(E_1)t^n \qquad \text{...[6.4.25]}$$

converges for

$$|t| < R \qquad \text{...[6.4.26]}$$

where R does not depend on E_1. This is Martin's main result. It should be compared with the shrinkage (C/E) of the analytic neighbourhood of $t = 0$ that corresponds to the Lehmann ellipse.

Further work by Martin (1966a, b) has extended the domain of analyticity of $F(E, t)$, but not in a way that gives any further improvement on high energy bounds. However, his work proceeds a significant part of the distance towards proving the cut-plane analyticity that is assumed by the Mandelstam representation.

7

The Pomeranchuk Theorems

These theorems relate to the asymptotic equality at high energy of cross sections for different processes. There are three types of result that were first indicated by Pomeranchuk (1956, 1958). The first was his suggestion that total cross sections tend to a constant at high energies. This result has never been proved from any basic theoretical assumptions although it is strongly suggested by experiment. The best theoretical result is the Froissart bound.

The second result is the asymptotic equality of particle-particle and of antiparticle-particle total cross sections. This is usually what is meant when one talks of the Pomeranchuk theorem. Various proofs of this theorem have been given but they all require assumptions additional to the basic axioms of quantum field theory. They are reviewed by Kinoshita (1966).

The third result is the equality of elastic differential cross sections. This is a consequence of crossing symmetry and is mathematically simpler than the Pomeranchuk theorem for total cross sections. There are useful reviews of these results by Greenberg (1964) and Van Hove (1964) and to some extent I will follow the latter's methods, which are based on earlier work by Meiman (1962). They involve essentially the same mathematics as was used in section 6 in discussing phase relations.

7.1. *Crossing symmetry and analyticity*

Consider the process $A+B \to A+B$. By crossing symmetry this is related to two other processes:

I	$A+B \to A+B$,	energy s,
II	$A+\bar{B} \to A+\bar{B}$,	energy u,
III	$A+\bar{A} \to B+\bar{B}$,	energy t.

We shall be concerned with the relations between processes I and II, for example when I denotes π^+p scattering and II denotes π^-p scattering. Then, as noted in section 6.1, for forward scattering, crossing symmetry gives

$$F_{\mathrm{I}}(-s-i\varepsilon, 0) = F_{\mathrm{II}}(s+i\varepsilon, 0). \qquad ...[7.1.1]$$

Hermitean analyticity gives

$$F_{\mathrm{I}}(-s+i\varepsilon, 0) = F_{\mathrm{I}}^*(-s-i\varepsilon, 0). \qquad ...[7.1.2]$$

For the non-forward amplitude we must use the crossing symmetric variable w instead of s,

$$w = s+\tfrac{1}{2}t-2m^2. \qquad ...[7.1.3]$$

Combining relations similar to [7.1.2] and [7.1.3] we obtain

$$F_{\mathrm{I}}(-w+i0, t) = F_{\mathrm{II}}^*(w+i0, t). \qquad ...[7.1.4]$$

It is simplest now to assume that these amplitudes have a definite functional form at infinity that excludes oscillations. We assume:

(a) for fixed t, $F_{\mathrm{I}}(w, t)$ is regular in $\mathrm{Im}\, w > 0$, and continuous on $\mathrm{Im}\, w = 0$,
(b) for fixed t, $F_{\mathrm{I}}(w, t)$ is polynomial bounded in w for large w,
(c) for fixed t, as $w \to \infty$,

$$\frac{F_{\mathrm{I}}(w+i0, t)}{w^{\alpha}(\log w)^{\beta}} \to C_1(t) \qquad ...[7.1.5]$$

where α and β depend on t.

Similar conditions on F_{II} give

$$\frac{F_{\mathrm{II}}(w+i0, t)}{w^{\alpha}(\log w)^{\beta}} \to C_2(t). \qquad ...[7.1.6]$$

Using [*7.1.4*] this gives

$$\frac{F_{\mathrm{I}}(-w+i0, t)}{(-w)^{\alpha}(\log(-w))^{\beta}} \to C_2^*(t) \exp(-i\pi\alpha). \qquad ...[7.1.7]$$

The Phragmén-Lindelöff theorem (§ 6.1) applied to the function on the left of [*7.1.5*] shows that

$$C_1(t) = C_2^*(t) \exp[-i\pi\alpha(t)]. \qquad ...[7.1.8]$$

The asymptotic equality of differential cross sections follows from this result (Logunov *et al.* 1963) since we have, as $w \to \infty$,

$$\left|\frac{F_{\mathrm{I}}(w+i0, t)}{F_{\mathrm{II}}(w+i0, t)}\right| \to \left|\frac{C_1}{C_2}\right| = 1. \qquad ...[7.1.9]$$

However this does *not* establish the asymptotic equality of total cross sections since these are related to $\mathrm{Im}\, F$ which depends on the phase. We consider this in the next section.

7.2. *Asymptotic equality of total cross sections*

With the special form for $F(w, t)$ assumed in section 7.1, we can prove the Pomeranchuk theorem on total cross sections. Note however that the form assumed already restricts the cross sections more strongly than the Froissart bound since if

$$F(s, t) \sim C(t) s^{\alpha(t)} (\log s)^{\beta(t)} \qquad \text{...[7.2.1]}$$

then

$$\frac{d}{dt} \log [\operatorname{Im} F(s, 0)] \sim \alpha(0) \log s. \qquad \text{...[7.2.2]}$$

From the inequality obtained in section 4.3, the total cross section is bounded by the logarithmic derivative

$$\sigma(s) < C\alpha(0) \log s, \qquad \text{...[7.2.3]}$$

which is one power of log s better than the Froissart bound.

Despite the special features of [*7.2.1*], a sum of terms of this type is a convenient form to illustrate the general principles that underlie the proofs of the Pomeranchuk theorem on total cross sections.
Consider the antisymmetric amplitude

$$F_-(s, t) = F_{\text{I}}(s, t) - F_{\text{II}}(s, t). \qquad \text{...[7.2.4]}$$

Strictly, we should use the variable w instead of s but it is not essential in this case for large s. Crossing symmetry for F_- gives

$$F_-(-s+i0, t) \sim -F_-^*(s+i0, t). \qquad \text{...[7.2.5]}$$

Note the additional minus sign compared with [*6.1.8*] for F_+ (the sum of F_{I} and F_{II}),

$$F_+(-s+i0, 0) = F_+^*(s+i0, 0). \qquad \text{...[7.2.6]}$$

We require to prove that

$$\frac{\operatorname{Im} F_{\text{I}}(s, 0)}{\operatorname{Im} F_{\text{II}}(s, 0)} \to 1 \quad \text{as } s \to \infty. \qquad \text{...[7.2.7]}$$

Equivalently we require

$$\frac{\operatorname{Im} F_-(s, 0)}{\operatorname{Im} F_+(s, 0)} \to 0 \quad \text{as } s \to \infty. \qquad \text{...[7.2.8]}$$

If this is not so then both numerator and denominator of [*7.2.8*] must be dominated by the same leading term, thus

$$\operatorname{Im} F_+(s, 0) \sim C_+ s^{\alpha} (\log s)^{\beta} \qquad \text{...[7.2.9]}$$

$$\operatorname{Im} F_-(s, 0) \sim C_- s^{\alpha} (\log s)^{\beta}. \qquad \text{...[7.2.10]}$$

To be definite, we will take $\alpha = 1$, (otherwise the total cross sections would decrease). From [7.2.6] and [7.2.9]

$$F_+(s, 0) \sim C_+ is\left[\log s - \frac{i\pi}{2}\right]^\beta. \qquad \ldots[7.2.11]$$

However, due to the minus sign in [7.2.5], we have

$$F_-(s, 0) \sim \frac{-C_- s}{(\beta+1)\pi/2}\,[\log s - i\pi/2]^{\beta+1}. \qquad \ldots[7.2.12]$$

Thus if $C_- \neq 0$,

$$\operatorname{Re} F_-(s, 0) \sim (\log s) \operatorname{Im} F_-(s, 0). \qquad \ldots[7.2.13]$$

This relation is a general one and, in some form, it occurs in all proofs of the Pomeranchuk theorem. If $C_- \neq 0$, $\operatorname{Re} F_-$ dominates $|F_+|$ and $|F_-|$ as $s \to \infty$. We will show that this leads to a contradiction with unitarity (Eden 1966b).

Unitarity requires that for both processes I and II

$$\frac{\operatorname{Im} F(s, 0)}{s} \sim \sigma(\text{total}) \geqq \sigma(\text{elastic}) \geqq \int_{-s}^{0} \left|\frac{F(s, t)}{s}\right|^2 dt. \qquad \ldots[7.2.14]$$

The elastic cross section is dominated for both processes by $\operatorname{Re} F_-(s, t)$, since from the form [7.2.1] near $t = 0$

$$F_-(s, t) \sim F_-(s, 0) \exp[t \log s]. \qquad \ldots[7.2.15]$$

Substituting from [7.2.1] and [7.2.12] into [7.2.14],

$$(\log s)^\beta \geqq C_- \frac{(\log s)^{2\beta+2}}{(\log s)} \geqq C_-(\log s)^{2\beta+1}. \qquad \ldots[7.2.16]$$

The inverse power of $\log s$ comes from the integration over t near $t = 0$.

It follows from [7.2.16] that if $\beta > -1$, then C_- must be zero. Thus if either total cross section tends to a constant, then the antiparticle total cross section must tend to the same constant. If either increases then their ratio tends to unity as $s \to \infty$.

More general forms for F or σ(total) involving oscillations require a special additional assumption in order to prove the Pomeranchuk theorem. This is usually of the type

$$\frac{\operatorname{Re} F(s, 0)}{(\log s) \operatorname{Im} F(s, 0)} \to 0 \quad \text{as } s \to \infty. \qquad \ldots[7.2.17]$$

However, if σ(total) increases as $s \to \infty$, the result that the cross sections' ratio tends to unity can be proved without the additional assumption [7.2.17]. Experimental results indicate for $\pi^+ p$ and $\pi^- p$ that

$$\frac{\sigma(\pi^+ p)}{\sigma(\pi^- p)} \sim 1 - \frac{C}{s^\alpha} \qquad \ldots[7.2.18]$$

where $\alpha \sim 0{\cdot}5$. This is a much stronger result than has yet been obtained by theoretical methods, except those based on special models in Regge theory.

Using isospin invariance the Pomeranchuk theorem has a number of important experimental consequences. For example charge exchange scattering cross sections tend to zero as the energy tends to infinity. These results lie outside the scope of this lecture series.

8

Partial Wave Amplitudes

8.1. *Bounds for large s, or for large l*

In section 4 we have given the derivation by Martin of the Froissart bound, which showed incidentally that

$$\operatorname{Im} f_l(s) < Cs^N \exp\{-l(t_0/k^2)^{\frac{1}{2}}\} \qquad ...[8.1.1]$$

where t_0 = constant. The inequality [*8.1.1*] applies also for the Greenberg-Low bound except that t_0 is bounded by the Lehmann ellipse and $t_0 = c/s$.

The derivation of [*8.1.1*] assumes temperedness of the amplitude

$$|\operatorname{Im} F(s, t)| < s^N \qquad ...[8.1.2]$$

for $t = t_0$, and that it remains regular in t out to t_0 at least. Taking $t_0 = 4m^2$, and assuming that poles have been subtracted out, $F(s, t)$ has been shown to satisfy

$$|F(s, t)| < s^{2-2\alpha}, \quad \text{for} \quad 0 \leqq t \leqq 4m^2 - \varepsilon \qquad ...[8.1.3]$$

where $\alpha = \alpha(\varepsilon) > 0$ for $\varepsilon > 0$. Hence we can use [*8.1.1*] with $N = 2$ giving

$$\operatorname{Im} f_l(s) < C \exp\{2 \log s - l(t_0/s)^{\frac{1}{2}}\} \qquad ...[8.1.4]$$

which decreases exponentially with l and becomes negligible for $l \gg L$, where

$$L \sim 2(s/t_0)^{\frac{1}{2}} \log s. \qquad ...[8.1.5]$$

Hence

$$\operatorname{Im} f_l(s) \to 0 \quad \text{as} \quad l \to \infty. \qquad ...[8.1.6]$$

Now

$$f_l(s) = \frac{\eta_l \exp(2i\delta_l) - 1}{2i} \qquad ...[8.1.7]$$

and

$$0 \leqq \eta_l \leqq 1. \qquad ...[8.1.8]$$

From [*8.1.6*] and [*8.1.7*]

$$\operatorname{Im} f_l(s) = \frac{1 - \eta_l \cos 2\delta_l}{2} \to 0. \qquad ...[8.1.9]$$

Thus for $l \to \infty$

$$\eta_l \to 1, \ \delta_l \to 0. \qquad \ldots[8.1.10]$$

It follows that as $l \to \infty$

$$\operatorname{Re} f_l(s) = \tfrac{1}{2}\eta_l \sin 2\delta_l \to 0 \qquad \ldots[8.1.11]$$

and $\operatorname{Re} f_l$, like $\operatorname{Im} f_l$, is negligible for $l \gg L$.

The same result can be obtained directly from the definition of $f_l(s)$ using the (experimental) fact that the scattering amplitude becomes negligible away from the forward peak at large s. Let the amplitude have a dominant term

$$F(s, t) \sim F(s) \exp\{\gamma(s)t\}, \qquad 0 > t > t_0, \qquad \ldots[8.1.12]$$

where $F(s)$ will be chosen to satisfy the Froissart bound, and t_0 is sufficiently large to make $F(s, t_0) \ll F(s)$. In practice, of course, γ would have to depend on both s and t, but the form [*8.1.12*] will be a good approximation if *either* $\gamma(s) \to \infty$, as $s \to \infty$, *or* $\gamma(s, t) \sim \gamma(s)$ for $0 < t < t_0(s)$, where, for example, $t_0(s) \sim \log s$. Substituting $t = -2k^2(1-\cos\theta)$ gives

$$F(s, t) \sim F(s) \exp[-\gamma k^2\theta^2]. \qquad \ldots[8.1.13]$$

The partial wave amplitude is

$$f_l(s) = \tfrac{1}{2}\int_{-1}^{1} d(\cos\theta)\, F(s, t(\theta))\, P_l(\cos\theta) \qquad \ldots[8.1.14]$$

$$\sim \tfrac{1}{2}\int_{0}^{\infty} \theta d\theta\, F(s) \exp[-\gamma k^2\theta^2]\, P_l(\cos\theta). \qquad \ldots[8.1.15]$$

We can make use of the formulae

$$\lim_{l\to\infty} P_l\left(1-\frac{x^2}{2l^2}\right) = J_0(x) \qquad \ldots[8.1.16]$$

$$\int_0^\infty \exp\{-y^2/2\}\, J_0(yb) y dy = \exp[-b^2/2]. \qquad \ldots[8.1.17]$$

These give, for large l, $l^2 > s\gamma(s)$

$$f_l(s) \sim \frac{F(s)}{s\gamma(s)} \exp[-2l^2/(s\gamma(s))] \to 0 \quad \text{as} \quad l \to \infty. \qquad \ldots[8.1.18]$$

This decrease is too fast since the bound [*8.1.4*] must be attained as $l \to \infty$, so that our approximation is not good in this limit.

For smaller values of l, $l^2 < s\gamma(s)$, we obtain from [*8.1.15*]

$$f_l(s) \sim f_0(s) \sim \tfrac{1}{2}\int_0^\infty \theta d\theta\, F(s) \exp[-\gamma s\theta^2/4] \qquad \ldots[8.1.19]$$

giving approximately

$$f_l(s) \sim \frac{F(s)}{s\gamma(s)}. \qquad \ldots[8.1.20]$$

Unitarity requires $|f_l(s)| < 1$, hence

$$|F(s)| < s\gamma(s). \qquad ...[8.1.21]$$

Since $F(s, t)$ must be tempered when $t > 0$, we have seen in section 4 that,

$$\gamma(s) < \log s. \qquad ...[8.1.22]$$

Hence, in this approximation,

$$|F(s)| < s \log s. \qquad ...[8.1.23]$$

This bound was derived earlier in section 4.3 by another method.

Now assume

$$F(s) \sim is\gamma'(s) \qquad ...[8.1.24]$$

where we must have $\gamma'(s) \leqq \gamma(s) \leqq \log s$. Then [*8.1.18*] gives for large l,

$$f_l(s) \sim i\{\gamma'(s)/\gamma(s)\} \exp\left[-2l^2/s\gamma(s)\right]. \qquad ...[8.1.25]$$

Note that one cannot deduce that a Gaussian form holds out to $l \to \infty$ since the form [*8.1.12*] for $F(s, t)$ is only approximate.

We next consider various possibilities (note that they are model dependent):

(a) If $\gamma'(s) \sim \gamma(s)$, only L partial waves will contribute significantly to the forward amplitude

$$F(s, 0) \sim \sum_0^\infty (2l+1) f_l(s) \qquad ...[8.1.26]$$

where $L^2 \sim s\gamma(s)$.
Each partial wave

$$f_l(s) \sim f_0(s) \sim \frac{i\gamma'(s)}{\gamma(s)} \sim \text{const.} \qquad ...[8.1.27]$$

giving

$$F(s, 0) \sim L^2 \sim s\gamma(s) \qquad ...[8.1.28]$$

which is of course necessary for consistency.

(b) If

$$\frac{\gamma'(s)}{\gamma(s)} \sim f_0(s) \to 0 \quad \text{as} \quad s \to \infty \qquad ...[8.1.29]$$

then we still have $L^2 \sim s\gamma(s)$, and L partial waves contribute to the series giving the forward amplitude, but each partial wave tends to zero. The forward amplitude is now

$$F(s, 0) \sim is\gamma(s) f_0(s). \qquad ...[8.1.30]$$

If the total cross section is constant, we deduce that the partial wave amplitude is

$$f_0(s) \sim \frac{1}{\gamma(s)} \qquad ...[8.1.31]$$

where

$$\gamma(s) = \left[\frac{d}{dt}\log F(s,t)\right]_{t=0}. \qquad ...[8.1.32]$$

The width of the diffraction peak, in a plot of $d\sigma/dt$ against t, is $1/\gamma(s)$.

From [*8.1.31*] we see that the width cannot *increase* with s if the total cross section is constant. Correspondingly, from [*8.1.27*] we see that if the cross section increases like $\log s$, then the width must decrease like $1/\log s$; i.e. as indicated by Regge theory.

In a discussion related to that given above, MacDowell and Martin (1964) obtain the following inequality

$$\left.\frac{d}{dt}\log F(s,t)\right|_{t=0} > \frac{1}{9}\left[\frac{\sigma_{\text{total}}}{4\pi}\left(\frac{\sigma_{\text{total}}}{\sigma_{\text{el. im.}}}\right)-\frac{1}{k^2}\right] \qquad ...[8.1.33]$$

where

$$\sigma_{\text{el. im.}} = \frac{4\pi}{k^2}\sum(2l+1)(\operatorname{Im} f_l)^2. \qquad ...[8.1.34]$$

This bound can either be used for direct comparison with experiment or for checking consistency of theoretical models.

8.2. *Partial wave phase shifts at high energy*

Crossing symmetry does not apply to partial wave amplitudes, so the methods of Meiman (1962), Van Hove (1964) and Khuri and Kinoshita (1965c) must be applied first to the full amplitude to obtain its phase. Then using the narrowness of the diffraction peak we can evaluate the partial wave phase shifts. We will use the methods of sections 6.1, 6.2 and 8.1, and consider typical behaviour of amplitudes that are consistent with known bounds, crossing, analyticity in s, and a narrow diffraction peak.

For the total cross section we will consider two possibilities

(a) $$\sigma(\text{total}) = \sigma_a \sim C_a + C_a' E^{\beta-1} \qquad ...[8.2.1]$$

corresponding to

$$\operatorname{Im} F_a(E,0) \sim C_a E + C_a' E^{\beta} \qquad ...[8.2.2]$$

where $\beta < 1$.

(b) $$\sigma(\text{total}) = \sigma_b \sim C_b \log E + C_b' E^{\beta-1} \qquad ...[8.2.3]$$

corresponding to

$$\operatorname{Im} F_b(E,0) \sim C_b E \log E + C_b' E^{\beta}. \qquad ...[8.2.4]$$

The amplitudes in the forward direction that give these cross sections and satisfy hermitean analyticity and crossing symmetry are

$$F_a(E,0) \sim C_a(iE)+C'_a\left\{\exp\left(-\frac{i\pi\beta}{2}\right)\right\}E^{\beta} \qquad \ldots[8.2.5]$$

$$F_b(E,0) \sim C_b(iE)(\log E - i\pi/2)+C'_b\left\{\exp\left(-\frac{i\pi\beta}{2}\right)\right\}E^{\beta}. \qquad \ldots[8.2.6]$$

For the example (a), the width $[\gamma(E)]^{-1}$ of the diffraction peak must not increase with E, but unitarity and temperedness permit any $\gamma(E)$ that satisfies

$$\text{constant} \leqq \gamma(E) \leqq C'' \log E. \qquad \ldots[8.2.7]$$

We will consider the two extreme possibilities

$$\gamma(E) = \gamma_a^0 = \text{constant} \qquad \ldots[8.2.8]$$

$$\gamma(E) = \gamma_a \log E. \qquad \ldots[8.2.9]$$

The second of these corresponds to the behaviour suggested by Regge theory. For example (b), there is only one possibility that is compatible both with unitarity and temperedness, namely

$$\gamma(E) = \gamma_b \log E. \qquad \ldots[8.2.10]$$

The non-forward amplitude is crossing symmetric in the variable w,

$$2mw = s-2m^2+\tfrac{1}{2}t. \qquad \ldots[8.2.11]$$

For fixed t, $|t| \ll s$,

$$w \sim E \sim s/2m. \qquad \ldots[8.2.12]$$

This gives the following asymptotic forms for the examples of non-forward amplitudes that we are considering:

(a, 1)

$$F_a^1(E,t) \sim \left[C_a(iE)+C'_aE^{\beta}\exp\left(-\frac{i\pi\beta}{2}\right)\right]\exp(\gamma_a^0 t) \qquad \ldots[8.2.13]$$

(a, 2)

$$F_a^2(E,t) \sim \left[C_a(iE)+C'_a(E)^{\beta}\exp\left(-\frac{i\pi\beta}{2}\right)\right]\exp[\{\gamma_a \log E - \tfrac{1}{2}i\pi\gamma_a\}t] \ldots[8.2.14]$$

(b)

$$F_b(E,t) \sim C_b(iE)(\log E - \tfrac{1}{2}i\pi)\exp[(\gamma_b \log E - \tfrac{1}{2}i\pi\gamma_b)t]$$
$$+C'_bE^{\beta}\exp\left(-\frac{i\pi\beta}{2}\right)\exp[(\gamma_b \log E - \tfrac{1}{2}i\pi\gamma_b)t]. \qquad \ldots[8.2.15]$$

The asymptotic form, for large E, of the partial wave amplitude is dominated by the non-forward amplitude near to $t = 0$, provided $l \ll L$, where $L^2 \sim E \log E$.

Then:

(a, 1)

$$f_l(E) \sim \frac{1}{\gamma_a^0}\left[iC_a + C_a' E^{\beta-1} \exp\left(-\frac{i\pi\beta}{2}\right)\right] \qquad \ldots[8.2.16]$$

(a, 2)

$$f_l(E) \sim \frac{\left[iC_a + C_a' E^{\beta-1} \exp\left(-\frac{i\pi\beta}{2}\right)\right]}{\gamma_a \log E - \frac{1}{2} i\pi\gamma_a} \qquad \ldots[8.2.17]$$

(b)

$$f_l(E) \sim \frac{iC_b}{\gamma_b} + \frac{C_b' E^{\beta-1} \exp\left(-\frac{i\pi\beta}{2}\right)}{\gamma_b(\log E - \frac{1}{2} i\pi)}. \qquad \ldots[8.2.18]$$

The phase shift $\delta_l(E)$ and the inelasticity coefficient $\eta_l(E)$ are obtained from

$$f_l(E) = \frac{\eta_l \exp(2i\delta_l) - 1}{2i}$$
$$= \frac{(1-\eta \cos 2\delta)i}{2} + \frac{\eta \sin 2\delta}{2} \qquad \ldots[8.2.19]$$

where η is independent of l for $l < L \sim (s \log s)^{\frac{1}{2}}$. Comparing real and imaginary parts we obtain:

(a, 1)

$$\delta \sim E^{\beta-1} \to 0, \quad \text{as} \quad E \to \infty \qquad \ldots[8.2.20]$$

$$(1-\eta) \sim \frac{2C_a}{\gamma_a^0} = \text{const.}, \quad \text{as} \quad E \to \infty. \qquad \ldots[8.2.21]$$

Note that the limit for δ is modulo $n\pi$, here and subsequently. Unitarity requires, if $C_a' > 0$,

$$0 < \frac{2C_a}{\gamma_a^0} < 1. \qquad \ldots[8.2.22]$$

If $C_a' < 0$, the cross section increases towards a constant and in this case equality, $2C_a/\gamma_a^0 = 1$, is possible.

(a, 2)

$$\delta \to 0, \quad \text{and} \quad \eta \to 1, \quad \text{as} \quad E \to \infty \qquad \ldots[8.2.23]$$

$$\delta \sim -\frac{C_a \pi}{(\log E)^2}, \quad C_a > 0 \qquad \ldots[8.2.24]$$

$$1 - \eta \sim \frac{2C_a}{\gamma_a \log E}. \qquad \ldots[8.2.25]$$

Note that the asymptotic phase is not dependent on the leading correction term, and that the inelasticity coefficient $\eta \to 1$ (the elastic value).

(b)

$$\delta \sim \frac{E^{\beta-1}}{\log E} \to 0 \quad \text{as} \quad E \to \infty \qquad \text{...[8.2.26]}$$

$$(1-\eta) \sim \frac{2C_b}{\gamma_b} = \text{constant}, \quad \text{as} \quad E \to \infty \qquad \text{...[8.2.27]}$$

From unitarity

$$0 < \frac{2C_b}{\gamma_b} < 1. \qquad \text{...[8.2.28]}$$

The asymptotic forms of the partial wave cross sections are given by [3.3.36-38] and δ_l, η_l given above. For $l < L$, they are

(a, 1)

$$\sigma^l_{\text{total}} \sim \frac{(2l+1)\pi}{k^2}\left(\frac{2C_a}{\gamma^0_a}\right) \qquad \text{...[8.2.29]}$$

$$\sigma^l_{\text{elastic}} \sim \frac{(2l+1)\pi}{k^2}\left(\frac{2C_a}{\gamma^0_a}\right)^2 \qquad \text{...[8.2.30]}$$

(a, 2)

$$\sigma^l_{\text{total}} \sim \frac{(2l+1)\pi}{k^2}\left(\frac{2C_a}{\gamma_a \log E}\right) \qquad \text{...[8.2.31]}$$

$$\sigma^l_{\text{elastic}} \sim \frac{(2l+1)\pi}{k^2}\left(\frac{2C_a}{\gamma_a \log E}\right)^2 \qquad \text{...[8.2.32]}$$

(b)

$$\sigma^l_{\text{total}} \sim \frac{(2l+1)\pi}{k^2}\left(\frac{2C_b}{\gamma_b}\right) \qquad \text{...[8.2.33]}$$

$$\sigma^l_{\text{elastic}} \sim \frac{(2l+1)\pi}{k^2}\left(\frac{2C_b}{\gamma_b}\right)^2. \qquad \text{...[8.2.34]}$$

In no case does the phase shift have any effect on the leading term in the partial wave cross sections. The cross sections are obtained from those above by multiplying by L^2,

$$L^2 \sim s \log s. \qquad \text{...[8.2.35]}$$

The most important unsolved problem about the partial wave amplitude is its asymptotic behaviour for complex values of energy. It has been proved to all orders in perturbation theory, for equal masses, that

$$g_l(z) = \frac{1}{k} f_l(z) \qquad \text{...[8.2.36]}$$

is analytic in the complex z-plane cut along ($-\infty$ to 0) and ($4m^2$ to ∞), with z the square of the energy, and $4k^2 = z-4m^2$. If

$$|g_l(z)| \to 0 \quad \text{as} \quad |z| \to \infty \qquad \ldots[8.2.37]$$

for all directions in the cut z-plane, this g satisfies the dispersion relation

$$g_l(z) = \frac{1}{\pi}\int_{4m^2}^{\infty} \frac{\text{Im } g_l(x)\,dx}{x-z} + \frac{1}{\pi}\int_{-\infty}^{0} \frac{\text{Im } g_l(x)\,dx}{x-z}. \qquad \ldots[8.2.38]$$

The conditions under which solutions exist to this dispersion relation that are consistent with unitarity on the right hand cut ($4m^2$, ∞) have been studied by Frye and Warnock (1965), and by many other authors. They establish conditions that must be satisfied by Im $g_l(x)$ on the left hand cut ($-\infty$, 0) so that a unitary solution exists.

The difficulty about the asymptotic behaviour of $g_l(z)$ that casts doubt on the assumption [*8.2.37*] is that for $z \sim -s$, the integral

$$g_l(-s) = \frac{1}{2k}\int_{-1}^{1} d\mu\, P_l(\mu)\, F\{-s, t(\mu)\} \qquad \ldots[8.2.39]$$

takes F into a domain where $-s$ and t both become large ($u \to +\infty, t \to +\infty$). This is a region where Regge branch cuts appear to cause difficulties and where we may even have

$$F(-s, t) \sim s^{t^{\frac{1}{2}}} \qquad \ldots[8.2.40]$$

with $t \to +\infty$ within the range of integration of [*8.2.39*]. It may happen that cancellations could make the partial wave amplitude much smaller than the amplitude for $s \to -\infty$, but this is a very difficult question.

If the Mandelstam representation is assumed to be valid with a finite number of subtractions, then $g_l(z)$ will satisfy a dispersion relation (possibly with some subtractions). In this case a number of properties of $g_l(z)$ can be established.

Jin and Martin (1964) have studied the discontinuity of the partial wave amplitude on the left hand cut. They establish that this discontinuity changes sign at least l times. Thus the left hand cut becomes increasingly complicated as $l \to \infty$, even when the simplifying assumption is made of a polynomial bound in z.

9

Summary of Bounds

There is an excellent detailed summary of bounds by Martin (1966b), some of which are listed in this section. The basic assumptions required for each bound are those of axiomatic quantum field theory unless otherwise noted.

9.1. *Upper bounds* (*reference section 4*)

The forward amplitude,

$$| F(s, 0) | \leqq Cs \log^2 (s/s_0) \qquad \ldots[9.1.1]$$

reference, Froissart (1961), Greenberg and Low (1961), Martin (1963a).

The total cross section

$$\sigma(\text{total}) \leqq C' \log^2 (s/s_0) \qquad \ldots[9.1.2]$$

$$C' \leqq 4\pi \,/\, (\text{Pion mass})^2 \qquad \ldots[9.1.3]$$

reference, Jin and Martin (1964), and for arbitrary spins reference Yamamoto (1963), Hara (1963), Cornille (1964).

Fixed momentum transfer squared,

$$| F(s, t) | < s(\log s)^{\frac{3}{2}} \qquad t < \varepsilon < 0 \qquad \ldots[9.1.4]$$

$$| F(s, t) | < s^{2-\varepsilon} \qquad 0 < t < 4\mu^2 \qquad \ldots[9.1.5]$$

reference, Martin (1963a) and Jin and Martin (1964).

Fixed angle,

$$| F(s, \cos\theta) | < Cs^{\frac{3}{4}} (\log s)^{\frac{3}{2}} \,/\, \sin^{\frac{1}{2}} \theta \qquad \ldots[9.1.6]$$

reference, Froissart (1961), Martin (1963a).

Diffraction peak, at $t = 0$,

$$\frac{d}{dt} \log \text{Im}\, F(s, 0) > \frac{1}{9}\left[\frac{(\sigma^{\text{total}})^2}{4\pi\sigma_{\text{el}}} - \frac{1}{k^2}\right] \qquad \ldots[9.1.7]$$

reference, MacDowell and Martin (1964).

Diffraction peak at $t = 0$,

$$\frac{d}{dt} \log \text{Im}\, F(s, 0) < C \log^2 s \qquad \ldots[9.1.8]$$

reference section 4.3.

9.2 *Lower bounds* (*reference section 5*)

Elastic cross sections,

$$\sigma(\text{elastic}) > \frac{C(\sigma^{\text{total}})^2}{(\log s)^2} \qquad \ldots[9.2.1]$$

reference, Martin (1963c).

Forward amplitude, (on average),

$$| F(s, 0) | > s^{-2} \quad \text{(assuming no unphysical cut)} \qquad \ldots[9.2.2]$$

reference, Jin and Martin (1964), see also Sugawara (1965).

Fixed angle, assuming Mandelstam representation,

$$| F(s, \cos\theta) | > \exp-[C(\theta)s^{\frac{1}{2}} \log s] \qquad ...[9.2.3]$$

$$C(\theta) \sim \theta^{\frac{1}{2}} \quad \text{for small } \theta \qquad ...[9.2.4]$$

reference, Cerulus and Martin (1963), Kinoshita (1964).

Form factors,

$$| F(t) | > \exp\{-[C\, | t |^{\frac{1}{2}}]\} \qquad ...[9.2.5]$$

reference, Martin (1965c).

BIBLIOGRAPHY

BARTON, G. 1965. *Dispersion Techniques in Field Theory* (Benjamin, New York).
BOAS, R. P. 1954. *Entire Functions* (Academic Press, New York).
CHEW, G. F. 1966. *The Analytic S-Matrix* (Benjamin, New York).
CHEW, G. F. and JACOB, B. 1963. *Strong Interactions and S-Matrix Theory* (Benjamin, New York).
DEWITT, C. and OMNES, R. (Editors). 1960. *Dispersion Relations and Elementary Particles* (Wiley, New York).
EDEN, R. J. 1967. *High Energy Collisions of Elementary Particles* (Cambridge University Press).
EDEN, R. J., LANDSHOFF, P. V., OLIVE, D. I. and POLKINGHORNE, J. C. 1966. *The Analytic S-Matrix* (Cambridge University Press).
FRAUTSCHI, S. C. 1963. *Regge Poles and S-Matrix Theory* (Benjamin, New York).
MOORHOUSE, R. G. (Editor). 1963. *Strong Interactions and High Energy Physics.* (Proceedings of the Scottish Universities Summer School; Oliver and Boyd, Edinburgh and London).
OMNES, R. and FROISSART, M. 1963. *Mandelstam Theory and Regge Poles* (Benjamin, New York).
TITCHMARSH, E. C. 1939. *Theory of Functions* (Oxford University Press).

REFERENCES

BLANKENBECLER, R. and GOLDBERGER, M. L. 1962. *Phys. Rev.* **126.** 766.
BROS, J., EPSTEIN, H. and GLASER, V. 1965. *Comm. Math. Phys.* **1**, 240.
CASTILLEJO, L., DALITZ, R. and DYSON, F. 1956. *Phys. Rev.* **101**, 453.
CERULUS, F. and MARTIN, A. 1963. *Phys. Lett.* **8**, 80.
CHEW, G. F. and FRAUTSCHI, S. C. 1961. *Phys. Rev. Lett.* **7**, 394, and **8**, 41.
COLLINS, G. 1966. *Stony Brook Conference Proceedings.*
CORNILLE, H. 1964. *Nuovo Cim.* **31**, 1101.
EDEN, R. J. 1961. *Phys. Rev.* **121**, 1567.
EDEN, R. J. 1965. Lectures on High Energy Scattering, University of Maryland, Physics Department.
EDEN, R. J. 1966a. *Phys. Lett.* **19**, 695.
EDEN, R. J. 1966b. *Phys. Rev. Lett.* **16**, 39.
EDEN, R. J. 1966c. Consistency Conditions on Models for High Energy Scattering, *J. Math. Phys.* Feb. 1967.
EDEN, R. J. and ŁUKASZUK, L. 1966. (To be published in *Nuovo Cim.*)
FOLEY, *et al.* 1965a. *Phys. Rev. Lett.* **14**, 862.
FOLEY, *et al.* 1965b. *Phys. Rev. Lett.* **15**, 45.
FROISSART, M. 1961. *Phys. Rev.* **123**, 1053.
FRYE, G. and WARNOCK, R. L. 1965. *Phys. Rev.* **130**, 478.

GOLDBERGER, M. L. 1955. *Phys. Rev.* **99**, 979.
GOLDBERGER, M. L. 1961. (See Bibliography—DeWitt and Omnes).
GREENBERG, O. W. 1964. *Lectures at Istanbul Summer School* (Editor—F. Gürsey, published by Gordon & Breach).
GREENBERG, O. W. and LOW, F. E. 1961. *Phys. Rev.* **124**, 2047.
GRIBOV, V. N. 1962. *Sov. Phys. JETP* **14**, 478 (published in Russian 1961).
HARA, Y. 1964. *Phys. Rev.* **136** B, 507.
HEPP, K. 1964. *Helv. Phys. Acta*, **37**, 639.
JIN, Y. S. and MARTIN, A. 1964. *Phys. Rev.* **135** B, 1369 and 1375.
JIN, Y. S. 1966. Brown University preprint.
JIN, Y. S. and MACDOWELL, S. W. 1965. *Phys. Rev.* **138** B, 1279.
KHURI, N. and KINOSHITA, T. 1965a. *Phys. Rev. Lett.* **14**, 84.
KHURI, N. and KINOSHITA, T. 1965b. *Phys. Rev.* **137** B, 720.
KHURI, N. and KINOSHITA, T. 1965c. *Phys. Rev.* **140** B, 706.
KINOSHITA, T. 1964. *Phys. Rev. Lett.* **12**, 256.
KINOSHITA, T. 1966. Cornell preprint (*Perspectives in Modern Physics*).
KINOSHITA, T., LOEFFEL, J. J. and MARTIN, A. 1964. *Phys. Rev.* **135** B, 1464.
LEADER, E. 1963. *Phys. Lett.* **5**, 75.
LEHMANN, H. 1958. *Nuovo Cim.* **10**, 579, and *Supp. Nuovo Cim.* **14**, 153 (1959).
LEVINSON, N. 1949. *Dan. Mat. Fys. Medd.* **25**, No. 9.
LINDENBAUM, S. J. 1965. *Proceedings of the Oxford Conference on Elementary Particles.*
LOGUNOV, A. A., NGUYEN VAN HIEN, TODOROV, I. T. and KRUSTALEV, O. A. 1963. *Phys. Lett.* **7**, 69.
MACDOWELL, S. W. and MARTIN, A. 1964. *Phys. Rev.* **135** B, 960.
MANDELSTAM, S. 1958. *Phys. Rev.* **112**, 1344.
MANDELSTAM, S. 1963. *Nuovo Cim.* **30**, 1148.
MARTIN, A. 1963a. *Phys. Rev.* **129**, 1432.
MARTIN, A. 1963b. (See Bibliography—Moorhouse).
MARTIN, A. 1963c. *Nuovo Cim.* **29**, 993.
MARTIN, A. 1965a. CERN preprint 65/1699/5-TH.630 (*Nuovo Cim.* 1966).
MARTIN, A. 1965b. CERN preprint 65/1857/5-TH.637.
MARTIN, A. 1965c. *Nuovo Cim.* **37**, 671.
MARTIN, A. 1966a. *Nuovo Cim.* **44**, 1219.
MARTIN, A. 1966b. *Stony Brook Conference Proceedings.*
MEIMAN, N. N. 1962. *Zhur. Elsp. i Teoret. Fiz.* **43**, 2277; (English translation 1965: *Sov. Phys. JETP*, **16**, 1609 (1963).
MORRISON, D. R. O. 1966. *Stony Brook Conference Proceedings.*
POLKINGHORNE, J. C. 1963. *Phys. Lett.* **7**, 217.
POMERANCHUK, I. YA. 1956. *Sov. Phys. JETP* **3**, 306.
POMERANCHUK, I. YA. 1958. *Sov. Phys. JETP* **7**, 499.
POMERANCHUK, I. YA. and OKUN, L. B. 1956. *Sov. Phys. JETP* **3**, 307.
REGGE, T. 1958. *Nuovo Cim.* **14**, 951.
REGGE, T. 1960. *Nuovo Cim.* **18**, 947.
SELOVE, W. *et al.* 1966. *Stony Brook Conference Proceedings.*
SUGAWARA, M. and KANEZAWA, A. 1961. *Phys. Rev.* **123**, 1895.
SUGAWARA, M. and TUBIS, A. 1963. *Phys. Rev.* **130**, 2127.
SUGAWARA, M. 1965. *Phys. Rev. Lett.* **14**, 336.
SYMANZIK, K. 1960. *J. Math. Phys.* **1**, 249, Appendix B.
VAN HOVE, L. 1964. *Rev. Mod. Phys.* **36**, 655. (See also CERN lectures, CERN 65-22.)
VAUGHN, H. T., AARON, R. and AMADO, R. D. 1961. *Phys. Rev.* **124**, 1258.
WIGHTMAN, A. 1960. (See Bibliography—DeWitt and Omnes).
WU, T. T. and YANG, C. N. 1965. *Phys. Rev.* **137** B, 708.
YAMAMOTO, K. 1963. *Nuovo Cim.* **27**, 1277.

2

VERY HIGH-ENERGY SCATTERING OF STRONGLY INTERACTING PARTICLES

L. Van Hove

CERN—Geneva

1

Introduction and Survey of Experimental Data

These lectures will be mainly concerned with two-body collisions of hadrons at high energy. Such collisions are of the type

$$A+B \to C+D \qquad ...[1]$$

A, B, C and D being hadrons, i.e. strongly interacting particles. They can be mesons, baryons and antibaryons. For the few choices of incident particles A, B for which actual collision experiments can be carried out, a considerable body of experimental information is now available. Depending on the case, the outgoing particles C and D can be stable, long-lived or short-lived hadrons, including mesonic, baryonic or antibaryonic resonances. In many cases they are detected through their decay products.

It has up to now been impossible to obtain detailed information concerning two-body collisions from cosmic ray experiments. The experimental results available to date have been obtained with high-energy accelerators, one of the initial particles, let us say A, being contained in a primary or secondary beam originating from the accelerator, and the other, say B, being essentially at rest in a target. Let $p^A = (p_0^A, \mathbf{p}^A)$ be the four-momentum of the beam particle. With accelerators operating at present the accessible energy interval is

$$p_0^A \leqq 33 \text{ GeV} = 33\times 10^9 \text{ eV} \qquad ...[2]$$

for protons, and correspondingly lower energies for other hadrons. The target particle B is usually a proton or a neutron. Representing its four-momentum by

$$p^B = (p_0^B, \mathbf{p}^B), \quad p_0^B \simeq m_B, \quad \mathbf{p}^B \simeq 0$$

where m_B is the mass of B, we have for the square of the c.m. energy (c.m. stands for "centre-of-mass")

$$s = (p^A+p^B)^2 = (p_0^A+p_0^B)^2 - |\mathbf{p}^A+\mathbf{p}^B|^2 \simeq 2m_B p_0^A + m_A^2 + m_B^2. \quad ...[3]$$

For $m_B = m_p$ = proton mass, the interval [2] becomes

$$s \lesssim 64\ (\text{GeV})^2. \quad ...[4]$$

Future extensions of the intervals in [2] and [4] can be expected to come from at least three sources: higher energy accelerators, intersecting beam devices for hadrons, and appropriate developments of cosmic ray experimentation. In a limited sense, an extension is already possible with existing accelerators, when B is a nucleon bound in a target nucleus and when one takes advantage of its Fermi motion. From recent measurements, using incident protons of 30 GeV energy on bound nucleons, one collision in about 10^7 gives a c.m. energy $s^{\frac{1}{2}}$ larger than 14 GeV.[1] In such cases, however, it is clear that the remaining part of the target nucleus plays an important dynamical role which must seriously affect the collision, making it more complex than a two-body collision, and our knowledge of nuclear structure is insufficient to allow for an accurate calculation of the necessary corrections.

Difficulties of similar kind but of much greater severity are encountered when one tries to extract information on collisions of virtual hadrons from collisions of real ones. Thus, one has advocated the use of incident pions hitting target protons in order to study pion-pion collisions, invoking the common assertion that the proton is surrounded by a cloud of virtual pions which are in some sense bound to it. However, the concept of virtual hadron is extremely ill-defined, and our knowledge of strong interaction dynamics is at present incapable of giving it an accurate mathematical content. It is therefore impossible to obtain anything but the crudest information concerning collisions involving virtual particles, and this information is of little help in our attempts to extract from the experimental data an accurate picture of strong interactions at high energies. In the following considerations, A and B are supposed to be real, free hadrons undergoing no interaction before they collide with each other, and so are C and D after the collision.

1.1. *Kinematical variables and cross-section formulae*

In terms of the four-momenta p^A, p^B, p^C, p^D, we define as usual the kinematical invariants of reaction [1]

$$s = (p^A+p^B)^2,\ t = (p^A-p^C)^2,\ u = (p^A-p^D)^2 \quad ...[5]$$

where the notation is the same as in [3] (signature of metric $+---$). As is well known the following relations are satisfied

$$(p^A)^2 = m_A^2,\ (p^B)^2 = m_B^2,\ (p^C)^2 = m_C^2,\ (p^D)^2 = m_D^2$$
$$p^A+p^B = p^C+p^D \quad ...[6]$$
$$s+t+u = M^2 \equiv m_A^2+m_B^2+m_C^2+m_D^2.$$

The units used are such that $\hbar = c = 1$. The c.m. momenta in the initial $(A+B)$ and final $(C+D)$ states will be denoted by k and k' respectively. They satisfy

$$s^{\frac{1}{2}} = (k^2+m_A^2)^{\frac{1}{2}}+(k^2+m_B^2)^{\frac{1}{2}} = (k'^2+m_C^2)^{\frac{1}{2}}+(k'^2+m_D^2)^{\frac{1}{2}}. \qquad ...[7]$$

The c.m. scattering angle θ, defined as the angle between the momenta of A and C in the c.m. system, is given by

$$t = [(k^2+m_A^2)^{\frac{1}{2}}-(k'^2+m_C^2)^{\frac{1}{2}}]^2-k^2-k'^2+2kk'\cos\theta. \qquad ...[8]$$

Equations [7] and [8] are special cases of the relativistically invariant definitions [5].

The six quantities s, t, u, k, k', θ are functions of two independent parameters, for which one may for example choose s and t, or k and θ. In addition, the collision depends on the spin orientations of those particles which have non-vanishing spin, so that the collision amplitude is a function of the form

$$T(s, t, \lambda_A, \lambda_B, \lambda_C, \lambda_D) \qquad ...[9]$$

where the λ_κ's ($\kappa = A, B, C, D$) label the various spin orientations of the particles. For the λ_κ's one often chooses the helicities of Jacob and Wick.[2] We take [9] to represent the relativistically invariant amplitude. Its relation to the S-matrix in an arbitrary reference system is given by

$$T(s, t, \lambda_A, \lambda_B, \lambda_C, \lambda_D)\,(p_0^A p_0^B p_0^C p_0^D)^{-\frac{1}{2}}\,\delta^4(p^A+p^B-p^C-p^D)$$
$$= i\langle \mathbf{p}^C, \lambda_C;\ \mathbf{p}^D, \lambda_D \mid 1-S \mid \mathbf{p}^A, \lambda_A;\ \mathbf{p}^B, \lambda_B\rangle \qquad ...[10]$$

with the single particle states normalized to

$$\langle \mathbf{p}', \lambda' \mid \mathbf{p}, \lambda\rangle = \delta_{\lambda'\lambda}\,\delta^3(\mathbf{p}'-\mathbf{p}). \qquad ...[11]$$

The differential cross-section for fully polarized incident and outgoing particles is

$$\frac{d\sigma}{dt} = (4\pi^3/k^2 s)\,|\,T(s, t, \lambda_A, \lambda_B, \lambda_C, \lambda_D)\,|^2. \qquad ...[12]$$

The special case of *elastic scattering* obtains when one of the outgoing particles (let us say C) is identical to A and the other (D) identical to B. All other cases of two-body collisions are called *inelastic*. The elastic scattering amplitude is written

$$T_{el}(s, t, \lambda_A, \lambda_B, \lambda'_A, \lambda'_B) \qquad ...[13]$$

$\lambda_{A,B}$ referring to the incident and $\lambda'_{A,B}$ to the outgoing particles. For s, t, u we keep the definitions [5], where $p^C = p'^A$ is now the outgoing momentum of A and $p^D = p'^B$ that of B. Through the optical theorem, which reads

$$\sigma_T(\lambda_A, \lambda_B) = (8\pi^2/ks^{\frac{1}{2}})\,\mathrm{Im}\,T_{el}(s, 0, \lambda_A, \lambda_B, \lambda_A, \lambda_B), \qquad ...[14]$$

T_{el} is related to the total cross-section $\sigma_T(\lambda_A, \lambda_B)$ for $A+B$ collisions.

We note that all our equations remain valid when the particles A, B are identical, and also when C, D are identical. The only modification then required for their validity is that the range of variation of the scattering angle θ should be restricted to $0 \leqq \theta \leqq \pi/2$ instead of the full range $0 \leqq \theta \leqq \pi$. For unpolarized incident particles, [*12*] and [*14*] have to be averaged over λ_A, λ_B.

1.2. *Summary of experimental information*

We attempt in this section a qualitative description of the experimental information on high-energy scattering, giving references to the most relevant literature. By high-energy region we shall understand the interval starting around

$$p_0^A \gtrsim 5 \text{ GeV}, \quad s \gtrsim 12(\text{GeV})^2 \qquad ...[15]$$

and reaching to the highest energies available with present accelerators, [*2*] and [*4*]. All measurements done up to now for two-body collisions of hadrons in this region have given differential cross-sections $d\sigma/dt$ which, at fixed t, behave like smooth, non-oscillating functions of s; the same is probably true for $d\sigma/dt$ as a function of s at fixed c.m. scattering angle θ. This smooth variation with s is in all cases very naturally extrapolated to represent a simple, non-oscillatory asymptotic behaviour for $s \to +\infty$, and it is under the assumption of validity of such extrapolations that the theoretical analysis of the high-energy scattering data is usually carried out. We shall adopt the same assumption here. We warn, however, that it may be wrong in several ways. Thus, oscillations in s may be present, either too slow to be visible in the limited interval $12(\text{GeV})^2 \lesssim s \lesssim 64(\text{GeV})^2$ now covered, or too fast to be detected with the resolution attained in available experiments; the possibility of such oscillations will be mentioned later (§ 2.1). Our description of the data ignores it.

The main feature of the existing data on two-body collisions is that in all cases the differential cross-section $d\sigma/dt$ tends rapidly to zero for increasing energy, with one notable exception which is *elastic scattering* $A+B \to A+B$ *for finite values of the momentum transfer* $(-t)^{\frac{1}{2}}$. We begin by discussing the latter case, for which good data are available.[3-6] For all choices of A, B which could be studied till now and at least in the interval

$$0 \leqq -t = -(p^A - p'^A)^2 \lesssim 1(\text{GeV})^2 \qquad ...[16]$$

(p'^A denotes the final four-momentum of A) the measured values of $d\sigma_{\text{el}}/dt$ taken for fixed t vary little with s for $s \to +\infty$. The data are compatible with the assumption that, for $s \to +\infty$ at fixed t, the elastic differential cross-section approaches a non-vanishing limit, i.e. that the amplitude verifies

$$s^{-1} \mid T_{\text{el}}(s, t, \lambda_A, \lambda_B, \lambda'_A, \lambda'_B) \mid \underset{s \to +\infty}{\longrightarrow} f(t, \lambda_A, \lambda_B, \lambda'_A, \lambda'_B) \qquad ...[17]$$

where f is finite for all values of its arguments, and is positive at least in the interval [*16*] and for at least some values of the helicities λ, λ'. Also the data on total cross-sections are compatible with the assumption that σ_T approaches a finite positive limit, which from [*14*] would mean that

$$s^{-1} \operatorname{Im} T_{el}(s, 0, \lambda_A, \lambda_B, \lambda_A, \lambda_B) \underset{s \to +\infty}{\longrightarrow} g(\lambda_A, \lambda_B) > 0 \qquad \ldots[18]$$

for at least some values of λ_A, λ_B. Of course, the data are also compatible with f or g being zero or infinity, but then the trend of $d\sigma_{el}/dt$ or σ_T toward a vanishing or infinite limit would be very slow at the energies measured.

At large energy, the shape of $d\sigma_{el}/dt$ as a function of t in the interval [*16*] is a very simple one, common to all cases studied. The experimental data are best parametrized in the form of the power series

$$\ln \frac{d\sigma_{el}}{dt} = a + bt + ct^2 + \ldots \qquad \ldots[19]$$

b and c are positive. The contribution of the second order term is very small in the interval [*16*]; one finds

$$0 \leqq c/b^2 \lesssim 0{\cdot}04. \qquad \ldots[20]$$

Further terms in the series [*19*] are negligible in the interval [*16*]. The parameters a, b vary little with energy, their behaviour being compatible with the assumption that they approach positive limits for $s \to +\infty$. The same may hold for c, but this parameter, giving only a very small contribution to [*19*], is not yet accurately measured.

One has no direct information at present on the relative contributions of non-spin-flip ($\lambda_A' = \lambda_A, \lambda_B' = \lambda_B$) and spin-flip ($\lambda_A' \neq \lambda_A$ and/or $\lambda_B' \neq \lambda_B$) amplitudes to elastic scattering. It is often assumed that at very high energy the interaction becomes *spin-independent*, which would mean that the spin-flip amplitudes become negligible compared to the non-spin-flip ones and that the latter become all equal to each other. Except for the obvious case where one of the particles has spin zero and where the scattering angle vanishes (forward scattering), there is no good justification for this spin independence assumption and experiment will have to decide. If one nevertheless accepts the assumption for very small t ($-t \lesssim 0{\cdot}05(\text{GeV})^2$), and if one considers the case where both A and B are charged, one can derive the phase of T_{el} from very small angle scattering data by making use of the interference between nuclear and Coulomb interactions. This has been done for pp[7-11] and for $\pi^{\pm}p$[12] collisions, with the result that the ratio

$$\rho = \operatorname{Re} T_{el}/\operatorname{Im} T_{el} \qquad \ldots[21]$$

is of order $-0{\cdot}3$ for pp and of order $-0{\cdot}2$ for $\pi^{\pm}p$ in the energy interval

$$10(\text{GeV})^2 \lesssim s \lesssim 50(\text{GeV})^2 \qquad \ldots[22]$$

and for $t \simeq 0{\cdot}01(\text{GeV})^2$, the region of maximum interference. The data on the ratio ρ are not yet reliable and accurate enough to draw definite conclusions as to its energy variation in the interval [22]. Within the large uncertainties ρ is about constant for pp and seems to approach zero for $\pi^{\pm}p$.

We proceed now with a brief summary of other cases of two-body collisions, i.e. elastic scattering with large momentum transfer and all types of inelastic collisions (i.e. collisions $A+B \rightarrow C+D$ with C and/or D different from A and/or B). For $s \gtrsim 10(\text{GeV})^2$ the most accurate experiments done so far concern pp[13] and πp[14] elastic scattering at large angles, backward πp and np elastic scattering[15,16,17] and a variety of inelastic processes: $\pi^- p \rightarrow \pi^0 n$,[18,19] $\pi^- p \rightarrow \eta n$,[20] $K^- p \rightarrow \bar{K}^0 n$,[21] $\bar{p}p \rightarrow \bar{n}n$[22] at low momentum transfer. All cross-sections $d\sigma/dt$ are found to go down rapidly with increasing energy. The decrease is of order s^{-1} to s^{-2} for $\pi^- p \rightarrow \pi^0 n$, $\pi^- p \rightarrow \eta n$, $K^- p \rightarrow \bar{K}^0 n$ and $\bar{p}p \rightarrow \bar{n}n$. One has a somewhat faster power decrease for πp and np backward elastic scattering. The decrease of $d\sigma/dt$ with s appears to be appreciably faster for large angle πp and pp elastic scattering; in fact, for c.m. scattering angles θ around 90^0, the drop of $d\sigma/dt$ with s can be described by an exponential, $\exp[-As^{\frac{1}{2}}]$ with $A \simeq 3(\text{GeV})^{-1}$. The slopes of the various differential cross-sections mentioned above show more structure and diversity than for elastic scattering at small t. In some cases there are pronounced dips which are of considerable interest. We must also mention that, beyond the experiments referred to before and at somewhat lower energies, numerous data on two-body inelastic reactions are available from bubble chamber work.[23] Finally, high-energy polarization experiments on $\pi^- p$ elastic[24] and charge-exchange scattering[25] and on pp elastic scattering[26] have recently been carried out and have given non-vanishing values for the polarization parameter.

1.3. *General remarks on theoretical methods*

The theoretical approach to very high energy scattering of hadrons is, in its essential features, rather similar to the theoretical framework of strong interaction physics at lower energies. There is no theory in the real sense of the word, which means that there is no mathematical formalism giving through well-defined and consistent procedures numerical predictions for the important experimental quantities (the most important one in our context being of course the scattering amplitude). There is, however, an incomplete but very useful mathematical framework, which is the *S-matrix formalism* as it can and has been abstracted from the known examples of renormalizable, relativistic quantum field theory. Despite the fact that more ambitious hopes have been expressed and that much work has been done to support them, it seems impossible to regard at present the *S*-matrix formalism as more than a mathematical framework. Even when supplemented by other general assumptions, it has not allowed the development of reliable computational methods having the predictive power of a true theory.

The framework of the *S*-matrix formalism, despite the limitations just mentioned, has proved to be well adapted to the task of analysing theoretically high-energy scattering phenomena. On the one hand, it allows to express very compactly the observable implications of the general *symmetry and analyticity principles* believed to hold true for strong interactions, which are C, P, T and Lorentz invariance, isospin and approximate unitary symmetry, as well as crossing symmetry and a certain amount of analyticity of the S-matrix elements in s and t. (The latter, as well as CPT and Lorentz invariance, are believed to hold not only for strong, but also for electromagnetic and weak interactions.) On the other hand, the S-matrix formalism provides the possibility of defining in a mathematically precise way various *dynamical models* which can then be used to analyse the experimental data in terms of the parameters or functions left undetermined in the models.

All the theoretical work on high-energy scattering combines to various degrees the use of specific dynamical models with the exploitation of the general symmetry and analyticity principles. In one class of investigations, briefly mentioned in sections 2.2 and 2.3, the latter principles, supplemented by a minimum of additional general assumptions, are used to derive relations between observable quantities. The most striking examples are the dispersion relation for forward pion-nucleon scattering and relations between the high energy behaviours of the reactions $A+B \rightarrow C+D$ and $\bar{C}+B \rightarrow \bar{A}+D$ ($\bar{C}$ is the antiparticle of C, and similarly for $\bar{A}$). The latter are extensions of the famous Pomeranchuk theorem on the equality of total cross-sections for $A+B$ and $\bar{A}+B$.[27] Up to now all such consequences of the symmetry and analyticity principles for which experimental tests were made have been found to agree with the facts, a circumstance which gives considerable confidence in the soundness of the general theoretical framework.

The situation is more complicated regarding the successes and failures encountered by dynamical models. The evaluation of the achievements of a model will clearly depend on the number and nature of the unspecified parameters or functions which it leaves free to fit the data. It will also depend on the simplicity and elegance of the model, in particular on its ability to incorporate in a natural way the general symmetry and analyticity principles mentioned above (thereby automatically achieving agreement with experiment on the observable consequences of these principles). Thus, *field theoretical models* (like for example the one-meson exchange models), which score rather well on the latter requirement, have little freedom in terms of unspecified parameters and functions. It is therefore not surprising that they encounter great difficulties in describing both the s and t dependence of the observed scattering amplitudes. In contrast, *optical models* have more than the freedom needed to fit the data, but they do not incorporate easily the symmetry and analyticity principles. A successful compromise is struck at present by the *Regge pole model* for high-energy two-body collisions. It elegantly embodies all accepted symmetry and analyticity principles, it has an ample number of free functions to adjust to the data, but it makes nevertheless some definite predictions which up to now agree with the facts (details on the Regge pole model are given in sections 2.4 to 2.6).

There is another general property of the S-matrix which we have not mentioned yet although it is a direct consequence of the most general principles of quantum physics. It is the *unitarity of the S-matrix*, and it provides the mathematical expression for the fact that the sum of probabilities of all possible final configurations in a collision is always equal to one. Its consequences for two-body collisions are of two types. It firstly implies inequalities involving two-body scattering amplitudes or cross-sections. Secondly, it gives relations connecting two-body scattering with certain bulk properties of the inelastic collisions (many of which have three or more hadrons in the final state). This relation is particularly important for elastic scattering at high energy if, as is suspected (§ 1.2), the elastic amplitude becomes almost purely imaginary for small t. The small angle elastic scattering is then almost purely shadow scattering of the inelastic collisions, and therefore reflects in a rather direct way some bulk properties of the latter. This allows one to discuss elastic scattering in terms of simple, qualitative features known or assumed to hold for the inelastic collisions (see section 3).

Considerable attention has been devoted lately to the consequences of the approximate SU_3 and SU_6 symmetries for high-energy two-body processes. The use of group-theoretical methods for this purpose encounters considerable difficulties for SU_6 symmetry, as is easily understood from the fact that the SU_6 symmetry concept clashes with relativistic effects. Another approach, which is more naive but turns out to be more constructive, starts from the quark model, i.e. the composite model of baryons and mesons suggested by the successful aspects of SU_6 symmetry. This *quark model approach to high-energy scattering* is in a less advanced state of development than the Regge pole model. Its results are nevertheless very suggestive and promising (see section 4).

2

High-energy Expansion of Two Particle Collision Amplitude at Finite Momentum Transfers. Regge Poles

2.1. *Form of the expansion*

We discuss the asymptotic behaviour of the scattering amplitude for the two-body collision

$$A+B \rightarrow C+D$$

for high energy, $s \rightarrow +\infty$, the momentum transfer

$$t = (p^A - p^C)^2$$

remaining fixed. We recall that the c.m. scattering angle θ between the directions of A and C then tends to zero. As mentioned in section 1.2, the experimental behaviour of the differential cross-section $d\sigma/dt$ under such conditions is a non-oscillatory one, with approximate constancy in s for small angle elastic scattering ($C = A$, $D = B$) and a pronounced decrease with s for other processes (large angle elastic scattering, $D = A \neq C = B$, as well as inelastic two-body collisions). This decrease seems to go with a power of s.

The behaviour just described is observed for momentum transfers in the interval [*16*] and for all energies measured above a limit as given in [*15*]. It is customary to assume this behaviour to be characteristic of the asymptotic variation of the scattering amplitude $T(s, t, \lambda_A, \lambda_B, \lambda_C, \lambda_D)$ at high energies ($s \to +\infty$). One often represents it theoretically by an asymptotic series expansion of form

$$T(s, t, \lambda_A, \lambda_B, \lambda_C, \lambda_D) \underset{s \to +\infty}{\cong} \sum_{j=1,2,\dots} c_j (s/s_j)^{\alpha_j} [\ln (s/s_j)]^{\beta_j} \qquad \dots[23]$$

where the parameters $c_j, s_j, \alpha_j, \beta_j$ can be functions of t and the helicities λ. In fact s_j, which has dimension [energy]^2, is largely arbitrary because changing its value only affects c_j and adds to the series new terms with lower powers of $\ln (s/s_j)$. The parameters c_j, α_j, β_j are dimensionless. While c_j must be allowed to be complex, α_j and β_j are usually taken real, with the condition

$$\alpha_{j+1} \leqq \alpha_j, \quad \beta_{j+1} < \beta_j \quad \text{if} \quad \alpha_{j+1} = \alpha_j. \qquad \dots[24]$$

This assumption is suggested by the fact that no oscillations have been found experimentally in the high-energy variation of the cross-sections. While this is undoubtedly a natural supposition, one must realize that present experiments exclude neither very fast nor very slow oscillations in some terms of [*23*]. As an example, we take forward elastic scattering without spin flip. Oscillations in the total cross-section, [*14*], could be caused by having one of the α_j complex and they would probably have remained undetected if either one of the following conditions is satisfied

$$|\operatorname{Im} \alpha_j| \gtrsim 10^3, \quad |\operatorname{Im} \alpha_j| \lesssim 1. \qquad \dots[25]$$

The first limit corresponds to fast oscillations in s with a period Δs of order $10^{-2}s$, the second to the case where the whole energy interval of [*2*] covers about a quarter oscillation. Since the actual values of $\alpha_1, \alpha_2, \dots$ are of order $1, \frac{1}{2}, \dots$ the first inequality may seem highly improbable, but the second one is in no way unreasonable. Nevertheless, unless stated otherwise, α_j and β_j will be taken real in the following.

Especially with this restriction, the ansatz [*23*] for the asymptotic behaviour of the amplitude T may appear much too restrictive as far as its phase is concerned, since so little is known experimentally about it. It implies that the phase of T has a well-defined limit for $s \to +\infty$. It is justified by the limited experimental evidence on the ratio ρ of [*21*] (see section 1.2) and by theoretical considerations presented later (§ 2.3). We stress nevertheless that it is in no way well established.

The inclusion of logarithmic terms in [23] is not required by present experimental results which can all be reproduced by means of powers of s only (i.e. all $\beta_j = 0$). They have, however, considerable theoretical interest (§ 2.4) and it may be hoped that in the long run experiment will be able to decide on their presence or absence. The meaning of this statement requires clarification because, evidently, it will always be possible to approximate a logarithmic term over a finite energy interval by a superposition of sufficiently many pure power terms. The point is that one wishes an asymptotic ansatz containing a small number of terms and covering with good accuracy all energies above a minimum of the order of equation [15]. From present evidence, one might hope to reach this goal with no more than a few terms (perhaps two, three or four) for each reaction and with a number of relations between the terms belonging to different reactions. (Such relations would follow from the general symmetry and analyticity principles mentioned in section 1.3, or would be characteristic of specific dynamical models.) Within this context, the analytic form of the terms in s becomes important and the presence of logarithmic factors (or other slowly varying functions, like ln (ln s) for example) may be necessary.

For theoretical reasons which will be mentioned later (see section 2.3), it is more convenient to replace the expansion [23] by

$$T(s, t, \lambda) \underset{s \to +\infty}{\cong} \sum_{j=1,2,\ldots} c_j (s_t/s_j)^{\alpha_j} \left[\ln (s_t/s_j) - \frac{i\pi}{2} \right]^{\beta_j} \qquad \ldots[26]$$

where s_t is defined as

$$s_t = s + \frac{t}{2} - \tfrac{1}{2}(m_A^2 + m_B^2 + m_C^2 + m_D^2) = s + \frac{t}{2} - \frac{M^2}{2}. \qquad \ldots[27]$$

Within the present experimental accuracy [26] has the same applicability as [23] and the comments just made concerning [23] apply to [26] without change. In particular the existing high-energy data can be described in the form [26] with a small number of terms and with $\beta_j = 0$, i.e. without logarithmic factors.

12.2. *Conservation of internal quantum numbers*

The conservation of isospin in the process $A+B \to C+D$ at large s and finite t is conveniently taken into account by separating the scattering amplitude $T(s, t, \lambda)$ into a sum of terms

$$T(s,t,\lambda) = \sum_I (I^A, I_3^A \mid I^C, I_3^C; I, I_3^A - I_3^C)\, T_I(s,t,\lambda)\, (I^B, I_3^B; I, I_3^D - I_3^B \mid I^D, I_3^D). \qquad \ldots[28]$$

Each term corresponds to exchange of isospin I between A and B, in the sense that when A becomes C it transfers isospin I to B which becomes D. The bracket symbols in [28] are Clebsch-Gordan coefficients for isospin.

The exchanged isospin I as defined in [28] is usually referred to as *isospin exchanged in the t channel* (see section 2.3). Because of isospin invariance, the same amplitudes T_I as occur in [28] will describe other reactions $A'+B' \to C'+D'$, where A' is a particle belonging to the same isospin multiplet as A and similarly for B', C', D'. Equation [28] can be used directly for the new reaction by making the necessary changes in the Clebsch-Gordan coefficients. As an example we quote the form taken by [28] for πp scattering:

$$T_{\pi^+p\to\pi^+p} = T_{I=0} - \frac{1}{(6)^{\frac{1}{2}}} T_{I=1}$$

$$T_{\pi^-p\to\pi^-p} = T_{I=0} + \frac{1}{(6)^{\frac{1}{2}}} T_{I=1} \qquad \ldots[29]$$

$$T_{\pi^-p\to\pi^0n} = -T_{\pi^0p\to\pi^+n} = -\frac{1}{(3)^{\frac{1}{2}}} T_{I=1}.$$

We note that [28] is only valid if isospin invariance holds exactly. This is supposed to be true for the strong interactions when electromagnetic and weak interaction effects are neglected. This approximation is certainly well founded for strong interaction processes having a sizable cross-section. It may become questionable for processes of very small cross-section, like, e.g., large angle scattering at very high energies (see section 1.2)—a possibility which only experiment will be able to explore.

When the isospin analysis [28] of T is applied to the high-energy expansion [26], it is clear that each term of [26] will make contributions to one or more of the T_I. The present evidence is that in general each term of [26] contributes to one amplitude T_I only. In other words the high-energy expansions of the various T_I do not seem to contain common powers. If this is so, each term j of the asymptotic series [26] contributes to exchange of a specific isospin I_j in the t channel. (In the opposite case, some terms of [26] would split between various values of the exchanged isospin.)

The situation is even simpler for the additive internal quantum numbers N and Y, i.e. baryon number and hypercharge. Given the reaction $A+B \to C+D$, the baryon number exchanged in the t channel has the unique value $N_A - N_C = N_D - N_B$, and similarly for the hypercharge. It can therefore trivially be said that each term in the expansion [26] contributes to exchange of $N_j = N_A - N_C$ and $Y_j = Y_A - Y_C$ in the t channel. For example

$$N_j = 0, \quad Y_j = -1 \quad \text{for} \quad \pi^- p \to K^0 \Lambda^0 \qquad \ldots[30]$$

$$N_j = -1, \quad Y_j = 1 \quad \text{for} \quad \pi^- p \to \Lambda^0 K^0. \qquad \ldots[31]$$

Notice that, since t is kept fixed as s increases, [30] differs from [31] by the fact that in the c.m. system the Λ^0 goes backward in [30] and forward in [31].

The procedure described above to take isospin conservation into account can be generalized in a straightforward way to SU_3 symmetry if the latter is assumed to hold exactly. The experimental situation indicates, however, that the SU_3 symmetry breaking is no less important in high energy collisions at small momentum transfer than in hadron spectroscopy, particularly in the mass splittings within SU_3 multiplets. (One has no experimental information on the degree of validity of unitary symmetry in large momentum transfer processes, a question which is obviously of the greatest interest.) When extending [28] to SU_3 one should therefore include the first order effects of symmetry breaking. This will give a formula of type

$$T = \sum_{\xi, I, Y} \gamma(A \mid C;\ \xi,\ YII_3)\, T_{\xi, YI}\, \gamma(B;\ \xi,\ YII_3 \mid D) \qquad \ldots[32]$$

where ξ labels the SU_3 representations ($\xi = 1, 8, 10, \overline{10}, \ldots$) and YII_3 the elements within a representation. The γ's are only SU_3 Clebsch-Gordan coefficients in the limit of exact SU_3; their actual values will contain symmetry breaking effects. Similarly, symmetry breaking implies that the amplitude $T_{\xi, YI}$ for exchange of ξ, YI in the t channel depends not only on ξ but also on Y, I. For the example of reactions [29], $T_{I=0}$ becomes a sum $T_{1,0,0}+T_{8,0,0}+T_{27,0,0}$, whereas $T_{I=1}$ becomes $T_{8,0,1}+T_{10,0,1}+T_{\overline{10},0,1}$.

As regards the use of [32] for the high-energy expansion [26] the situation is the same as described for the isospin case; in general each term of [26] contributes to a single amplitude $T_{\xi, YI}$ and therefore to exchange in the t channel of a specific element Y_j, I_j of a specific SU_3 representation ξ_j. We shall assume this to be the case and shall refer again later on to the complete set of exchanged internal quantum numbers N_j, ξ_j, Y_j, I_j belonging to each term of [26].

In our opinion, there is at present no fully satisfactory way of dealing with broken SU_6 symmetry for high-energy scattering. In these lectures (see section 4) we shall regard SU_6 properties as resulting from the dynamical assumption of a composite model of hadrons (quark model). This model allows one to obtain many of the broken SU_3 properties without using group theory.

2.3. *Analyticity and crossing symmetry between s and u channels*

The most important implications of the analyticity and crossing symmetry properties of the scattering amplitude for high energy and finite momentum transfer will now be described in terms of the asymptotic expansion [26]. They are generalizations of the well-known Pomeranchuk theorem on total cross-sections.[27]

The essential content of the principles of analyticity and crossing symmetry is that the scattering amplitude $T(s, t, \lambda)$ for $A+B \to C+D$ can be continued analytically in s and t along properly selected paths, and that proper linear combinations of the continuations of T or T^* (depending on the path) become identical to the scattering amplitudes of the crossed reactions

$$\bar{C}+B \to \bar{A}+D \qquad \ldots[33]$$

$$A+\bar{C} \to \bar{B}+D. \qquad \ldots[34]$$

The kinematical variables of the crossed reactions are given by equation [5] with the substitutions

$$p^A \rightarrow -p^{\bar{A}},\ p^C \rightarrow -p^{\bar{C}} \quad \text{for} \quad \bar{C}+B \rightarrow \bar{A}+D \qquad ...[35]$$

$$p^B \rightarrow -p^{\bar{B}},\ p^C \rightarrow -p^{\bar{C}} \quad \text{for} \quad A+\bar{C} \rightarrow \bar{B}+D \qquad ...[36]$$

so that $u^{\frac{1}{2}}$ is the c.m. energy of [33] and $t^{\frac{1}{2}}$ that of [34]. Accordingly the reactions $A+B \rightarrow C+D$, [33] and [34] are often said to belong to the *s, u and t channels* respectively (this explains the term *t* channel used in the previous section).

We are interested for the moment in the crossing relation between the *s* and *u* channel reactions. It is obtained by keeping *t* constant and continuing $T(s, t, \lambda)$ analytically in the variable *s* through the upper half complex *s*-plane toward such values on the real axis that

$$u = -s-t+m_A^2+m_B^2+m_C^2+m_D^2 \qquad ...[37]$$

takes a physically realizable value for the (c.m. energy)^2 of the *u* channel reaction [33], i.e., such that one can have

$$u = (p^{\bar{C}}+p^B)^2,\ t = (p^{\bar{C}}-p^{\bar{A}})^2 \qquad ...[38]$$

for physical values of the four-momenta involved. The complex conjugate $T^*(s, t)$ of this analytically continued $T(s, t)$ is related to the scattering amplitude $\bar{T}$ of [33] by the *crossing relation*

$$\bar{T}(u, t) = T^*(s, t) \qquad ...[39]$$

when *u*, *s* and *t* satisfy [37]. The crossing relation [39] is written down for the case of spinless particles. For particles with arbitrary spin, the crossing relation can be given a similar form

$$\bar{T}(u, t, \lambda'_{\bar{C}}, \lambda'_B, \lambda'_{\bar{A}}, \lambda'_D) = T^*(s, t, \lambda_A, \lambda_B, \lambda_C, \lambda_D) \qquad ...[40]$$

with *u*, *s*, *t* satisfying again [37] and with

$$\lambda'_{\bar{C}} = -\lambda_C,\ \lambda'_B = \lambda_B,\ \lambda'_{\bar{A}} = -\lambda_A,\ \lambda'_D = \lambda_D. \qquad ...[41]$$

The simple form [40] is only obtained for some special choices of definition of the spin amplitudes. It holds for the "generalized" helicity amplitudes defined, following Trueman and Wick,[28] in an arbitrary reference frame which is not supposed to change in the analytic continuation of *T* in *s*.[29]

(*Remark*: The generalized helicity amplitude depends on the reference frame and should therefore be written $T(p^Ap^Bp^Cp^D,\ \lambda_A\lambda_B\lambda_C\lambda_D)$, with explicit reference to the four-momenta. One writes similarly for the *u* channel amplitude $\bar{T}(p^{\bar{C}}p^Bp^{\bar{A}}p^D,\ \lambda'_{\bar{C}}\lambda'_B\lambda'_{\bar{A}}\lambda'_D)$. The crossing relation [40] is then

$$\bar{T}(p^{\bar{C}}p^Bp^{\bar{A}}p^D,\ \lambda'_{\bar{C}}\lambda'_B\lambda'_{\bar{A}}\lambda'_D) = T^*(-p^{\bar{A}}, p^B, -p^{\bar{C}}, p^D, -\lambda'_{\bar{A}}, \lambda'_B, -\lambda'_{\bar{C}}, \lambda'_D) \qquad ...[40']$$

where $T(p^A p^B p^C p^D, \lambda_A \lambda_B \lambda_C \lambda_D)$ has been continued analytically in p^A and p^C, the energies p_0^A, p_0^C retaining always a non-negative imaginary part. For a straightforward derivation of [*40'*] and [*40*], using the reduction technique of local field theory, see B. E. Y. Svensson, reference 29.)

It holds also for the "transversity amplitudes" of Kotanski, for which the spin quantization axis is taken to be normal to the scattering plane.[30,31]‡ It should be noted that [*40*] is valid only for specific choices of the phases of the spin states; otherwise it would contain a phase factor; the derivation makes explicit use of time-reversal invariance.[29]

For other choices of the spin amplitudes, [*40*] becomes more complicated in that the right-hand side is replaced by a linear combination of spin amplitudes. This is the case for the usual helicity amplitudes of Jacob and Wick[2] where the helicities are taken by definition to refer to the c.m. system of the collision. The crossing relation for this case, originally derived by Trueman and Wick[28] for another path of analytic continuation, has been given for continuation at fixed t in the upper half s-plane by Białas and Svensson,[32] to whose paper we refer for the explicit form of the relation. For the very simple and common case of spin 0 for A, C and spin $\frac{1}{2}$ for B, D, one can also write the crossing relations in the form

$$\bar{g}(u, t) = g^*(s, t), \; \bar{h}(u, t) = h^*(s, t). \qquad ...[42]$$

The functions g and h are related to the amplitude T of the s channel reaction by

$$T = \chi_D \left[g + i \frac{\sigma \cdot (\mathbf{k}_A \times \mathbf{k}_C)}{|\mathbf{k}_A \times \mathbf{k}_C|} h \right] \chi_B \qquad ...[43]$$

where $\mathbf{k}_A$, $\mathbf{k}_C$ are the three-momenta in the c.m. system, whereas χ_B, χ_D are the Pauli two-component spinors representing the spin orientations of particles B and D in their respective rest frames. The definition of $\bar{g}$, $\bar{h}$ is the same for the u channel reaction [*33*].

Remark: For unpolarized B, the cross-section and polarization formulae are then

$$\frac{d\sigma}{dt} = \frac{4\pi^3}{k^2 s} (|g|^2 + |h|^2), \; \sigma_T = \frac{8\pi^2}{k s^{\frac{1}{2}}} \operatorname{Im} g(s, 0),$$

$$P \equiv \left\langle \bar{\chi}_D \left[\frac{\sigma \cdot (\mathbf{k}_A \times \mathbf{k}_C)}{|\mathbf{k}_A \times \mathbf{k}_C|} \right] \chi_D \right\rangle = \frac{2 \operatorname{Im}(gh^*)}{|g|^2 + |h|^2}$$

where P must be averaged over the spin states χ_D of D generated by the collision for random spin states χ_B of B.)

After this description of the crossing relation between the s and u channels,

‡ The normal to the scattering plane is defined for example in the c.m. system; it remains unchanged when going over to any other reference system moving with respect to the c.m. system in a direction lying in the scattering plane, e.g. when going over to the rest system of an incoming or outgoing particle.

we return to high-energy scattering. From [*39*], [*40*] or [*42*], and the analogous relation obtained by permuting T and $\bar{T}$ (which is valid since the crossing relations are symmetric in T and $\bar{T}$), one can write T as a sum

$$T = T_+ + T_-, \quad T_\pm = \tfrac{1}{2}[T(s, t, \lambda) \pm \bar{T}(s, t, \lambda')] \qquad \ldots[44]$$

where the notation λ, λ' of equations [*40*] and [*41*] is used. The new amplitude T_+ (T_-) is even (odd) under crossing, i.e. it verifies the identity

$$T_\pm(u, t, \lambda') = \pm T_\pm^*(s, t, \lambda). \qquad \ldots[45]$$

$T_+(T_-)$ is said to be the *positive* (*negative*) *signature part of T.*

To apply the subdivision [*44*] to the high-energy expansion [*26*], one can again assume for simplicity that each term of [*26*] makes a contribution either to T_+, or T_-. If a term would contribute both to T_+ and T_-, it could be split accordingly into its two contributions; experiment up to now does not require such degeneracies. We can therefore say that each term j of the expansion [*26*] has a definite signature $\varepsilon_j = \pm 1$ ($+1$ for positive, -1 for negative signature). Using the usual assumptions of analyticity and slower than exponential growth of T in the upper half s-plane one can then demonstrate that the phase of the coefficient c_j is determined up to π by the exponent α_j and the signature ε_j. It is given by the formula

$$c_j = \pm |\, c_j \,| \exp\left[-\frac{i\pi}{2}(\alpha_j + \tfrac{1}{2}\varepsilon_j - \tfrac{1}{2})\right]. \qquad \ldots[46]$$

This important consequence of analyticity and crossing symmetry is a far reaching generalization of the Pomeranchuk theorem on total cross-sections.[27] It was derived in various stages of generality in a number of papers.[33-37] Its validity requires the special form adopted in [*26*] for the high-energy expansion, which uses s_t instead of s and ($\ln s_t - i\pi/2$) instead of $\ln s$ as expansion parameters.

To give an example of application of the phase rule [*46*], we consider the case of elastic scattering. Experiment suggests that at very high energy the differential cross-section $d\sigma_{el}/dt$ becomes approximately independent of s at fixed t. From [*12*] this implies that for T_{el} the leading term $j = 1$ of the expansion [*26*] has $\alpha_1 \simeq 1$ for $0 \leqq -t \lesssim 1(\text{GeV})^2$. Applying [*46*], one concludes that c_j is purely imaginary if the signature ε_1 of the leading term is $+1$, whereas it would be purely real if $\varepsilon_1 = -1$. Experiment suggests that the former case is by far the most likely to occur. As another example we take $\pi^- p \to \pi^0 n$. The crossed reaction [*33*] is $\pi^0 p \to \pi^+ n$. From isospin invariance, [*29*], we have

$$T_{\pi^0 p \to \pi^+ n} = -T_{\pi^- p \to \pi^0 n}. \qquad \ldots[47]$$

Hence $T = T_{\pi^- p \to \pi^0 n}$ is odd under crossing, $T = T_-$, the signature is odd, and [*46*] applies with the negative sign to each term of the expansion of the charge exchange amplitude, $T_{\pi^- p \to \pi^0 n}$.

2.4. *Interpretation of the high-energy expansion in terms of the t channel*

In the preceding paragraph we used crossing between the s and u channels. Let us consider now crossing between the s and t channels, i.e. between the original reaction and $A+\bar{C} \to \bar{B}+D$, [*34*]. The corresponding substitution of momenta is given by [*36*]. The scattering amplitude of [*34*] is denoted by

$$\tilde{T}(t, s, \tilde{\lambda}_A, \tilde{\lambda}_{\bar{C}}, \tilde{\lambda}_{\bar{B}}, \tilde{\lambda}_D), \; t = (p^A+p^{\bar{C}})^2, \; s = (p^A-p^{\bar{B}})^2. \qquad ...[48]$$

We continue it analytically in t and s, in the upper half t-plane, so as to keep

$$u = (p^A-p^D)^2 = M^2-(t+s), \; M^2 = m_A^2+m_{\bar{B}}^2+m_{\bar{C}}^2+m_{\bar{D}}^2 \qquad ...[49]$$

constant and so as to reach values of s larger than $(m_A+m_B)^2$. In full analogy with [*39*], (except that $\bar{C}$ plays the role of A, A of B and $\bar{B}$ of C), we obtain the crossing relation

$$T(s, t, \lambda_A, \lambda_B, \lambda_C, \lambda_D) = \tilde{T}^*(t, s, \tilde{\lambda}_A, \tilde{\lambda}_{\bar{C}}, \tilde{\lambda}_{\bar{B}}, \tilde{\lambda}_D) \qquad ...[50]$$

where $\tilde{T}^*$ is the complex conjugate of the analytically continued $\tilde{T}$, and where the spin orientations are related by

$$\tilde{\lambda}_A = \lambda_A, \; \tilde{\lambda}_{\bar{C}} = -\lambda_C, \; \tilde{\lambda}_{\bar{B}} = -\lambda_B, \; \tilde{\lambda}_D = \lambda_D. \qquad ...[51]$$

We have assumed the definition of spin amplitudes to be such as to give to the crossing relation this simple form, as discussed after equations [*40*] and [*41*]; the best choice consists in taking the transversity amplitudes of Kotansky.[30,31] They have over the generalized helicity amplitudes the advantage that the momenta enter only through the relativistic invariants s, t. The normal to the scattering plane is an invariant concept and can be kept the same when crossing from the s to the u and t channels.

In section 2.2 we discussed for the reaction $A+B \to C+D$ the internal quantum numbers exchanged in the t channel. It is clear that for the reaction $A+\bar{C} \to \bar{B}+D$ these internal quantum numbers are simply those of the $A+\bar{C}$ and $\bar{B}+D$ systems themselves. Thus, if $\tilde{T}_I$ is the scattering amplitude for the isospin I component of the $A+\bar{C} \to \bar{B}+D$ reaction, so that

$$\tilde{T}(t, s, \tilde{\lambda}) = \sum_I (I^A, I_3^A;\; I^C, I_3^{\bar{C}} \mid I, I_3^A+I_3^{\bar{C}})\, \tilde{T}_I(t, s, \tilde{\lambda})\, (I, I_3^{\bar{B}}+I_3^D \mid I^B, I_3^{\bar{B}};\; I^D, I_3^D) \qquad ...[52]$$

the analytic continuation of $\tilde{T}_I$, carried out as in [*50*], will give through

$$T_I = \tilde{T}_I^* \qquad ...[53]$$

the amplitude T_I of [*28*].

Also the signature concept, introduced for reaction $A+B \to C+D$ in [*44*], has a simple and important interpretation in terms of the t channel reaction $A+\bar{C} \to \bar{B}+D$. It can be obtained as follows. Equation [*50*] relates the scattering amplitude T of the s channel reaction $A+B \to C+D$ to the analytic continuation of the amplitude $\tilde{T}$ of $A+\bar{C} \to \bar{B}+D$. We write down the analogous equation for the scattering amplitude $\bar{T}$ of the u channel reaction $\bar{C}+B \to \bar{A}+D$. It is

$$\bar{T}(u, t, \lambda'_{\bar{C}}, \lambda'_B, \lambda'_{\bar{A}}, \lambda'_D) = \tilde{\bar{T}}^*(t, u, \tilde{\lambda}'_C, \tilde{\lambda}'_A, \tilde{\lambda}'_B, \tilde{\lambda}'_D). \qquad \text{...[54]}$$

The arguments in $\bar{T}$ are

$$u = (p^{\bar{C}}+p^B)^2, \quad t = (p^{\bar{C}}-p^{\bar{A}})^2.$$

The right-hand side of [54] is the complex conjugate of the analytic continuation, at constant $t+u$ and in the upper half t-plane, of the scattering amplitude $\tilde{\bar{T}}(t, u, \tilde{\lambda}'_C, \tilde{\lambda}'_A, \tilde{\lambda}'_B, \tilde{\lambda}'_D)$ of the reaction $\bar{C}+A \to \bar{B}+D$. The arguments in the amplitude $\tilde{\bar{T}}$ are

$$t = (p^{\bar{C}}+p^A)^2, \quad u = (p^{\bar{C}}-p^B)^2.$$

The relation between the λ's is similar to [51]

$$\tilde{\lambda}'_C = \lambda'_{\bar{C}}, \; \tilde{\lambda}'_A = -\lambda'_{\bar{A}}, \; \tilde{\lambda}'_B = -\lambda'_B, \; \tilde{\lambda}'_D = \lambda'_D. \qquad \text{...[55]}$$

We now remark that the reaction $\bar{C}+A \to \bar{B}+D$ is identical to the t channel reaction $A+\bar{C} \to \bar{B}+D$, except that the roles of the two incident particles are interchanged. Hence $\tilde{\bar{T}}$ is given in terms of $\tilde{T}$ by

$$\tilde{\bar{T}}(t, u, \tilde{\lambda}'_C, \tilde{\lambda}'_A, \tilde{\lambda}'_B, \tilde{\lambda}'_D) = \tilde{T}(t, M^2-t-u, \tilde{\lambda}'_A, \tilde{\lambda}'_C, \tilde{\lambda}'_B, \tilde{\lambda}'_D). \qquad \text{...[56]}$$

We proceed by considering the positive and negative signature parts $T_\pm$ of T, as defined in [44], and apply [50] and [54] to them. Notice first that [41], [51] and [55] give

$$\tilde{\lambda}'_C = \tilde{\lambda}_C, \; \tilde{\lambda}'_A = \tilde{\lambda}_A, \; \tilde{\lambda}'_B = \tilde{\lambda}_B, \; \tilde{\lambda}'_D = \tilde{\lambda}_D. \qquad \text{...[57]}$$

Hence we obtain from [44], [50], [54] and [56]

$$T_\pm(s, t, \lambda_A, \lambda_B, \lambda_C, \lambda_D) = \tilde{T}^*_\pm(t, s, \tilde{\lambda}_A, \tilde{\lambda}_C, \tilde{\lambda}_B, \tilde{\lambda}_D) \qquad \text{...[58]}$$

where the right-hand side is the analytic continuation, at constant $t+s$ and in the upper half t-plane, of

$$\tilde{T}_\pm(t, s, \tilde{\lambda}_A, \tilde{\lambda}_C, \tilde{\lambda}_B, \tilde{\lambda}_D)$$
$$= \tfrac{1}{2}[\tilde{T}(t, s, \tilde{\lambda}_A, \tilde{\lambda}_C, \tilde{\lambda}_B, \tilde{\lambda}_D) \pm \tilde{T}(t, M^2-t-s, \tilde{\lambda}_A, \tilde{\lambda}_C, \tilde{\lambda}_B, \tilde{\lambda}_D)]. \qquad \text{...[59]}$$

$\tilde{T}_+$ and $\tilde{T}_-$ are the even and odd parts of the amplitude $\tilde{T}$ for exchange of the incident particles A and $\bar{C}$ in the t channel, i.e. for exchange of s and $u = M^2-t-s$.

Consider further the partial wave expansion of $\tilde{T}$ for the t channel reaction $A+\bar{C} \to \bar{B}+D$, which we write in the form

$$\tilde{T}(t, s, \tilde{\lambda}) = \sum_{l=0,1,2\ldots} (2l+1)\, \zeta_l(t, \tilde{\lambda})\, D_l(\tilde{\lambda}, \cos\theta_t). \qquad \text{...[60]}$$

D_l is a rotation function depending on the definition of the spin amplitudes;

it is a linear superposition of Legendre polynomials $P_l(\cos\theta_t)$ and of their derivatives. The angle θ_t is the scattering angle of A to $\bar{B}$ in the c.m. system. It is related to s by the equivalent equations

$$s = [(k_t^2+m_A^2)^{\frac{1}{2}}-(k_t'^2+m_B^2)^{\frac{1}{2}}]^2-k_t^2-k_t'^2+2k_tk_t'\cos\theta_t \quad \text{...[61]}$$

$$u = M^2-t-s = [(k_t^2+m_C^2)^{\frac{1}{2}}-(k_t'^2+m_B^2)^{\frac{1}{2}}]^2-k_t^2-k_t'^2-2k_tk_t'\cos\theta_t \quad \text{...[62]}$$

k_t, k_t' being the c.m. momenta of initial and final states:

$$t = [(k_t^2+m_A^2)^{\frac{1}{2}}+(k_t^2+m_C^2)^{\frac{1}{2}}]^2 = [(k_t'^2+m_B^2)^{\frac{1}{2}}+(k_t'^2+m_D^2)^{\frac{1}{2}}]^2. \quad \text{...[63]}$$

Assume that either $m_A = m_C$, or $m_B = m_D$.‡ Exchange of s and u then simply means replacement of θ_t by $\pi-\theta_t$, so that $\tilde{T}_\pm$ is given by

$$\tilde{T}_\pm(t, s, \tilde{\lambda}) = \tfrac{1}{2} \sum_{l=0,1,2,\ldots} (2l+1)\,\zeta_l^\pm(t, \tilde{\lambda})\,[D_l(\tilde{\lambda}, \cos\theta_t) \pm D_l(\tilde{\lambda}, -\cos\theta_t)] \quad \text{...[64]}$$

where we have written $\zeta_l^\pm$ for the values of ζ_l entering $\tilde{T}_\pm$. Because of the fact that the forces in the s and u channels (e.g. the forces due to exchange of particles in these channels) are in general independent of each other, it is indeed natural to regard $\tilde{T}_+$ and $\tilde{T}_-$, and hence ζ_l^+ and ζ_l^-, as independent of each other. Because of parity invariance one can assume that D_l is even (odd) in $\cos\theta_t$ for l even (odd), as is the case for spinless particles when D_l is simply the Legendre polynomial P_l. For positive (negative) signature the sum [*64*] then contains only terms with even (odd) l, and ζ_l^+ (ζ_l^-) is only defined for even (odd) l.

One now introduces the essential concept of continuation of the orbital momentum l to complex values. Under proper assumptions, (e.g. when the Mandelstam representation holds for the scattering amplitude), one can follow the procedure developed by Regge[39] in non-relativistic potential scattering and obtain one and only one continuation of $\zeta_l^\pm(t, \tilde{\lambda})$ into an analytic function of the complex variable l, t and $\tilde{\lambda}$ being regarded as fixed parameters.§ One then applies to [*64*] the so-called "Sommerfeld-Watson transformation", transforming the sum into a contour integral (for its history see S. C. Frautschi, reference 40, p. 107),

$$\tilde{T}_\pm(t, s, \tilde{\lambda}) = \oint_C (2l+1)\,\zeta_l^\pm(t, \tilde{\lambda})\,\frac{[D_l(\tilde{\lambda}, -\cos\theta_t) \pm D_l(\tilde{\lambda}, \cos\theta_t)]}{4i\sin\pi l}\,dl. \quad \text{...[65]}$$

C is the contour encircling counter-clockwise the positive real semi-axis of the l-plane:

$$l = +\infty+i\varepsilon \to i\varepsilon;\; l = \varepsilon e^{i\phi},\; \phi = \tfrac{1}{2}\pi \to \tfrac{3}{2}\pi;\; l = -i\varepsilon \to +\infty-i\varepsilon$$

‡ The correct treatment of the limit $s \to +\infty$, t fixed, when $m_A \neq m_C$ and $m_B \neq m_D$ is much more complicated, although the main results may be the same (see reference 38).

§ The continuations of ζ_l^+ and ζ_l^- will in general be different functions of l. For details, see reference 40.

with ε positive infinitesimal. In [65] the even (odd) character of D_l in $\cos\theta_t$ for l even (odd) has been used. The analytic continuation of D_l in l can be obtained from the standard analytic continuation of P_l. (See E. J. Squires, reference 40, p. 4 and following.)

The next step is to assume that the analytic continuation in s and t leading from $\tilde{T}$ to T, [50] and [58], can be carried out in the right-hand side of [65], where [61] and [63] allow one to eliminate $\cos\theta_t$ in terms of s and t. One can then discuss the asymptotic behaviour of T_+, T_- and $T = T_+ + T_-$ for $s \to +\infty$ at fixed t by making use of the integral representation [65]. Indeed, for $s \to +\infty$, t fixed, one has $\cos\theta_t \to -\infty$; (remember that the analytic continuation in s and t must go through the upper half t-plane with u = real constant, so that $k_t k_t'$ becomes negative). Furthermore, one knows that for l real or complex,

$$\left.\begin{aligned} P_l(z) \underset{z\to\infty}{\simeq} & \frac{\Gamma(l+\frac{1}{2})}{\Gamma(l+1)} \frac{(2z)^l}{\pi^{\frac{1}{2}}} F\left(-\frac{l}{2}, -\frac{l-1}{2}, -l+\tfrac{1}{2}, z^{-2}\right)+ \\ & + \frac{\Gamma(-l-\frac{1}{2})}{\Gamma(-l)} \frac{(2z)^{-l-1}}{\pi^{\frac{1}{2}}} F\left(\frac{l+1}{2}, \frac{l}{2}+1, l+\tfrac{3}{2}, z^{-2}\right) \end{aligned}\right\} \quad \ldots[66]$$

and that the hypergeometric function F in [66] tends to unity for $z \to \infty$. The asymptotic behaviour is therefore obtained by opening up the contour C to the left until one reaches the line $\operatorname{Re} l = -\frac{1}{2}$ and it is controlled by those singularities of $\zeta_l^{\pm}(t, \tilde{\lambda})$ in the l-plane which have the largest values of $\operatorname{Re} l$ and satisfy $\operatorname{Re} l \geqq -\frac{1}{2}$. In succession of decreasing $\operatorname{Re} l$, they give rise to the successive terms of the asymptotic expansion [26].

If these singularities are simple poles, they give for $s \to +\infty$ terms of type [26] without logarithmic factors. This follows from [66]. Thus, in the special case where $D_l = P_l$, a simple pole of $\zeta_l^{\pm}$ at $l = \alpha_j$ with $\operatorname{Re}\alpha_j > -\frac{1}{2}$ gives a term in [26] with the power $s_t^{\alpha_j}$, plus other terms with powers $s_t^{\alpha_j-1}, s_t^{\alpha_j-2}, \ldots$. The term in $s_t^{\alpha_j}$ originates from the occurrence, in the pole contribution to [65], of the combination

$$P_{\alpha_j}(-\cos\theta_t) \pm P_{\alpha_j}(\cos\theta_t) \simeq \frac{\Gamma(\alpha_j+\frac{1}{2})}{\Gamma(\alpha_j+1)} \frac{|2\cos\theta_t|^{\alpha_j}}{\pi^{\frac{1}{2}}} (1 \pm \exp(i\pi\alpha_j)) \quad \ldots[67]$$

for $\cos\theta_t \to -\infty$. The sign of the exponent in $\exp(i\pi\alpha_j)$ is obtained from the fact that the analytic continuation of [61] in s and t, for large s, leads to the value

$$\cos\theta_t = -|\cos\theta_t| + i\varepsilon, \qquad \varepsilon > 0.$$

After insertion of $\tilde{T}_\pm$ in [58], the resulting contribution to $T_\pm$ contains

$$\frac{\Gamma(\alpha_j+\frac{1}{2})}{\Gamma(\alpha_j+1)} \frac{|2\cos\theta_t|^{\alpha_j}}{\pi^{\frac{1}{2}}} \frac{1 \pm e^{-i\pi\alpha_j}}{\sin\pi\alpha_j}. \quad \ldots[68]$$

We note that the phase agrees with the one given in [46] on the basis of much more general arguments, so that we conclude that the residue of $\zeta_l^{\pm}$ in the pole $l = \alpha_j$ is real. From [*61*]-[*62*], under the restrictive assumption already mentioned that $m_A = m_C$ or $m_B = m_D$, we note that

$$s_t = \frac{s-u}{2} = 4k_t k_t' \cos\theta_t \qquad ...[69]$$

which explains again the particular role played by s_t in asymptotic expansions.

If some singularities of $\zeta_l^{\pm}$ in the complex plane for $\mathrm{Re}\, l \geqq -\frac{1}{2}$ are more complicated than simple poles, e.g. if they are branch cuts, the asymptotic expansion of T will contain terms of type [*26*] with logarithmic factors. Such terms are also likely to arise as a consequence of the integration along the line $\mathrm{Re}\, l = -\frac{1}{2}$. The theoretical situation concerning branch cuts is rather unclear. From the work of Amati, Fubini and Stanghellini[41] and of Mandelstam,[42] one expects that poles cannot be the only singularities of $\zeta_l^{\pm}$ in l and that branch cuts must exist. One has also a tentative mechanism through which, if poles exist, cuts are generated in positions determined by the poles. (See e.g. the Addendum II in S. C. Frautschi, reference 40, p. 185.) All such considerations are uncertain however, because they could well be affected by unforeseen cancellations between singularities in various contributions to the partial wave amplitudes. As mentioned in section 2.1, the matter is also open from the experimental side, in the sense that there is as yet no experimental indication for or against logarithmic factors in the asymptotic expansion [*26*].

The present attitude consists in *assuming* that the leading terms of the expansion [*26*] originate from simple poles of the partial wave amplitudes $\zeta_l^{\pm}(t, \tilde{\lambda})$ in the $\mathrm{Re}\, l > -\frac{1}{2}$ region of the complex l-plane, and to assume that these poles have the properties which can be inferred from consideration of the likely behaviour of the t channel reaction.‡ This gives the *Regge pole model* description of high-energy scattering, as proposed by several authors in 1962, especially Chew and Frautschi,[43] and Gribov and Pomeranchuk.[44] (For a more detailed historical survey see S. C. Frautschi, reference 40, p. 144.) The main features of this model will be reviewed in the next section, a brief comparison with experiment being given in section 2.6.

2.5. *The Regge pole model*

The assumption formulated at the end of the previous section leads to an asymptotic expansion of form

$$T(s, t, \lambda) \underset{s \to +\infty}{\cong} \sum_{j=1,2,\ldots} c_j (s_t/s_j)^{\alpha_j} \qquad ...[70]$$

for the scattering amplitude T of the s channel reaction $A+B \to C+D$, the α_j being given by simple poles in l of the partial wave amplitude $\zeta_l^{\pm}$ of the

‡ This behaviour is usually assumed to be a reasonable relativistic generalization of the behaviour established by Regge to hold for the non-relativistic Schrödinger equation with a superposition of Yukawa potentials (reference 39).

t channel reaction $A+\bar{C} \to \bar{B}+D$ in the way described above, [67]-[69]. This assumption leads to a series of very interesting predictions which we now list:

(a) *Universality property of the exponents* α_j.
(b) *Factorization property of residues.*

A priori, the α_j could depend on all parameters determining $\zeta_l^{\pm}(t, \tilde{\lambda})$ as a function of l, i.e. on t, the four particles $A, \bar{C}, \bar{B}, D$, the signature $\pm$ and the spin orientations $\tilde{\lambda}$. In the t channel, it is however possible to regard $\zeta_l^{\pm}$ as a matrix element

$$\zeta_l^{\pm} = \langle A, \tilde{\lambda}_A;\ \bar{C}, \tilde{\lambda}_{\bar{C}} \mid Z_J^{\pm}(t) \mid \bar{B}, \tilde{\lambda}_{\bar{B}};\ D, \tilde{\lambda}_D \rangle \qquad \text{...[71]}$$

of a matrix $Z_J^{\pm}(t)$, where J is the total angular momentum, which in each particular case can be expressed as a function of l and $\tilde{\lambda}$. For the particles $A, \bar{C}, \bar{B}, D$ and their spin orientations $\tilde{\lambda}$, all choices compatible with the signature and the internal quantum numbers in the t channel can be taken. Poles of $\zeta_l^{\pm}$ in the complex l-plane correspond to poles of the matrix $Z_J^{\pm}(t)$ in the complex J-plane, and the residue of $\zeta_l^{\pm}$ at a pole in l will be given by the residue of the matrix element [71] at the corresponding pole in J. Concerning the poles $J_j^{\pm}(t)$ of $Z_J^{\pm}(t)$ in J (the so-called *Regge poles*) the following predictions can be made immediately: the position of the poles depends exclusively on the value of t, on the signature and on the internal quantum numbers in the t channel. Since a simple pole of a matrix is usually a pole of only one of its eigenvalues, the residues of the matrix element [71] have the form of a product‡

$$R_j^{\pm}(t) = [\gamma_j^{\pm}(t;\ A, \tilde{\lambda}_A;\ \bar{C}, \tilde{\lambda}_{\bar{C}})]^* \, [\gamma_j^{\pm}(t;\ \bar{B}, \tilde{\lambda}_{\bar{B}};\ D, \tilde{\lambda}_D)]. \qquad \text{...[72]}$$

Except for inessential complications due to the fact that J, l and α may differ in the case of particles with spin, the above properties of the poles $J_j^{\pm}$ of $Z_J^{\pm}$ apply to the poles of $\zeta_l^{\pm}$, and hence to the exponents α_j in [70]. In fact, even in presence of spin, and with the normalization of amplitude here used, see [12], one finds $J_j^{\pm}(t) = \alpha_j$. We therefore conclude that the exponents α_j in [70] depend only on the value of t, on the signature and on the internal quantum numbers in the t channel (*universality property*); the residues of $\zeta_l^{\pm}$ have the product form [72] (*factorization property*). From now on we shall write $\alpha_j^{\pm}(t)$ instead of α_j, thereby stressing that α_j should in general depend on t.

(c) *Relation of* $\alpha_j^{\pm}(t)$ *to particles*

A pole of the matrix $Z_J^{\pm}(t)$ in the J-plane at fixed t, if its position depends on t, will in general also be a pole of $Z_J^{\pm}(t)$ in the t-plane at fixed J. If J has a non-negative integral or half-integral value which is physically compatible

‡ That $Z_J^{\pm}$ has a complete set of eigenvectors and eigenvalues follows from unitarity of the S-matrix in the t channel. The factorization property [72] was first recognized by M. Gell-Mann[45] and by V. N. Gribov and I. Ya. Pomeranchuk[44].

with the signature and the internal quantum numbers in the t channel,‡ a pole of $Z_J^{\pm}(t)$ at $t = t_0$ in the t-plane will correspond to a particle of mass $t_0^{\frac{1}{2}}$, with spin J and the same quantum numbers, where one should suppose

$$\text{Re}\, t_0 > 0, \; |\text{Im}\, t_0| \ll \text{Re}\, t_0.$$

If $\text{Im}\, t_0 = 0$ the particle is stable, otherwise it is a resonance. The parity of this particle or resonance is related to the signature.‡

Conversely, as advocated by Chew, Frautschi,[43] Blankenbecler and Goldberger,[46] for every particle X one can assume that the matrix $Z_J^{\pm}(t)$ with the same internal quantum numbers and with the signature allowed by the parity of the particle‡ has a Regge pole $J_j^{\pm}(t)$ such that

$$J_j^{\pm}(m_X^2) = J_X \qquad \ldots[73]$$

m_X, J_X being the mass and spin of X. The function $J_j^{\pm}(t)$ is also called the *Regge trajectory* of X. Several particles can fall on the same trajectory by satisfying [*73*] for the same $J_j^{\pm}(t)$. In this case, because the signature is fixed, their spins J will differ by 2, 4 . . . , while their parities and internal quantum numbers must be the same. Such "Regge recurrences" are now well known for baryons and are likely to exist also for mesons. For baryons we quote as a single example the nucleon resonances of isospin $\frac{3}{2}$ with masses 1236 MeV and 1924 MeV. Their spins and parities are $\frac{3}{2}^+$ and $\frac{7}{2}^+$ (the positive parity corresponding to negative signature in the pion-nucleon system).

We therefore conclude that the same Regge trajectories $J_j^{\pm}(t)$ can be expected to pass through the known particles, as in [*73*], and to determine the exponents $\alpha_j^{\pm}(t) = J_j^{\pm}(t)$ of the asymptotic expansion [*70*].

(d) *Reality and positive slope of* $\alpha_j^{\pm}(t)$

As long as t is positive one can regard $Z_J^{\pm}(t)$ as describing a system of energy $t^{\frac{1}{2}}$ and angular momentum J. Hence, if $J_0 = J_j^{\pm}(t_0)$ with t_0 positive, $t_0^{\frac{1}{2}}$ is an eigenvalue of the Hamiltonian generalized to the value J_0 of the angular momentum, a value which will in general be non-physical. Conversely one can regard $t_0^{\frac{1}{2}}$ as given and determine J_0 as an eigenvalue of the angular momentum variable J entering the generalized Hamiltonian.§ As long as $t_0^{\frac{1}{2}}$ is below the threshold of the continuum of scattering states, the eigenvalue $t_0^{\frac{1}{2}}$ is real if J_0 is real, because real J_0 makes the generalized Hamiltonian Hermitian. Conversely, if $t_0^{\frac{1}{2}}$ is a given real energy below the continuum threshold, the eigenvalue J_0 of J will also be expected to be real.

‡ Considering as in [*64*] the equal mass case, one notes that the signature is related to the space parity in the c.m. system of the t channel, because the replacement $\cos\theta_t \to -\cos\theta_t$ corresponds to inversion of the space coordinates of the incident particles A and $\bar{C}$ in that system. This is particularly simple to see on the example $A = \pi^+$, $\bar{C} = \pi^-$. Positive signature means even parity and $J = 0, 2, \ldots$, negative signature giving odd parity and $J = 1, 3, \ldots$.

§ This generalized Hamiltonian is very simple to write down in non-relativistic Schrödinger theory with a central potential. One takes the radial wave equation and introduces a general value of J in the centrifugal potential term $J(J+1)r^{-2}$.

This leads to the expectation that $J_j^{\pm}(t)$, and therefore $\alpha_j^{\pm}(t)$, is real for t positive below the continuum threshold. It is tempting to assume that the same property would continue to hold for negative values of t, as are relevant for the $A+B \to C+D$ reaction. While this assumption seems reasonable when the t channel system $A+\bar{C}$ behaves as a boson,[47] it was shown by Gribov to be wrong when $A+\bar{C}$ behaves like a fermion[48]: the $J_j^{\pm}(t)$ for $t < 0$ become complex and occur in pairs which are complex conjugate to each other. Hence, if the reaction $A+B \to C+D$ is such that bosons are exchanged in the t channel, the Regge pole model predicts reality of the exponents α_j in the expansion [*70*]. If fermions are exchanged, the exponents become complex for $t < 0$, so that oscillations could occur in the asymptotic behaviour. According to Kinoshita[47] such oscillations are not expected in the cross-section but in the polarization.

A word of caution is needed for the case of fermion exchange. All simple reactions of this type, like

$$\pi^{\pm}+p \to p+\pi^{\pm}, \; \pi^{-}+p \to n+\pi^{0}, \; \pi^{-}+p \to \bar{p}+d, \; p+p \to \pi^{+}+d$$

correspond to the unequal mass case ($m_A \neq m_C$ and $m_B \neq m_D$ in our general notation) for which a correct treatment of fermions is not yet available. (The recent discussion by Goldberger and Jones[38] is carried out for spinless particles.) This, however, in no way decreases the interest of their careful experimental investigation.

Returning to our previous discussion of the Regge poles $J_j^{\pm}(t)$ for t positive below the continuum threshold, it is clear that one there expects the slope $dJ_j^{\pm}/dt$ to be positive. Indeed, $t_0^{\frac{1}{2}}$ can be regarded as an energy eigenvalue of a generalized Hamiltonian with angular momentum $J_0 = J_j^{\pm}(t_0)$. Increasing J_0 means increasing the centrifugal potential, and this will increase the energy eigenvalue $t_0^{\frac{1}{2}}$, thereby showing that $dJ_j^{\pm}/dt$, and hence $d\alpha_j^{\pm}/dt$ *must be positive*. These properties are likely to remain valid for $t < 0$, at least for boson trajectories. As we mentioned before fermion trajectories become complex for $t < 0$, but the imaginary part will be small for small t.

We finally note that all considerations under (c) and (d) only apply to *moving poles*, i.e. when $J_j^{\pm}(t)$ effectively varies with t. There has been much argument in the literature on whether *fixed poles* or more generally *fixed singularities* (i.e. singularities independent of t) can exist or not. The matter is essentially undecided in view of the difficulty that, if there exist branch cuts moving with t (which is likely), they can move across a fixed singularity and force the latter to move from one sheet to another of the Riemann surface in l.

(e) *Reality and zeros of residues*

It is generally believed that the residue of the partial wave amplitude [*71*] at a Regge pole $J_j^{\pm}(t)$ is a real quantity for values of $t = (p^A - p^C)^2$ as are taken in the $A+B \to C+D$ reaction. In the simple case of $D_l = P_l$ considered to derive [*68*], we have already concluded to the reality of the residue.

The argument there used applies generally when $J_j^{\pm}(t)$ and $\alpha_j^{\pm}(t)$ are real and shows that reality of the residue then follows from analyticity of T at fixed t (§ 2.3). For the fermion exchange case the same is probably true to the approximation that the $\alpha_j^{\pm}(t)$ would be almost real as long as $|t|$ is small.

Another interesting property of residues is that they may be expected to vanish for special values of $t = (p^A - p^C)^2$ in the scattering process $A+B \to C+D$ at high energy. Consider the residue $R_j^{\pm}(t)$ of the partial wave amplitude [71] at $J = J_j^{\pm}(t)$, and let $\alpha_j^{\pm}(t) = \alpha_j$ be the exponent of the corresponding term in the expansion [70]. The coefficient c_j of this term is essentially the product of R_j and of an expression of type [68] resulting from the last factor in the integrand of [65]. Now [68] has a pole for $\alpha_j = \alpha_j^{\pm}(t)$ equal to an even (odd) integer when the signature is even (odd). If the asymptotic expansion [70] is valid, the residue $R_j^{\pm}(t)$ must then vanish for the corresponding value of t so that the scattering amplitude T remains finite. It has been conjectured[49-51] that $R_j^{\pm}(t)$ would also vanish for some values of t for which $\alpha_j^{\pm}(t)$ equals an odd (even) integer when the signature is even (odd). Since [68] is then finite, this would mean that the j-th term of the asymptotic expansion would vanish for such t, an effect which would be observable. The present version of this conjecture is that $R_j^{\pm}(t)$ would only vanish if, in addition to the above circumstance, either one of the following conditions would be satisfied:

(i) the value J' of $J_j^{\pm}(t)$ at the t value in question‡ is negative *or*

(ii) J' being non-negative, a Feynman diagram for one-particle exchange in $A+B \to C+D$, consisting solely of exchange of a particle of spin J' in the t channel, gives contribution zero to the particular amplitude $T(s, t, \lambda_A, \lambda_B, \lambda_C, \lambda_D)$ considered.

(f) *Analyticity properties of poles and residues*

In addition to the above properties or conjectures, one has some information on the analytic properties in t of the Regge poles $J_j^{\pm}(t)$ and residues $R_j^{\pm}(t)$ of the partial wave amplitude [71]. Either from analogy with non-relativistic Schrödinger theory, or from consideration of the Mandelstam representation for the relativistic amplitude, one expects that $J_j^{\pm}(t)$ and $R_j^{\pm}(t)$ are real analytic functions of t if $A+\bar{C}$ is a boson, or of $t^{\frac{1}{2}}$ if $A+\bar{C}$ is a fermion, with a cut from the lowest threshold in the t channel to $+\infty$. (See E. J. Squires, reference 40, section 4.1, and also S. C. Frautschi, M. Gell-Mann and F. Zachariasen[43] and A. Barut and D. E. Zwanziger.[52])

2.6. *Comparison with experiment*

We end this section with a brief discussion of the experimental situation in terms of the asymptotic expansion [70] and of the Regge pole model. (A

‡ Since the corresponding value of l is integral, J' is an integer (half-integer) for boson (fermion) exchange in the t channel.

more detailed review and references will be found in the various contributions of reference 53.)

For elastic scattering, the experimental facts suggest that the leading term of the expansion has, for all reactions, the exponent

$$\alpha_1 \simeq 1,$$

leading to constant σ_T and $d\sigma_{el}/dt$ in the high-energy limit, that its signature is positive and that its t channel quantum numbers are those of the vacuum. In the Regge pole model, $\alpha_1 = \alpha_1^+(t) \simeq 1$ is referred to as the Pomeranchuk trajectory. It appears to be universal, its slope $d\alpha_1^+/dt$ is vanishing or small, it can be associated with either of the 2^+ mesons having $I = Y = 0$ (f_0 with mass 1250 MeV, f_0' with 1525 MeV).

The deviations of σ_T and $d\sigma_{el}/dt$ from constancy can be explained by introducing further terms of the expansion, which have the t channel quantum numbers of an SU_3 nonet of mesons of odd signature and of an octet or a nonet of mesons of even signature. If the slope $d\alpha_j^{\mp}/dt$ for these terms is given a value of order $1(\text{GeV}/c)^{-2}$, the corresponding Regge trajectories can be extrapolated as approximately straight lines through the well-known nonet of 1^- mesons (for the odd signature ones) and the suspected nonet of 2^+ mesons (for the even signature ones).[49, 54]

Charge exchange reactions give information on the terms of the expansion with charge ± 1 and zero hypercharge in the t channel. Among the trajectories mentioned, only the odd signature term, i.e. the ρ trajectory $\alpha_\rho^-(t)$, contributes to $\pi^- p \to \pi^0 n$. This has allowed the experimental determination of its shape, which was indeed found to be very straight for $-1(\text{GeV/c})^2 \lesssim t \leqq 0$ and to have the slope $\sim 1(\text{GeV/c})^{-2}$ which makes it pass through the spin 1 of the ρ meson when extrapolated as a straight line to $t = m^2$. This corresponds to an appreciable shrinkage of the peak and is the most remarkable success of the Regge pole model to date. The prediction on vanishing residues described in the previous section (end of point (d)) applies to the spin-flip part of the $\pi^- p \to \pi^0 n$ amplitude. It is predicted to vanish at the t value where $\alpha_\rho^-(t)$ vanishes, which is $t = \simeq 0{\cdot}6(\text{GeV/c})^2$. It is striking that the cross-section has a marked minimum for this value of t. Only one aspect of $\pi^- p \to \pi^0 n$ is not explained in terms of the ρ trajectory alone. It is the non-vanishing polarization at 6 GeV,[25] which must be explained by interference between the ρ trajectory term and another term in the amplitude. At present, many choices are possible for this term, but it cannot belong to the two nonets mentioned above.

The reaction $K^- p \to \bar{K}^0 n$ is dominated by the ρ trajectory and its positive signature counterpart, which is supposed to belong to the A_2 meson of $I = 1$, $Y = 0$ with mass 1290 MeV. (This trajectory assignment requires the spin parity of A_2 to be 2^+, a point which is not completely settled.) The A_2 trajectory passes through the A_2 meson at spin $J = 2$ if it is given the slope of $1(\text{GeV/c})^{-2}$ mentioned above. This allowed the theoretical prediction of the reaction $\pi^- p \to \eta n$, which is entirely dominated by the A_2 trajectory,[55] in good agreement with the later experimental findings.[20]

On the whole the Regge pole analysis of meson-nucleon forward reactions has been extremely successful and it will be of great importance to carry it through to a higher degree of precision ($\pi^- p \to \pi^0 n$ is the only reaction where the tests of the Regge model have reached the truly quantitative stage). Backward $\pi^\pm p$ scattering can also be successfully described in terms of two very straight trajectories of opposite signature, one belonging to the nucleon and one to $N^*_{\frac{3}{2}\frac{3}{2}}$ (1238 MeV). The data[15,16] show that the latter is more weakly coupled than the former. Furthermore, a dip in $d\sigma/dt$ for $\pi^+ p \to p\pi^+$ can be accounted for in terms of the prediction in the previous section (end of point (d)), which is here applicable to the residue of the nucleon trajectory.[53]

The situation for charge-exchange reactions with nucleons and anti-nucleons is less satisfactory. The results available on $np \to pn$,[15-17] and $\bar{p}p \to \bar{n}n$[22,56] cannot be explained in terms of the two nonets of trajectories mentioned above.[57,70] At least one further trajectory is needed,[57] or one has to resort to other explanations, like the long-range part of the one-pion exchange proposed by Byers[59] in the framework of the coherent droplet model.[60]

3

Diffraction as Shadow Scattering

3.1. *Unitarity condition and overlap function*

It is likely that the scattering amplitude T_{el} for elastic scattering $A+B \to A+B$ becomes almost purely imaginary at high energy for small momentum transfers. Since the differential cross-section $d\sigma_{el}/dt$ becomes extremely small for all but small values of t, one concludes that elastic scattering, whenever appreciable, is essentially shadow scattering caused by the existence of inelastic processes, and it can therefore be approximately expressed in terms of bulk properties of the inelastic processes. The present section discusses some implications of this general property, which is an immediate consequence of the unitarity of the S-matrix.

The unitarity condition in operator form is

$$S^\dagger S = 1, \text{ or } i(T^\dagger - T) = T^\dagger T \text{ for } T = i(1-S). \qquad ...[74]$$

We take its matrix element between the states $\langle \mathbf{p}'^A, \mathbf{p}'^B |$ and $| \mathbf{p}^A, \mathbf{p}^B \rangle$ of the $A+B$ system, neglecting spins altogether as would be allowed if the spin-flip amplitudes become negligible compared to the non-spin-flip ones at very high energy. In the matrix element of $T^\dagger T$ we separate the elastic and the inelastic intermediate states. The formulae become particularly simple in terms of the partial wave amplitudes $i\eta_l$, which are defined by

$$T_{el}(s, t) = \frac{is^{\frac{1}{2}}}{4\pi k} \sum_l (2l+1)\eta_l P_l(\cos\theta) \qquad ...[75]$$

with the same notation as in section 1.1. One finds

$$\text{Re}\,\eta_l - \tfrac{1}{2}|\eta_l|^2 = f_l \qquad ...[76]$$

where f_l is determined by the inelastic collisions according to the following rule. For initial state $|\boldsymbol{p}^A, \boldsymbol{p}^B\rangle$ of the $A+B$ system, consider the inelastic part of the final state

$$|\psi(\boldsymbol{p}^A, \boldsymbol{p}^B)\rangle = PS|\boldsymbol{p}^A, \boldsymbol{p}^B\rangle \qquad ...[77]$$

where S is the S-matrix and P the projection operator on all inelastic states, i.e. on all states other than $A+B$. We then have

$$\langle\psi(\boldsymbol{p}'^A, \boldsymbol{p}'^B)|\psi(\boldsymbol{p}^A, \boldsymbol{p}^B)\rangle = (p_0^A p_0^B)^{-1}\,\delta^4(p^A+p^B-p'^A-p'^B)\,F(s, t) \qquad ...[78]$$

$$F(s, t) = \frac{s^{\frac{1}{2}}}{2\pi k} \sum_l (2l+1) f_l P_l(\cos\theta) \qquad ...[79]$$

p^A, p^B are as usual the four-momenta of A and B, p'^A and p'^B are other possible values of these variables, with

$$\begin{aligned} (p^A)^2 &= (p'^A)^2 = m_A^2,\ (p^B)^2 = (p'^B)^2 = m_B^2 \\ s &= (p^A+p^B)^2,\ t = (p^A-p'^A)^2 \\ s^{\frac{1}{2}} &= (k^2+m_A^2)^{\frac{1}{2}}+(k^2+m_B^2)^{\frac{1}{2}} \\ t &= -2k^2(1-\cos\theta). \end{aligned} \qquad ...[80]$$

The relativistically invariant function $F(s, t)$ has a simple significance. Taking two initial states $|\boldsymbol{p}^A, \boldsymbol{p}^B\rangle$, $|\boldsymbol{p}'^A, \boldsymbol{p}'^B\rangle$ with $p^A+p^B = p'^A+p'^B$, i.e. with same c.m. system and same c.m. energy, $F(s, t)$ measures the amount of overlap between the wave functions $|\psi(\boldsymbol{p}^A, \boldsymbol{p}^B)\rangle$ and $|\psi(\boldsymbol{p}'^A, \boldsymbol{p}'^B)\rangle$ of the corresponding inelastic final states. This function, which is called the *overlap function*,[61] describes the bulk properties of inelastic collisions which are relevant for determining the resulting shadow. Indeed, F fixes the f_l through [79], and f_l determines η_l through [76] when T_{el} is purely imaginary, i.e. η_l purely real.‡ Equation [76] then gives two possible values of η_l

$$\eta_l = 1 \pm (1-2f_l)^{\frac{1}{2}} \qquad ...[81]$$

which are real because unitarity of the S-matrix applied to the l-th partial wave for elastic $A+B$ scattering gives

$$|1-\eta_l| \leqq 1 \quad \text{or} \quad 0 \leqq f_l \leqq \tfrac{1}{2}. \qquad ...[82]$$

‡ This remains true when T_{el} has a non-negligible real part if the phase of T_{el} can be determined on independent theoretical grounds, e.g. from the asymptotic expansion [26] and the phase rule [46] as discussed in section 2.

The sign ambiguity in [81] is easily lifted by a continuity argument: experiment suggests that at high energy η_l and f_l tend to continuous functions of the impact parameter $b = lk^{-1}$, that η_l never exceeds 1, and that $\eta_l \to 0$ for $l \to \infty$. Continuity then requires one to choose the negative sign

$$\eta_l = 1-(1-2f_l)^{\frac{1}{2}}. \qquad ...[83]$$

For qualitative considerations one can approximate [83] by the first term of its power expansion:

$$\eta_l \simeq f_l, \quad \text{or} \quad T_{el}(s, t) \simeq \frac{i}{2} F(s, t) \qquad ...[84]$$

because experiment shows that, except at small l, higher order corrections are small. This is due to the almost exponential shape of the diffraction peak, see [19] and [20], which implies that the ratio σ_{el}/σ_T of integrated elastic to total cross-sections is always small ($\lesssim 0{\cdot}2$). One therefore concludes that in first approximation the shape of the diffraction peak, i.e. of $d\sigma_{el}/dt$ at small $|t|$, is directly given by the t dependence of the overlap function $F(s, t)$ defined in [78]. Experimentally, the most common inelastic collisions at high energy appear to produce an appreciable number of secondaries with small transverse momentum but rather large longitudinal momentum in the c.m. system. This is often referred to as the "*jet*" structure of common high-energy collisions. The jet structure immediately implies that $F(s, t)$ should drop steeply with t and therefore predicts a similar steep drop for $d\sigma_{el}/dt$. It is difficult to put such considerations in more quantitative terms before more detailed information becomes available on common inelastic collisions. The work done in this direction will be briefly reviewed in section 3.3. It is clear, however, that the problem is of considerable interest, because it is on the t dependence of differential cross-sections at high energy that existing theoretical models have made the least progress.‡

We end this section with a few cross-section formulae:

$$\left.\begin{aligned} \sigma_T &= 2\pi k^{-2} \sum_l (2l+1) \,\mathrm{Re}\, \eta_l, \quad \sigma_{el} = \pi k^{-2} \sum_l (2l+1)\,|\eta_l|^2 \\ \sigma_{in} &= \sigma_T - \sigma_{el} = 2\pi k^{-2} \sum_l (2l+1) f_l = 4\pi^2 k^{-1} s^{-\frac{1}{2}} F(s, 0) \end{aligned}\right\} \qquad ...[85]$$

3.2. *Inequalities between cross-sections*

In the same qualitative vein as in the discussion after equation [84], consideration of the overlap function allows one to predict on simple, intuitive grounds a number of inequalities between cross-sections which are in agreement with experiment. Consider first as an example pp inelastic collisions at high energy. Their main characteristic is the jet structure mentioned above, especially the small values of all transverse momenta. It suggests the assumption that the outgoing particles can be roughly divided into two groups, one

‡ Thus, the Regge pole model makes only very few predictions on the t dependence of the residue functions.

containing the particles which go forward in the c.m. system (*forward group*) and one containing the particles going backward (*backward group*). In each of these groups the *leading particle*, i.e. the particle with largest c.m. momentum, is almost always a nucleon or a nucleon isobar.[62] In addition, antinucleon and strange particle production is rare. Hence, both the forward and the backward groups have most of the time their total baryon number and strangeness equal to those of a proton. In fact, although few measurements are available, it seems likely that also the total charge of each group would be mostly equal to the proton charge.

These properties can be formulated more generally in terms of exchange. Call A the incident particle, B the target particle, A' the forward group and B' the backward group (A' and B' are in general many-particle systems, i.e. neither single particles nor resonances). In analogy with two-body reactions, the process $A+B \rightarrow A'+B'$ can be called the s channel reaction, and $A+\bar{A}' \rightarrow \bar{B}+B'$ the t channel one. The properties mentioned above then mean that t channel exchange of baryon number, strangeness and charge is very rare in the bulk of inelastic collisions at high energy, as we know it to be also in two-body collisions. This would probably hold also for isospin, i.e. for all internal quantum numbers conserved by strong interactions. The complete inelastic final state ψ_{AB} of an $A+B$ collision is thereby split approximately as

$$\psi_{AB} \simeq \psi_{AB}^{(0)}+\psi_{AB}^{(1)} \qquad ...[86]$$

where $\psi_{AB}^{(0)}$ refers to those final states for which vanishing internal quantum numbers are exchanged in the t channel, and $\psi_{AB}^{(1)}$ to the final states where non-vanishing exchange took place. Using the definition [*78*] one obtains a corresponding separation of the overlap function into two parts

$$F_{AB}(s, t) \simeq F_{AB}^{(0)}(s, t)+F_{AB}^{(1)}(s, t). \qquad ...[87]$$

The expected absence of t channel exchange of non-vanishing quantum numbers at high energy then implies

$$F_{AB}^{(1)}/F_{AB}^{(0)} \rightarrow 0 \quad \text{for} \quad s \rightarrow \infty. \qquad ...[88]$$

Compare now e.g. pp and np collisions. Any final state $p'+p''$ belonging to $\psi_{pp}^{(0)}$ will correspond to a final state $n'+p''$ belonging to $\psi_{np}^{(0)}$, simply by relating n' to p' through the same isospin rotation as turns n into p, and vice-versa. Hence

$$F_{pp}^{(0)}(s, t) = F_{np}^{(0)}(s, t). \qquad ...[89]$$

The same does not hold, however, for the final states $p'+p''$, $n'+p''$ contained in $\psi_{pp}^{(1)}$, $\psi_{np}^{(1)}$. There will be more such states of the $n'+p''$ type because charge can be exchanged in more ways between n and p than between p and p. Since each particular final state contributes a positive number to the overlap function at $t = 0$, see [*78*], we conclude

$$F_{pp}^{(1)}(s, 0) < F_{np}^{(1)}(s, 0). \qquad ...[90]$$

Remembering [*85*], this gives an inequality between inelastic cross-sections.‡

$$\sigma_{\text{in}}(pp) < \sigma_{\text{in}}(np). \qquad ...[91]$$

Similarly, considering $\bar{p}p$ and $\bar{n}p$, the more numerous possibilities of charge exchange in the first than in the second reaction gives the prediction

$$\sigma_{\text{in}}(\bar{p}p) > \sigma_{\text{in}}(\bar{n}p) \qquad ...[92]$$

whereas the abundant possibility of nucleon exchange (mainly annihilation) in $\bar{n}p$ means that

$$\sigma_{\text{in}}(\bar{n}p) > \sigma_{\text{in}}(np). \qquad ...[93]$$

In deriving the last inequality, we have assumed that

$$F^{(0)}_{\bar{n}p}(s, t) = F^{(0)}_{np}(s, t).$$

One expects indeed in general that

$$F^{(0)}_{\bar{A}B}(s, t) = F^{(0)}_{AB}(s, t) \qquad ...[94]$$

corresponding to the fact that only the charge conjugation number $C = +1$ would be exchanged in the t channel at very high energy.§

Entirely similar considerations apply to meson-baryon collisions. They give the following inequalities:

$$\sigma_{\text{in}}(\pi^- p) > \sigma_{\text{in}}(\pi^+ p) \qquad ...[95]$$

$$\sigma_{\text{in}}(K^- p) > \sigma_{\text{in}}(\bar{K}^0 p) \qquad ...[96]$$

$$\sigma_{\text{in}}(\bar{K}^0 p) > \sigma_{\text{in}}(K^0 p) \qquad ...[97]$$

$$\sigma_{\text{in}}(K^0 p) > \sigma_{\text{in}}(K^+ p). \qquad ...[98]$$

All except [*97*] follow from considerations of charge exchange. Equation [*97*] uses again [*94*] and relies on the fact that strangeness exchange is possible in many more inelastic states for $\bar{K}^0 p$ than for $K^0 p$, because hypercharge $Y = 0$ hyperons are numerous whereas no $Y = 2$ baryon is known so far with certainty.

All inequalities above are due to the presence of t channel exchange of non-vanishing internal quantum numbers. As expressed in [*88*] this exchange becomes rare at high energy, so that in each inequality [*91*]-[*98*] the two inelastic cross-sections become equal for $s \to +\infty$, just as the corresponding elastic scattering cross-sections $d\sigma_{\text{el}}/dt$ have the same asymptotic property for the same reasons.

‡ Such inequalities are more readily compared with experiment for total cross sections σ_T, but the smallness of $\sigma_{\text{el}}/\sigma_T$ and the similarity of shape of all diffraction peaks (they differ mainly in width and height, not so much in shape) implies that σ_T will obey the same inequalities as σ_{in}.

§ The analogous property for two-body processes is that the leading term of the expansion [*26*] for elastic scattering has positive signature, which implies that the corresponding value of C exchanged in the t channel is $+1$.

To present the above considerations, we have assumed that the outgoing particles in most inelastic collisions can be subdivided into a forward and a backward group. Our conclusions, however, have a somewhat greater validity. What is required in order to compare $\sigma_{in}(A_1B)$ with $\sigma_{in}(A_2B)$ is that the final states of, say, inelastic A_1+B collisions have analogues of equal amplitude in A_2+B, whereas some final states of A_2+B have no analogue in A_1+B. This property is clearly more general than that assumed in our discussion. How exactly it would be implemented in nature is impossible to say at present, because our experimental knowledge of inelastic final states containing more than two particles or resonances is extremely limited. In fact it is bound to progress very slowly in view of the great complications inherent in the analysis of many-particle systems. It would be desirable, however, to exhibit in some way the type of properties described in the present paragraph.

3.3. *The t dependence of the overlap function*

When the real part of the elastic scattering amplitude is neglected, the t dependence of the overlap function $F(s, t)$ can be deduced from the observed differential cross-section, [*19*]. If one goes beyond the approximation [*84*] and uses [*83*], one finds that the t dependence of F is closer to an exact exponential

$$F(s, t) = F(s, 0) \exp (b't) \qquad ...[99]$$

than is the case for $d\sigma_{el}/dt$. In fact, using [*99*], one calculates a differential cross-section having for small $|t|$ the shape [*19*] with $c/b^2 \simeq 0{\cdot}02$, whereas at large t it behaves as $\exp(-b''(-t)^{\frac{1}{2}})$. (See the last two papers quoted in reference 51 and W. N. Cottingham and R. F. Peierls, reference 63, where actual πp and pp data are fitted.) It appears therefore that the simple t dependence [*99*] is a good approximation to the facts. It probably holds over a range of t such that F decreases by more than an order of magnitude.

One can ask whether the simple exponential shape of F can be given a qualitative explanation. A possibility is that the most common inelastic final states might have many degrees of freedom excited, with weak correlations between them. In the extreme case where even the correlations imposed by the conservation laws would be neglected, the inelastic final state [*77*] would then have the form of a product

$$\psi(\mathbf{p}^A, \mathbf{p}^B) = \prod_{i=1}^{N} \psi_i(\mathbf{p}^A, \mathbf{p}^B) \qquad ...[100]$$

where $\psi_1, \ldots \psi_N$ would refer to various groups of degrees of freedom, N being a rather large number at high energy. Equation [*78*] then implies

$$F(s, t) = \prod_{i=1}^{N} F_i(s, t), \qquad ...[101]$$
$$F_i(s, t) \propto \langle \psi_i(\mathbf{p}'^A, \mathbf{p}'^B) \mid \psi_i(\mathbf{p}^A, \mathbf{p}^B) \rangle$$

where the constraint of energy-momentum has been dropped from the formulae (the effect of this constraint will of course be the smaller, the larger the value of N). All $F_i(s, t)$ are real functions decreasing with t. If the number N of factors is large the t decrease of the product is necessarily of exponential type

$$F(s, t) \simeq \prod_i F_i(s, 0)(1+a_i(s)t) \simeq [\prod_i F_i(s, 0)] \exp [t \sum_i a_i(s)]. \quad ...[102]$$

This type of consideration has been carried out in more detail, taking into account momentum and energy conservation, for a special "uncorrelated jet" model.[64] It is impossible to say at present whether it provides the correct explanation for the exponential shape of F. While the multiplicity of particles produced in high-energy collisions seems appreciable, information is lacking completely on possible correlations among them.

We mention in closing this section an investigation by Kokkedee.[65] He studies $\bar{p}p$ collisions by considering the overlap function as composed of two parts, one referring to inelastic collisions with annihilation and one without, corresponding roughly to the terms $F^{(1)}$ and $F^{(0)}$ in [87]. The rapid energy variation still shown by the $\bar{p}p$ system at the highest energies mentioned, compared with the approximate constancy of the pp and np systems, suggests that the contribution of $F^{(1)}$ is sizable but decreases rapidly with energy. Kokkedee proposes a model for $F^{(1)}$ which corroborates this fact and accounts well for the data.

4

Quark Model of High-Energy Processes

4.1. *SU_6 symmetry, current algebras and quark model*

Whereas there is no particular difficulty in using the methods of section 2 to analyse the effect of approximate SU_3 symmetry on high-energy reactions, it is more difficult to adapt them to the discussion of SU_6 symmetry which involves spin in an essential way; one then encounters all the complications of relativistic spin-orbit coupling. Still, very soon after its introduction, SU_6 led to a remarkable prediction concerning differences between total cross-sections, the Johnson-Treiman relations[66]

$$\tfrac{1}{2}[\sigma_T(K^+p)-\sigma_T(K^-p)] = \sigma_T(\pi^+p)-\sigma_T(\pi^-p) = \sigma_T(K^+n)-\sigma_T(K^-n), \quad ...[103]$$

which are in reasonable agreement with experiment. On the other hand, the agreement is very poor for other relations like for example the triangular inequality for forward scattering[67]

$$\left[\frac{d\sigma}{dt}(K^-p \to \pi^-\Sigma^+)\right]^{\frac{1}{2}} \geq \left[\frac{d\sigma_{el}}{dt}(\pi^-p)\right]^{\frac{1}{2}} - \left[\frac{d\sigma_{el}}{dt}(K^-p)\right]^{\frac{1}{2}} \quad ...[104]$$

although it follows from SU_3 symmetry alone and does not require SU_6. In fact, the left-hand side of [*104*] is found experimentally to decrease rapidly with energy, whereas the right-hand side stays positive and almost constant. Clearly, symmetry breaking is very important in some cross-sections or combinations of cross-sections while others are hardly affected by it. This type of difficulty affects of course also the theoretical analysis of SU_3 and SU_6 symmetry in high-energy reactions. The problem is usually, in deriving successful relations between cross-sections, not to use arguments which would lead to further relations in disagreement with experiment.

This situation is not very different for the one encountered in the SU_6 symmetry analysis of electromagnetic and weak interactions of hadrons. Also here a consistent and successful treatment is lacking by group-theoretical methods, and one has therefore developed other approaches, mainly those based on current algebras[68] and on the quark model.[69] The latter is a dynamical model treating the hadrons as composite particles formed of SU_3 triplets called quarks. It gives rise to very successful but admittedly very crude calculations, based on a non-relativistic treatment of the quarks inside the hadrons but by-passing and leaving unanswered the problems connected with the binding force between quarks. The current algebra method makes a less explicit use of quarks. It postulates for the currents the commutation rules which hold when the current operators are given the simplest form in terms of quark fields. These commutation rules are then used to derive sum rules for observable properties of hadrons, without any further reference to quarks. Also the current algebra method has achieved many successes, the most remarkable being the Adler-Weisberger sum rule which gives in terms of pion-nucleon total cross-sections the absolute value of the ratio between axial-vector and vector coupling constants of beta decay.[70]

One of the most attractive features of the quark model is that it predicts in a very direct way relations between certain properties of nucleons and analogous properties of mesons. The best known example is the correct relation between the proton magnetic moment and the partial width of the decay $\omega^0 \to \pi^0 + \gamma$.[71] As we shall see later on, an even more striking prediction is obtained for high-energy scattering, where the quark model gives for total cross-sections:

$$\lim_{s \to +\infty} \sigma_T(\pi p) \,/\, \sigma_T(pp) = \tfrac{2}{3}. \qquad \ldots[105]$$

Extrapolating measured πp and pp cross-sections to constant values at infinite energy gives $\sigma_T(\pi p) \simeq 22$ mb, $\sigma_T(pp) \simeq 36$ mb, in close agreement with [*105*]. It is the qualitative success of such simple predictions and the fact that they do not presuppose exact SU_3 symmetry which make the quark model so attractive and fruitful in the present stage of theoretical developments. Most of its predictions can be obtained in slightly modified form from less radical assumptions by means of the current algebra method, but the derivation is then more complex and the input depends more on the particular property considered.‡ Thus we mention a very recent extension of the current algebra method to high-energy scattering by Cabibbo, Horwitz, Kokkedee and Ne'eman.[73] These authors set out to obtain [*105*] and other consequences of the quark model for high-energy scattering by introducing scalar and vector currents for the coupling of the particles to two nonets of Regge poles, the scalar (vector) currents being coupled to the even (odd) signature poles. The commutation rules of these "strong currents" are derived from the quark model but no further use is made of quarks. One obtains the remarkable results that, while [*105*] and many other successful relations at $t = 0$ are derived, a fit to the s dependence of total cross-sections requires all σ_T to approach zero as s^a, $a \simeq -0{\cdot}07$ for $s \to \infty$.§

The subsequent discussion will be devoted to the quark model approach to high-energy scattering. While we shall review only a few among the many relations which have been derived for the forward scattering amplitude and its imaginary part, i.e., the total cross-section,[74] we shall pay more attention to the t dependence of the amplitude at small t and to the consistency of the basic additivity assumption.[75]

4.2. *Additivity assumption for scattering amplitude*

The basic assumption underlying all applications of the quark model[69] is one of *additivity*, stating that some particular property of a hadron is a sum of terms belonging to the quarks and antiquarks composing that hadron. In weak or electromagnetic interactions, the matrix element of the (weak or electromagnetic) current operator J_μ between two hadronic states A, A' is written as a sum of terms:

$$\langle A' \mid J_\mu \mid A \rangle = \sum_i \langle A' \mid J_\mu^i \mid A \rangle \qquad \ldots[106]$$

where J_μ^i acts on the i-th quark‖ in A and leaves the other quarks of A unaffected. The structure of $\langle A' \mid J_\mu^i \mid A \rangle$ is always assumed to be of the form

$$\langle A' \mid J_\mu^i \mid A \rangle = \langle Q_i^{A'} \mid J_\mu^i \mid Q_i^A \rangle f_i^{A'A}(\Delta p_\mu). \qquad \ldots[107]$$

$\langle Q_i^{A'} \mid J_\mu^i \mid Q_i^A \rangle$ represents the current matrix element between quark states

‡ Another method to derive some of the consequences of the quark model without assuming quarks is provided by the concept of universality, which is an assumption that all relevant currents or couplings are dominated by the known mesons. This has been studied in detail by Freund. For application of universality to high-energy scattering see reference 72.

§ It should be said that the model of Cabibbo *et al.* does not reproduce equation [*105*] when SU_3 breaking is taken into account. The asymptotic ratio $\sigma_T(\pi p)/\sigma_T(pp)$ can then take any value depending on the F/D ratio for the baryon coupling to the even signature octet, which is a free parameter; see the second paper of reference 73. The situation in the quark model is different because πp and pp scattering, which involves only the $I = \frac{1}{2}$ quarks, is not influenced by SU_3 breaking which first appears when the $I = 0$ quark is also involved.

‖ To be more precise one should say here the i-th quark or antiquark, but we shall usually omit the explicit distinction between quark and antiquark. For mesons the index i runs over one quark and one antiquark, for baryons over three quarks.

Q_i^A, $Q_i^{A'}$ calculated for the same value of the momentum transfer $p_\mu^{A'} - p_\mu^A = \Delta p_\mu$ as occurs in $\langle A' \mid J_\mu \mid A \rangle$. It depends on A, A' only through the spin and SU_3 variables of Q_i^A, $Q_i^{A'}$. The form factor $f_i^{A'A}$ represents the overlap between the quark wave function of hadron A' and the wave function generated by action of J_μ^i on the quark wave function of A. When A' is identical to A and J_μ^i does not flip either spin or SU_3 state of Q_i one has

$$f_i^{AA}(0) = 1. \qquad ...[108]$$

Equations [*106*] and [*107*] describe the contents of the additivity assumption. Equation [*106*] states the additivity proper and [*107*] describes the approximation which allows one to compare properties of, say, baryons and mesons. Indeed $\langle Q_i^{A'} \mid J_\mu^i \mid Q_i^A \rangle$ is supposed to depend only on Δp_μ, on the spin and on the SU_3 states of quark Q_i in A and A', whereas $f_i^{A'A}$ describes the effect of the quark wave functions of A and A'.‡

We now generalize the additivity assumption to the two-body scattering process

$$A + B \rightarrow A' + B'. \qquad ...[109]$$

A and A' are supposed to have the same quark composition, i.e. to be both baryons or both mesons; the same holds for B and B'. As usual s will denote the square of the c.m. energy. We write Δp_μ for the four-momentum transfer

$$\Delta p_\mu = p_\mu^{A'} - p_\mu^A = p_\mu^B - p_\mu^{B'}. \qquad ...[110]$$

Let Q_i denote the quarks in A ($i = 1, 2, 3$ if A is a baryon; $i = 1, 2$ if A is a meson; as before we use the term quark both for quarks and antiquarks). Similarly, let Q_j denote the quarks in B. The additivity assumption now means that the process [*109*] takes place through scattering of a single quark Q_i in A on a single quark Q_j in B, the remaining quarks in A and B remaining unaffected. In terms of the collision matrix

$$T = i(1 - S) \qquad ...[111]$$

this can be written

$$\langle A'B' \mid T \mid AB \rangle = \delta^4(p^{A'} + p^{B'} - p^A - p^B) \sum_{ij} \langle A'B' \mid T_{ij} \mid AB \rangle \qquad ...[112]$$

where the operator T_{ij} acts on the i-th quark of A and the j-th quark of B, but leaves the other quarks unaffected. Equation [*112*], which is analogous to [*106*], must be supplemented by an equation analogous to [*107*]. We write it in the form

$$\langle A'B' \mid T_{ij} \mid AB \rangle = \langle Q_i^{A'} Q_j^{B'} \mid T_{ij} \mid Q_i^A Q_j^B \rangle f_i^{A'A}(\Delta p_\mu) f_j^{B'B}(-\Delta p_\mu). \qquad ...[113]$$

‡ The separation between spin and space coordinates of quarks is assumed possible as is the case when the quark states in hadrons are non-relativistic and Δp_μ is not too large. The non-relativistic assumption is desirable anyway if the quark model is to reproduce the SU_6 multiplet structure of hadrons, especially their spin values.

$\langle Q_i^{A'} Q_j^{B'} \mid T_{ij} \mid Q_i^A Q_j^B \rangle$ describes the collision of quarks Q_i, Q_j with momentum transfer Δp_μ, bringing them from the spin and SU_3 states Q_i^A, Q_j^B which they have in A, B to the spin and SU_3 states $Q_i^{A'}$, $Q_j^{B'}$. As in [107] the form factors $f_i^{A'A}$ represent the overlap between the quark wave function of A' and the wave function generated by the action of T_{ij} on Q_i, the remaining quarks of A being unaffected, and similarly for $f_j^{B'B}$. (See previous footnote.) The form factors satisfy [108] for $A' = A$, $B' = B$ and those forward collisions where Q_i and Q_j keep their spin and SU_3 states unchanged.

The dependence of the various factors in [113] on the incident energy $s^{\frac{1}{2}} = p_0^A + p_0^B$ deserves special attention. It is normal to consider that for given Δp_μ the form factors $f_i^{A'A}, f_j^{B'B}$ are independent of s. The same can be assumed reasonably for $\langle Q_i^A Q_j^B \mid T_{ij} \mid Q_i^A Q_j^B \rangle$ in the case of elastic scattering at very high energy ($A' = A$, $B' = B$, $s \to +\infty$) if all cross-sections become constant. In that case, indeed, one can assume that the covariant amplitude [13] becomes proportional to s, so that, from the normalization [10], the matrix element [112] becomes s independent. The same can then naturally be assumed for [113]. In other words, the fact that the $A+B$ collision becomes energy independent at very high energy is accounted for by the assumption that the same holds for the $Q_i + Q_j$ collisions. In this simple case one can avoid the question of which energy to attribute to the $Q_i + Q_j$ collision when the energy $s^{\frac{1}{2}}$ of the $A+B$ collision is given.

This question must be faced, however, for elastic scattering at finite values of s and at all s for inelastic processes ($A' \neq A$ and/or $B' \neq B$) with decreasing cross-sections. Since the quark model is regarded as mainly characterized by additivity, it seems reasonable to say that the properties carried by the hadrons are the sum of those carried by the constituting quarks, and that this holds not only for baryon number, isospin, hypercharge and spin, but also for momentum and energy. Thus if A has four-momentum p_μ^A, its constituting quarks Q_i will have four-momenta

$$p_\mu^i = c_i^A p_\mu^A, \quad \sum_i c_i^A = 1 \qquad \text{...[114]}$$

and it is natural to assume all c_i^A to be approximately equal so that

$$c_i^A \simeq c_A, \; c_A = \tfrac{1}{2} \text{ for mesons, } c_A = \tfrac{1}{3} \text{ for baryons.} \qquad \text{...[115]}$$

It is clear that [114] is not compatible with the relation

$$(p^i)^2 = m_Q^2 \gg m_A^2 \qquad \text{...[116]}$$

which would hold for a free quark of large mass m_Q. The properties discussed here are those of quarks as they are supposed to behave when bound together in the hadrons, and these quarks have then an effective mass m_Q^{eff} which [114] and [115] suggest to be of order

$$m_Q^{\text{eff}} \simeq \tfrac{1}{3} m_B \simeq \tfrac{1}{2} m_M \qquad \text{...[117]}$$

where m_B, m_M are average values of the baryon and meson masses respectively. Such a low effective mass for the quark is also suggested by the value of the baryon magnetic moments under the additivity hypothesis.[71,76] In fact, all successful applications of the quark model use the additivity property for bound quark properties, and are quite independent of what a free quark would be if it existed. The essential assumption, however, is that the bound quark properties are the same in baryons and mesons, and it is this assumption which is tested when comparing baryons with mesons in the framework of the quark model.

We remark also in this connection that all applications of the model refer to small momentum transfer properties, so that they deal with bound quark properties averaged over lengths and times $\gtrsim m_B^{-1}$. On a shorter time scale, quarks may have other properties of oscillating nature.‡ This may hold in particular for the Fermi momentum of the quarks inside the hadrons; they may be appreciable but give a small contribution to the time-averaged quantities p_μ^i of [*114*].

We shall take [*112*] and [*113*] as the explicit formulation of the additivity assumption for high-energy scattering with the understanding that the $\langle Q_i^{A'} Q_j^{B'} \mid T_{ij} \mid Q_i^A Q_j^B \rangle$ are the same for different hadron collisions when the resulting quark-quark collision energy given by [*114*]-[*115*] is the same. This energy is given by

$$s_{ij} = (p^i+p^j)^2 \simeq (c_A m_A)^2 + (c_B m_B)^2 + 2c_A c_B (p^A p^B). \qquad ...[118]$$

At high energy it is well approximated by

$$s_{ij} \simeq c_A c_B (s - m_A^2 - m_B^2) \simeq c_A c_B s_0 \qquad ...[119]$$

where s_0 is the value at $t = 0$ of s_t, [*27*]. This means that to compare baryon-baryon, meson-baryon and meson-meson collisions, they must be taken at energies in the following proportions

$$\frac{s_0^{BB}}{9} = \frac{s_0^{MB}}{6} = \frac{s_0^{MM}}{4}. \qquad ...[120]$$

The agreement with experiment is indeed better if the comparison between baryon-baryon and meson-baryon cross-sections is done in this way.

4.3. *Relations between total cross-sections*

The additivity assumption, equations [*112*]-[*113*], implies many relations between scattering amplitudes. We shall present a few of them here in the form in which they have been first derived,[74] i.e. under neglect of the spin variables. As was shown recently by Itzykson and Jacob,[77] spin variables can be taken into account; several relations derived with neglect of spin are

‡ One can speculate that the fast oscillations would be due to the strong binding force to which quarks are subjected in the hadrons. Also charge and baryon number can oscillate, so that their fractional values $\pm\frac{1}{3}$, $\pm\frac{2}{3}$ could be averages over integral values.

recovered for the spin average quantities and new relations can be obtained. In the few cases where experimental tests are possible with present data, the agreement is surprisingly good in view of the expected crudeness of the model.

Consider first forward elastic processes, i.e.,

$$A' = A, \; B' = B, \; \Delta p_\mu = 0. \qquad \text{...[121]}$$

All form factors reduce to 1, as shown by [*108*]. We use the quark composition of baryons and mesons, with spins neglected

$$\left.\begin{aligned} \pi^+ &= Q_p\bar{Q}_n, \quad K^+ = Q_p\bar{Q}_\lambda, \\ p &= Q_pQ_pQ_n, \quad n = Q_pQ_nQ_n \end{aligned}\right\} \qquad \text{...[122]}$$

where Q_p, Q_n, Q_λ denote the $I = \frac{1}{2}$, $I = 0$ quarks respectively. Using the abbreviations

$$\left.\begin{aligned} \langle AB \mid T \mid AB\rangle &= (AB), \\ \langle Q_iQ_j \mid T_{ij} \mid Q_iQ_j\rangle &= (Q_iQ_j), \end{aligned}\right\} \qquad \text{...[123]}$$

we obtain immediately

$$\left.\begin{aligned}
(pp) &= 4(Q_pQ_p)+4(Q_pQ_n)+\ (Q_nQ_n) \\
(np) &= 2(Q_pQ_p)+5(Q_pQ_n)+2(Q_nQ_n) \\
(\bar{n}p) &= 2(\bar{Q}_pQ_p)+4(\bar{Q}_nQ_p)+\ (\bar{Q}_pQ_n)+2(\bar{Q}_nQ_n) \\
(\bar{p}p) &= 4(\bar{Q}_pQ_p)+2(\bar{Q}_pQ_n)+2(\bar{Q}_nQ_p)+\ (\bar{Q}_nQ_n) \\
(\pi^+p) &= 2(Q_pQ_p)+\ (Q_pQ_n)+2(\bar{Q}_nQ_p)+\ (\bar{Q}_nQ_n) \\
(\pi^-p) &= 2(\bar{Q}_pQ_p)+\ (\bar{Q}_pQ_n)+2(Q_nQ_p)+\ (Q_nQ_n) \\
(K^+p) &= 2(Q_pQ_p)+\ (Q_pQ_n)+2(\bar{Q}_\lambda Q_p)+\ (\bar{Q}_\lambda Q_n) \\
(K^+n) &= \ (Q_pQ_p)+2(Q_pQ_n)+\ (\bar{Q}_\lambda Q_p)+2(\bar{Q}_\lambda Q_n) \\
(K^-p) &= 2(\bar{Q}_pQ_p)+\ (\bar{Q}_pQ_n)+2(Q_\lambda Q_p)+\ (Q_\lambda Q_n) \\
(K^-n) &= \ (\bar{Q}_pQ_p)+2(\bar{Q}_pQ_n)+\ (Q_\lambda Q_p)+2(Q_\lambda Q_n).
\end{aligned}\right\} \qquad \text{...[124]}$$

A simple consequence is

$$\begin{aligned} \tfrac{1}{3}[(pp)+(np)+(\bar{n}p)+(\bar{p}p)] &= [(\pi^+p)+(\pi^-p)] \\ &= [2(Q_pQ_p)+3(Q_pQ_n)+(Q_nQ_n)+2(\bar{Q}_pQ_p)+(\bar{Q}_pQ_n)+2(\bar{Q}_nQ_p)+(\bar{Q}_nQ_n)]. \end{aligned} \qquad \text{...[125]}$$

At very high energy, the absence of exchange scattering and annihilation implies that all amplitudes in the same square bracket become equal, so that

$$\tfrac{4}{3}(pp) = 2(\pi^+p) = 12(Q_pQ_p). \qquad \text{...[126]}$$

Through the optical theorem, [*105*] follows. Applying the optical theorem directly to [*125*] one obtains

$$\frac{2[\sigma_T(\pi^+p)+\sigma_T(\pi^-p)]}{\sigma_T(pp)+\sigma_T(np)+\sigma_T(\bar{n}p)+\sigma_T(\bar{p}p)} = \tfrac{2}{3}. \qquad \ldots[127]$$

Taking the πp cross-sections at 13 GeV and the nucleon-nucleon and anti-nucleon-nucleon ones at 20 GeV, as proposed by the rule [*120*], the experimental ratio is $0{\cdot}59 \pm 0{\cdot}05$. It would be $0{\cdot}54 \pm 0{\cdot}05$ if all cross-sections were taken at the same energy, 13 GeV or 20 GeV.

As another example we consider the Johnson-Treiman relations [*103*]. From [*124*] one obtains:

$$\begin{aligned}(K^+p)-(K^-p) &= 2(\delta_{pp}-\delta_{\lambda p})+\delta_{pn}-\delta_{\lambda n}\\ (\pi^+p)-(\pi^-p) &= 2(\delta_{pp}-\delta_{np})+\delta_{pn}-\delta_{nn}\\ (K^+n)-(K^-n) &= 2(\delta_{pn}-\delta_{\lambda n})+\delta_{pp}-\delta_{\lambda p}\end{aligned} \qquad \ldots[128]$$

with the notation

$$\delta_{ab} = (Q_aQ_b)-(\bar{Q}_aQ_b); \quad a, b = p, n, \lambda. \qquad \ldots[129]$$

Isospin invariance gives

$$\delta_{pp} = \delta_{nn}, \; \delta_{pn} = \delta_{np}, \; \delta_{\lambda p} = \delta_{\lambda n}. \qquad \ldots[130]$$

Hence

$$(K^+p)-(K^-p) = (\pi^+p)-(\pi^-p)+(K^+n)-(K^-n). \qquad \ldots[131]$$

As shown by Lipkin and Scheck,[74] the corresponding relation for total cross-sections is in very good agreement with experiment for $s \gtrsim 15(\text{GeV})^2$. To obtain the Johnson-Treiman relations [*103*] one needs further

$$\delta_{\lambda p} = \delta_{np} \qquad \ldots[132]$$

which holds if the quark-quark and quark-antiquark amplitudes obey SU_3 symmetry. It is therefore satisfactory that [*131*] for σ_T, which is not affected by symmetry breaking, is better satisfied by the data than the full Johnson-Treiman relations.

As a last remark on total cross-sections, we note that the relations discussed above begin to disagree with the facts when s decreases below a value of order $10(\text{GeV})^2$. This shows that the additivity assumption for scattering amplitudes holds only in the high-energy region.

4.4. *Differential cross-sections and consistency of the additivity assumption*

On differential cross-sections we shall discuss the implications of the quark model with additive scattering amplitudes for the case of elastic scattering at very high energy.‡ This will allow us to obtain, at least in two

‡ Many other cases of differential cross-sections, including spin phenomena, are studied by Itzykson and Jacob.[77]

limiting cases, an understanding of why the additivity assumption can be a good approximation. At very high energy, it is natural to assume that the quark-quark elastic scattering becomes purely diffractive and has an energy independent cross-section. Hence in [*113*] applied to elastic scattering ($A' = A$ and $B' = B$) we put

$$\langle Q_i^A Q_j^B \mid T_{ij} \mid Q_i^A Q_j^B \rangle = ig_{ij}(t) = ig_{ji}(t), \quad t = (\Delta p)^2 \qquad \text{...[133]}$$

with $g_{ij}(t)$ real. Isospin symmetry gives

$$g_{pp}(t) = g_{nn}(t), \quad g_{\lambda p}(t) = g_{\lambda n}(t) \qquad \text{...[134]}$$

where the indices p, n, λ refer to the quarks Q_p, Q_n, Q_λ. Neglecting SU_3 symmetry breaking effects would give in addition

$$g_{\lambda\lambda}(t) = g_{pp}(t), \quad g_{pn}(t) = g_{\lambda n}(t). \qquad \text{...[135]}$$

The absence of charge exchange at very high energy implies the further relation

$$g_{pp}(t) = g_{pn}(t). \qquad \text{...[136]}$$

Similar relations hold for antiquark-quark scattering. We list them with the necessary assumptions ($\bar{p}, \bar{n}, \bar{\lambda}$ refer to $\bar{Q}_p, \bar{Q}_n, \bar{Q}_\lambda$):

(i) isospin and charge conjugation:

$$g_{\bar{p}p} = g_{\bar{n}n}, \; g_{\bar{p}n} = g_{\bar{n}p}, \; g_{\bar{\lambda}p} = g_{\bar{\lambda}n}, \; g_{\bar{p}\lambda} = g_{\bar{n}\lambda}, \; g_{\bar{\lambda}p} = g_{\bar{p}\lambda} \qquad \text{...[137]}$$

(ii) SU_3 $\quad g_{\bar{\lambda}\lambda} = g_{\bar{p}p}, \; g_{\bar{p}n} = g_{\bar{\lambda}n}$ ...[*138*]

(iii) no charge exchange: $g_{\bar{p}p} = g_{\bar{n}p}$. ...[*139*]

Finally, $\bar{Q}_i Q_j$ and $Q_i Q_j$ elastic scattering become identical at very high energy,‡ which means

$$g_{\bar{p}p}(t) = g_{pp}(t), \; g_{\bar{\lambda}p}(t) = g_{\lambda p}(t), \; g_{\bar{\lambda}\lambda}(t) = g_{\lambda\lambda}(t). \qquad \text{...[140]}$$

Using these various relations, except the ones requiring SU_3 symmetry, we obtain for very high energy

$$\langle AB \mid T \mid AB \rangle = ig_{AB}(t) f^A(t) f^B(t) \qquad \text{...[141]}$$

with §

$$f^A(t) = \sum_i f_i^{AA}(\Delta p_\mu), \quad f^B(t) = \sum_j f_j^{BB}(-\Delta p_\mu) \qquad \text{...[142]}$$

‡ This is the generalized Pomeranchuk theorem for energy-independent diffraction scattering; see reference 78, or in section 2 of these lectures, for the properties of the leading term of the asymptotic expansion [*26*] for elastic scattering, especially the fact that it has positive signature.

§ To separate the form factors f from the scattering amplitudes g when strange particles are involved one relies on the fact that $f_i^{AA}(t)$ is independent of i. This is obvious if A is a meson because it contains two quarks only, and hence one relative momentum. It is valid for hyperons to the degree that their wave function is symmetric in the three quarks.

$$\left.\begin{array}{lll} g_{AB}(t) = g_{pp}(t) & \text{for } A, B = \pi, p, n, \bar{p}, \bar{n} & \\ g_{AB}(t) = \tfrac{1}{2}[g_{\lambda p}(t)+g_{pp}(t)] & \text{for } A = K, \bar{K} & \text{and } B = p, n, \pi \\ g_{AB}(t) = \tfrac{1}{3}[g_{\lambda p}(t)+2g_{pp}(t)] & \text{for } A = \Lambda, \Sigma & \text{and } B = p, n, \pi \\ g_{AB}(t) = \tfrac{1}{3}[2g_{\lambda p}(t)+g_{pp}(t)] & \text{for } A = \Xi & \text{and } B = p, n, \pi \\ g_{AB}(t) = g_{\lambda p}(t) & \text{for } A = \Omega^{-} & \text{and } B = p, n, \pi. \end{array}\right\} \quad ...[143]$$

With SU_3 symmetry all $g_{AB}(t)$ reduce to $g_{pp}(t)$.

A number of consequences of the above relations have been discussed elsewhere by Kokkedee and the present author.[74,75] No accurate experimental tests are available with present data, but the model predicts a great degree of universality for high-energy diffraction of all hadrons, which is indeed strongly suggested by existing measurements. In our further considerations we limit ourselves to scattering of pions (π) and nucleons (N). With SU_3 symmetry, our formulae would apply to all mesons and baryons. With the notation

$$g(t) = g_{pp}(t) \quad ...[144]$$

we have instead of [*141*]

$$\langle AB \mid T \mid AB\rangle = ig(t)f^{A}(t)f^{B}(t). \quad ...[145]$$

This gives the factorization property

$$\langle AB \mid T \mid AB\rangle^2 = \langle AA \mid T \mid AA\rangle \,.\, \langle BB \mid T \mid BB\rangle \quad ...[146]$$

which determines for example the $\pi\pi$ diffraction scattering from known data

$$\langle \pi\pi \mid T \mid \pi\pi\rangle = \frac{f^{\pi}(t)}{f^{N}(t)}\langle \pi N \mid T \mid \pi N\rangle = \frac{\langle \pi N \mid T \mid \pi N\rangle^2}{\langle NN \mid T \mid NN\rangle}. \quad ...[147]$$

Equation [*145*] shows that the t dependence of hadron-hadron diffraction scattering is controlled by two types of functions, the hadron form factors $f^{A}(t), f^{B}(t)$ and the asymptotic quark-quark amplitude $g(t)$. The effects of these functions cannot be separated on the basis of hadron-hadron diffraction data alone. We therefore discuss two extreme cases between which a continuum of intermediate situations can be imagined. They will allow a better understanding of the validity of the additivity assumption.

In the first extreme case, *the t dependence of the diffraction peak is supposed to be mainly determined by the form factors*, in the sense that $g(t)$ remains practically equal to $g(0)$ over the whole width of the diffraction peak, the slope of which is determined by the factor $f^{A}(t)f^{B}(t)$ in [*145*]. In other words, the size of quarks as it manifests itself in their high-energy diffraction is supposed smaller than the measured size of hadrons. The minimum size of quarks is obtained by assuming that they behave as black spheres. Their radius R_Q is then given by

$$\sigma_T(QQ) = 2\pi R_Q^2 \quad ...[148]$$

where $\sigma_T(QQ)$, the total quark-quark cross-section, is in our normalization

$$\sigma_T(QQ) = 4\pi^2 g(0). \qquad ...[149]$$

(*Remark*: We notice that at very high energy, t being finite, the matrix elements $\langle A'B' \mid T \mid AB\rangle$, $\langle Q_i^{A'} Q_j^{B'} \mid T_{ij} \mid Q_i^A Q_j^B\rangle$ used in equations [*112*] and [*113*] are the same for all coordinate systems in which the incident momenta $\boldsymbol{p}$, $\boldsymbol{p}'$ are collinear, have directions opposite to each other and are both very large in absolute value. This is seen from [*10*] by remarking that in these systems one has

$$(p_0^A p_0^B p_0^C p_0^D)^{\frac{1}{4}} \approx \tfrac{1}{2}(p_0 p_0' - \boldsymbol{p}\cdot\boldsymbol{p}') = \text{invariant}$$

where p_0, p_0' are the incident energies. The c.m. system of the $A+B$ collision is such a system both for the $A+B$ collision and for the resulting $Q_i + Q_j$ collisions. The cross-sections for the latter can therefore be calculated by direct application of equations [*10*], [*12*] and [*13*], the matrix elements [*10*] of $T = i(1-S)$ being the ones used throughout in section 4.)

But from [*145*], [*142*] and [*108*], one has

$$\sigma_T(pp) = 36\pi^2 g(0) = 9\sigma_T(QQ) \qquad ...[150]$$

so that we find

$$R_Q^2 = \frac{\sigma_T(pp)}{18\pi}. \qquad ...[151]$$

This is indeed much smaller than the size of proton (or pion) as determined from their diffraction, not only because of the factor 9 in [*150*], but also because the exponential shape of the former corresponds to an absorption which is only strong at zero impact parameter and decreases as a Gaussian. In the notation of section 3.1, one has in fact approximately

$$\eta_l \simeq \eta_0 \exp\left(-\frac{l^2}{2bk^2}\right), \qquad ...[152]$$

$b \simeq 9(\text{GeV/c})^2$ being the width of the high-energy pp or πp diffraction peaks as defined in [*19*]. The parameter η_0 is close to 1 for pp, of order 0·7 for πp.

Our first extreme case leads us to a picture of hadrons composed of small size quarks separated from each other by distances large compared to their size. In this picture the success of the additivity assumption for quark scattering amplitudes becomes a consequence of the small size of the quarks which makes multiple scattering effects rare. The picture also suggests another interesting property. The hadron-hadron elastic scattering has now a t dependence given by

$$\frac{d\sigma_{\text{el}}}{dt} = \left(\frac{d\sigma_{\text{el}}}{dt}\right)_{t=0} [f^A(t) f^B(t)]^2. \qquad ...[153]$$

But in our picture of small quarks forming large hadrons, the form factor $f^A(t)$ should apply to all hadron properties carried by the quarks, including the electromagnetic and weak ones. For f^A, f^B in [*153*] one could then take the electromagnetic form factors. It is remarkable that this conjecture is in excellent agreement with experiment for the case $A = B =$ proton, the only one where it can be tested at present.[75] It holds in the interval $0 \leqq -t \lesssim 1(\text{GeV/c})^2$ where $d\sigma_{el}/dt$ drops by a factor 10^{-3}. The conjecture allows to derive from πp diffraction scattering the electromagnetic form factor of the pion, which is found to be very similar to the proton one.

We consider now the second extreme case where *the t dependence of the diffraction peak is supposed to be mainly determined by the quark-quark amplitude* $g(t)$, the hadron form factors $f^{A,B}(t)$ in [*145*] remaining approximately equal to $f^{A,B}(0)$ over the whole diffraction peak. The resulting picture is opposite to the previous one. It is now the quark itself which has a large size for diffraction, while the various quarks inside one hadron overlap almost exactly with each other in space so as to give a weak t dependence to the form factors $f^{A,B}(t)$. All hadrons have then, except for a constant factor, the same diffraction scattering in the limit of very high energy, namely

$$\langle AB \mid T \mid AB \rangle = i n_A n_B g(t) \qquad \ldots[154]$$

where $n_{A,B} = f^{A,B}(0)$ is the number of quarks composing A, B. This picture is compatible with the data because an extrapolation of the pp, $\bar{p}p$, $\pi^{\pm}p$ diffraction peaks to infinite energy indeed indicates that their shapes become remarkably close to each other. The quark-quark diffraction scattering is now entirely determined by the equations

$$\sigma_T(QQ) = \tfrac{1}{9}\sigma_T(pp), \quad \frac{d\sigma_{el}}{dt}(QQ) = \tfrac{1}{81}\frac{d\sigma_{el}}{dt}(pp) \qquad \ldots[155]$$

and the corresponding absorption coefficient η_l is given approximately by [*152*] with $\eta_0 \simeq \frac{1}{9}$. While the quark has a large size for diffraction, it has only very weak absorption, reaching a maximum of about 10% at zero impact parameter. In the present picture the success of the additivity assumption for scattering amplitudes is a consequence of this low absorption or high transparency. While the quarks overlap in space, their diffractive interaction is so weak that again multiple processes involving scattering on more than one quark in the same hadron are very unlikely. This would not remain true, however, for scattering at large t because the fast drop of $g(t)$ with t would then favour multiple processes.

The weak nature of the diffractive interaction between quarks obtained in the present case suggests that it can be treated in first Born approximation, each small volume in one quark interacting with each small volume in the other one, and the total scattering amplitude being the sum over all such contributions. This suggests

$$g(t) = g(0)[F(t)]^2 \qquad \ldots[156]$$

where $F(t)$ is the form factor of the quark describing how its diffractive interaction strength is distributed over space.‡ If the same form factor $F(t)$ describes the distribution of electromagnetic and weak interactions in the quark, it must be identified with the electromagnetic form factors of the proton and the pion because the $f^{A,B}(t)$ do not contribute to the latter in our approximation. As mentioned after [*153*] the experimental facts are compatible with this identification. Needless to say, in the picture just discussed, the quark is even more difficult to visualize as a separate particle than was the case with the previous picture of small-size quarks.

The previous considerations have shown how the additivity assumption can be understood to work in two extreme cases. It is reasonable to expect that intermediate cases would combine elements of the two extremes but would again lead to the conclusion that multiple scattering on various quarks in the same hadron would only contribute small effects in hadron-hadron collisions.

4.5. *Relation to Regge pole model and inelastic collisions in quark model*

We end with a few remarks comparing the quark model approach to high energy scattering with the Regge pole model (§2) and the considerations of section 3 on shadow scattering.

A few of the quark model predictions are qualitatively similar to features found in the Regge pole model. Thus, the factorization property contained in [*145*] and [*146*] for high energy diffraction scattering and total cross-sections would also be obtained from Regge pole theory, because of the factorization property (§ 2.5, point (b)) applied to the residue of the Pomeranchuk pole ($\alpha_1 \simeq 1$) which dominates elastic scattering. A more direct and more general consequence of the additivity assumption [*112*]-[*113*] is that for $s \to \infty$, t finite, the quantum numbers exchanged in the t channel are those of the quark-antiquark system with positive and negative signatures, corresponding to the nonets required in Regge pole theory to describe the meson-nucleon reactions. In fact, one can say that the quark-antiquark system taken for positive and negative values of $t = m^2$ (m = mass of the system) provides a dynamical model for the Regge trajectories of the mesons, and that the matrix $Z_J^{\pm}(t)$ of equation [*71*] refers entirely to this $Q\bar{Q}$ system. This accounts then directly for the observed universality of the exponents α_j and for their relations to particles [section 2.5, points (a) and (c)]. The analogies mentioned are of course only partial and, while the Regge pole and quark models of high-energy scattering have these points in common, their full implications would be rather different in many respects. It is very likely, however, that both are no more than approximate descriptions of a truly complex class of strong interaction phenomena.

‡ If one quark would be point-like and the other one spread out, the Born approximation would give $g(t) = g(0)F(t)$. The additional factor $F(t)$ in [*156*] accounts for the fact that the two quarks are spread out.

The principal ideas of the quark model for high-energy scattering at small momentum transfer can be extended to a large class of inelastic phenomena by allowing A', B' in [*112*] and [*113*] to be all states obtained by transferring a not too large amount Δp_μ of energy-momentum, and possibly some isospin or hypercharge from a quark Q_i of A to a quark Q_j of B. The states A', B' would then not necessarily be resonances but would be excited systems which would decay through virtual creation of quark-antiquark pairs liberated as mesons.‡ The inelastic final states $A'+B'$ could coincide with those discussed in section 3.2, A' and B' forming a more precise characterization of the "forward" and "backward" groups of particles there considered. The occurrence of the "leading" particle there described would be due to the fact that the excitation energy and momentum of A' (or B') radiated away in the form of mesons would have a preferential direction in space, namely the direction of motion of the particle B (or A) which has hit A (or B) and excited it to A' (or B'). This direction is of course backward for the incident particle A, and forward for the target particle B. The effect discussed is rather easily visualized when A (B) is a baryon. The quark-quark collision then slows down one quark only in A (B), the two others tending to continue with their initial momentum. The latter two quarks will keep the first one bound to them with a rather high forward (backward) momentum, the remaining energy of A' (B') being radiated away as mesons with much lower momentum. We also remark that this type of directional excitation is common in many-particle systems. It occurs for example when a crystal is excited by a fast particle incident on it; the phonons excited in the crystal will have a direction of propagation strongly correlated to the direction of motion of the incident particle.

Also the separation [*86*] of the final state in non-exchange and exchange parts $\psi_{AB}^{(0)}$, $\psi_{AB}^{(1)}$ is easily connected to [*112*] and [*113*]. It results from the fact that the Q_iQ_j collision can transfer vanishing or non-vanishing values of isospin and hypercharge. The only states missing are the ones contributed to $\psi_{AB}^{(1)}$ by annihilation when A = antibaryon, B = baryon. They cannot be generated by a single quark-quark collision but require triple quark exchange. As a consequence, their possible separation into forward and backward groups cannot be interpreted in the same terms as before. More generally, all processes involving baryon exchange will require a separate treatment in the quark model, also for example backward meson-nucleon scattering.

The further analysis of the annihilation effects in antibaryon-baryon collisions throws in fact some doubt on whether the quark model with additive amplitudes, when applied to total cross-sections and elastic scattering [sections 4.3 and 4.4], treats properly that contribution to the antibaryon-baryon elastic scattering amplitude which is due to the shadow of annihilation. In section 4.4 we have treated hadron-hadron and quark-quark elastic scattering

‡ We do not consider Δp_μ to become large enough for ejection of free quarks, although this could be allowed to happen if the latter had integral charge and baryon number.

as purely diffractive, as indicated by the reality of g_{ij} in [*133*]. But the shadow of antibaryon-baryon annihilation cannot be obtained in an additive way from the shadow of antiquark-quark annihilation, whereas additivity can hold for the shadow of the non-annihilation part of $\psi_{AB} = \psi_{AB}^{(0)} + \psi_{AB}^{(1)}$. It may therefore be that our considerations require a further term in the antibaryon-baryon elastic amplitudes and total cross-sections. This is also suggested by the data, because [*127*] is in closer agreement with the data if the annihilation contribution to $\sigma_T(\bar{n}p)$ and $\sigma_T(\bar{p}p)$ is removed.

A last remark concerns the shape of the diffraction peak. In section 3 we have related it through unitarity to the overlap function of inelastic collisions, whereas in section 4.4 it was suggested to be given in terms of the hadronic form factor. This then implies a relation between form factor and overlap function. We do not know at present whether it can be given a particular significance, nor whether the approximate exponential shape of the overlap function is better understood in terms of the inelastic collisions, as proposed in section 3.3, or in terms of the form factor.

REFERENCES

1. D. E. Dorfan, J. Eades, L. M. Lederman, W. Lee, C. C. Ting, P. Piroué, Stuart Smith, J. L. Brown, J. A. Kadyk and G. H. Trilling. 1965. *Phys. Rev. Lett.* **14**, 995.
2. M. Jacob and G. C. Wick. 1959. *Ann. Phys.* **7**, 404.
3. A. N. Diddens, E. Lillethun, G. Manning, A. E. Taylor, T. G. Walker and A. M. Wetherell. 1962. *Phys. Rev. Lett.* **9**, 108, 111.
4. K. J. Foley, S. J. Lindenbaum, W. A. Love, S. Ozaki, J. J. Russel and L. C. L. Yuan. 1963. *Phys. Rev. Lett.* **10**, 376, 543; 1963. **11**, 425, 503.
5. D. Harting, P. Blackall, B. Elsner, A. C. Helmholz, W. C. Middelkoop, B. Powell, B. Zacharov, P. Zanella, P. Dalpiaz, M. N. Focacci, S. Focardi, G. Giacomelli, L. Monari, J. A. Beaney, R. A. Donald, P. Mason, L. W. Jones and D. O. Caldwell. 1964. *Phys. Lett.* **8**, 288; 1965. *Nuovo Cim.* **38**, 60.
6. K. J. Foley, R. S. Gilmore, S. J. Lindenbaum, W. A. Love, S. Ozaki, E. H. Willen, R. Yamada and L. C. L. Yuan. 1965. *Phys. Rev. Lett.* **15**, 45.
7. A. E. Taylor, A. Ashmore, W. S. Chapman, D. F. Falla, W. H. Range, D. B. Scott, A. Astbury, F. Capocci and T. G. Walker. 1965. *Phys. Lett.* **14**, 54.
8. K. J. Foley, R. S. Gilmore, R. S. Jones, S. J. Lindenbaum, W. A. Love, S. Ozaki, E. H. Willen, R. Yamada and L. C. L. Yuan. 1965. *Phys. Rev. Lett.* **14**, 74.
9. L. Kirillova, V. Nikitin, V. Pantuev, V. Sviridov, L. Strunov, M. Khachaturyan, L. Khristov, M. Shafranova, Z. Korbel, L. Rob, S. Damyanov, A. Zlateva, Z. Zlatanov, V. Iordanov, Kh. Kanazirsky, P. Markov, T. Todorov, Kh. Chernev, N. Dalkhazhav and D. Tuvdendorzh. 1965. *Sov. J. Nucl. Phys.* **1**, 379.
10. G. Bellettini, G. Cocconi, A. N. Diddens, E. Lillethun, J. P. Scanlon, J. Walkers, A. M. Wetherell and P. Zanella. 1965. *Phys. Lett.* **14**, 164.
11. G. Bellettini, G. Cocconi, A. N. Diddens, E. Lillethun, J. P. Scanlon and A. M. Wetherell. 1965. *Phys. Lett.* **19**, 341, 705.
12. K. J. Foley, R. S. Gilmore, R. S. Jones, S. J. Lindenbaum, W. A. Love, S. Ozaki, E. H. Willen, R. Yamada and L. C. L. Yuan. 1965. *Phys. Rev. Lett.* **14**, 862.
13. G. Cocconi, V. T. Cocconi, A. D. Krisch, J. Orear, R. Rubinstein, D. B. Scarl, B. T. Ulrich, W. F. Baker, E. W. Jenkins and A. L. Read. 1965. *Phys. Rev.* **138**, B165.

14. J. Orear, R. Rubinstein, D. B. Scarl, D. H. White, A. D. Krisch, W. R. Frisken, A. L. Read and H. Ruderman. 1965. *Phys. Rev. Lett.* **15**, 309.
15. W. R. Frisken, A. L. Read, H. Ruderman, A. D. Krisch, J. Orear, R. Rubinstein, D. B. Scarl and D. H. White. 1965. *Phys. Rev. Lett.* **15**, 313.
16. S. W. Kormanyos, A. D. Krisch, J. R. O'Fallon, K. Ruddick and L. G. Ratner. 1966. *Phys. Rev. Lett.* **16**, 709.
H. Brody, R. Lanza, R. Marshall, J. Niederer, W. Selove, M. Shocket and R. Van Berg. 1966. *Phys. Rev. Lett.* **16**, 828, 968.
17. J. L. Friedes, H. Palevsky, R. L. Stearns and R. J. Sutter. 1965. *Phys. Rev. Lett.* **15**, 38.
G. Manning, A. G. Parham, J. D. Jafar, H. B. van der Raay, D. H. Reading, D. G. Ryan, B. D. Jones, J. Malos and N. H. Lipman. 1966. *Nuovo Cim.* **41**, 167.
18. I. Mannelli, A. Bigi, R. Carrara, M. Wahlig and L. Sodickson. 1965. *Phys. Rev. Lett.* **14**, 408.
19. A. V. Stirling, P. Sonderegger, J. Kirz, P. Falk-Vairant, O. Guisan, C. Bruneton, P. Borgeaud, M. Yvert, J. P. Guillaud, C. Caversasio and B. Amblard. 1965. *Phys. Rev. Lett.* **14**, 763; 1966. *Phys. Lett.* **20**, 75.
20. O. Guisan, J. Kirz, P. Sonderegger, A. V. Stirling, P. Borgeaud, C. Bruneton, P. Falk-Vairant, B. Amblard, C. Caversasio, J. P. Guillaud and M. Yvert. 1965. *Phys. Lett.* **18**, 200.
21. P. Astbury, G. Finocchiaro, A. Michelini, C. Verkerk, C. H. West, W. Beusch, B. Gobbi, M. Pepin, M. A. Pouchon and E. Polgar. 1965. *Phys. Lett.* **16**, 328.
22. G. Finocchiaro, A. Michelini, W. Beusch, W. E. Fischer, B. Gobbi and E. Polgar. 1966. Paper presented at the Stony Brook Conference on High-Energy Two-Body Reactions, Stony Brook, N.Y., April, 1966.
23. D. R. O. Morrison. 1966. Review paper presented at the Stony Brook Conference on High-Energy Two-Body Reactions, Stony Brook, N.Y., April, 1966.
24. M. Borghini, G. Coignet, L. Dick, L. di Lella, A. Michalowicz, P. C. Macq and J. C. Oliver. 1966. *Phys. Lett.* **21**, 114.
25. P. Sonderegger. 1966. Paper presented at the Stony Brook Conference on High-Energy Two-Body Reactions, Stony Brook, N.Y., April, 1966.
26. L. Holloway. 1966. Paper presented at the Stony Brook Conference on High-Energy Two-Body Reactions, Stony Brook, N.Y., April, 1966.
27. I. Ya. Pomeranchuk. 1958. *Zhur. Eksp. i Teoret. Fiz.* **34**, 725; (English translation: 1958. *Sov. Phys. JETP* **7**, 499).
28. T. L. Trueman and G. C. Wick. 1964. *Ann. Phys.* **26**, 322.
29. B. E. Y. Svensson. 1967. *Arkiv för Fysik.* **33**, 93.
30. A. Kotanski. 1966. *Acta Phys. Polonica.* **29**, 699.
31. A. Kotanski. 1966. Jagellonian University, Cracow, preprint TPJU-4/66.
32. A. Białas and B. E. Y. Svensson. 1966. *Nuovo Cim.* **42A**, 672.
33. L. Van Hove. 1963. *Phys. Lett.* **5**, 252.
34. A. A. Logunov, Nguyen Van Hieu, I. T. Todorov and O. A. Khrustalev. 1963. *Phys. Lett.* **7**, 69.
35. A. Białas and E. Białas. 1965. *Nuovo Cim.* **37**, 1686.
36. L. Van Hove. 1965. *High-Energy Physics and Elementary Particles* (International Atomic Energy Agency, Vienna), p. 179.
37. L. Van Hove. 1966. *1965 Tokyo Summer Lectures in Theoretical Physics* (Syokabo, Tokyo and Benjamin, New York). Part II, p. 17.
38. M. L. Goldberger and C. E. Jones. 1966. *Phys. Rev.* **150**, 1269.
D. Z. Freedman and J. M. Wang. 1967. *Phys. Rev.* **153**, 1596.
39. T. Regge. 1959. *Nuovo Cim.* **14**, 951; 1960, **18**, 947.
A. Bottino, A. M. Longoni and T. Regge. 1962. *Nuovo Cim.* **23**, 954.
40. E. J. Squires. 1963. *Complex angular momenta and particle physics.* (W. A. Benjamin, New York and Amsterdam.)
S. C. Frautschi. 1963. *Regge poles and S-matrix theory.* (W. A. Benjamin, New York and Amsterdam.)
R. Omnès and M. Froissart. 1963. *Mandelstam theory and Regge poles.* (W. A. Benjamin, New York and Amsterdam.)

41. D. Amati, S. Fubini and A. Stanghellini. 1962. *Phys. Lett.* **1**, 29.
42. S. Mandelstam. 1963. *Nuovo Cim.* **30**, 1127 and 1148.
43. G. F. Chew and S. C. Frautschi. 1962. *Phys. Rev. Lett.* **8**, 41.
S. C. Frautschi, M. Gell-Mann and F. Zachariasen. 1962. *Phys. Rev.* **126**, 2204.
44. V. N. Gribov and I. Ya. Pomeranchuk. 1962. *Phys. Rev. Lett.* **8**, 343 and 412.
45. M. Gell-Mann. 1962. *Phys. Rev. Lett.* **8**, 263.
46. R. Blankenbecler and M. L. Goldberger. 1962. *Phys. Rev.* **126**, 766.
47. T. Kinoshita. 1962. CERN Report 62-33, (Appendix A).
48. V. N. Gribov. 1962. *Proceedings of the 1962 International Conference on High-Energy Physics at CERN*, Ed. by J. Prentki, (CERN, Geneva), p. 547.
49. R. J. N. Phillips and W. Rarita. 1965. *Phys. Rev.* **139**, B1336; **143**, 1390(E).
50. G. Höhler, J. Baacke, H. Schlaile and P. Sonderegger. 1966. *Phys. Lett.* **20**, 79.
51. F. Arbab and C. B. Chiu. 1966. *Phys. Rev.* **147**, 1045.
L. L. Wang. 1966. *Phys. Rev. Lett.* **16**, 756.
G. F. Chew. 1966. Contribution to the Stony Brook Conference on High-Energy Two-Body Reactions, Stony Brook, N.Y., April, 1966.
52. A. O. Barut and D. E. Zwanziger. 1962. *Phys. Rev.* **127**, 974.
53. *Proceedings of the Stony Brook Conference on High-Energy Two-Body Reactions*, Stony Brook, N.Y., April, 1966.
54. V. Barger and M. Olsson. 1966. *Phys. Rev. Lett.* **16**, 545 and 952; *Phys. Rev.* **148**, 1428.
55. R. J. N. Phillips and W. Rarita. 1965. *Phys. Rev.* **140**, B200; **143**, 1390(E).
56. P. Astbury, G. Brautti, G. Finocchiaro, A. Michelini, D. Websdale, C. H. West, E. Polgar, W. Beusch, W. E. Fischer, B. Gobbi and M. Pepin. 1966. *Phys. Lett.* **23**, 160.
57. H. Høgaasen and W. Fischer. 1966. CERN preprint TH. 672, June 1966, and *Phys. Lett.* **22**, 516.
58. J. L. Friedes, private communication.
59. N. Byers. 1966. UCLA preprint, June.
60. N. Byers and C. N. Yang. 1966. *Phys. Rev.* **142**, 976.
61. L. Van Hove. 1962. Physikertagung Stuttgart (Mosbach-Baden Physik Verlag, 1963), p. 84; *Nuovo Cim.* **28**, 798 (1963); *Rev. Mod. Phys.* **36**, 655 (1964).
A. Zichichi, Ed. 1964. *Strong, Electromagnetic and Weak Interactions.* (W. A. Benjamin, New York and Amsterdam), p. 84.
62. A. Białas. 1964. *Nuovo Cim.* **33**, 972.
63. W. N. Cottingham and R. F. Peierls. 1965. *Phys. Rev.* **137**, B147.
64. L. Van Hove. 1964. *Rev. Mod. Phys.* **36**, 655.
65. J. J. J. Kokkedee. 1966. *Nuovo Cim.* **43A**, 919.
66. K. Johnson and S. B. Treiman. 1965. *Phys. Rev. Lett.* **14**, 189.
67. H. Harari and H. J. Lipkin. 1964. *Phys. Rev. Lett.* **13**, 208.
68. M. Gell-Mann. 1962. *Phys. Rev.* **125**, 1067; 1964. *Physics* **1**, 63.
69. M. Gell-Mann. 1964. *Phys. Lett.* **8**, 214.
G. Zweig. 1964. CERN preprints TH. 401 and TH. 402, unpublished.
70. S. L. Adler. 1965. *Phys. Rev. Lett.* **14**, 1051; 1965. *Phys. Rev.* **140**, B736.
W. I. Weisberger. 1965. *Phys. Rev. Lett.* **14**, 1047.
71. W. E. Thirring. 1965. *Phys. Lett.* **16**, 335; 1966. *Acta Phys. Austr. Suppl.* II, 205 and *Suppl.* III, 294.
C. Becchi and G. Morpurgo. 1965. *Phys. Lett.* **17**, 352; 1965. *Phys. Rev.* **140**, B687.
72. P. G. O. Freund. 1965. *Phys. Rev. Lett.* **15**, 929; 1966. *Nuovo Cim.* **43A**, 1171.
73. N. Cabibbo, L. Horwitz and Y. Ne'eman. 1966. *Phys. Lett.* **22**, 336.
N. Cabibbo, L. Horwitz, J. J. J. Kokkedee and Y. Ne'eman, 1966. *Nuovo Cim.* **45**, 275.
74. E. M. Levin and L. L. Frankfurt. 1965. *Zhur. Eksp. i Teoret. Fiz. Pisma v Redak.* **2**, 105; (English translation: 1965. *JETP. Lett.* **2**, 65).
H. J. Lipkin and F. Scheck. 1966. *Phys. Rev. Lett.* **16**, 71 and many later papers by H. J. Lipkin and collaborators.
J. J. J. Kokkedee and L. Van Hove. 1966. *Nuovo Cim.* **42**, 711.
J. J. J. Kokkedee. 1966. *Phys. Lett.* **22**, 88.
75. J. J. J. Kokkedee and L. Van Hove, Ref. 74.

L. Van Hove. 1966. *Proceedings of the Stony Brook Conference on High-Energy Two-Body Reactions*, Stony Brook, N.Y., April, 1966.
76. N. N. Bogoliubov *et al.* 1965. Dubna preprint D-1968.
77. C. Itzykson and M. Jacob. 1966. Saclay preprint. (July, 1966.)
78. L. Van Hove. 1963. *Phys. Lett.* **5**, 252.

3

SOME FEATURES OF ELECTROMAGNETIC INTERACTIONS

RICHARD WILSON

Harvard University

1

INTRODUCTION

I propose, in these lectures, to give a general outline which, I believe, will be appropriate to an experimenter. This will involve the presentation of enough theory to explain why the experiments are important but no great detail will be given. At the same time I wish to show the interest in the phenomena of electromagnetic interactions.

The first topic will be the study of the proton and the neutron by electron scattering. This is a classical study and it is already well worn, though by no means exhausted. In order to introduce this, I shall assume that you already know the major details of how to calculate a cross-section. I shall go into one or two details which illustrate the particular features of electron scattering. To outline the experiments and their present interpretation, I shall use methods which are perhaps unconventional from the theoretical point of view, but, nevertheless, are correct and give some insight to experimenters.

I refer, in electron scattering, to several recent review articles.[1,2,3,4,5,6]

I shall follow a calculation by Hand which is outlined in reference 7. The whole method of calculation is based on finding the square of the current operator interaction which involves a little calculation which I shall present.

2

ELASTIC ELECTRON SCATTERING

2.1. *Kinematics and definitions of form factors*

We choose the parameters according to the diagram shown in Fig. 1. p_i, p_f, P_i, and P_f are the four-momenta of the incident and final electron and proton respectively. q is the four-momentum transfer,

$$q^2 = (p_i - p_f)^2 = (P_i - P_f)^2. \qquad ...[1]$$

We choose the convention that q^2 is positive for electron-proton scat-

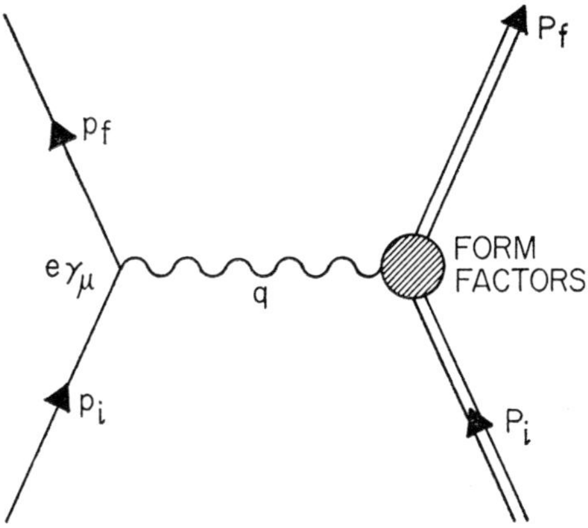

FIG. 1

Diagram for *ep* scattering in the one photon exchange approximation, showing the definition of kinematic quantities.

tering. We allow the final nucleon to have a mass different from the nucleon mass (M) to include inelastic scattering.

If we allow only one photon exchange, the transition amplitude has as its main factor the interaction $|j_\mu J^\mu|$ where j_μ is the matrix element of the electron current and J^μ is the matrix element of the proton current, both evaluated in the same reference frame. There are several standard ways of evaluating this, but I shall choose a way which gives some results of utility later.

We shall be concerned with four reference frames. (See Fig. 2.) The first two are "Breit" or "brick wall" frames. In each of these, there is no energy transferred. In the first frame we have $\boldsymbol{p}_i^{**} = -\boldsymbol{p}_f^{**}$. We call this the electron's Breit frame and quantities are double starred.

In the second, $\boldsymbol{P}_i^* = -\boldsymbol{P}_f^*$ and quantities are single starred.

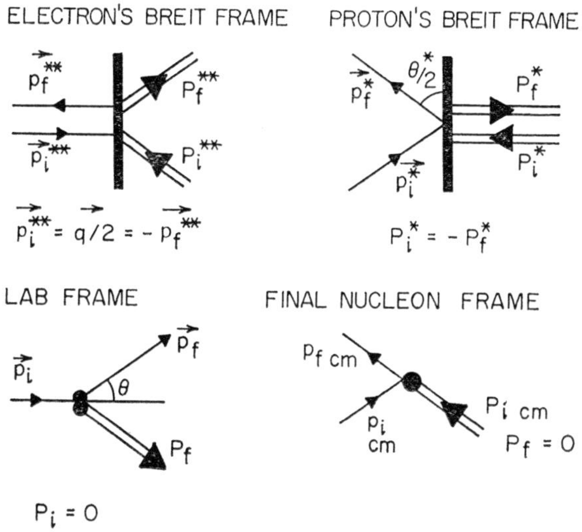

FIG. 2. Some reference frames of interest.

The laboratory frame, in which $\boldsymbol{P}_i = 0$, is also useful as is the final nucleon c.m. frame in which $\boldsymbol{P}_f = 0$.

We shall first evaluate j_μ in the electron's Breit frame, using *rest frame* spinors. j_μ represents

$$\langle \bar{u}(p_f) \mid e\gamma_\mu \mid u(p_i) \rangle$$

where u is a four component spinor; for $u(0)$ we write $\begin{pmatrix} \chi \\ 0 \end{pmatrix}$ where χ is a two component Pauli spinor. Then we find:

$$j_0^{**} = e\chi_f^* \chi_i \qquad \text{...[2]}$$

$$\boldsymbol{j}^{**} = ie\chi_f^* \left(\boldsymbol{\sigma} \times \frac{\boldsymbol{q}^{**}}{2m} \right) \chi_i. \qquad \text{...[3]}$$

From these equations we note several things.

1. For the usual momentum transfers, $q^2/4m^2 \gg 1$ and we may neglect j_0, hence the statement that the electron scatters primarily through its spin. (A pion of the same mass would not scatter as much.)

2. $\boldsymbol{j}$ is purely transverse and corresponds to circular polarization. In the limit $m \to 0$, it is well known that we have helicity conservation and the scattered electron retains the helicity of the original electron. The vanishing of both j_0 and j_z clearly gives this.

3. We note that the current of a charged particle with no spin is down by order $(m^2/q^2)^{\frac{1}{2}}$. Thus we can say that an electron scatters by its magnetic moment.

Now we carry out a Lorentz transformation to evaluate the electron current in the proton's Breit frame. For simplicity we shall drop m^2 in comparison with q^2. Then

$$j_0^* = \frac{ie\beta q^{**}}{2m(1-\beta^2)^{\frac{1}{2}}} (\chi_f^* \sigma_y \chi_i) \qquad \text{...[4]}$$

$$j_z^* = 0 \qquad \text{...[5]}$$

$$j_y^* = \frac{ieq^{**}}{2m} (\chi_f^* \sigma_x \chi_i) \qquad \text{...[6]}$$

$$j_x^* = \frac{ieq^{**}}{2m(1-\beta^2)^{\frac{1}{2}}} (\chi_f^* \sigma_y \chi_i). \qquad \text{...[7]}$$

Before we proceed to multiply this with the proton current, we note:

1. j_0^* has crept in again, but now with a phase factor i.
2. j_x^* and j_y^* are now unequal, corresponding to a polarization of the electromagnetic field (virtual photon) of the electron.

We define the effective linear polarization of the virtual photon as

$$P \equiv \frac{|j_y|^2-|j_x|^2}{|j_y|^2+|j_x|^2} = \frac{1/(1-\beta^2)-1}{1/(1-\beta^2)+1} = \frac{\beta^2}{2-\beta^2} = \frac{\cos^2(\theta^*/2)}{2-\cos^2(\theta^*/2)} = \frac{\cot^2(\theta^*/2)}{2+\cot^2(\theta^*/2)}. \quad ...[8]$$

This is clearly of interest to inelastic scattering, a subject to which we shall return later.

Now the proton current, even evaluated in its own Breit frame, is not as simple as the electron current. We must allow for structure. We are limited in what we may do by the requirements of gauge invariance, Lorentz invariance and time-reversal symmetry. We find that the most general form possible is[5, 6]

$$J_0^* = eG_E(q^2)X_f^*X_i \quad ...[9]$$

$$\mathbf{J}^* = ieG_M(q^2)X_f^*\left(\frac{\boldsymbol{\sigma}\times\mathbf{q}^*}{2M}\right)X_i \quad ...[10]$$

where $G_E(q^2)$ and $G_M(q^2)$ are arbitrary functions of q^2.

Now we evaluate $j_\mu J^\mu$.

$$\left.\begin{aligned} j_\mu J^\mu = e^2\left(\frac{q^2}{4m^2}\right)^{\frac{1}{2}} \Bigg\{ &\frac{iG_E\beta}{(1-\beta^2)^{\frac{1}{2}}}(X_f^*X_i)(\chi_f^*\sigma_y\chi_i) \\ &-G_M\left(\frac{q^2}{4M^2}\right)^{\frac{1}{2}}\frac{1}{(1-\beta^2)^{\frac{1}{2}}}(X_f^*\sigma_yX_i)(\chi_f^*\sigma_y\chi_i) \\ &-G_M\left(\frac{q^2}{4M^2}\right)^{\frac{1}{2}}(X_f^*\sigma_xX_i)(\chi_f^*\sigma_x\chi_i)\Bigg\}. \end{aligned}\right\} \quad ...[11]$$

When we square and average over spin directions the cross terms vanish and using $\cot^2(\theta^*/2) = \beta^2/(1-\beta^2)$ we get

$$\sum_{\text{spins}}|j_\mu J^\mu|^2 = \frac{q^2}{4m^2}\{G_E^2\cot^2(\theta^*/2)+(q^2/4M^2)G_M^2[2+\cot^2(\theta^*/2)]\}. \quad ...[12]$$

Transforming to the lab. system we find, with $\tau = q^2/4M^2$,

$$\sum|j_\mu J^\mu|^2 = \frac{q^2}{4m^2}\left\{G_E^2\frac{\cot^2(\theta/2)}{1+\tau}+\tau G_M^2\left[2+\frac{\cot^2(\theta/2)}{1+\tau}\right]\right\} \quad ...[13]$$

and the cross-section becomes

$$\frac{d\sigma}{d\Omega} = \frac{\alpha^2 r_e^2 m^2}{4E^2\sin^2(\theta/2)}\cdot\frac{1}{1+\dfrac{2E}{M}\sin^2(\theta/2)}\left\{G_E^2(q^2)\frac{\cot^2(\theta/2)}{1+\tau} + \tau G_M^2(q^2)\left[2+\frac{\cot^2(\theta/2)}{1+\tau}\right]\right\}. \quad ...[14]$$

The interaction of the proton current with an electromagnetic field breaks up into two simple terms in the Breit frame; J_0^* interacts with the electric field and $\boldsymbol{J}^*$ interacts with the magnetic field. Hence the two form factors $G_E(q^2)$ and $G_M(q^2)$ are called the electric and magnetic form factors. At the limit $q^2 \to 0$, we clearly get the electric charge and magnetic moment respectively, normalized to unity and to the static moment in Bohr magnetons.

The separation also corresponds to a separation in terms of no spin flip or spin flip of the proton; these two facts were first pointed out by Yennie, Levy and Ravenhall.[8] It was not until it was realized[2] that the scattering cross-section formula was particularly simple with these form factors that they achieved general acceptance and the present notation.

If the current J_μ had been produced by a static charge distribution, then $\boldsymbol{J}^* = 0$ and

$$J_0^* = e \int \rho(\boldsymbol{x})\, e^{i\boldsymbol{q}\cdot\boldsymbol{x}}\, d^3x = ef(q^2) \qquad \text{...[15]}$$

is given by the Fourier transform of the charge distribution, and similarly for a static distribution of magnetic moment.

It is tempting, therefore, to call $G_E(q^2)$ and $G_M(q^2)$ the Fourier transforms of the charge and magnetic moment distributions of the proton, but this is overly naïve. The Breit frame is obtained from the frame in which the proton is initially at rest by a Lorentz transformation with velocity $(q/2M)$. This is variable as q is varied; so the Breit frame is a different frame for each value of q^2 and the meaning of the Fourier transform of $G_E(q^2)$ is far from clear.

We can introduce the Dirac and Pauli form factors associated with these parts of the proton current by the definitions

$$\left.\begin{aligned} \kappa F_2(q^2) &= \frac{G_M(q^2) - G_E(q^2)}{1+\tau} \\ F_1(q^2) &= \frac{\tau G_M(q^2) + G_E(q^2)}{1+\tau} \end{aligned}\right\} \qquad \text{...[16]}$$

and we introduce isoscalar and isovector forms

$$\left.\begin{aligned} G_{MV} &= G_{Mp} - G_{Mn} \\ G_{MS} &= G_{Mp} + G_{Mn}. \end{aligned}\right\} \qquad \text{...[17]}$$

In his otherwise excellent book,[6] Gourdin uses, without adequate comment, a notation different from that now used by most experimenters. Thus he normalizes the Pauli form factor to the anomalous magnetic moment with $F_2(0) = \kappa$, whereas we, and most authors, use $F_2(0) = 1$. He uses

$$G_0(q^2) = G_E(q^2)/(1+\tau)^{\frac{1}{2}}, \qquad \text{...[18]}$$

$$G_1(q^2) = G_M(q^2)/(1+\tau)^{\frac{1}{2}}. \qquad \text{...[19]}$$

However, for inelastic scattering, he defines

$$\sigma_L(q^2, W^2) = \sigma_L(q^2, K) = \sigma_0(q^2, K) \qquad \ldots[20]$$

and

$$\sigma_T(q^2, W^2) = \sigma_T(q^2, K), \qquad \ldots[21]$$

as does Hand.

2.2. *Experiments on elastic ep scattering*

2.2a. *General.* We must now discuss the experiments on elastic *ep* scattering. Our aim is clearly to determine $G_E(q^2)$ and $G_M(q^2)$. From the cross-section formula we notice the way to do this. At 180°, the cross-section involves $G_M(q^2)$ alone. Indeed, G_M^2 dominates at angles sufficiently large that

$$2\tau(1+\tau)G_M^2 > \cot^2(\theta/2). \qquad \ldots[22]$$

At forward angles, G_M^2 and G_E^2 are mixed together and we measure $G_E^2+\tau G_M^2$. At momentum transfers

$$q^2 > 1(\text{BeV/c})^2 > 25\text{f}^{-2},$$

for which $\tau > 0{\cdot}2$, the G_E^2 term, being, as we shall see, down by $1/\mu^2 = 1/(2{\cdot}79)^2$ from G_M^2, gives less than the G_M^2 term and becomes increasingly hard to find.

Our recipe then is, at each value of momentum transfer q^2, to measure the cross-section at the smallest possible angle that the available energy will allow us. We have

$$q^2 = 4EE'\sin^2(\theta/2) \simeq E^2\theta^2. \qquad \ldots[23]$$

Then we measure, at the same q^2, at the largest angle that counting rate and experimental convenience will allow. If we are limited by counting rate, and the solid angle of the spectrometer is a fixed parameter, then a simple calculation shows that 180° is not usually the right angle to measure, but that it is best to measure between 100° and 150° depending on q^2.

Sometimes measurements are made at intermediate angles and the part of the cross-section given by $(4m^2/q^2)\,|j_\mu J^\mu|^2$ is plotted against $\cot^2(\theta/2)$, in what should be a straight line fit. This checks the validity of the assumption of one photon exchange. Such a check is of very limited use, as we shall see later.

2.2b. *Experimental methods.* The precautions and procedures to be followed in accurate (1% to 2%) electron scattering measurements were first made clear in the work at Orsay.[12] This was followed by the Harvard-Stanford collaboration[9] and now by extensive measurements at Stanford[10] and experiments just beginning at CEA-Harvard. The most accurate experiments at this time seem to be those of Drickey and Hand.[9]

In order to measure the cross-section we must attend to certain details:

1. Measure the energy, direction and intensity of the beam.
2. Measure the target density and thickness.
3. Measure the solid angle of acceptance of the scattered particles and the efficiency of detection.

We shall now go through these, showing how various laboratories differ. There are now five laboratories in this field: Stanford (MK III),[9,10] Orsay,[12] Cornell, CEA[13,14] and DESY, with SLAC and NINA soon to join. A typical layout (Orsay[12]) is shown in Fig. 3.

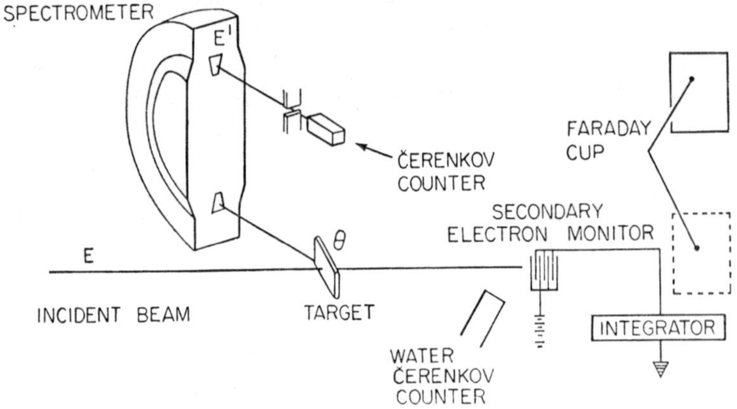

FIG. 3. Experimental arrangement for *ep* scattering, as used at Orsay.

2.2c. *Beam energy*. The beam energy may be measured by passing it through a known bending magnet ahead of the apparatus. This is done by a floating wire at Stanford[10] to an accuracy of 0·2%. At CEA this was done by a magnet with a field measured to better than 0·1% and the synchrotron, thereby, calibrated for subsequent use.[11] At Orsay,[12] and in the earlier CEA[13] measurements, the spectrometer used was calibrated to an accuracy of 0·2% by wire measurements (Orsay) and magnetic field measurements (CEA) and used to infer the incident energy.

The cross-section, without form factors, varies as $(1/E)^2$. Including form factors it varies roughly as $(1/E)^4$ for constant angle. The new CEA measurements of 0·1% in energy give, therefore, ½% uncertainty in cross-section.

2.2d. *Beam direction*. At the linear accelerator and the external beam of CEA the direction is determined by passing the beam through fluorescent screens or glass slides which are surveyed in place; an accuracy of 1 minute of arc (0·3 milliradians) is usual. The stability of the beam direction is not always that good. At the CEA, for example, it wanders about 1 minute. The new CEA measurements are at small angles—10° to 20°. The cross-section varies as $(1/\theta)^8$ so that this can give up to 1% error if careful monitoring is not done.

2.2e. *Intensity monitor.* The best intensity monitor is a Faraday cup. This is made of lead and large enough to catch 99·99% of the energy in a shower; the collected charge is then measured. Problems can arise due to low energy electrons knocked backwards from the cup or forward from the entrance window. These are bent out by a magnet and it is verified that these cause no trouble by varying the relative DC potential between the cup and the outside case. The efficiency can then be taken to be $100 \pm 0{\cdot}2\%$. Such cups are used at Orsay, Stanford, and the CEA. Their design has been well discussed.[17]

At the CEA a further check is used, as suggested by Hague.[16] The beam intensity may be measured by passing the beam through a toroidal coil and measuring the e.m.f. induced. At the CEA this is used as a null method and the Faraday cup signal is returned through the same coil. Technical problems still remain, but so far the Faraday cup is checked to 1%.[18]

At Cornell, and at the early CEA measurements, the beam stayed in the synchrotron and use of a Faraday cup was not possible. They measured the bremsstrahlung from the target, using a quantameter of R. R. Wilson.[19] The quantameter is, in itself, absolute (to 3%), and it has also been compared to this accuracy with a Faraday cup both at Stanford and the CEA. We need also to know the bremsstrahlung cross-section from hydrogen, which is probably known only to 4%. Thus these internal beam measurements are inherently less accurate.

2.2f. *Target.* Early targets (Orsay and Cornell) were of $(CH_2)^n$; the density and length measurement is trivial but background problems enter. Now, hydrogen targets are used. Care must be taken to measure the target length when *cold* because of the contraction of the target cup. The density of the target may be inferred from its temperature using NBS tables; at CEA, we use the hydrogen itself as a vapour pressure thermometer.

The main problem, so far inadequately discussed in any paper, arises in the understanding of bubbling in the target. If all the heat input to the target leaves by evaporation alone, it is easy to calculate that there are large density changes. Fortunately, observation through transparent target cups shows that bubbling is mostly on the surfaces and is not as severe as the postulated case.

At the CEA we encourage convection in the cup, the only other good method of heat transfer, by having two tubes to the cup from the target cooler. The problems get larger with these cups as the beam intensity increases. At 0·1 μA the heat input from the beam equals that by radiation from the outside.

2.2g. *Solid angle.* This subject is hardly discussed in the literature so I propose to discuss it here in some detail.

Backgrounds prevent the solid angle being defined by an aperture. The aim is to make this aperture thin, so that it defines a solid angle well, but also thick enough to stop particles outside the aperture (see Fig. 4). For electrons, a radiation length of tungsten is 3 mm; so a thickness of 1 cm is adequate. At a distance of L cm we get a *maximum* uncertainty of $(2/L)$ in the solid angle. By sloping the sides, as in Fig. 4, this can be reduced; a finite size target increases the uncertainty again. In Fig. 4 I draw two trajectories which must be discussed in detail. For the large double focussing spectrometers at Orsay and Stanford the trajectories drawn will almost always not

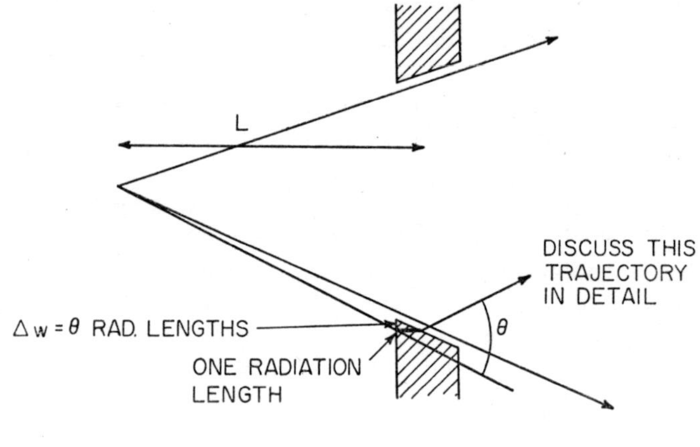

θ = MULTIPLE SCATTERING ANGLE FOR 1 RADIATION LENGTH

FIG. 4. This shows the possible uncertainties in the determination of solid angle.

reach the counter. At the CEA they might. On the other hand, the distances are more important; at Stanford only 50 cm; at Orsay 100 cm; at the CEA 200 cm. In all, an accuracy of 0·5% can be achieved.

2.2h. *Spectrometers.* Following the defining aperture there is always a magnet to sweep out unwanted particles and to define the momentum of the scattered particles. This magnet is a necessary nuisance because one has to be sure that all desired scattered electrons which pass through the defining aperture reach the detector, and a necessity because radiative corrections must be considered.

At both Stanford[10] and Orsay,[12] a large (180° bend) spectrometer has been used and the final detector is a single, simple lucite Cerenkov counter (see Fig. 3).

At the synchrotrons of Cornell, CEA[13] and DESY, there have been changes. The energy is higher; π meson fluxes are appreciable; at the same time simpler and cheaper spectrometers have been used; the discrimination against background has not been so clean and has involved shower detectors and Cerenkov counters. The vastly superior duty cycle of the synchrotrons has made the use of the complex counter system possible.

2.2i. *Detectors.* At low momentum transfers (Orsay and Stanford) the detector has always been a lucite Cerenkov counter. The efficiency for

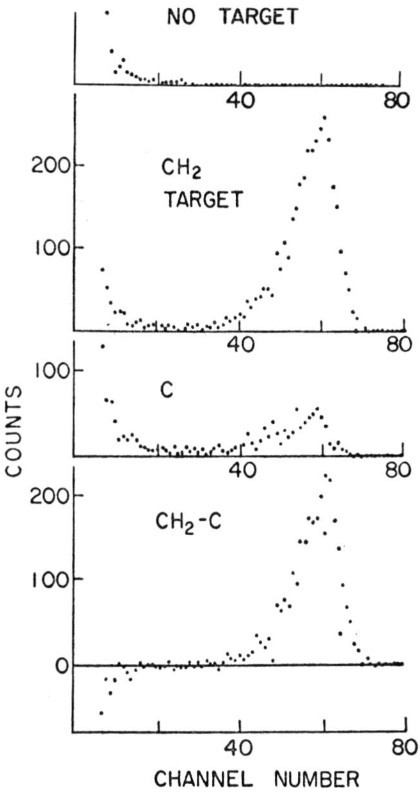

FIG. 5. Pulse height distribution from the Cerenkov counter used at Orsay. The efficiency is obtained by extrapolating the bottom curve to zero.

detecting electrons is obtained by examining the pulse height distribution. It is assumed that it consists of a peak and a tail, and that the efficiency is 100% if the peak is extrapolated to zero. This is made clear in Fig. 5 where one of the Orsay pulse height distributions is shown. The extrapolation to zero is assumed to fall between a straight line to the origin and a horizontal straight line. Errors are assigned with these limits. By this means an efficiency of $99{\cdot}5+0{\cdot}3\%$ was established for this particular run.

At Cornell and Harvard a large flux of π mesons was present, which penetrated absorbers and shielding to reach the counters. The spectrometers, as noted above, are simpler and could not separate them. Although kinematic calculations tell us that no π meson can have a momentum as great as an elastically scattered electron, the background junk must be separated and, also, we want to study *inelastic* processes.

Accordingly, two counters are used; a threshold gas Cerenkov counter and a shower counter to identify the electrons. On each of these, the efficiency is determined in the same way as shown above (Fig. 5) for the lucite Cerenkov counter. In addition, the energy is measured in a scintillation counter bank.

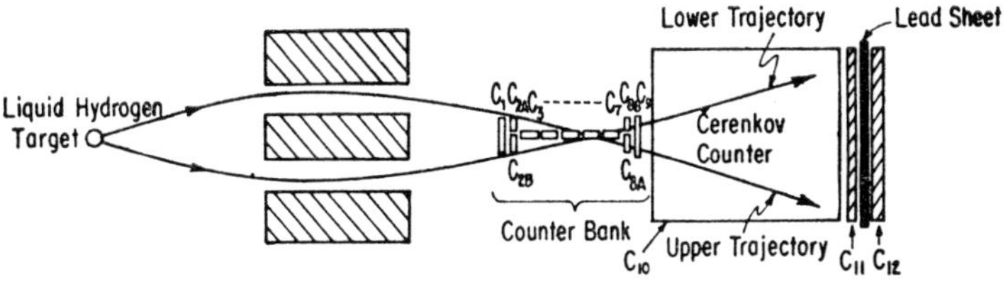

FIG. 6. Schematic diagram of the counter arrangement for high-energy electron detection at CEA.

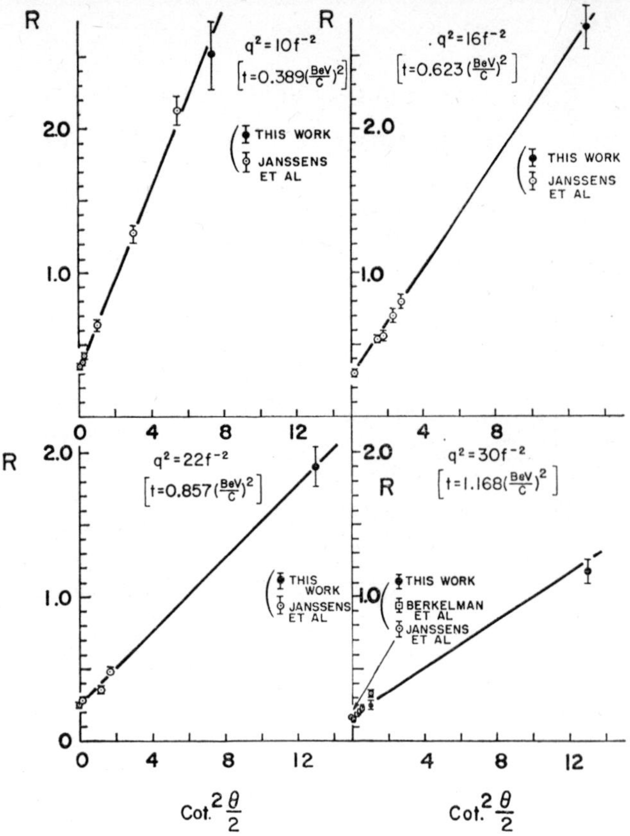

FIG. 7

R plotted against $\cot^2 (\theta/2)$ for $q^2 = 10$ to $30\ f^{-2}$.

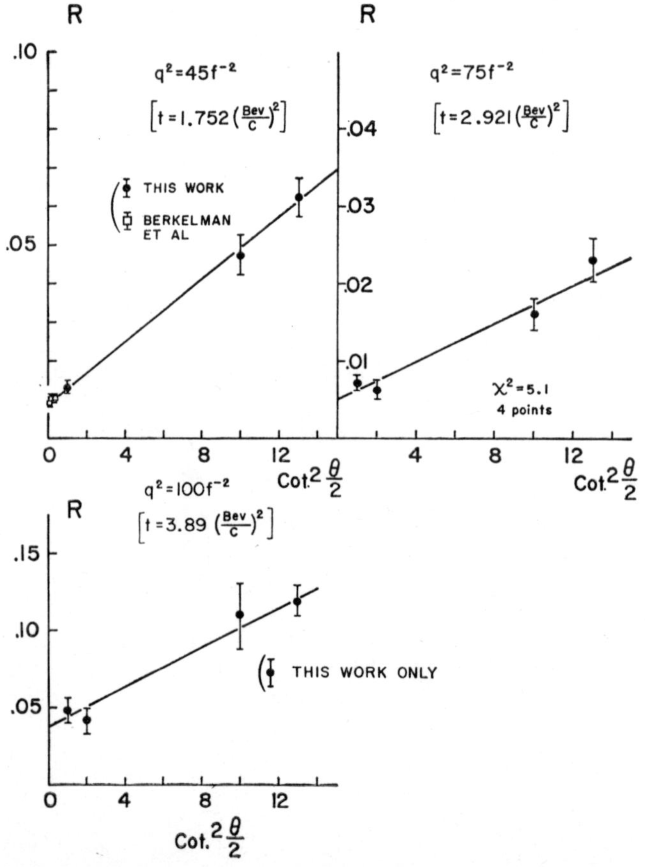

FIG. 8

R plotted against $\cot^2 (\theta/2)$ for $q^2 = 45$ to $100\ f^{-2}$.

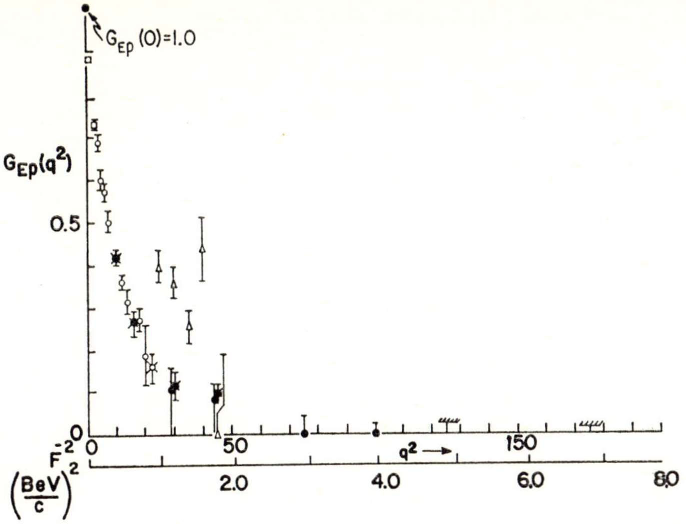

FIG. 9*a*

Electric form factor of the proton $G_{E_p}(q^2)$, plotted against q^2.

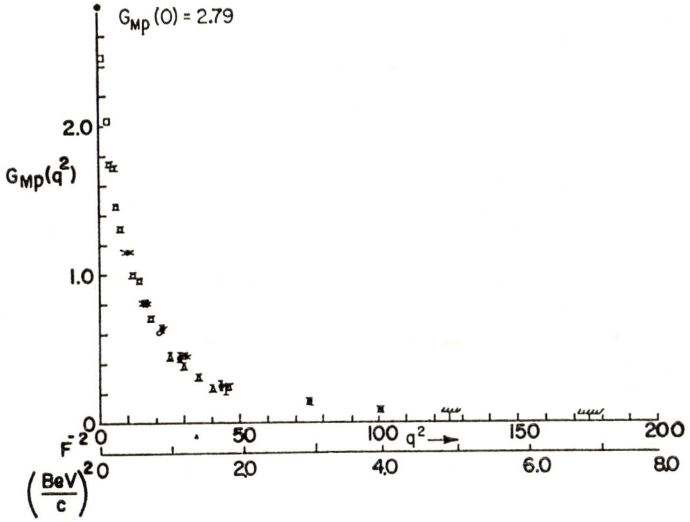

FIG. 9*b*

Magnetic form factor of the proton $G_{M_p}(q^2)$, plotted against q^2.

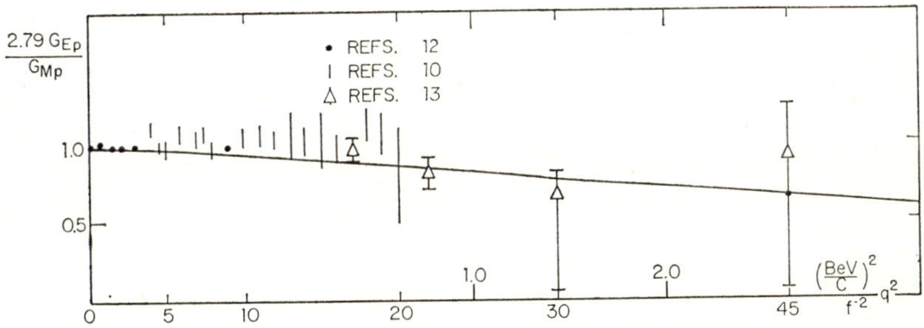

FIG. 9*c*. 2·79 G_{E_p}/G_{M_p}, plotted against q^2. The solid line is the fit from the 4-pole fit (form factors given by equations [*64*], [*65*], [*67*] and [*68*]).

The system (Fig. 6) is, therefore, quite complicated and it is harder to get the efficiency to high precision.

2.2j. *Radiative corrections.* Before the experimental cross-sections can be compared to theory we must discuss the radiative correction. I shall adopt a point of view which is adequate for an experiment and refer others to the relevant literature.[20]

No electron-proton collision is purely elastic. The probability of emitting one or more low energy photons is unity. Thus we measure a cross section for a collision in which we integrate over inelastic processes. Specifically, in most experiments to date, we integrate over those collisions where the scattered energy is less than that for purely elastic kinematics by up to an amount ΔE. This allows emission of photons up to an energy Δk, where, unfortunately, due to the kinematics, Δk is a function of angle.

The theoretical cross-section derived earlier has no direct meaning. However, the renormalization methods of quantum electrodynamics give us a procedure for deriving an ideal experimental cross-section to compare with this theory. We take

$$\sigma_{\text{exp}}(\text{ideal}) = \sigma_{\text{exp}}\, e^{\delta}$$

where

$$\delta = -\frac{2\alpha}{\pi}\left(A + B\log\frac{\Delta E}{E}\right); \quad B = \log(q^2/m^2) - 1. \qquad \ldots[24]$$

The A term does not affect experimental matters and I now ignore it (though of course it must be included somewhere in the theory/experiment comparison). The $\log(\Delta E)/E$ term comes from integration over the $(dk)/k$ of the bremsstrahlung spectrum.

Now if we do not define ΔE properly we can have an error. The error is smaller if we make $(\Delta E)/E$ larger; however, we need a small ΔE to exclude inelastic (in the sense of nucleon excitation) scattering. Also, we must be sure that ΔE is larger than the experimental momentum resolution and that any second order terms are taken into account. This is all straightforward; however, the evaluation of e^{δ} is, due to computational complexities, uncertain to 1% or so.

2.3. *Experimental results*

We now discuss the cross-sections, corrected for the radiative processes above, so that they are not actually experimental cross-sections. This is, however, the normal "break" between theoretical and experimental papers. It is conventional to plot the quantity

$$R(q^2) = \frac{G_E^2 + \tau G_M^2}{1+\tau}\cot^2(\theta/2) + 2\tau G_M^2 \qquad \ldots[25]$$

versus $\cot^2(\theta/2)$, or $R\tan^2(\theta/2)$ versus $\tan^2(\theta/2)$, for constant q^2. The former has the advantage that the physically possible point $\theta = 180°$ is included. So far, all such plots have given straight lines and to this extent verify the assumption of one photon exchange which led to equation [*14*]. But we must be careful. We shall see later that this is a very poor way to see multi-photon exchange effects. For the moment, however, we shall assume theoretically one photon exchange. Although the values of the form factors could be changed beyond their errors if we are wrong, no major conclusion will be altered.

In Fig. 7 and Fig. 8 we present such plots of R versus $\cot^2(\theta/2)$. In Figs. 9a and 9b we present G_{Ep} and G_{Mp} as functions of q^2. In Fig. 9c we present the ratio $G_{Mp}/\mu G_{Ep}$ as a function of q^2. This parameter seems close to unity—an exciting relation first noticed some five years ago in the interpretation of the Orsay data.

Note the precise low momentum transfer data from Harvard-Stanford[9] and Orsay[12] which are seen as dots. (It was the $q^2 = 2{\cdot}98 f^{-2}$ point which started the interest in this relation). Note also the medium momentum transfer data from Stanford,[10] which are *systematically* different and when averaged give $G_M/\mu G_E$ higher than 1, and the low precision high momentum transfer data from Harvard.[13]

We notice also the significant feature that *both* G_E and G_M tend to zero as $q^2 \rightarrow \infty$. This result is referred to as the absence of a *core* in the proton. It is not surprising but it prevents naïve explanations of proton structure which were current six years ago. These two facts and, of course, the slopes of G_E and G_M at the origin, are the most important ones and all that the theory can presently assimilate.

Recent (August 1966) data from DESY suggest that $G_M(q^2)$ may be lower than Harvard data at high q^2. This is surprising to me but makes no changes in the present interpretation.

Note added in proof: It is now confirmed that the cross-sections of reference 13 are indeed high at large momentum transfers by up to a factor of 3. $G_M(q^2)$ is reduced by up to a factor of 1·7. Although figures 9*a*, 9*b* and 22 must be amended slightly, no theoretical conclusions are altered.

2.4. *Annihilation processes*

The process of Fig. 1 can be viewed by turning the page on its side to give the crossed reaction

$$p+\bar{p} \rightarrow e^+ + e^-$$

or

$$e^+ + e^- \rightarrow p+\bar{p}.$$

We again make the one photon exchange approximation and we realize we are studying a similar process, but in a different kinematic region. In particular the four-momentum transfer q^2 is now timelike (negative in our metric).

We found the Breit frame (which we defined by putting $\boldsymbol{P}_i = -\boldsymbol{P}_f$) very useful in the discussion of elastic scattering. If we make a similar definition for the same pair of particles (nucleons) we now have $\boldsymbol{P}_p = -\boldsymbol{P}_{\bar{p}}$ and this is clearly the centre of mass system. We expect, therefore, that the centre of mass system, defined as the system where the momentum transfer is zero, has the same sort of significance that the Breit frame has for scattering.

Following the usual formulae we get

$$\left(\frac{d\sigma}{d\Omega}\right)_{\text{c.m.}} = \frac{\alpha^2}{16M^2}\left(\frac{1+\tau}{\tau}\right)^{\frac{1}{2}} \{\sin^2\theta\, G_E^2(q^2) + (1+\cos^2\theta)\, G_M^2(q^2)\} \quad \ldots [26]$$

where θ is the angle of emission of the outgoing particles with respect to the incoming direction. $G_E(q^2)$ and $G_M(q^2)$ are again the coefficients of the current operator, as in equations [*9*] and [*10*].

The method that has been used is to use a beam of antiprotons on stationary protons. There is a large background of pions from the two-body final state

$$\bar{p}+p \rightarrow \pi^+ + \pi^-$$

and from the more probable many body final states.

The experimental procedure is as follows. The main problem is to obtain an intense beam of antiprotons, free from pions if possible to avoid further background. This is obtained by electromagnetic separators both at CERN and at the AGS. The beam intensity is measured by counting the beam in coincidence with the final state products. (Alas, we wish in vain for an antiproton intensity high enough that dead time effects need consideration.) The two-body decay processes are identified by a kinematic coincidence between the two outgoing particles.

We still have to separate the electron positron final state. This has been done by examining the shower produced. $\pi^\pm$ mesons can produce a shower by first making a charge-exchange collision and then the π^0 produces a shower. But this does not occur every time. Moreover, when the shower from a π meson is examined in a spark chamber separated by lead shower forming plates it appears different from one caused by an electron. It is possible experimentally to set, therefore, an upper limit of 10^{-3} on the detection efficiency for one charged π meson and 10^{-6} for two π mesons. This is enough.

A similar reaction to

$$\bar{p}+p \rightarrow e^+ + e^-$$

is

$$\bar{p}+p \rightarrow \mu^+ + \mu^-.$$

Since a muon and electron are similar the nucleon form factors are the same.

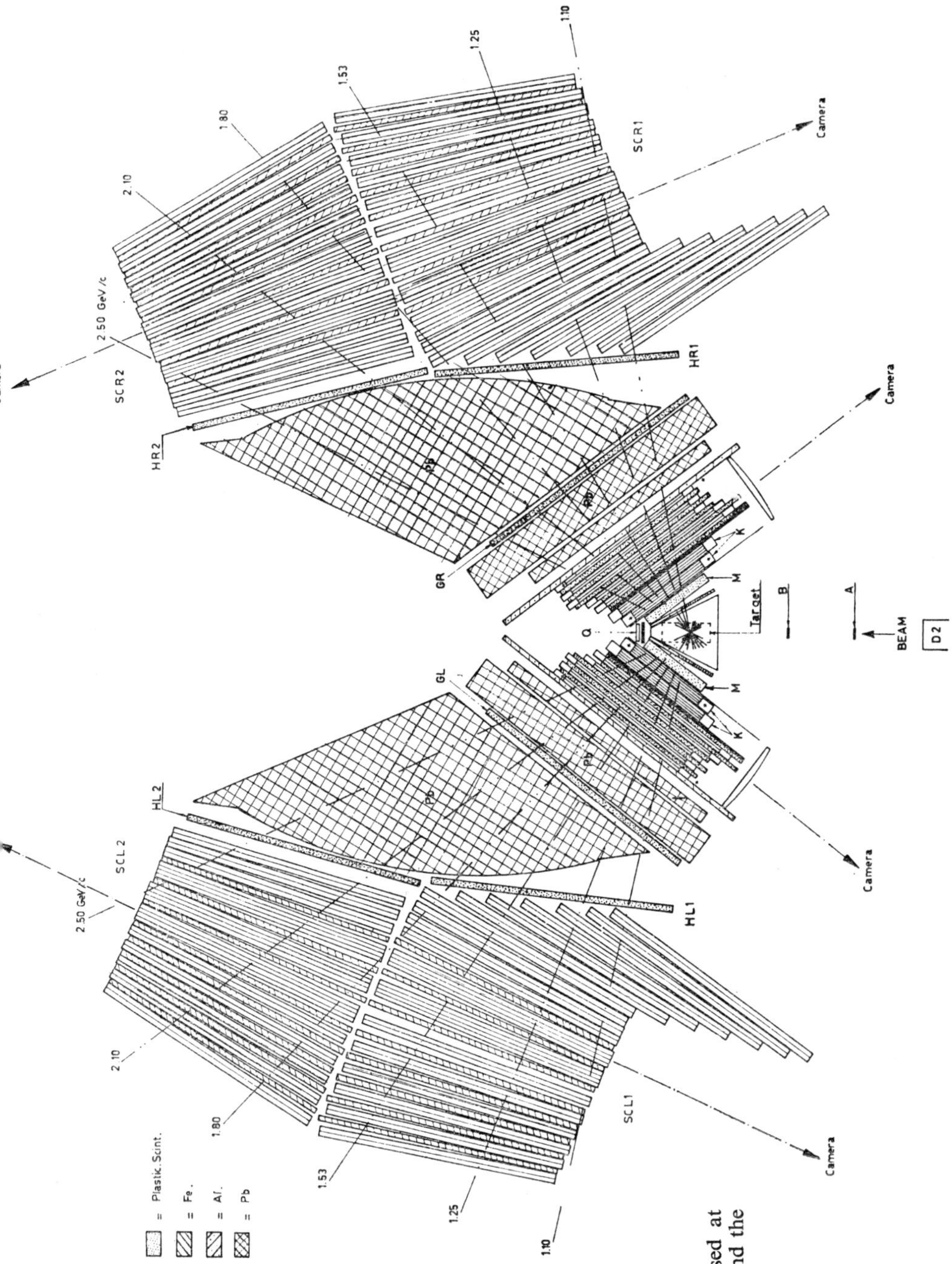

FIG. 10

Experimental apparatus used at CERN in an attempt to find the reaction $\bar{p}+p \rightarrow$ leptons.

A muon is distinguished from a π meson by its long range and lack of nuclear interaction.

Two experiments have been done. The one at CERN[21] (Fig. 10) has measured no events but has given upper limits (90% confidence level) of

$$\sigma_T(pp \rightarrow e^+e^-) \leqq 1{\cdot}2 \times 10^{-33} \text{ cm}^2$$

$$\sigma_T(pp \rightarrow \mu^+\mu^-) \leqq 0{\cdot}6 \times 10^{-33} \text{ cm}^2$$

at $q^2 = -6{\cdot}8(\text{BeV/c})^2$.
The other at the AGS[22] gives $\sigma_T(p\bar{p} \rightarrow e^+e^-) \leqq 10^{-33}$ cm^2 at $q^2 = -6{\cdot}8$ (BeV/c)^2 with no events and $\sigma_T(p\bar{p} \rightarrow e^+e^-)$ at $q^2 = -5{\cdot}1(\text{BeV/c})^2$ with two possible events.

We then deduce that

$$G_E \leqq 0{\cdot}2 \qquad \text{...[27]}$$

$$G_M \leqq 0{\cdot}2 \qquad \text{...[28]}$$

at $q^2 = -6{\cdot}8(\text{BeV/c})^2$ and $q^2 = -5{\cdot}1(\text{BeV/c})^2$.

2.5. *Threshold constraint*

We have the relations

$$G_M = F_1 + \kappa F_2 \qquad \text{...[29]}$$

$$G_E = F_1 - \tau\kappa F_2 \qquad \text{...[30]}$$

where F_1 and F_2 are the Dirac and Pauli form factors. Most people find it unpalatable to let F_1 and F_2 have singularities at the threshold for the annihilation reaction and so at $\tau = -1$ we have the result

$$G_M = G_E. \qquad \text{...[31]}$$

Alternatively, we note the angular distribution at threshold of $e^+ + e^- \rightarrow \bar{p} + p$ should be that of an s wave—i.e. isotropic. Looking at equation [26], we find $G_M^2(-4M^2) = G_E^2(-4M^2)$; a similar argument for the polarization puts the phases equal.

3

Neutron Form Factors

When we come to the task of finding neutron form factors, we find four main sources of information:

1. Scattering of low energy neutrons by high Z elements.
2. Elastic electron-deuteron scattering.
3. Quasi-elastic electron-deuteron scattering.
4. Inelastic electron-proton scattering.

We shall discuss the last later.

3.1. *Neutron-electron interactions*

The scattering of thermal neutrons has two parts: coherent scattering by the nucleus and scattering by the atomic electrons. In order to measure this scattering we must separate these two, which we do by the fact that they have appreciably different *atomic* form factors. We define the coherent nuclear amplitude $a_{n(coh)}$ and the electron-neutron amplitude a_{en}.

The first such measurement was by Fermi and Marshall[23] who measured the angular distribution in the scattering from krypton—a rare gas with no complicating chemical or liquid effects. The aim is to find the interference term between these amplitudes because the electron-neutron amplitude is very small by itself.

The most recent and accurate measurement of this type is by V. E. Krohn.[24] The quantity measured is

$$\frac{\sigma_{45}}{\sigma_{135}} = 1+\frac{2a_{en}a_{n(coh)}}{(\sigma_T/4\pi)}\left[\left(\int nFd\lambda\right)_{45}-\left(\int nFd\lambda\right)_{135}\right] \qquad \ldots[32]$$

where the integrals are over the neutron wavelength weighted by the atomic form factors. There is a calculable spurious effect, due to the centre of gravity motion, about equal in size to the desired one.

Melkonian *et al.*[25] measured the *total* cross-section from liquid bismuth. Whereas the neutron-nucleus cross-section is constant with energy, the neutron-electron interaction falls with increasing energy because of the atomic form factor. They measure:

$$\sigma = \sigma_n+2Z(\sigma_n\sigma_{en})^{\frac{1}{2}}\bar{f}+Z^2\sigma_{en}\langle f^2\rangle \qquad \ldots[33]$$

where $\bar{f}$, the average over the form factor, is a calculable function of energy. The choice of liquid bismuth is for the following reasons:

1. No coherent effects.
2. No spin incoherence.
3. Large Z, giving large neutron-electron scattering.
4. Large A, giving small c.m. motion (Doppler correction).
5. Only one isotope.
6. Availability.

Corrections are needed for:

(i) The energy dependent $(1/v)$ neutron capture, which is measured in a pile oscillator.

(ii) The Doppler effect $\delta\sigma/\sigma = 1+(kT)/(2\mu E_0)$, where μ is the reduced mass.

(iii) The liquid diffraction cross-section, which is the biggest and worst. This is based on a theory of Placzek. Otto Halpern[26] doubts the reliability of this correction.

Then we come to the third experiment. This neutron-electron scattering experiment[27] uses the technique of a liquid mirror. The nuclear cross-sections for oxygen and bismuth are measured from total cross-sections at high (thermal) energy and the total amplitudes are compared by scattering an epithermal neutron beam from a liquid oxygen/bismuth interface. The measurement of the angle of total external reflection gives the difference between the two amplitudes and hence the neutron-electron interaction.

These measurements are all in fair agreement. They measure[28] $(dG_{En})/(dq^2)$. It is usual and interesting to express the results in terms of $(\kappa/4M^2)$—a term due to the anomalous moment alone, *without* any variation in the Dirac form factor F_{1n}, and with, therefore, $F_{1n} \equiv 0$.

The results are shown in Table I. They are all in rough agreement and are certainly very distinct from zero. This is important because other experiments and theories suggest $G_{En}(q^2)$ is everywhere zero.

TABLE I

Author	Method	$\left(\frac{dG_{En}}{dq^2}\right)/(\kappa/4M^2)$	$\frac{dG_{En}}{dq^2}\left(\frac{\text{BeV}}{c}\right)^{-2}$
Krohn	K, A, Xe Scattering	0·846±0·050	0·459±0·02
Melkonian	Liquid Bi	1·062±0·034	0·575±0·019
Hughes	Bi/O mirror	0·946±0·090	0·512±0·019

3.2. *Elastic ed scattering*

We next come to the elastic electron-deutron scattering experiments. We note first that the separation of the scattering cross-section into two parts, which we carried out for electron-proton scattering, is in fact a characteristic of the vector (spin 1) character of the electromagnetic field being exchanged. It holds also for scattering from particles of any spin provided that the incoming and outgoing spins are not measured, as shown clearly, for example, by Durand.[29] At higher momentum transfers the quadrupole moment, with its form factor, enters.

The calculations are based on what always used to be called the "impulse approximation" (following Chew) but now the term "spectator approximation" is coming into use.

We have to *add* the amplitudes for scattering from the neutron and proton, multiplied by a phase factor because of the deuteron size; clearly we get roughly

$$G \sim (G_{Ep}+G_{En})F_d. \qquad \ldots[34]$$

There are, however, refinements which the experts[30, 31, 32, 33, 34] take into account. Probably the clearest description is in Gourdin's book.[6] Gourdin calculates directly in the Breit frame and obtains (in our notation)

$$d\sigma = d\sigma_{\text{Mott}}\left\{\frac{G_{Ed}^2(q^2)}{1+\tau}+\tfrac{8}{9}\tau^2\,\frac{G_{Qd}^2(q^2)}{1+\tau}+\tfrac{2}{3}\tau^2 G_{Md}^2\left[\frac{1}{1+\tau}+2\tan^2(\theta/2)\right]\right\} \qquad \ldots[35]$$

where

$$G_{Ed}(q^2) = 2G_{ES}(q^2)F_{Ed}(q^2) \qquad ...[36]$$

$$G_{Qd}(q^2) = 2G_{ES}(q^2)F_{Qd}(q^2) \qquad ...[37]$$

$$G_{Md}(q^2) = 2\frac{M_d}{M}G_{MS}(q^2)F_M(q^2) \qquad ...[38]$$

$$F_{Ed}(q^2) = \int_0^\infty [u^2(r)+w^2(r)]j_0\left(\frac{qr}{2}\right)dr \rightarrow 1 \text{ as } q^2 \rightarrow 0 \qquad ...[39]$$

$$F_{Qd}(q^2) = \frac{6M_d^2}{q^2}\int_0^\infty [u(r)w(r)\sqrt{2}-\tfrac{1}{2}w^2(r)]j_2\left(\frac{qr}{2}\right)dr \rightarrow M_d^2Q \text{ as } q^2 \rightarrow 0. \qquad ...[40]$$

Now how do we know the features of the deuteron ground state wave functions $u(r)$ and $w(r)$? From nucleon-nucleon scattering[35] we know that the asymptotic forms are given by the binding energy, i.e.

$$u(r) \text{ and } w(r) \rightarrow e^{-\gamma r} \text{ as } r \rightarrow \infty, \qquad ...[41]$$

where

$$\gamma^2 = 2M\varepsilon/\hbar^2. \qquad ...[42]$$

Also, if we expand the above expression for F_{Ed} we find the next term is given by $r_t(-\varepsilon, -\varepsilon)$, the *np* triplet effective range.

The first of these is a direct and the second an almost direct consequence of experiment. Moreover, the higher energy data may be approximated by a static potential so that higher terms are fairly well determined too—the next term by the twin requirements that there be a repulsive core and that the *np* interaction at large distances be given by one pion exchange. *Most* theorists calculate with a Hamada potential or something like it, but you must realize that at low q^2, F_{Ed} is determined fairly directly from experiment and that many of the potentials (e.g. Hamada's) do not fit exactly the experimental parameters. With this caveat, and a note on the small relativistic and exchange corrections to equation [34], we proceed to discuss experiments. We need now consider only five.[36,37,38,39,40]

At low q^2 ($q^2 < 2\text{f}^{-2}$) we consider the experiments to measure G_{ES}. Since electron-proton scattering is measured with the same apparatus we measure

$$G_{ES}/G_{Ep} = 1+G_{En}/G_{Ep} \qquad ...[43]$$

and some systematic errors cancel. Fig. 11[38] shows this. The measured quantity is

$$\frac{G_{Ed}^2+(q^4/18M_d^4)\,G_{Qd}^2}{(G_{Ep})_{\text{exp}}^2}$$

and is compared to the theory supposing $G_{En} = 0$. The two potentials assumed differ primarily in the values for the triplet effective range. The potential GK no. 8,[41] we find (see Table II), has a triplet scattering length (which experimental quantity is closely related to the effective range) too high by one standard deviation. For the Hamada-Johnson potential this scattering length is too low by a factor of 2 or 3. It is noteworthy that the differences are smaller than the experimental errors. Table II summarizes the data. Provided we realize that, up to $q^2 = 3\text{f}^{-2}$, the primary uncertainty is the triplet scattering length, we could in principle "correct" the results to any desired effective range using the formulation of Wilson.[35]

The conclusion of looking at the neutron form factors is that there is a discrepancy from a smooth extrapolation of the low energy result

$$\frac{dG_{En}}{dq^2} = 0{\cdot}018\text{f}^2. \qquad \ldots[44]$$

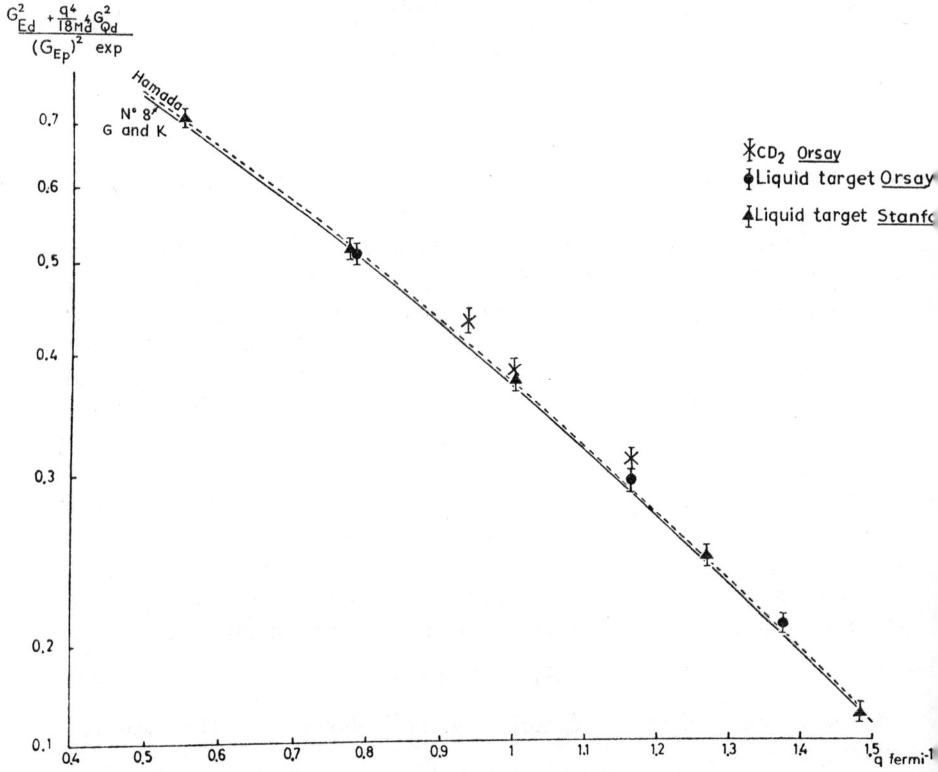

FIG. 11. $[G_{Ed}^2+(q^4/18M_d^4)G_{Qd}^2]/(G_{Ep}^2)_{\text{exp}}$ versus q^2. The triangles are the Harvard/Stanford results (Ref. 36) and the circles and crosses, the Orsay results (Ref. 37, 38). The solid line is a calculation from the Glendinning and Kramer potential No. 8 (Ref. 41) and the dotted line from the Hamada potential.

TABLE II

$q^2(\text{f}^{-2})$	$q^2\left(\frac{\text{BeV}}{c}\right)^2$	G_{En}		Reference
0·606	0·0236	$-0·003 \pm 0·014$	GK No. 8	38
1·351	0·0525	$+0·009 \pm 0·014$		38
1·964	0·0764	$+0·015 \pm 0·018$		38
3·0		$+0·02 \pm 0·015$	Hamada Potential	37
4·0		$-0·015 \pm 0·017$		37
5·0		$-0·001 \pm 0·015$		37
6·0		$+0·01$		37
0·3		$+0·005 \pm 0·006$	Hamada Potential	9
0·6		$-0·007 \pm 0·007$		9
1·0		$+0·002 \pm 0·005$		9
1·6		$+0·002 \pm 0·008$		9
2·2		$-0·002 \pm 0·009$		9

But we know (it will be shown later) that $G_{En} \simeq 0$ at $q^2 \to \infty$ so that a more reasonable extrapolation would be $F_{1n}(q^2) = 0$. All the fits to the data presented later fall close to this. This gives $G_{En} \simeq 0·04$ at $q^2 = 3\text{f}^{-2}$. There is still a definite experimental discrepancy for this value of G_{En} of about 2 or 3 standard deviations at several points. However, the theory is probably uncertain.

Note added in proof: A forthcoming paper by Gross finds agreement by making relativistic corrections and by using the Feshbach-Lomon wave-function.[42]

The magnetic scattering from the deuteron gives, in the non-relativistic application of the impulse approximation,

$$G_{Md}(q^2) = 2\frac{M_d}{M}\int_0^\infty \left\{ [G_{MS}(u^2+w^2) - \tfrac{3}{2}(G_{MS} - \tfrac{1}{2}G_{ES})w^2] j_0\left(\frac{qr}{2}\right) \right.$$

$$\left. + [(1/2^{\frac{1}{2}})(G_{MS})w(u+w/2^{\frac{1}{2}}) - \tfrac{3}{18}G_{ES}w^2] j_2\left(\frac{qr}{2}\right)\right\} dr. \qquad \ldots[45]$$

When $q^2 = 0$, we find this reduces to

$$\mu_D = (\mu_p + \mu_n)(1 - \tfrac{3}{2}P_D) + \tfrac{3}{4}P_D \qquad \ldots[46]$$

where

$$P_D = \frac{\int w^2 dr}{\int (u^2 + w^2) dr} \qquad \ldots[47]$$

is the percentage D state. Now it is well known that this leads to $P_D \simeq 4\%$. This result is non-relativistic and is widely disbelieved because relativistic corrections are clearly non-negligible. It has been common to talk about a 7% D state. The reasons are:

(i) The potential models which fit nucleon-nucleon scattering give this value.

(ii) It gives a better fit to photodisintegration of the deuteron at 70 MeV ($^3D \rightarrow {}^3P$ states are enhanced giving a more isotropic angular distribution).

(iii) The elastic *ed* form factor $G_Q(q^2)$, which dominates small angle scattering at high momentum transfers, tends to depend directly on P_D and suggests $\simeq 7\%$.

(iv) Coherent π^0 production from deuterium also gives such a value.

However, recent work of Feshbach and Lomon,[42] using a boundary condition model at the edge of a hard core to describe the nucleon-nucleon interaction, finds $P_D \simeq 4\%$, fits photo-disintegration and generally destroys most of these arguments. Their work suggests the fault may be in the potential description. So experiments leading to P_D are exciting.

Many people have attempted to calculate meson exchange corrections to equation [*46*]. One dominant term has been picked out,[43] the effect of the $\pi\rho\gamma$ coupling. Drell and Adler find a change

$$\Delta\mu_d \simeq 2\%, \qquad \ldots[48]$$

but they admit this is only one of several terms. Other methods of calculation[49] give the same sort of result but the opposite sign. It is clearly an open field.

At the higher momentum transfer we present in Figs. 12a and 12b the

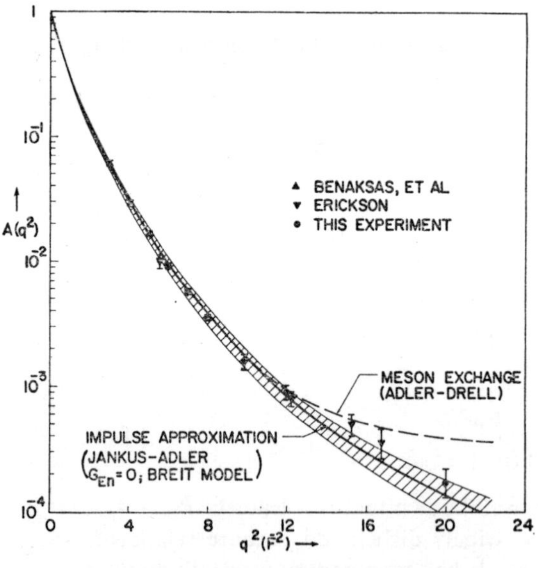

FIG. 12*a*. The quantity $A(q^2)$ in the equation $\sigma = \sigma_{\text{NO SPIN}}\,[A(q^2)+B(q^2)\tan^2(\theta/2)]$. The data is that of Refs. 39 and 36, and the theory uses the Breit potential. The Adler-Drell[43] exchange correction is shown.

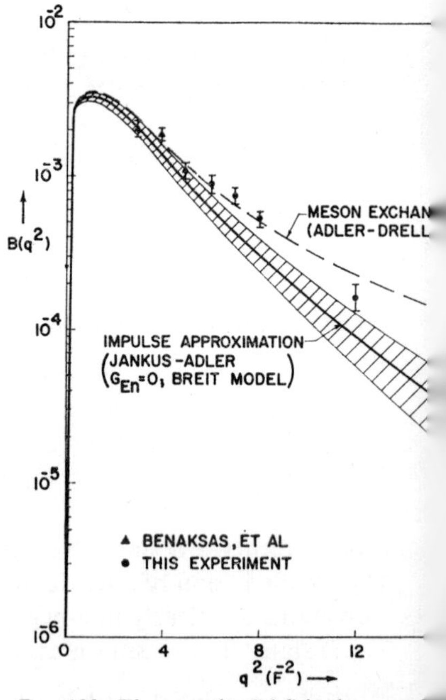

FIG. 12*b*. The quantity $B(q^2)$ in the equati $\sigma = \sigma_{\text{NO SPIN}}\,[A(q^2)+B(q^2)\tan^2(\theta/2)]$.

results of Buchanan and Yearian. The general features of the theory still hold and it is clear that the deuteron ground-state wave function derived from nucleon-nucleon scattering data describes the data well with G_{En} still equal to 0. This is not quite a trivial comment because in principle we can fit the effective range expansion, for example, to all orders and by choosing the interaction obtain *any* answer for *ed* scattering. But the interaction would then be very strange; it would not be short range.

3.3. *Electron-deuteron inelastic scattering*

The primary source of information on neutron form factors is the inelastic scattering of electrons from deuterons. The theory is again the impulse approximation, but now we work at sufficiently large transfers of momentum to single nucleons that we can regard them as approximately free. They are not, however, stationary, but will have a distribution in momentum given by the deuteron ground-state wave function in momentum space. This is, of course, as well known as that in co-ordinate space. We take the S and D state Fourier transforms separately and recombine.

The experiments fall into three classes.

(i) Non-coincidence detection of the electron.

(ii) Coincidence detection of electron with proton.

(iii) Coincidence detection of electron in coincidence with the neutron.

The first of these includes several recent experiments. (See references 44, 45, 46, 47.)

Elastic electron-proton scattering at an angle θ gives a definite scattered electron energy (with the exception of the radiative tail). The scattered electron energy from the deuteron is distributed about that from the proton because of the momentum distribution (primarily that of the component along the beam direction). This spread varies from 10% full width at half maximum at $q^2 = 3{\cdot}8\text{f}^{-2}$ and $\theta = 60°$ to 5% at $\theta = 60°$ and higher energies. Unlike the case for non-relativistic kinematics, the fractional spread is *constant* for constant θ as the energy increases (at $\theta = 20°$, $\frac{1}{2}$%; at $\theta = 90°$, 8%). This leads to a natural upper limit on the possible momentum transfers.

Now we expect, from the naïve argument above, that, if we integrate over this peak

$$\sigma_{ed} = \sigma_{en} + \sigma_{ep}. \qquad \ldots[49]$$

Moreover, we can calculate the shape of the peak and, therefore, if we do not detect all of it, we can make a correction. Radiative corrections clearly distort the peak shape slightly in a calculable way[48] but we must include the radiation from the proton line, as done by Meister and Yennie.[20]

Now, the problems of this method are threefold:

(i) We find that $\sigma_{en} \leqq \frac{1}{2}\sigma_{ep}$. Therefore, to determine σ_{en}, we must obtain σ_{ed} to three times the accuracy desired. This is statistically difficult, though fortunately some systematic experimental errors can cancel in a ratio measurement.

(ii) For the same reason as in (i) the corrections to the theory of the scattering are multiplied by three in finding σ_{en}. So are their uncertainties.

(iii) At high incident energies the first nucleon resonance (M = 1238 MeV) is excited (approximately equal in probability to leaving an unexcited nucleon). The scattered electron energy differs from that for elastic scattering (see equation [*95*]) by

$$\frac{\Delta E}{E} = \frac{K}{E} \qquad \text{...[50]}$$

$$= \frac{330 \text{ MeV}}{E} \qquad \text{...[51]}$$

which gets *smaller* as E increases, *in contrast* to the effect of the deuteron momentum distribution. Thus the quasi-elastic peak can only be separated from the quasi N^* production peak below $q^2 = 70\text{f}^{-2}$ (as shown in Fig. 13 from Dunning *et al.*[47]). I know of no simple way around this problem.

Now we come to the coincidence methods. In the first, we detect the electron and a high energy (> 20 MeV) proton in coincidence. This is a quasi *ep* scattering. According to our naïve ideas, we should find, after correction for the tails of the momentum distribution, that

$$\frac{\sigma_{ep}\text{ (deuterium)}}{\sigma_{ep}\text{ (hydrogen)}} = 1. \qquad \text{...[52]}$$

Fig. 14 shows such a plot from Dunning *et al.*[47] Of course, we must be sure to pick up *all* the recoil protons. A good measurement of the angular distribution of these recoil protons (now in progress at Harvard) should be a good test of the theory and possibly, if we are very lucky, add to our information about the ground-state wave functions of the deuteron.

Now we can count from deuterium

$$\frac{\text{electron events without coincidence}}{\text{electron events with coincidence}} = \frac{\sigma_e - \sigma_{ep}}{\sigma_{ep}} = \frac{\sigma_{en}}{\sigma_{ep}}. \qquad \text{...[53]}$$

This ratio is free from the bad statistical problems of the non-coincidence experiment. Moreover, it is better by a factor of 3, theoretically, because we no longer subtract a large number (σ_{ep}) from an uncorrected (σ_{ed}).

The ability to obtain numbers independent of the theory is improved even more because the final state interactions tend to depress or accentuate both numerator and denominator together. If there were no charge exchange, this would be exactly true; with charge exchange, it would be true for $\sigma_{ep} = \sigma_{en}$. Thus we expect only about $\frac{1}{4} = \frac{1}{2} \times \frac{1}{2}$ of the final state interaction correction applicable to σ_{ep} or σ_{en} independently to be applicable to the ratio. This method needs a good duty cycle and is, therefore, *particularly* suitable for synchrotrons but *not* for linear accelerators. Fig. 15 presents some very recent Harvard data, with statistical errors only.

The third method, also a coincidence method, involves the measure of *en* coincidences directly.[50] This method, like the last, avoids both statistical and theoretical problems. However, we must use a neutron counter. This is very background sensitive and it is difficult to measure its efficiency. Thus at the moment this method is used at low momentum transfers only. Table III shows the results obtained.

There have been many calculations giving refinements to the simple theory stated above. We note that again, because of the generality of the angular

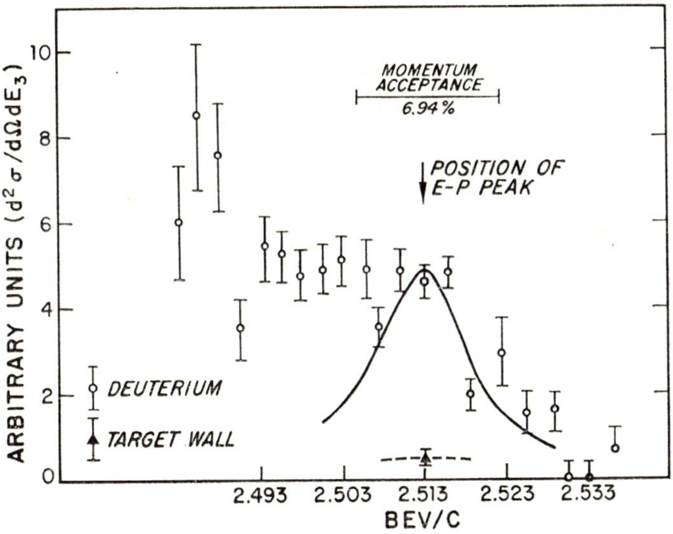

FIG. 13. Inelastic electron momentum spectrum from deuterium at $q^2 = 2{\cdot}92$ (BeV/c)^2 and $\theta = 31°$.

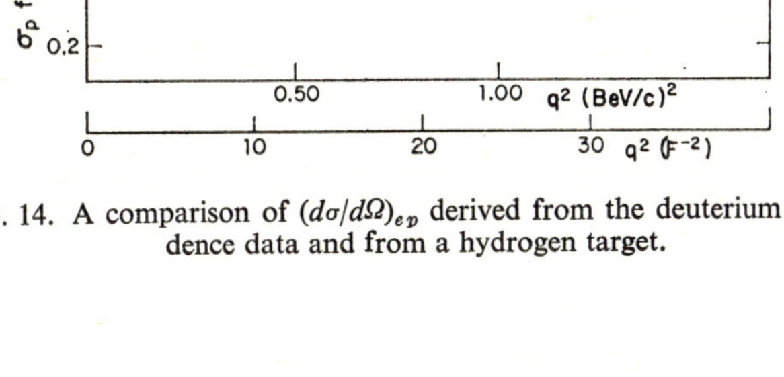

FIG. 14. A comparison of $(d\sigma/d\Omega)_{ep}$ derived from the deuterium coincidence data and from a hydrogen target.

TABLE III‡

The neutron form factors are given as a function of the four-momentum transfer q^2. For comparison the values for G_{Mn} from Hughes *et al.*[45] are given.

$q^2(f^{-2})$	G_{En}^2 This expt.	G_{Mn} This expt.	G_{Mn} Hughes[45]
5·5	$+0{\cdot}0017 \pm 0{\cdot}0070$	....	....
10·0	$-0{\cdot}0037 \pm 0{\cdot}0141$	$-0{\cdot}825 \pm 0{\cdot}064$	$-0{\cdot}825 \pm 0{\cdot}081$[a]
14·5	$-0{\cdot}0055 \pm 0{\cdot}0088$	$-0{\cdot}638 \pm 0{\cdot}042$	$-0{\cdot}650 \pm 0{\cdot}072$[a, b]

‡ Table I of reference 50.
a Error includes 5% theoretical uncertainty.
b This value has been determined by extrapolation.

distribution, we can separate into longitudinal and transverse parts. The final state interactions will clearly be different in these two parts. With no spin flip (longitudinal scattering) we ask for interaction in ^3S, ^1P states and with spin flip (transverse scattering) in ^1S, ^3P, etc. states.

The corrections have been evaluated by various authors who do not quite agree.[51, 52, 53, 54] The corrections which remain uncertain are related to calculations relating to the D state part of the wave function. These lead to uncertainties of 1% or 2%. The final state interaction corrections (assuming S states alone) of Nuttall and Whippman seem the best and are also about 2% to 3%. The calculations of McGee[53] for the D state give a peculiar kinematic peak at high momentum transfers (which experimentally does not occur).

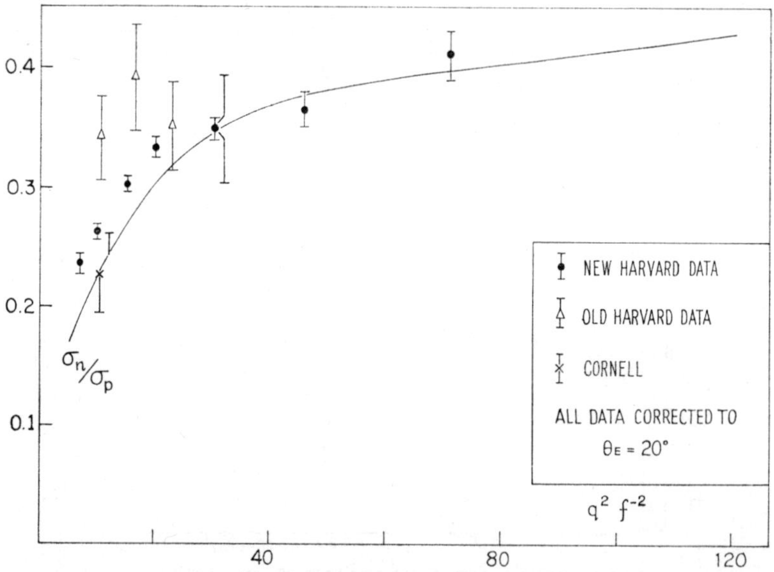

FIG. 15. σ_{en}/σ_{ep} for small angle electron scattering; the new Harvard points are preliminary and include only statistical errors. The line is from equations [*71*], [*72*] and [*73*].

In the theory of the deuteron, the whole question of relativistic corrections remains uncertain. Some calculations—such as McGee's—predict certain phenomena which seem physically implausible. Bosco and collaborators[54A] find a different sign for the correction (using a dispersion theory approach) than do Durand and his associates[51,53] using a potential theory. In principle the two methods should agree. (Bosco[54B] in another calculation on *np* capture claimed a similar improvement over a potential calculation but included fewer terms than the refined potential calculations.[35,55A] He may have done the same here.) Therefore only non-relativistic calculations are used to understand the data (but with relativistic kinematics), and Nuttall and Whippman's correction. The data are summarized in Figs. 15, 16 and 17.

Before we come to the theory of the form factors, we note that they proceed very smoothly and that there remain some statistical, and other, errors. See ref. 55 for a discussion of the systematic discrepancies between the data. We note that G_{Ep}, G_{En}, G_{Mp}, G_{Mn} all seem to go to zero at high momentum transfers. We also have good values for dG/dq^2 at $q^2 = 0$.

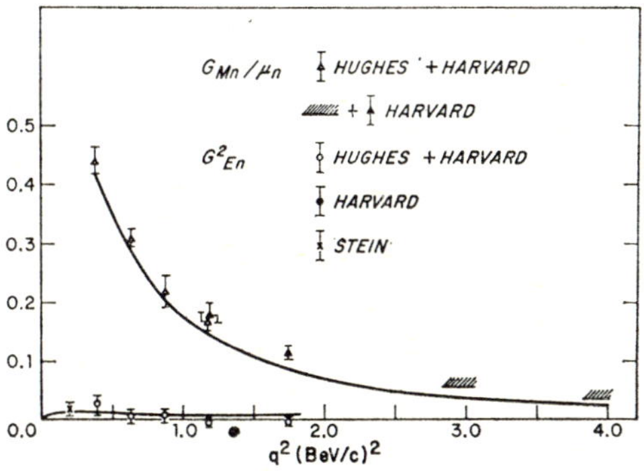

FIG. 16. The neutron electromagnetic form factors.

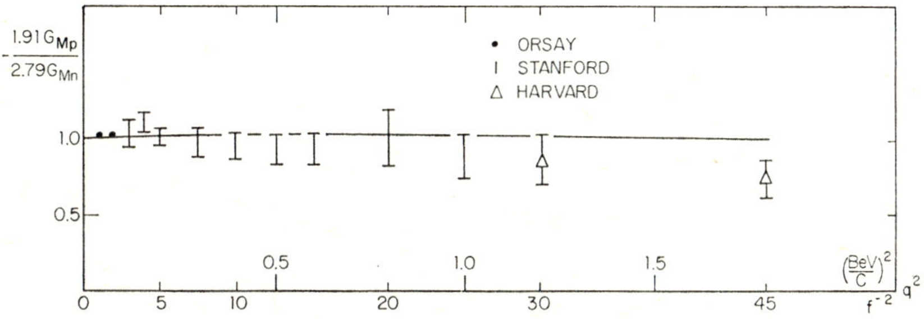

FIG. 17. The ratio ($-1{\cdot}91\ G_{Mp}/2{\cdot}79\ G_{Mn}$) versus q^2. The solid line is the 4-pole fit of equations [*64*], [*65*], [*67*] and [*68*].

4

THEORY OF FORM FACTORS

4.1. *Pion form factor*

Before we discuss the theory of *nucleon* form factors, let us discuss that of the pion. We assume $F_\pi(q^2) \to 0$ as $q^2 \to \infty$. Then we can write a simple dispersion relation[56] for $F_\pi(q^2)$ assuming that the only singularity is a cut from $q^2 = -4M_\pi^2$ to $q^2 = -\infty$. We have the following domain of analyticity (see Fig. 18).

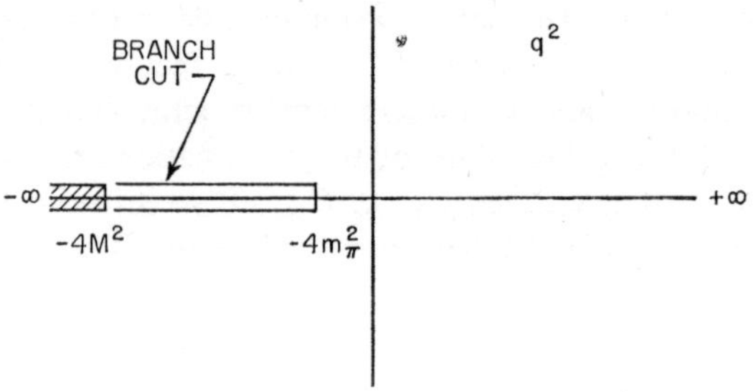

FIG. 18. Domain of analyticity of the pion (and nucleon) form factors.

The form factor at a point q^2 can be found by applying Cauchy's theorem to the function

$$\frac{F(q'^2)}{q^2 - q'^2}$$

where q^2 is now considered complex.
Then

$$F(q^2) = \frac{1}{2\pi i} \int \frac{F(q'^2)\, dq'^2}{q'^2 - q^2} \qquad \text{...[54]}$$

where the contour is the real axis and semicircle at infinity (where $F(q'^2)$ vanishes by experiment). Then,

$$F(q^2) = \frac{1}{2\pi i} \lim_{\varepsilon \to 0} \int_{-\infty}^{+\infty} \frac{F(q'^2)\, dq'^2}{q'^2 - q^2 - i\varepsilon}. \qquad \text{...[55]}$$

Using

$$\frac{1}{x - i\varepsilon} = P\left(\frac{1}{x}\right) + i\pi\delta(x) \qquad \text{...[56]}$$

we have

$$F(q^2) = \frac{1}{i\pi} P \int_{-\infty}^{\infty} \frac{F(q'^2)\, dq'^2}{q'^2 - q^2}. \qquad \text{...[57]}$$

This gives

$$\text{Re}\, F(q^2) = \frac{P}{\pi}\int_{-4M_\pi^2}^{-\infty} \frac{\text{Im}\, F(q'^2)dq'^2}{q^2-q'^2} \qquad ...[58]$$

since $\text{Im}\, F(q^2) = 0$ for $q^2 > -4M_\pi^2$.

Now if the $\pi\pi$ interaction is dominated by a resonance, we can approximate $\text{Im}\, F(q^2)$ by a resonance. We find a Breit-Wigner form for $F(q^2)$

$$F(q^2) \simeq \frac{M_\rho^2}{M_\rho^2+q^2-i\Gamma M_\rho}, \qquad ...[59]$$

neglecting a small term in the numerator.

We know of only one such resonance, the ρ meson, with parameters

$$\left.\begin{aligned} M_\rho &= 765\pm3 \text{ MeV} \\ \Gamma_\rho &= 124\pm4 \text{ MeV.} \end{aligned}\right\} \qquad ...[60]$$

We will assume that this is the correct form for the π form factor. It must be considered an approximation to a correct treatment of the $\pi\pi$ interaction, but looks like a use of a diagram (see Fig. 19a). Indeed Gell-Mann *et al.*[57] talk about the coupling constants in such a diagram, although, as drawn, it is not even gauge invariant. Gell-Mann[58] justifies this as a limit of a dispersion calculation. We note that this is an unsubtracted relation; this is because of the Harvard experiments that say that all form factors tend to zero as $q^2 \to \infty$.

4.2. *Nucleon isovector form factors*

Now for the isovector nucleon form factors we again expect the ρ meson to be dominant. We may then guess:

$$2G_{MV}(q^2) = (\mu_p-\mu_n)\frac{M_\rho^2}{M_\rho^2+q^2} \qquad ...[61]$$

$$2G_{EV}(q^2) = \frac{M_\rho^2}{M_\rho^2+q^2} = 1-\frac{q^2}{M_\rho^2}+\ldots \qquad ...[62]$$

although we do *not* know whether we should express G_M and G_E or F_1 and F_2 in this simple form. However, it is easy to see that this simple approach gets into trouble very fast.[2,15,55] Firstly, equations [61] and [62] do *not* satisfy $G_{EV} = G_{MV}$ at $\tau = -1$, which is necessary if we wish to keep F_{1V} and F_{2V} finite. Secondly, the slope is

$$2\frac{dG_{EV}}{dq^2} = -\frac{1}{M_\rho^2}, \qquad ...[63]$$

which is only one-half the measured slope.

There are suggestions to get around this dilemma:

(i) Assume another vector meson (ρ') at a mass 900 MeV-1100 MeV. This is unsatisfactory because it should have been found. (See references 55, 69.) It should also fit somewhere in $SU(6)$ but it doesn't. This fit gave

$$2G_{EV} = \frac{2{\cdot}35}{1+q^2/14{\cdot}9} - \frac{1{\cdot}85}{1+q^2/19{\cdot}7} \qquad \text{...[64]}$$

$$2G_{MV} = \frac{8{\cdot}27}{1+q^2/14{\cdot}9} - \frac{5{\cdot}91}{1+q^2/19{\cdot}7}. \qquad \text{...[65]}$$

(ii) Put in an *ad hoc* form factor $f_{\rho NN}$ to fit. This should be slightly different in the two spin states. Clearly, it is about $1/(1+q^2/\Lambda^2)$ where $\Lambda \sim$ 900-1100 MeV. Cocho[59] and Massam[60] have adopted this approach. It is effectively the same as the first because of the algebraic identity

$$\frac{(B^2-A^2)}{(q^2+A^2)(q^2+B^2)} = \frac{1}{(q^2+A^2)} - \frac{1}{(q^2+B^2)}. \qquad \text{...[66]}$$

(iii) Try to understand the terms we have left out. These are diagrams of the form shown in Fig. 19b.

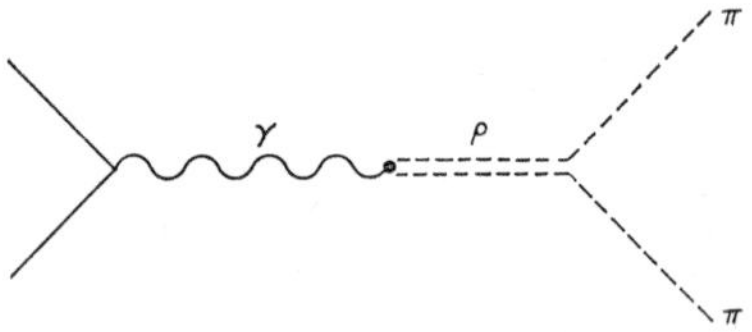

FIG. 19*a*

This illustrates the equation by which the pion-pion interaction (ρ meson) dominates the pion form factor.

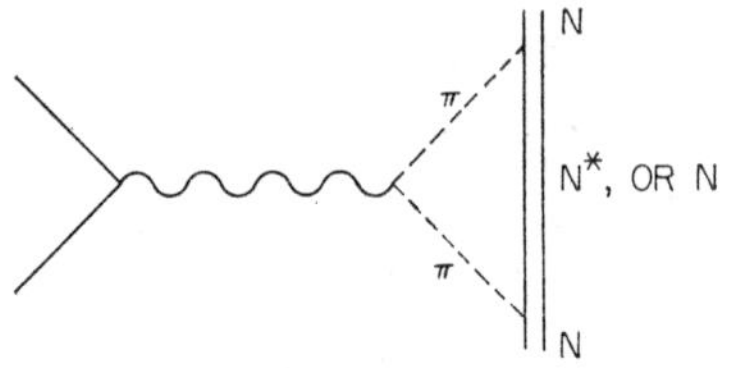

FIG. 19*b*

This illustrates processes not included in the vector model (but included in reference 61).

Now Furuichi and Watanabe[61] have calculated such a diagram, using the dispersion theory (CGLN) for the ρNN^* vertex, using the constants (including an arbitrary constant for high energy effects) given by Bowcock *et al.*[62] They claim a good fit. (See Figs. 20a and 20b.) This is to my mind an important observation and is probably the solution to the problem.

4.3. *Nucleon isoscalar form factors*

I now leave the isovector form factors and come to the isoscalar form factors. We now have no simple diagram (Fig. 21) to describe the non-resonant part of the interaction. Historically, it was hoped five years ago that the ω meson would describe the data but we have the same problem as

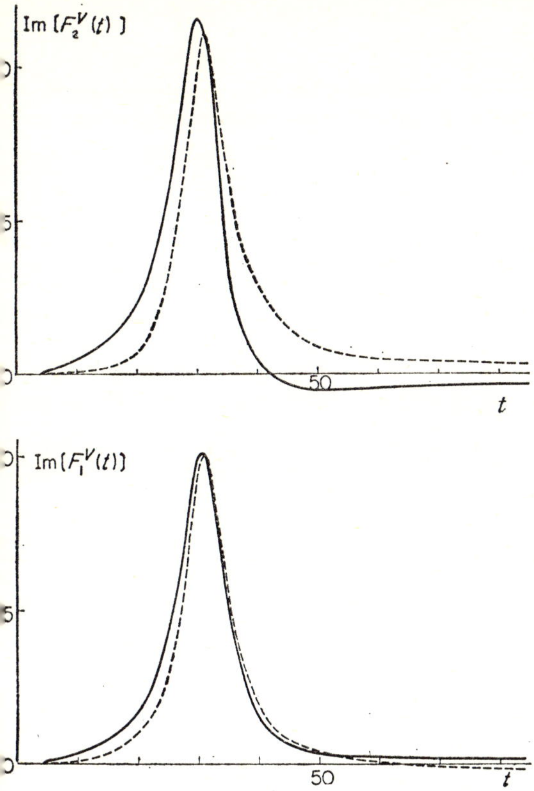

FIG. 20*a*

The form factors in the time-like region from the resonance contribution (dotted) and the *total* contribution (solid), as given in reference 61.

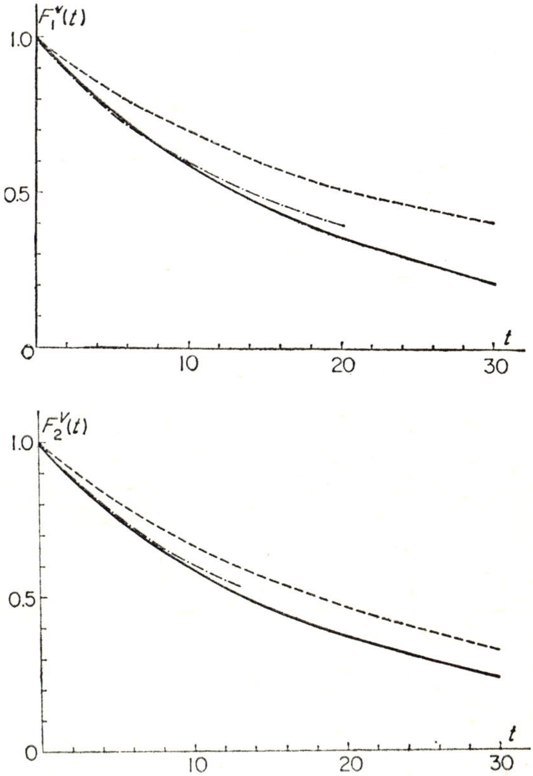

FIG. 20*b*

The form factors in the space-like region from the resonance contribution (dotted) and the *total* contribution (solid), as given in reference 61.

in the isovector mesons. Then Hand, Miller and Wilson[2] introduced an *ad hoc* additional term. This was later identified with the ϕ meson and a fine fit develops which together with the *ad hoc* isovector fit above, fits all the data.

$$2G_{ES} = \frac{1 \cdot 21}{1+q^2/15 \cdot 7} - \frac{0 \cdot 71}{1+q^2/26 \cdot 7} \qquad \ldots [67]$$

$$2G_{MS} = \frac{1 \cdot 09}{1+q^2/15 \cdot 7} - \frac{0 \cdot 65}{1+q^2/26 \cdot 7}. \qquad \ldots [68]$$

The trouble is that theorists do not like the numbers one gets. In particular, the ϕ and ω must contribute in opposite sign to the form factors, if there is no other contributor, in contrast to the $SU(3)$ prediction.

We should note that in deriving fits such as equations [*64*], [*65*], [*67*] and [*68*], we must fit to *cross-sections* (not form factors). This is more

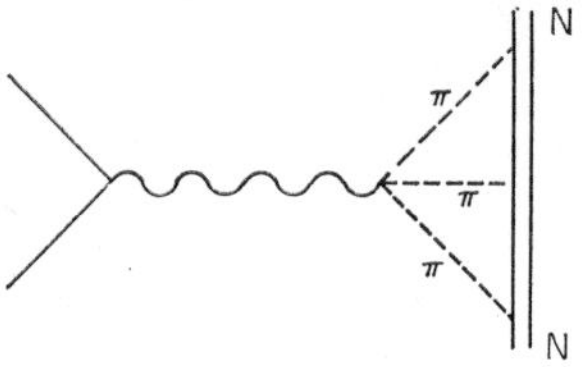

FIG. 21

Diagram for the isoscalar form factors for uncorrelated π mesons which, unlike Fig. 19*b*, cannot be simple calculated.

restrictive, but we get coupled errors. The fit quoted above has a χ^2 of 400 with 200 degrees of freedom; the failure to fit well comes from systematic differences between the data of different laboratories. One reason for the high value of χ^2 is that the estimates of systematic errors were *not* taken into account in the fit.

An ω_0 is supposed to be a member of the same octet as the ρ and ϕ_0 a singlet at the same mass, and the coupling of the ω_0 to gamma rays, to be used with a single diagram such as Fig. 21, can be derived from the $\rho\gamma$ coupling. The ϕ_0 does *not* couple to gamma rays. The coupling of the real ω and ϕ is obtained from the $\omega_0\gamma$ coupling using the theory of $\omega\phi$ mixing.[63] The mixing angle in turn is derived from a necessity to fit a mass formula of assumed form. Thus, we get

$$f_{\omega\gamma} = f_{\rho\gamma} \cos\theta \qquad \ldots [69]$$

$$f_{\phi\gamma} = f_{\rho\gamma} \sin\theta. \qquad \ldots [70]$$

Now Okubo has come up with a formula for relating $f_{\omega NN}$ and $f_{\phi NN}$ to $f_{\rho NN}$ (which are each two quantities, one each for spin flip and no spin flip). Thus, however uncertain, we have an $SU(3)$ prediction. To make this fit the data, however, we have the same problem as for the isovector form factors. Roughly speaking, the effect of the mixed ω and ϕ on the form factors is the same as of a single unmixed ω_0 of the same octet as the ρ meson. Cocho[59] and Zichichi[60] add an extra multiplicative term $(1+(q^2/\Lambda^2))^{-1}$ to the isoscalar form factors as already discussed for the isovector form factors. This can have no such simple justification, such as Furuichi and Watanabe gave for the corresponding isovector term, so it remains *ad hoc*. It is interesting, however, that it is the same size as the corresponding term for the isovector case (as first shown by Hand, Miller and Wilson[2]).

We now return to general symmetry suggestions. We noted that, at least approximately,

$$\frac{G_{Mp}}{\mu_p} = \frac{G_{Mn}}{\mu_n} \qquad ...[71]$$

$$\frac{G_{Mp}}{\mu_p} = G_{Ep} \qquad ...[72]$$

$$G_{En} \simeq 0. \qquad ...[73]$$

These have suggested to theorists that there may be some symmetry rule. I am no expert on $SU(6)$, though I gather that some versions predict all three of these; however S. Coleman tells me that the only *consistent* version of $SU(6)$ fails to predict the second.

Of course, the symmetry laws probably do not predict *exact* rules, so the fact that the vector meson fits to form factors do not fit [*71*], [*72*] and [*73*] exactly need not disturb us.

A recent paper from Trieste[64] suggests, in their version of $SU(6)$, that the simple pole description of form factors is only true if multiplied by an additional term

$$(1+(q^2/4M^2))^{7/2} \simeq 1+\tfrac{7}{8}(q^2/M^2). \qquad ...[74]$$

This has the same size (Λ = 1000 MeV) as the *ad hoc* term introduced earlier. However, the exact form also makes G_E and G_M *vanish* at $q^2 = -4M^2$ avoiding completely the problem of making them equal at this point.

4.4. *Relation to pp scattering*

The size of the proton as measured by electron scattering is given by the first term in the expansion of the form factor

$$G(q^2) = 1-\tfrac{1}{6}\langle r^2\rangle q^2 \qquad ...[75]$$

where $\langle r^2\rangle^{\frac{1}{2}} = 0{\cdot}813$f, for both electric and magnetic form factors G_E and G_M. It is interesting to note that this is close to the size of the proton as deduced from *pp* scattering

$$\sigma(\theta)/\sigma(0) = 1-(2/6)\langle r^2\rangle q^2. \qquad ...[76]$$

The factor 2 is put in to allow for the mutual interactions of the two meson clouds. We see from *pp* scattering data that

$$\langle r^2\rangle^{\frac{1}{2}} = 0{\cdot}8-1{\cdot}0\text{f}. \qquad ...[77]$$

The near equality of the r.m.s. radii derived from electron-nucleon and nucleon-nucleon scattering has been observed before.

We now turn to a conjecture of Wu and Yang.[65] They note that at large momentum transfers *pp* scattering may be expressed as

$$\sigma(\theta)/\sigma(0) = \exp(-P_T/P_0) \qquad ...[78]$$

where P_T is the transverse momentum and $P_0 = 0{\cdot}15$ BeV/c. They attribute the exponential to a sticking probability of the proton. They guess that for *ep* scattering a similar relation holds for the form factors, but we must take the fourth root. Thus, with q^2 in $(\text{BeV/c})^2$, we have

$$G_{Ep}(q^2) = G_{Mp}(q^2)/\mu_p = \exp[-(q^2)^{1/2}/0{\cdot}6] \qquad ...[79]$$

which is valid at large momentum transfers only.

The fit to the form factors G_{Ep} and G_{Mp} is shown in Fig. 22. The model is not expected to fit near $q^2 = 0$ and, indeed, the analyticity properties are not quite right. $G(q^2)$ should have no imaginary part in the region $-4M_\pi^2 < q^2 < 0$, in contrast to equation [79]. We could make the analytic properties work out correctly by writing

$$G_{Ep}(q^2) = \{\exp[-(q^2+4M_\pi^2)^{1/2}/0{\cdot}6]\}/\exp(-2M_\pi/0{\cdot}6). \qquad ...[80]$$

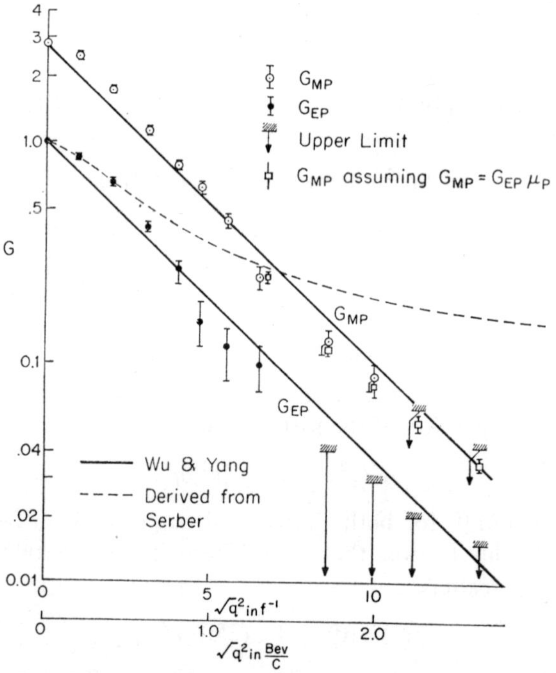

FIG. 22. The form factors G_{Mp} and G_{Ep}, compared with the simple model of Wu and Yang (equation [79]) and a model derived from Serber's fit to *pp* scattering (equation [84]). We see that the form factors *do not* fall as fast as we would predict.

It fits quite well at higher momentum transfers though there is a tendency for the model to fall *faster* than the data. This is accentuated if we insist that $G_{Mp} = \mu_p G_{Ep}$ and therefore have no error in the form-factor separation. The new DESY data, if correct, fit [*79*] better than the Harvard data.

The formula [*79*] gives an oscillating behaviour for the form factor in the time-like region; this is equivalent to choosing many vector resonances whose coupling constants have the same magnitude but whose signs alternate as a function of q^2.

Another approach can be based on the analysis of the *pp* scattering by Serber.[66] Serber notes that the amplitude is mainly imaginary and that therefore many cancellations can occur. He finds that the major features of the experiment can be summarized by assuming an imaginary interaction

$$V(r) = i\hbar c \exp(-\lambda r)/r, \text{ with } \lambda = 2{\cdot}12\,\mathrm{f}^{-1}. \qquad \ldots[81]$$

We may now guess at the consequences of this approach for *ep* scattering. We assume that this imaginary potential is caused by a cloud of strongly interacting particles, interacting at short range, and that most of these particles have a charge. Then we might attribute a charge and magnetic moment distribution equal to the potential

$$\rho(r) = \exp(-\lambda r)/r \qquad \ldots[82]$$

$$G(q^2) = \lambda^2/(q^2+\lambda^2). \qquad \ldots[83]$$

Again, however, we must allow for the fact that in *pp* scattering there are two interacting clouds and that the form factor of equation [*83*] is the product of the two nucleon form factors. We must thus take the square root of [*83*] to obtain

$$G_{Mp}(q^2)/\mu_p = G_{Ep}(q^2) = \left(\frac{\lambda^2}{q^2+\lambda^2}\right)^{1/2} = 1-\tfrac{1}{2}q^2/\lambda^2+\ldots, \qquad \ldots[84]$$

which gives a radius of $0{\cdot}67 f^{-1}$ which is about right but falls off too slowly at high momentum transfer. This is shown in Fig. 22.

The difference in the behaviour at high momentum transfers from the idea of Wu and Yang is that Serber attributes the cancellations giving the small *pp* cross-section at high momentum transfer entirely to the phase, which is imaginary; these cancellations do not then occur in *ep* scattering which has a real amplitude. It is clear that this connection is still confused and that more precise data may lead to further understanding.

It might be thought that the annihilation experiments prevent our acceptance of the ideas of Wu and Yang, but their ideas represent an average; they merely demand several resonances and that the *average* form factor is unity. The annihilation experiments are at two points only and could fall accidentally between two resonances. However, as we shall now see, there exists an experiment[68] which rules out this possibility.

5

FURTHER EXPERIMENTS ON FORM FACTORS

5.1. *Leptonic decays of vector mesons*

It would be nice to test the interactions $\rho \to \gamma$, $\phi \to \gamma$ and $\omega \to \gamma$, and to search for other mesons and couplings. This has been done in two cases—for the ρ and the ω meson. There are two experiments[67, 68] measuring the $\rho\gamma$ coupling. The apparatus is shown in Fig. 23.

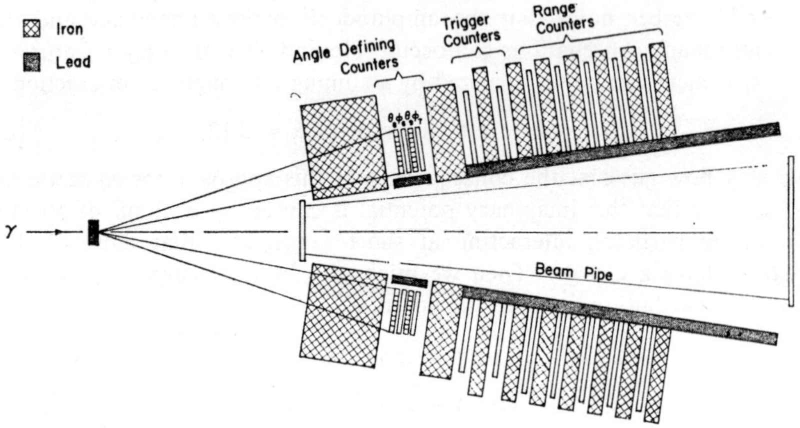

FIG. 23. Arrangement of apparatus for μ pair experiments (reference 67).

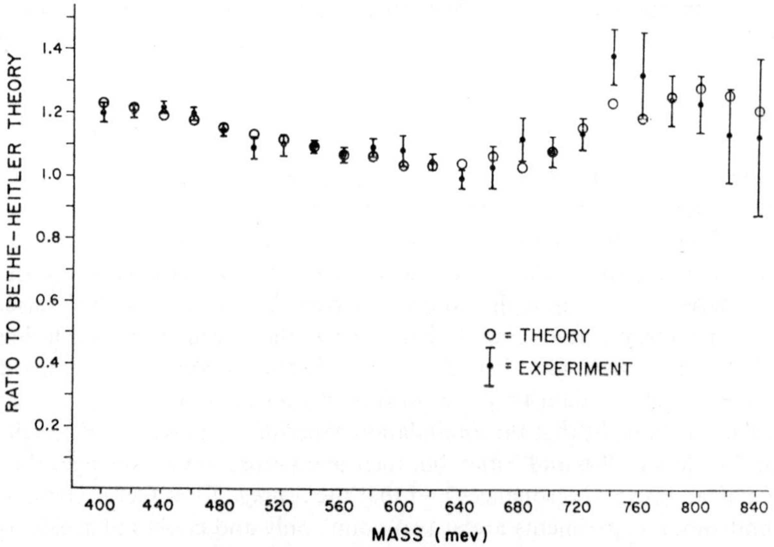

FIG. 24. Invariant mass spectrum of μ pairs. Both experiment and theory are normalized to Bethe-Heitler Theory (reference 67).

In the first of these ρ mesons are photoproduced and the decay into two μ mesons is found by measuring the mass of the 2μ system from the μ meson range and opening angle. This experiment is a by-product of an experiment to measure the photoproduction of μ pairs by the ordinary Bethe-Heitler process at large angles. This will be discussed as a test of quantum electrodynamics later, though we note here that the mass of the 2μ system is the mass of the virtual muon line in such a calculation. The mass plot is shown in Fig. 24. It peaks at the ρ mass although the ρ decay muons are superimposed on a background of Bethe-Heitler produced μ pairs.

Fortunately, there should be no interference terms, according to theory, because the Bethe-Heitler amplitude is wholly real and the ρ photoproduction amplitude imaginary. There is other data[69] on the photoproduction of ρ mesons. This is obtained by examining the π pair production which is, incidentally, a dominant photoproduction process. This π pair production peaks with little background at the ρ mass. We, therefore, derive the result

$$B = 4{\cdot}4^{+1{\cdot}6}_{-0{\cdot}7} \times 10^{-5}. \qquad \ldots[85]$$

In Wehmann *et al.*'s experiment,[68] 12 BeV π mesons were allowed to fall on carbon (in a carbon plate spark chamber), see Fig. 25. The triggering event was the observation of two high energy muons. Then the range of the muons was measured by their range in steel together with the opening angle measured in the carbon plate spark chamber. The mass resolution is 8%.

Fig. 26 shows the mass plots for all events with a small (~5%) background, from π mesons decaying in flight, not yet subtracted. We note the ρ peak, with no sign of a ρ' meson above the peak. If this were to have the properties desired to fit isovector form factors we would expect the imaginary part of the form factors to change sign as in Fig. 27a and the contribution to

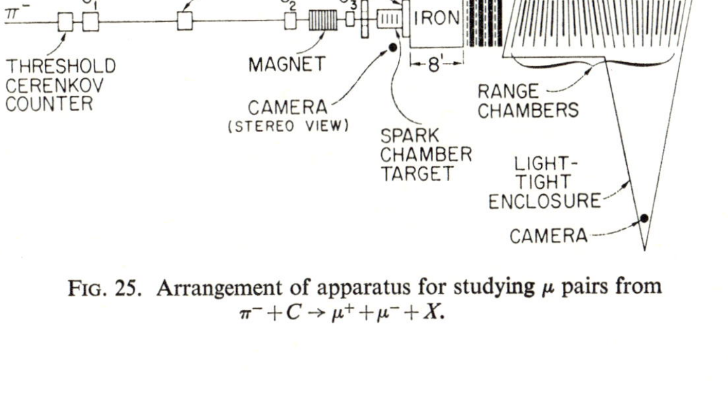

FIG. 25. Arrangement of apparatus for studying μ pairs from $\pi^- + C \rightarrow \mu^+ + \mu^- + X$.

this experiment to be as in Fig. 27b. For these events, where no other particle comes from the vertex, we can say that there was energy conservation. Now we can use the information on the beam energy (±1%) and the inaccurate range determination may be ignored. Then we get overall 4%, which is better.

We can compare with two experiments on production of ρ mesons by 12 BeV pions[69A, 69B] finding the branching ratio

$$B = (4{\cdot}3 \pm 1{\cdot}4) \times 10^{-5}. \qquad \text{...[86]}$$

These two are to be compared with the theoretical prediction from [*59*], which is the pion form factor dominated by the ρ meson;

$$B = \frac{\alpha^2}{36}(M_\rho/\Gamma_\rho)^2(0{\cdot}95) = 5{\cdot}6 \times 10^{-5} \qquad \text{...[87]}$$

for $\Gamma_\rho = 124$ MeV, $M_\rho = 765$ MeV. If we are to assume the existence of a ρ' meson (see equations [*64*] and [*65*]), or a failure of the photon propagator (see section 9.9), or a variation of the $\rho\gamma$ coupling with momentum transfer, the predicted value of B increases to between 10^{-4} and 2×10^{-4}.

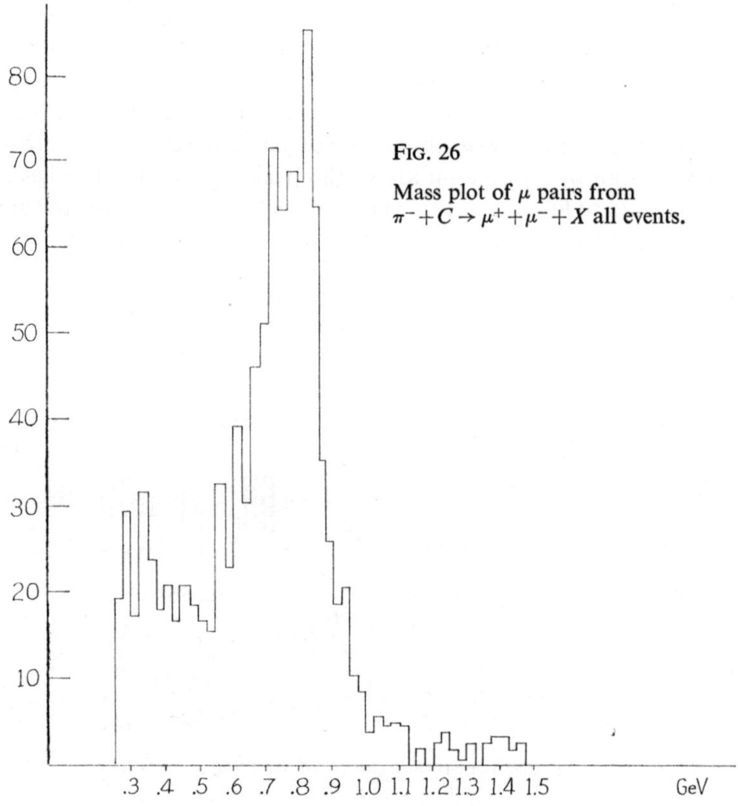

FIG. 26

Mass plot of μ pairs from $\pi^- + C \rightarrow \mu^+ + \mu^- + X$ all events.

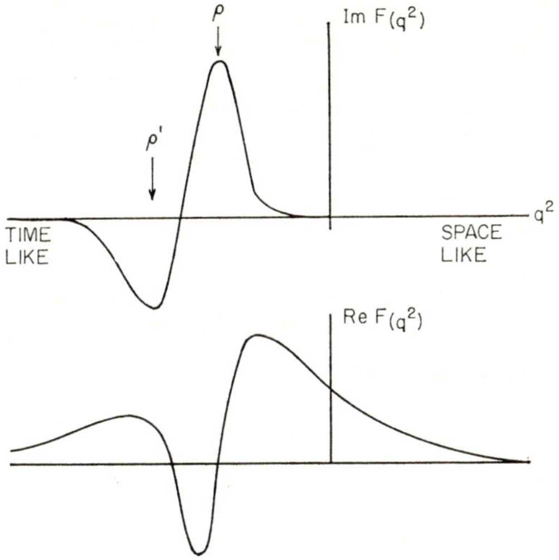

FIG. 27*a*

This illustrates qualitatively how two resonances ρ and ρ' should appear in the form factor.

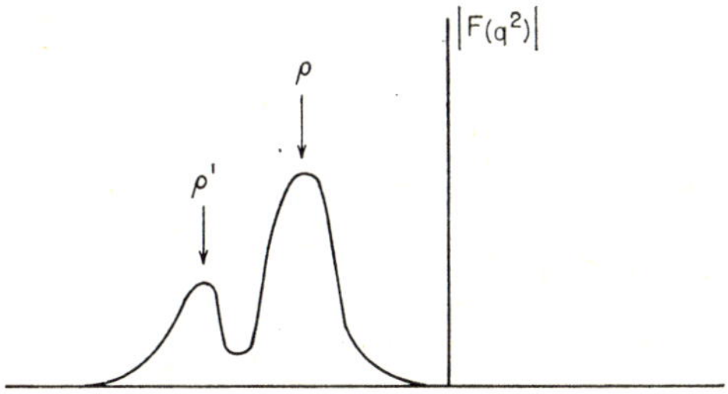

FIG. 27*b*

This illustrates how two resonances ρ and ρ' should appear in the mass plot.

Another powerful feature of this experiment is that no appreciable sign is found of a $J^P = 1^-$ resonance above the ρ meson. This excludes the possibility that the Wu-Yang theory is right and that the annihilation experiments accidentally sat between two resonances.

The twin factors: (i) the correct branching ratio, (ii) no μ pairs above the peak, confirm the ρ dominance of the pion form factor and force us to search in such a way as Furuichi and Watanabe, for things other than new resonances in the nucleon form factors. Indeed, the second of these experiments, if viewed as a one pion exchange peripheral process (as the ρ production experi-

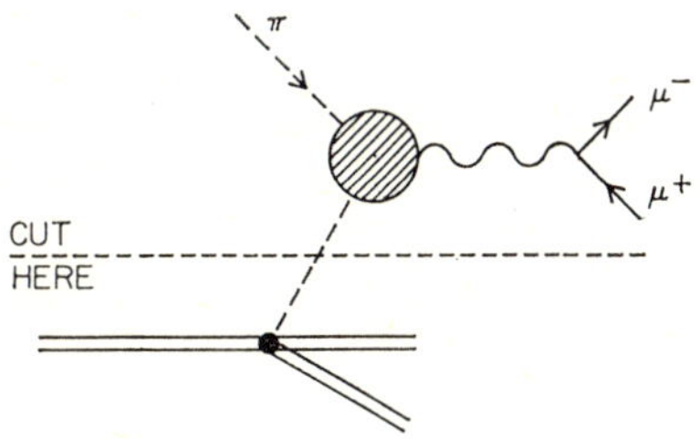

FIG. 28

This shows how the μ pair experiment measures the π form factor if we assume one pion exchange.

ments suggest), tells us *directly* the pion form factor in the time-like region, as we see from the diagram, Fig. 28.

D. M. Binnie *et al.* find a branching ratio for the ω to two electrons which lies between 5×10^{-5} and 60×10^{-5}. This also agrees with the $SU(3)$ prediction from the ρ dominance of 10×10^{-5}. This needs to be improved to be useful and we want to find the ϕ branching ratio too. This can be done most generally by colliding beam experiments which we shall now discuss. Indeed, these last two experiments[67, 68] are merely nibbling away at the colliding beam field where good experiments are imminent.

6

Colliding Beam Devices

6.1. *Colliding beams for form factors*

The question of colliding beam experiments now naturally arises. At this point I will discuss collisions between electrons and positrons. It is clearly an interesting thing to do. The experiments just discussed on the ρ, ω and ϕ decays into leptons are nibbling away at the field that properly belongs to colliding beams. For example, we avoid a discussion of the process of producing the ρ meson if we study the inverse process

$$e^+ + e^- \to \rho^0 \to \pi^+ + \pi^- .$$

Provided we can adequately monitor the colliding beams, we can measure the $\rho \to \gamma$ coupling directly. Alternatively, this is a *direct* measure of the pion form factor in this time-like region without any assumptions of one pion exchange as made by Wehmann *et al.*[68] A calculation of many of the processes one can measure is given by Cabibbo and Gatto[72] and in a recent review by Gatto.[73]

6.2. *Technical status of colliding beams*

Let us start with the present experimental status. In Table IV a list of the leptonic colliding beams projects on which work is now proceeding is given.

The colliding beams were, I believe, first suggested—as were many other things—by Wideroe about 25 years ago. Then, in 1956, G. K. O'Neill[74] suggested proton-antiproton colliding beams. Although at CEA Ramsey and myself considered them, the time was not ripe and the first active storage ring was proposed in 1958 by O'Neill and built at Stanford.

In order to keep electrons continuously circulating in a ring it is necessary to provide RF power to make up for the energy radiated by synchrotron radiation. At low energy and intensities this is easy, but at high energy and intensities this can be very expensive and is a practical limit to the intensity available. This radiation, however, performs a useful task in that vertical oscillations are damped and the beams become small.

TABLE IV

Location	Energy (each beam)	Luminosity	Working
e^-e^- rings			
Stanford (Mk III)	300 to go to 500 MeV	$\sim 10^{29}$ cm^{-2} sec^{-1}	1964
Novosibirsk (WEP I)	150 MeV	$\sim 10^{28}$ cm^{-2} sec^{-1}	1964
e^+e^- rings			
Frascati-Orsay (Ada)	250 MeV	$\sim 10^{27}$ cm^{-2} sec^{-1}	1962
Orsay	510 MeV	$\sim 10^{29}$ cm^{-2} sec^{-1}	1966
Novosibirsk (WEPP II)	750 MeV	$\sim 10^{28}$ cm^{-2} sec^{-1}	1965-6
Frascati (Adone)	1500 MeV	$\sim 10^{29}$ cm^{-2} sec^{-1}	1966-7
Cambridge (By-Pass)	3-4 BeV	$\sim 10^{31}$ cm^{-2} sec^{-1}	1968-9
Moscow (Lebedev)	250 MeV	$\sim 10^{27}$ cm^{-2} sec^{-1}	1968
Stanford (SLAC)	3 BeV	$\sim 10^{32}$ cm^{-2} sec^{-1}	Proposal

The value of a colliding beam for experiments is primarily determined by the luminosity—which is simply defined as the counting rate divided by the cross-section for the process being counted. The luminosity clearly depends

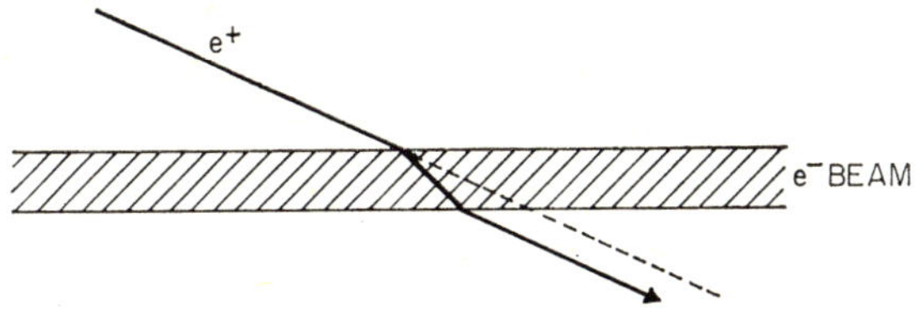

FIG. 29. This shows how the electrostatic attraction of a positron by the electron beam changes a trajectory and hence the betatron frequency ν.

directly upon the interacting currents, the size of the beam and the length, height and width (l, h and w) of the interaction region;

$$L \propto \frac{i_1 i_2 l}{hw}. \qquad \ldots[88]$$

Now it was at first thought that the vertical height h, of the beam could be made very small by the damping due to the synchrotron radiation. However, there are limits. At low energies, if we allow h to get too small, an instability, called the Touschek effect, is produced, which was first found in

Ada. This is transfer of energy from betatron oscillations to longitudinal (synchrotron) oscillations by electron-electron collisions in the same beam resulting in the loss of the particles.

At higher energies and intensities other effects occur.

1. There is a basic space charge limit due to the defocusing of a particle in one beam by the other. The change in trajectory is illustrated in Fig. 29. If the defocusing is enough to change the betatron wavelength from a stable to an unstable region, the beam will be lost. Clearly this effect *also* increases as i_1 and as the beam size. This was calculated first by O'Neill for the e^-e^- case and by Amman and Ritson.[76] We can relate the luminosity to the change in betatron frequency by a simple relation

$$L = \frac{i}{e}\frac{1}{r_e}\frac{\gamma}{\beta}\Delta\nu. \qquad ...[89]$$

i/e is the rate of particles crossing the interaction region, r_e is the classical electron radius, $\gamma = E/mc^2$, and β is the β function of the magnet system which is smaller the stronger the focusing.

In order to increase L, without increasing $\Delta\nu$, we can increase i (but now L varies only linearly as i because we change other beam parameters to keep $\Delta\nu$ constant) or we can change the focusing characteristics. The strong focusing machines have a higher luminosity for the same current as given by the small β factor. Clearly the *stronger* the focusing at the interaction region the smaller may be the beam for the same $\Delta\nu$, and hence the greater luminosity. Unfortunately, the strength of focusing available is limited by many practical matters. The first design, by Courant, Livingston and Snyder, of a strong focusing accelerator had $\nu = 100$ and an aperture of one inch diameter, but we must inject into such a machine and we need a larger aperture. Adone, for an example, will run with $\nu \simeq 3{\cdot}3$. The vastly improved luminosity planned with the CEA By-Pass[77] (which converts the CEA into a storage ring) is obtained by realizing that the aperture is *not* necessary after injection because the beam has damped down (by synchrotron radiation) to a small size. So after injection the beam may be switched to a stronger focusing region. This scheme can, of course, be used with other rings.

2. There are many instabilities which can occur. See for example the report of a Summer Study at Stanford.[75] The beam induces currents in the walls of the vacuum chamber. These currents, because of the resistance of the walls, will persist and produce fields (wake fields) after the beam has passed. These fields can then act on the beam during the next revolution. The phase is such that an oscillation can be set up. This oscillation is coherent for all particles in the beam and is clearly worse for thin beams and high currents.

Fortunately, we now believe that all these instabilities can be controlled by various means. This is not the place for detail but we can summarize. Any coherent instability can be controlled by sensing the instability with pickup electrodes and applying a feedback in the right phase. This has been done at MURA and BNL. The incoherent instabilities are assumed to be connected with the change $\Delta\nu$, and experimental data can probably be transferred between different storage rings for similar $\Delta\nu$. Although we no longer believe we can allow ν to change all the way between half integral resonances ($\Delta\nu \simeq \frac{1}{2}$), we believe $\Delta\nu$ can be as great as 0·05.

6.3. *List of experiments for colliding beams*

The excitement of the storage ring physics arises from the fact that many experiments are fundamental and direct. After the electrons and positrons annihilate, only energy, angular momentum 1, and negative parity are left.

Some exciting experiments are:

1. A study of the resonances ω, ρ, ϕ, and any other 1^- resonances which couple to photons, in the most direct way possible and a search for other resonances.
2. Measurements of the π form factor directly in the time-like region.
3. Measurements of p, n, Λ, Ξ etc. form factors if the energy is high enough to produce these in pairs. (The cross-section for producing these may be low, however, because the form factors will all be similar to proton form factors, which are known to be small.)
4. A search for the intermediate vector boson W.
5. A search for higher mass leptons μ^*, e^*.
6. A study of quantum electrodynamics.

Once the beams are colliding, these experiments are easy. The cross-sections for the processes vary as $F^2(q^2)/E^2$ where $F(q^2)$ is a form factor, $q^2 = 4E^2$ and E is the energy of one of the colliding beams. At 500 MeV, a luminosity of $L = 10^{29}$ cm^{-2} sec^{-1} gives reasonable counting rates—a few a minute. At a resonance $F(q^2)$ becomes larger as M/Γ, and the cross-section as $(M/\Gamma)^2$, as we see at once from [59]. The narrow ϕ resonance is much enhanced. Thus the study of the $J^P = 1^-$ resonances ρ, ω and ϕ is easy compared with the studies just described.

At higher energies, not only does the cross-section fall as $1/E^2$ but also $F(q^2)$ falls if our present belief that there are no more 1^- resonances is correct. For example, for the CEA ring we expect, for $L = 10^{31}$ cm^{-2} sec^{-1} at 3 BeV, the following rates

$e^+ + e^- \rightarrow e^+ + e^-$	5/min over 4π
$\rightarrow \mu^+ + \mu^-$	1/min
$\rightarrow \gamma + \gamma$	20/hr
$\rightarrow$ (excited lepton pairs)	1/min
$\rightarrow W^+ + W^-$	5/min, if mass is low enough
$\rightarrow \pi^+ + \pi^-$	1/hr, or lower
$\rightarrow p + \bar{p}$	1/hr, or lower.

An apparatus to distinguish all of these processes except the last two is shown in Fig. 30. The $e^{\pm}$ are distinguished by a shower, the $\mu^{\pm}$ by their range and the W by its decay and non-collinear tracks. This, incidentally, is the cleanest known method of finding a W if the mass is right, because the π background is low. To distinguish the last two processes—and pairs of all

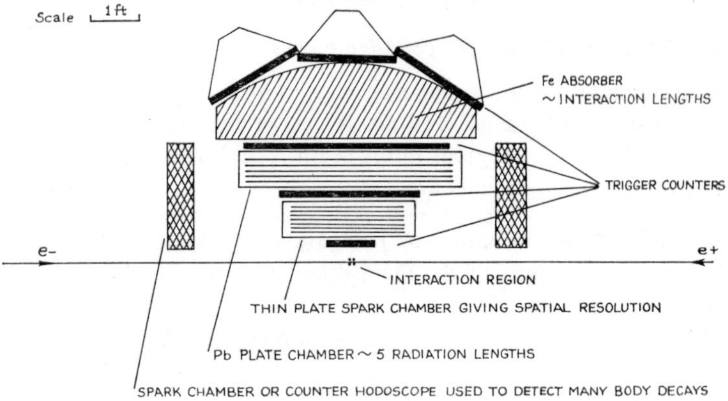

FIG. 30. Diagramatic representation of a detector which has a cylindrical symmetry around the axes of the beams for a simple colliding beam experiment.

other strongly interacting particles—will be hard and will need a magnetic field. A magnetic field detector is likely to cost 2 million dollars.

In short, we can hope for good data on the ω, ρ and ϕ mesons from storage rings at Orsay, Frascati and Novosibirsk within a year or two, and we can also hope for better tests of QED than now are available.

7

INELASTIC ELECTRON-PROTON SCATTERING

7.1. *Kinematics and form factors*

The study of atoms by inelastic electron scattering was begun by Franck and Hertz fifty years ago. This method has been applied to nuclei and we now wish to apply it to nucleons. This was first done by Panofsky and Allton.[78] The theory of the inelastic scattering parallels that of the elastic scattering given earlier and we here again follow Hand's treatment.[7] (For alternative treatments see references 80, 80A, 97, 144 and 6.)

We will for the moment confine ourselves to observation of the outgoing electron. Again, we calculate $(j_\mu J^\mu)$ in the proton's Breit frame. The calculation of J^μ is now complicated for it is a function of all the spins and momenta of the final state particles over which we are, for the moment, summing. We are therefore averaging over independent rotations of the initial and final states about the momentum transfer axis. Then terms proportional to $J^0\boldsymbol{J}$

average to zero and we have, as in the case of elastic scattering (for $m^2 \ll q^2$)

$$|j_\mu J^\mu|^2 = \frac{e^2 q^2}{4M^2}\{\overline{J^{02}}\cot^2(\theta^*/2)+\tfrac{1}{2}\overline{J^2}[1+\operatorname{cosec}^2(\theta^*/2)]\}. \qquad ...[90]$$

The bars denote averages and we note that the cross term in $J_0 \boldsymbol{J}$ only averages to zero if J_0 and $\boldsymbol{J}$ are both real, or if we average over initial states. As we shall see when we discuss time reversal, J_0 and $\boldsymbol{J}$ can be complex if time reversal is not valid. We are then left with a term in σ_y (see equation [*11*]), the component of spin of the initial proton which is normal to the scattering plane.

Then we get the cross-section (for $m^2 \ll q^2$)

$$\frac{d^2\sigma}{d\Omega dE'} = \Gamma_T(q^2, K)\sigma_T(q^2, K)+\Gamma_0(q^2, K)\sigma_0(q^2, K) \qquad ...[91]$$

where

$$\Gamma_T = \frac{\alpha}{4\pi^2}\frac{K}{q^2}\frac{E'}{E}\left[2+\frac{\cot^2(\theta/2)}{1+\tau'}\right]; \; \sigma_T = \frac{1}{2K}|\boldsymbol{J}|^2 \qquad ...[92]$$

and

$$\Gamma_0 = \frac{\alpha}{4\pi^2}\frac{K}{q^2}\frac{E'}{E}\left[\frac{\cot^2(\theta/2)}{1+\tau'}\right]; \; \sigma_0 = 1/K|J^0|^2, \qquad ...[93]$$

and we have put $\tau' = (q_0^2/q^2)$, and $\cot^2(\theta/2) = (1+\tau')\cot^2(\theta^*/2)$. The virtual photon energy K is defined by

$$K = q_0-(q^2/2M) = \varepsilon(1+\varepsilon/2M) = (M^{*2}-M^2)/2M \qquad ...[94]$$

where ε is the excitation energy of the nucleon and M^* is the mass of the excited system. The kinematics tells us that the scattered particle energy is given by

$$E' = \frac{E}{1+(2E/M)\sin^2(\theta/2)}\left(1-\frac{K}{E}\right) \qquad ...[95]$$

which tells us that the resolution $(\Delta E')/E'$ needed to separate a resonance of width ΔK in photoproduction is dependent only on the incident particle energy and not on the scattering angle. We noted earlier in discussion of quasi-elastic electron-deuteron scattering measurements how the different dependences of this, and the spread of the quasi-elastic peak of the electrons as E was varied, puts an upper limit to the easy neutron form factor measurements.

We now note the type of information that can be given by simple experiments of the Franck and Hertz (or Panofsky-Allton) type.

Firstly, the definition of $\sigma_T(q^2, K)$ in [*92*] yields $\sigma_T(0, K) = \sigma_\gamma(K)$ or the photoproduction cross-section. Then the factors have the interpretation of the number of virtual photons per MeV/ster. of the scattered electron. The separation of the kinematic factors between Γ and σ is arbitrary, provided that the limit at $q^2 = 0$ has the physical meaning provided above. It is convenient, however, to make the separation such that

$$\sigma_T = \sigma_T(q^2, K) = \sigma(0, K)F(q^2) = \sigma_\gamma(K)F(q^2) \qquad \text{...[96]}$$

where σ_γ is the photoproduction cross-section and $F(q^2)$ is a form factor defined by the above relation. This factorization is not possible to do exactly in terms of nucleon form factors (as can be seen by a detailed examination of the FNW theory [110]), but it holds approximately.

The separation above into scalar and transverse matrix elements holds only if the total production of electrons is measured and we integrate over outgoing pions and protons. If the outgoing mesons in coincidence with an electron are measured, and in particular if the azimuthal dependence of these mesons is measured, an interference term is produced. This will be ignored for the purposes of the immediately following discussions.

Note that here we use a transverse/scalar separation, whereas, for example, Dalitz and Yennie use a transverse/longitudinal separation. The two [109] are related by the requirement of gauge invariance on the electron and proton current operators,

$$j_\mu q^\mu = 0 = J_\mu q^\mu \qquad \text{...[97]}$$

giving

$$j_z J^z = j_0 J^0 \left(\frac{q_0^2}{|\boldsymbol{q}|^2}\right). \qquad \text{...[98]}$$

The longitudinal cross-section is given by $j_0 J^0 - j_z J^z$. The properties of electron scattering are especially easy to understand in the Breit frame (where $q_0^2 = 0$). In this frame j_z is zero (to order q/m). However, if we were considering electron-positron annihilation, it would be preferable to use a gauge transformation to put $j_0 J^0 = 0$. We are interested in three frames of reference (the lab. frame, the centre-of-mass frame and the Breit frame), which are obtained from each other by Lorentz transformations along the momentum transfer direction q. The separation remains correct in any of these frames.

The previous section establishes a connection with photoproduction which is conceptually very useful. It enables us to consider extending to electroproduction any process which has been usefully studied in photoproduction. The extension gives a factor of 2 in parameters—σ_0 as well as σ_T—and an infinite extension also, $q^2 = 0$ to $q^2 = \infty$. Because photoproduction is already a big field, electroproduction is bigger; it remains to narrow it down to a few regions which are interesting, possible to study, or are otherwise partially understood by the author. These are classified below. *All* involve detection of a scattered electron.

1. Study of the total cross-sections $\sigma_T(q^2, K)$ and $\sigma_0(q^2, K)$ as functions of q^2 and K.

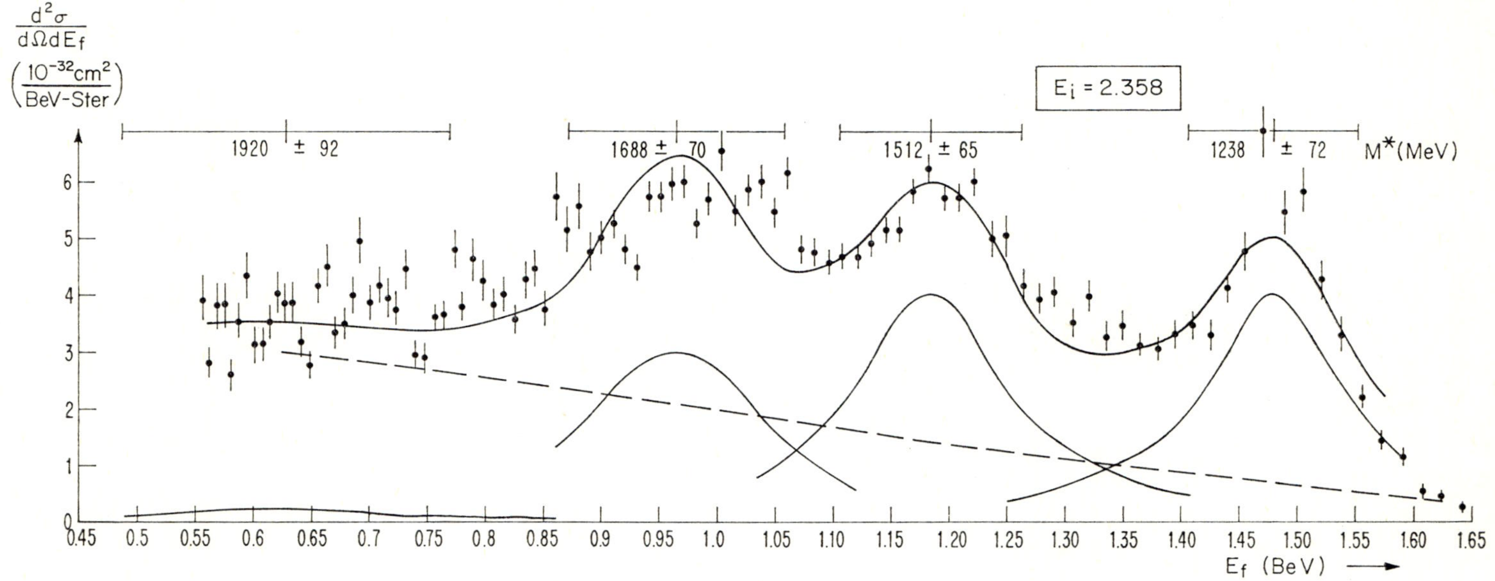

FIG. 31

Inelastic scattering of electrons at 31°. The experimental data has been corrected for backgrounds and radiative corrections.

2. Study of the nucleon resonance regions $K = 300$, 700, 1000 MeV etc. with a π or two in coincidence with the scattered electron.
3. Study of peripheral interactions (Drell process) by electroproduction.

Some of these are considered below.

7.2. *Detection of electron only—multipole terms*

The minimum aim is to measure $(d^2\sigma)/(d\Omega dE')$ and to deduce σ_0 and σ_T from [*91*]. This study is what has been started by Panofsky and Allton[78] and by Hand,[81] Lynch[81A] and Cone.[79] Let us consider the uses of such a study. First, a total gamma cross-section is hard to obtain. An absorption measurement, as is possible for neutrons, measures primarily electronic absorption processes such as pair production. We shall discuss later a possible experiment in this area. There are similar problems in measuring $\sigma(q^2, K)$ at low q^2; the analogue of the problem of separating the gamma ray, nuclear absorption and pair production is the problem of separating the inelastic electron-proton scattering from the radiative tail of the elastic scattering. The lower limit of $q^2 > 2\,\mathrm{f}^{-2}$ seems reasonable.

An experiment is shown in Fig. 31. We see unequivocally the excitation of three known resonances at $M^* = 1238$ MeV, 1512 MeV and 1688 MeV and a probable excitation of the 1920 MeV resonance. The procedure is simply to put an electron-detecting-and-pion-rejecting detector at the focus of the spectrometer. At the higher energies, at CEA, a threshold gas Cerenkov counter and a lead plate/scintillator shower counter were both used and no appreciable pion contamination was present.

How do we interpret these experiments? The general theory is beautifully described by Bjorken[80] in an entirely different notation from this (or any other) paper. It is similar to the theory for nuclear excitations which has been used extensively in the past.[83A] We note the principal features here. We can define a threshold behaviour valid when the product $(r\boldsymbol{q}_c)$ of the three-momentum transfer $\boldsymbol{q}_c$ in the excited nucleon's rest frame and r, a characteristic radius of the interactions, is close to zero. For nuclei, we can approach this limit, but we are far from it for the nucleon.
Then

$$\sigma_T = (\boldsymbol{q}/K)^{2l-2}\sigma_\gamma F(q^2) \qquad ...[99]$$

and

$$\sigma_0 = \left(\frac{q^2}{q_0^2}\right)\left(\frac{\boldsymbol{q}}{K}\right)^{2l-2}\sigma_\gamma \frac{l}{l+1} F(q^2) \qquad ...[100]$$

for electric multipole and longitudinal multipole radiation *except* monopole, and

$$\sigma_T = (\boldsymbol{q}/K)^{2l}\sigma_\gamma F(q^2) \qquad ...[101]$$

for magnetic multipoles. $F(q^2)$ is a form factor. For any resonance we have

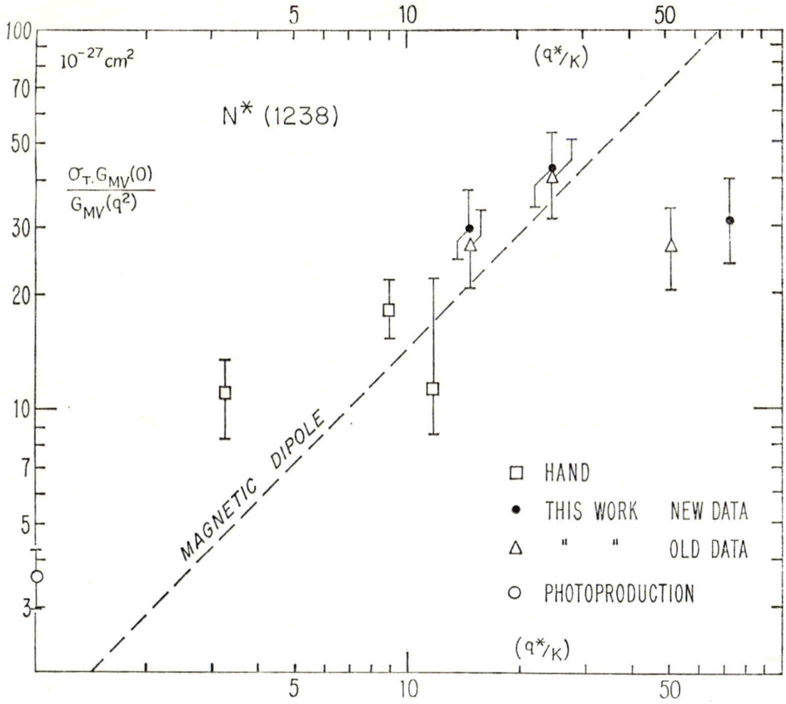

FIG. 32. The excitation of the resonance at $M = 1238$ MeV versus q^2.

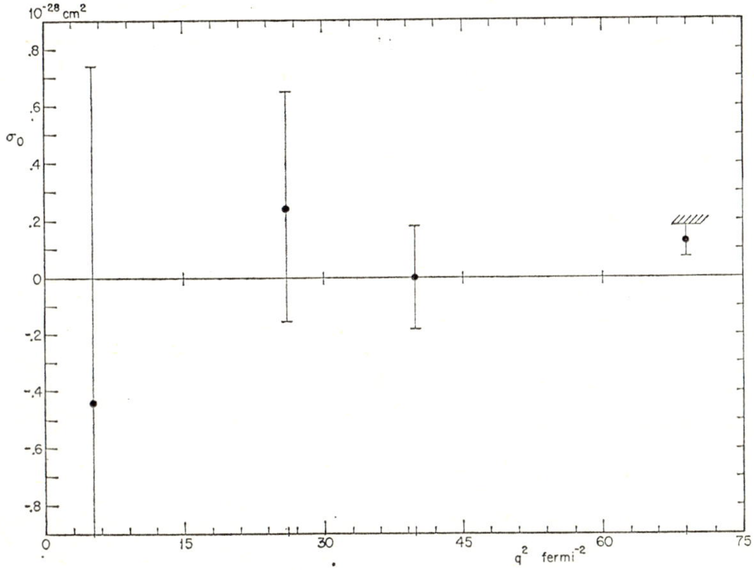

FIG. 33. σ_0 versus q^2 for $M = 1238$ MeV.

three excitations; one electric, one magnetic and one longitudinal. We can understand this threshold behaviour non-relativistically as the first terms in the expansion of

$$| \int j_l(qr)\rho(r)r^2dr |^2.$$

Let us study the 1238 MeV resonance. It is an M1 (abnormal parity) transition with a small $E2$ mixture and a small $L2$ associated with the $E2$. For the $M1$ transition we find *only* transverse excitations. Experimentally this is the case (Figs. 32 and 33). Since the transition goes by an isotopic spin flip ($T = \frac{1}{2}$, $J = \frac{1}{2}$ to $T = \frac{3}{2}$, $J = \frac{3}{2}$) as well as ordinary spin flip, the form factor is $F(q^2) = G_{MV}(q^2)/G_{MV}(0)$.

There are more precise calculations using dispersion theory, of which we mention those of Zagury,[84] Adler[85] and Loubaton[70, 86] which refine the earlier dispersion treatment of CGLN.[87] In their dominant (resonant) term they confirm the threshold behaviour above, but differ slightly at high momentum transfers. This is a remarkable triumph of theory. Using πN phase shifts they predict the total photoproduction cross-section to 10% with no adjustable parameters. (The angular distribution is not so well fitted but it depends on interference terms with small phases.) From there on, they follow the threshold behaviour falling a factor of 3 lower at $q^2 = 90\,\mathrm{f}^{-2} = 4(\mathrm{BeV/c})^2$. (But this is to be expected from the next terms in the non-relativistic expansion of $j_l(qr)$.) Figs. 34 and 35 show fits up to this highest momentum transfer. No longitudinal terms have been found in the total cross-section (Fig. 33).

With this triumph of theory we proceed to the 1512 resonance. This is excited only at 30° and calculation suggests that it is primarily a longitudinal excitation. Unfortunately the threshold behaviour as described by Bjorken[80] and Jones[80A] is inadequate. σ_0 becomes infinite when $q_{0c} = 0$, which happens in the experimental region. Being uncertain what to do, we have replaced q_{0c} by K (which is correct at the threshold anyway) and get:

$$\frac{\sigma_0}{\sigma_T} = \frac{q^2}{K^2}\frac{l}{l+1}. \qquad \ldots[102]$$

We are also uncertain about what form factor to use but they are all similar.

Assuming this formula for the 1512 resonance we may plot (Fig. 36a) $\sigma_T/(G_{MV}^2)$ versus q^2 for this resonance. The threshold prediction fails. Maybe we are exciting the S resonance at 1550 MeV (which we could not distinguish from the 1512 resonance). This would vary as $(q^*)^4$. Fig. 36b shows a fit for the 1688 MeV resonance which is very good.

7.3. *Radiative corrections*

7.3a. *General.* The radiative correction calculation for the inelastic scattering is, in principle, the same as for the elastic case but becomes numerically very complex. Some of the scattered electrons at the position we expect from excitation of the N^* come from elastic *ep* scatterings accompanied by radiation. Thus we must understand the radiative tail from the elastic peak, at large photon energies, to evaluate the $N^*(1238)$ cross-section. Then we must understand *both* the radiative tails from the elastic scattering *and* those from the $N^*(1238)$ to evaluate the $N^*(1512)$ cross-section. It is

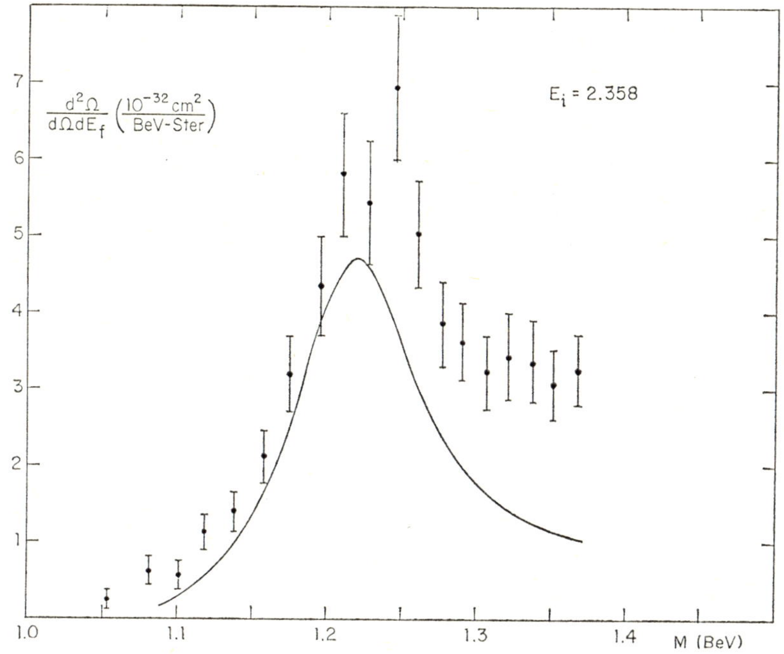

FIG. 34. The differential cross-section at $q^2 = 40\ f^{-2}$ versus theory (Adler) near $M = 1238$ MeV.

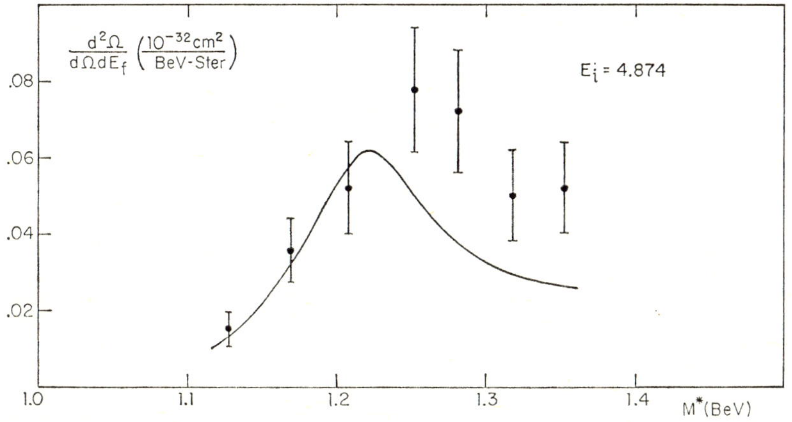

FIG. 35. The differential cross-section at $q^2 = 90\ f^{-2}$ versus theory near $M = 1238$ MeV.

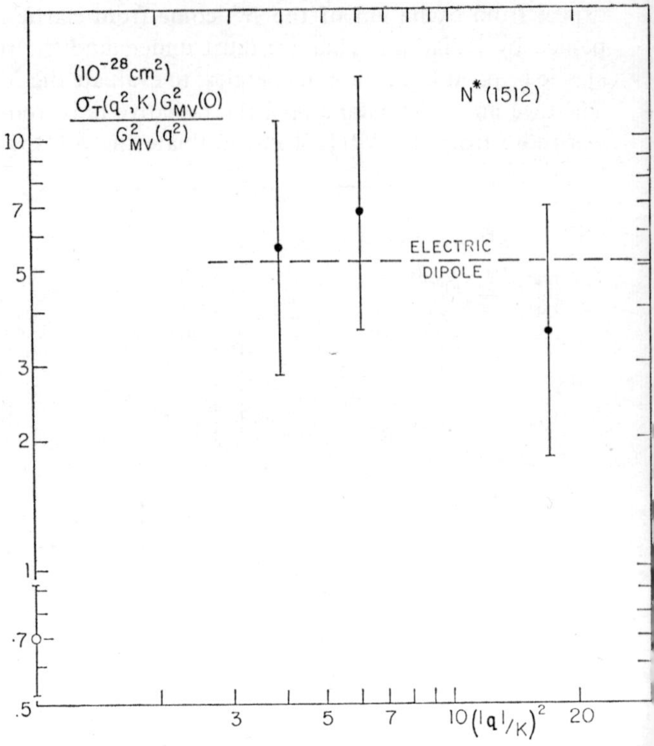

FIG. 36*a*

σ_T/G_{MV}^2 for the resonance $M = 1512$ MeV, assuming σ_0/σ_T given by the threshold formula.

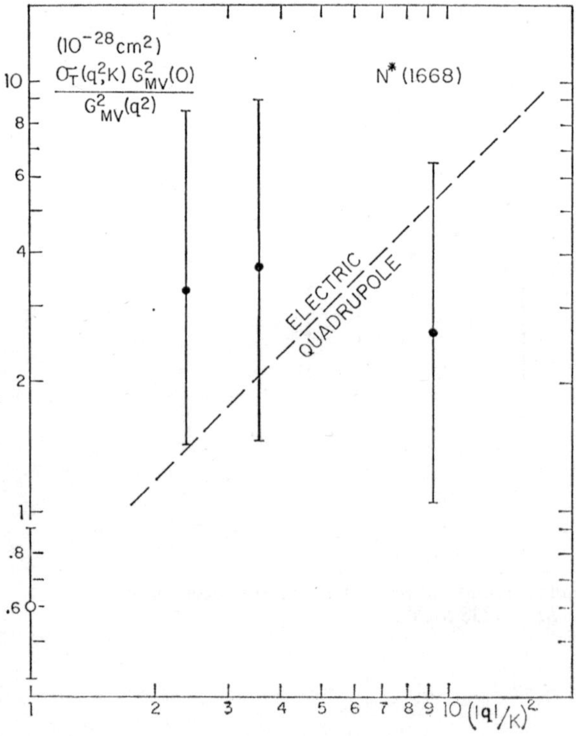

FIG. 36*b*

σ_T/G_{MV}^2 for the resonance M = 1668 MeV, assuming σ_0/σ_T given by the threshold formula.

thus an iterative process. Moreover, it is a two dimensional iteration as we shall now see. We refer to two methods of calculation, by Perez y Jorba[82] and Bjorken.[83] We write the total measured inelastic cross-section in terms of "bare" cross-sections as follows.

Write $(d^2\sigma)/(d\Omega dE')$ at an incident energy E as $\sigma(E, E')$ and let $X(E, E')$ be the probability of an electron of energy E radiating and leaving a scattered energy E'. Then we find

$$\left(\frac{d^2\sigma}{d\Omega dE'}\right)_{\text{meas.}} = \sigma(E, E')_{\text{meas.}}$$

$$= \sigma(E, E')_{\text{bare}}(1-\delta) + \int \sigma(E, E'')_{\text{bare}} X(E'', E')\,dE'' + \int X(E, E'')\sigma(E'', E')_{\text{bare}}\,dE'' \quad \ldots[103]$$

where δ is an ordinary radiative correction (radiating out of the detection interval); the second term represents radiation *after* scattering and the third radiation *before* scattering. Radiation both before and after is ignored.

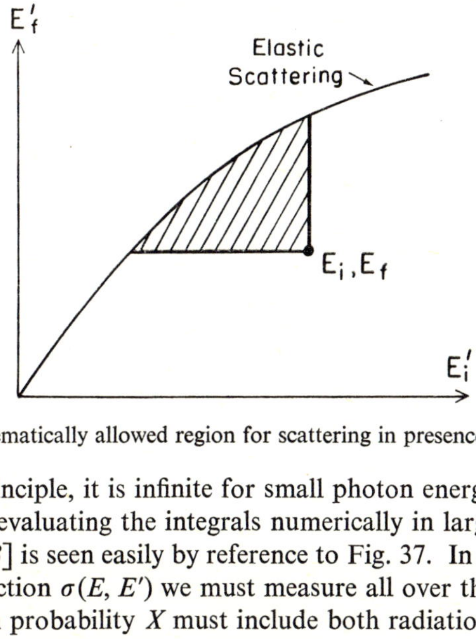

FIG. 37. Kinematically allowed region for scattering in presence of radiation.

(Although, in principle, it is infinite for small photon energies we can avoid this problem by evaluating the integrals numerically in large enough steps.) The form of [*103*] is seen easily by reference to Fig. 37. In order to evaluate the bare cross-section $\sigma(E, E')$ we must measure all over the shaded area.

The radiation probability X must include both radiation in any physical radiator placed in the beam and *virtual* processes. Thus we can approximate

$$X(E'', E')\,dE'' = \frac{1}{E''-E'}\,dE''\left[t\log 2 + \frac{\alpha}{\pi}(\log(q^2/m^2)-1)\right] = \frac{\varepsilon\,dE''}{E''-E'} \quad \ldots[104]$$

where t is the physical thickness in radiation lengths, placed both in the incident *and* the scattered beam. Panofsky[78] first noticed that one can include real and virtual processes together.

Using a better approximation for X and including occasions where radiation occurs *both* before and after collision, Bjorken inverted equation [*103*], using the techniques of a transform, to get

$$\sigma(E, E')_{\text{bare}}$$
$$= -E' \frac{\partial^2}{\partial E \partial E'} \int_0^E dE'' \int_{E'}^{\infty} \frac{dE'''}{E'''} \psi \left[\frac{E}{E'''}, (1-\varepsilon)\right] \sigma(E'', E''')_{\text{meas.}} \psi \left[\frac{E'''}{E'}, (1-\varepsilon)\right] \qquad \ldots[105]$$

where $E\psi(E, E')$ is the probability of an electron of energy E radiating and ending up with an energy E'. $\psi \propto (1-\varepsilon E'/E)$.

This is, in principle, easier to compute (though less easy to understand) than [*103*], however, care must still be taken because of the divergence at one limit of the integration. So far we have obtained sensible answers only from the Perez y Jorba[82] method.

7.3b. *Minimizing radiative background.* Now we ask a qualitative general question.

We are concerned with whether or not the inelastic counts with nucleon excitation stand out above the radiative tail. In order to see this easily, we rearrange [*91*] into a cross-section differential in the virtual photon energy

$$\frac{d^2\sigma}{d\Omega dK} = \frac{d^2\sigma}{d\Omega dE'} \left|\frac{dE'}{dK}\right| = \eta \frac{d^2\sigma}{d\Omega dE'} \qquad \ldots[106]$$

where $\eta = \dfrac{E'}{E}$ (for elastic scattering). Hence

$$\frac{d^2\sigma}{d\Omega dK} = \frac{\alpha K}{4\pi^2} \frac{\eta}{4E^2 \sin^2(\theta/2)} \left\{\sigma_T \left[2 + \frac{\cot^2(\theta/2)}{1+\tau'}\right] + \sigma_0 \frac{\cot^2(\theta/2)}{1+\tau'}\right\} \qquad \ldots[107]$$

where $\tau' = q_0^2/q^2 = (K+q^2/2M)^2/q^2$.

We can note the similarity, even in constants, with the elastic scattering formula derived earlier (equation [*14*]) which we here repeat for convenience:

$$\frac{d\sigma}{d\Omega} = \frac{\alpha^2 r_e^2 m^2 \eta}{4E^2 \sin^2(\theta/2)} \left\{\tau G_M^2 \left[2 + \frac{\cot^2(\theta/2)}{1+\tau}\right] + G_E^2 \frac{\cot^2(\theta/2)}{1+\tau}\right\} \qquad \ldots[108]$$

where $\tau = q_0^2/q^2 = q^2/(4M^2)$ for elastic scattering.

If the radiated γ-ray energy is small compared with the incident energy, we can write the radiative tail[20]

$$\frac{d^2\sigma}{d\Omega dK} = \frac{d\sigma_{\text{elastic}}}{K d\Omega} \frac{2\alpha}{\pi} [\log(q^2/m^2) - 1]. \qquad \ldots[109]$$

The ratio of the inelastic scattering to the radiative tail is given by

$$R = \frac{\sigma_I}{\sigma_R} = AK^2 \frac{\{\text{Inelastic}\}}{\{\text{Elastic}\}} \qquad \text{...[110]}$$

where

$$A = \frac{8\pi^2\alpha^2 r_e^2 m^2}{\log(q^2/m^2)-1}$$

is almost a constant and the braces are those of [*107*] and [*108*].

At low q^2, G_E dominates the elastic scattering and equals unity, σ_T tends to a constant σ_γ, and σ_0 tends to 0 for most models. Therefore

$$\frac{\sigma_I}{\sigma_R} \to AK^2\sigma_\gamma\left[(\theta^2/2)+\frac{q^2}{K^2}\right], \qquad \text{...[111]}$$

which tells us that it is difficult to find inelastic scattering at low momentum transfers. Assuming that σ_T varies roughly as $q^2G_M^2$ the ratio will increase with increasing momentum transfer until both q_0^2/q^2 becomes less than unity and G_M begins to dominate the elastic scattering, i.e. for $q^2 > 10\,\text{f}^{-2}$ ($\simeq \frac{2}{5}(\text{BeV/c})^2$) for low lying resonances; higher momentum transfers are necessary for the higher resonances.

If the radiated energy is larger in comparison with the incident energy, σ_R is larger than given by [*109*] (see for example, reference 81). The form of [*110*] and [*111*], tells us at once that if we choose E and θ conveniently, so that σ_R reaches its lower limit of [*109*], the inelastic scattering will be most apparent. This happens at small angles and high energies for a given q^2. This suggests that we work at the highest possible energy, i.e. SLAC.

7.4. *Neutron-proton mass difference*

Apart from the interest in mapping $\sigma_T(q^2, K)$ and $\sigma_0(q^2, K)$, there is some interest in evaluating integrals of σ_T and σ_0. Thus Cottingham[88] has expressed the electromagnetic mass of the nucleon in terms of these parameters. Also, by a unitary relation, one may be able to obtain some idea of the imaginary part of the elastic *ep* scattering amplitude (due of course to multiphoton terms), in the same way that one can express the imaginary part of the forward amplitude in terms of the total cross-section. In terms of σ_T and σ_0 we find for the electromagnetic self mass

$$\Delta M = \Delta M^E + \Delta M^I \qquad \text{...[112]}$$

with

$$\Delta M^I = -\left(\frac{1}{2\pi}\right)^3 \int_0^\infty dq^2 \int_c^\infty dq_0 \frac{K}{(q^2+q_0^2)^{\frac{1}{2}}}\left[\frac{(1-x)(2+x)}{1+x}\right]\sigma_T + \left(\frac{1}{2\pi}\right)^3 \int_0^\infty dq^2 \int_c^\infty dq_0 \frac{K}{(q^2+q_0^2)^{\frac{1}{2}}}\left[\frac{1+x+x^2}{1+x}\right]\sigma_0 \qquad \text{...[113]}$$

where $x = (\tau'/(1+\tau'))^{\frac{1}{2}}$ with $\tau' = (q_0^2)/q^2$, so that $x < 1$, and

$$c = [2M\mu+\mu^2+q^2]/(2M),$$

where μ is the pion mass. The expression for ΔM^E is similar, but includes the elastic form factors.

In this type of experiment, μ scattering may be better than electron scattering. The radiative correction at a given q^2 is given by a term $\log(q^2/m^2)$; at $q \simeq 250$ MeV/c the improvement with μ mesons is thus $\log(10^6/4)/\log 4$, or a factor of 20.

Now if we wish to study the higher nucleon resonances at $M^* \simeq 3$ BeV by electrons we note that the photoproduction of these resonances stands out only $\frac{1}{2}\%$ above the general background of peripheral processes.[89] We obviously need to enhance this, and note the factor $(\boldsymbol{q})^{2l}$ in [*99*]. Since these nucleon resonances have probably a high spin, we have hopes of an enhancement, but we note that when $q^2 = 0$, $|\boldsymbol{q}| \simeq 2$ BeV and

$$\boldsymbol{q}^2 = q^2 + \frac{1}{4M^{*2}}[M^{*2}-M^2+q^2]^2. \qquad ...[114]$$

To increase $|\boldsymbol{q}|$ we need to go to a value of q^2 of about 9(BeV/c)^2 (and still retain small angles for the radiative subtraction). This again indicates a job for SLAC.

7.5. *Coincidence experiments—pion form factors*

Since the excitation of the $N^*(1238)$ has been so successfully fitted by theory we are encouraged to ask for more details. Let us consider the coincidence experiments.

We noted in [*6*] and [*7*] that $|j_x|$ and $|j_y|$ were different. This leads to an effective transverse polarization

$$P = \frac{|j_x|^2 - |j_y|^2}{|j_x|^2 + |j_y|^2} = \frac{\Gamma_0}{\Gamma_T}. \qquad ...[115]$$

(Berkelman[90,91] uses ε for this quantity.)

Then if we consider the direction of momentum transfer $\boldsymbol{q}$ to be the direction of the virtual photon, we can ask for the angular distribution of the pions in the decay of the $N^*(1238)$. Since we are now looking at the outgoing particles, we have longitudinal-transverse interference with a $\cos\phi$ dependence and the polarization gives a $\cos 2\phi$ dependence.

Thus we find a differential cross-section of the form

$$\frac{d^3\sigma}{d\Omega dE' d\Omega_\pi} = \sigma_T\Gamma_T + \sigma_0\Gamma_0 + \sigma_I[\Gamma_T(\Gamma_T+\Gamma_0)]^{\frac{1}{2}}\cos\phi + \Gamma_0\sigma_P\cos 2\phi \qquad ...[116]$$

where σ_P is the polarized cross-section and σ_I an interference cross-section. σ_P is a new quantity but σ_I can be expressed in terms of the others. We can now count the full 3 independent measurable quantities given by one photon exchange, from the four components of j_μ minus one for gauge invariance.

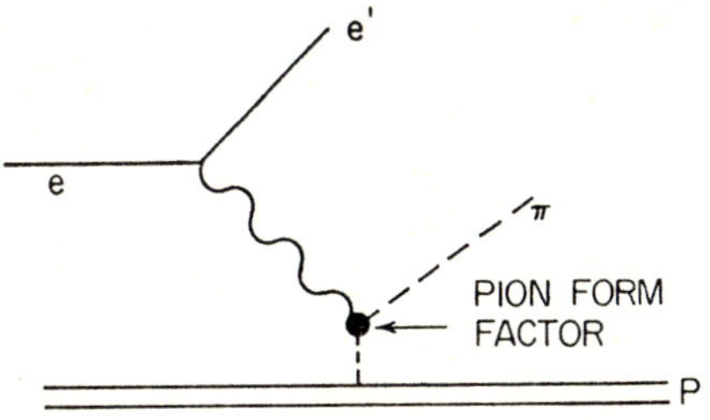

FIG. 38

Diagram showing how the pion form factor can enter in inelastic *ep* scattering.

Now let us consider the process at $\theta_\pi = 0$. An examination of the theory shows that the nonresonant diagram, Fig. 38, which comes from the pion form factor, appears in σ_0 only and dominates σ_0. Berkelman has examined the cross-section at several electron angles and plotted it as a function of $\Gamma_0/\Gamma_T = P = \varepsilon$ in Fig. 39. He finds a clear dependence. Although we must regard the conclusions as tentative, he finds a form factor

$$F_\pi = 0{\cdot}76 \pm 0{\cdot}14 \quad \text{at} \quad q^2 = 2{\cdot}96\,\text{f}^{-2} \qquad \ldots[117]$$

whence the r.m.s. radius $\langle r_\pi^2 \rangle^{\frac{1}{2}} = 0{\cdot}70 \pm 0{\cdot}20$ f. The prediction for ρ-meson dominance is 0·62 f.

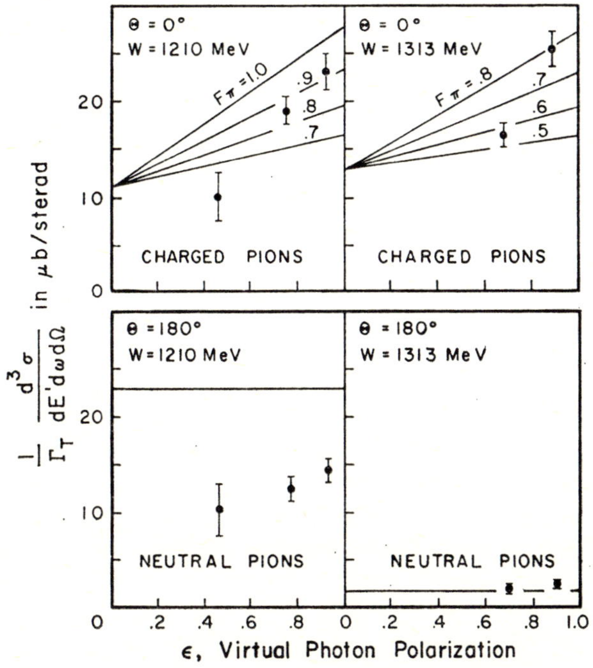

FIG. 39. The cross-section for inelastic *ep* scattering versus $\varepsilon = \Gamma_0/\Gamma_T$. The lines are FNW theory calculations.

Experiments are in progress at CEA to measure the cross-section at several θ_π near 0°. The CGLN theory for photoproduction is notoriously bad at 0° because the small (non-resonant) phases are very important and no one quite knows how to handle them. In his other paper[90] Berkelman found that there was indeed a term in cos 2ϕ, in agreement with data from photoproduction by polarized photons. This is also shown in Fig. 39.

The theory is not yet good enough to extend to higher virtual photon energies. At these higher energies, we would hope that the Drell process (an extension of the diagram of Fig. 38) dominates and we can derive the pion form factor again. This is now still theoretically *more* uncertain than the lower energy predictions. But this is clearly a field of great scope for synchrotrons.

7.6a. *Photon tagging and virtual photon tagging*. The inelastic scattering is very general and one can do all the experiments with photons over again as a function of momentum transfer. Of course, life is too short to do this, so the previous discussions have been based on picking up one or two interesting regions. In two respects, electron experiments are simpler than photon experiments since:

(i)—The energy of the virtual photon is well defined.

(ii)—There is one polarization of the virtual photon.

This leads naturally to the integration over all outgoing electron directions (but not high energies) and thus having a high intensity of tagged virtual photons. The number of virtual photons is obtained[144] by integrating over equation [*91*] (or more precisely its equivalent when m^2 is not neglected or else the answer is infinite) assuming the form factors are constant, because small angles and momentum transfers in fact dominate the integral. If we omit σ_0, because σ_0/σ_γ varies as q^2/q_0^2 and is small at small momentum transfers, then we find the cross-section is

$$\frac{d\sigma}{dE'} = \frac{R_{\text{eff}}\,\sigma_\gamma}{k} \qquad \text{...[118]}$$

where

$$R_{\text{eff}} = \frac{\alpha}{\pi}\left\{\frac{E^2+E'^2}{E^2}\log\left(\frac{2EE'+m^2}{m(E-E')}\right)-\frac{(E+E')^2}{2E^2}\log\left(\frac{E+E'}{E-E''}\right)-\frac{E'}{E}\right\}. \qquad \text{...[119]}$$

This is the number of events given by the gamma rays produced from a target of radiation length R_{eff}. It is therefore called the virtual radiator and has a value of about 0·02 to 0·03. This is close to the well known Weiszacker-Williams approximation.[92,93,144]

So now we can compare three processes. Real photon tagging,[94,95] and virtual photon tagging with electrons and muons.[96] The advantage in tagging of virtual photons is the improvement in the background. This is particularly clear when a μ meson beam can be used.

A gamma ray tagging experiment with real photons (Case 1) is shown schematically in Fig. 40 and one with the same electrons producing virtual photons (Case 2) in Fig. 41. The muon case (Case 3) also follows Fig. 41.

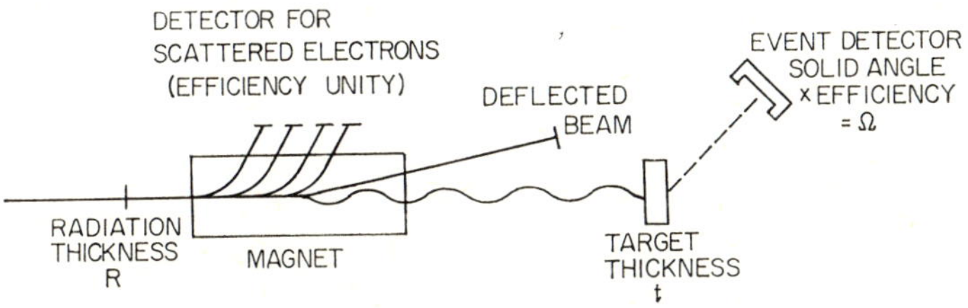

FIG. 40. This illustrates the tagging of real photons.

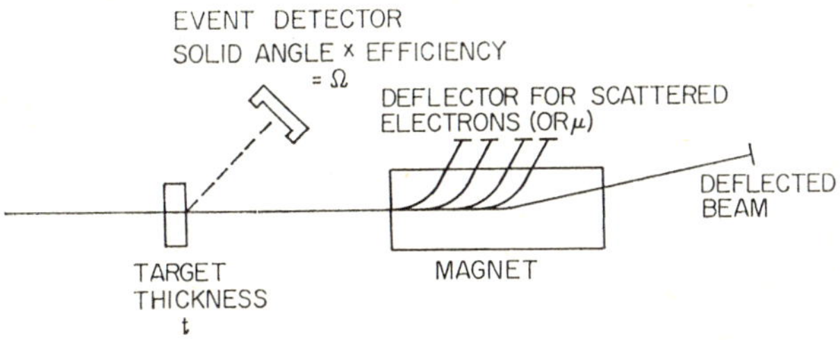

FIG. 41. This illustrates the tagging of virtual photons.

Since it will transpire that beam intensity is *not* a limitation on these experiments, we shall consider the background random coincidence rate between the scattered electron counters and the "event" detector as a number of interest, and the event rate of desired events produced by a gamma ray in a small interval dk, as a number to be kept constant.

We now present in tabular form a comparison between the random rates in three cases, bearing in mind that the event rate and not beam intensity is the independent variable. We note also that the principle counting rate accompanies nuclear events; bremsstrahlung is lower than this by a factor of about 4. This causes the difference in the rate S between cases 2 and 3.

	Case 1	*Case 2*	*Case 3*
$E =$	$N\frac{R}{X_0}\frac{\Delta k}{k}\,t\sigma(k)\,\frac{\Omega L}{A}$	$N\frac{R_{\text{eff}}}{X_0}\frac{\Delta k}{k}\,t\sigma(k)\,\frac{\Omega L}{A}$	$N\frac{R_{\text{eff}}}{X_0}\frac{\Delta k}{k}\,t\sigma(k)\,\frac{\Omega L}{A}$ ...[120]
$T =$	$N\frac{R}{X_0}\,t\,\frac{\Omega L}{A}\int_0^{k(\max)}\frac{\sigma(k)dk}{k}$	$N\frac{R_{\text{eff}}}{X_0}\,t\,\frac{\Omega L}{A}\int_0^{k(\max)}\frac{\sigma(k)dk}{k}$	$N\frac{R_{\text{eff}}}{X_0}\,t\,\frac{\Omega L}{A}\int_0^{k(\max)}\frac{\sigma(k)dk}{k}$ ...[121]

$$S = N\frac{R}{X_0}\frac{\Delta k}{k} \qquad N\frac{t}{X_0}\frac{\Delta k}{k} \qquad N\frac{t}{X_0}\frac{\Delta k}{k}4\frac{M_e^2}{M_\mu^2} \qquad \ldots[122]$$

$$C = TS \qquad \ldots \qquad \ldots \qquad \ldots[123]$$

$$= N^2\frac{R^2}{X_0^2}t\frac{\Omega L}{A}\frac{\Delta k}{k}\int\frac{\sigma(k)dk}{k} \qquad \ldots \qquad \ldots \qquad \ldots[124]$$

$$= \frac{E^2\int_0^{k(max)}\frac{\sigma(k)dk}{k}}{t\Omega(L/A)\frac{\Delta k}{k}[\sigma(k)]^2} \qquad \frac{E^2\int_0^{k(max)}\frac{\sigma(k)dk}{k}}{R_{eff}\Omega(L/A)\frac{\Delta k}{k}[\sigma(k)]^2} \qquad 4\frac{M_e^2}{M_\mu^2}\frac{E^2\int_0^{k(max)}\frac{\sigma(k)dk}{k}}{R_{eff}\Omega(L/A)\frac{\Delta k}{k}[\sigma(k)]^2} \qquad \ldots[125]$$

The symbols used in the above equations are shown in Table V.

TABLE V

Physical quantity	Symbol	Units
Beam intensity	N	sec^{-1}
Gamma ray energy	k	
Radiation length of material of radiator	X_0	g cm^{-2}
Thickness of target	t	g cm^{-2}
Thickness of radiator	R	g cm^{-2}
Number of molecules in a mole.	$L = 6 \times 10^{23}$	g^{-1}
Atomic weight	A	
Cross-section (maybe partial) for event of interest, which can trigger event counter	$\sigma(k)$	cm^2
Solid angle multiplied by efficiency for event of interest	Ω	(fraction of 4π)
Count rate in scattered electron detector	S	sec^{-1}
Event rate of desired events	E	sec^{-1}
Total count rate in event detector	T	sec^{-1}
Background coincidence rate	C	sec^{-1}
Effective radiator for virtual photons (Equation [119])	R_{eff}	g cm^{-2}
Total resolving time for coincidence (including duty cycle)	τ	sec

We note that these equations [*120*]-[*125*] are identical for the two cases (1 and 2) if R_{eff} equals t. If $t < R_{eff}$ tagging of virtual photons is preferable. For a liquid hydrogen target this corresponds to a target 20 cm long. With targets longer than this tagging of real photons is preferable if only electron beams are available (but one then gets large radiative corrections). If muon beams are available, tagging of virtual photons is always preferable.

We note that the count rate in the detector for scattered electrons (S) is given mainly by the probability for emitting gamma rays (except for case 3) and *not* by a reaction of interest.

The random coincidence rate, then, represents the number of times *two* scattered electrons appear in the time interval for detection and we cannot therefore identify the gamma ray energy unambiguously. If we have detectors for the whole range of scattered electron energies, these confusing events can be identified at once by noting that two particles exist and hence can be ignored; otherwise, the usual delayed count must be made and the events subtracted from the undelayed, with the usual statistical problem.

We note, in passing, a feature about equations [*120*] to [*125*] which have prevented frequent use of tagged photons. If either the solid angle Ω or the cross-section σ is small, C becomes large and experiments impossible. Until recently,[95] only small solid angle detectors could be used. The advent of large solid angle detectors (spark chambers) makes the technique of great utility.

7.6b. *Illustrative numbers.* Consider, for example, the measurement of a partial cross-section for strange particle production by 1-6 BeV photons in intervals (dk/k) of 0·05; over this range $\sigma(k)$ is about 5μb. We use a target of liquid hydrogen 20 cms long (=1·4 g/cm^2) which is about equal to R_{eff} ($R_{\text{eff}}/X_0 \simeq 0{\cdot}023$). $\tau = 10^{-6}$ seconds (10^{-8} seconds with a duty cycle 1 in 100). We assume that the event counter detects no events with gamma ray energies less than 1 BeV ($\sigma(k) = 0$ for $k < 1$ BeV).

Then for cases 1 and 2 we obtain $C = 70\ E^2$/sec. So if E is one event in ten minutes, $C/E = \frac{1}{6}$ which is a tolerable random rate.

The requisite beam intensity, N, can be found from equation [*122*] and is $3{\cdot}4 \times 10^6$ electrons per second (case 2) and the same for case 1 if the radiator thickness is the same as the target thickness in radiation lengths ($R/X_0 = t/X_0$).

For muons we find, for 1 event in ten minutes, $C/E = 10^{-5}$ which is negligible. The requisite muon rate is about a factor two greater (7×10^6/sec) due to the change in R_{eff}.

Electron accelerators produce beams of about 10^{12}/sec, so our earlier assumption that electron beam intensities are no limit is confirmed. Muon beams have intensities of the order of 10^6/sec, so this is not now true for muons. We could increase the count rate for muons by using a longer target or larger beam. Clearly, we could use muon beams up to 10^{11}/sec, and event rates in our example of 40/sec, without bad random rates. We could not, however, use larger electron beams. The comparison between muons and electrons is *more* favourable to muons if an accelerator (for example SLAC) with a poorer duty cycle than 10^{-2} is considered. For a duty cycle of 2×10^{-4}, τ becomes 5×10^{-5} secs and for $E = 1/10$ minutes $C/E = 8$ for cases 1 and 2 —an impossible situation. For muons, however, C/E remains tolerable at 5×10^{-4}.

As a second illustrative example, we consider a total gamma proton cross-section excluding pair production and other purely electrodynamic processes. This has not yet been done, but could surmount the problems for total γ cross-section measurement discussed earlier. The cross-section is about 100 μb from 1 to 6 BeV. This could also involve taking spark chamber pictures of each event. We can identify the event by either of two methods.

Firstly, we can count any charged particle at a large angle, such that low momentum transfer pair production is excluded. Secondly, we can use an anticoincidence method. The event detector is a total energy gamma ray shower counter in the forward direction which can detect either gamma rays or electron positron pairs produced in the absorber. The event is then identified by the presence of a scattered electron and the absence of a gamma ray or electron positron pair.

Again consider a hydrogen target 200 cm long and tag photons in intervals dk/k of 0·05 from 2 BeV to 6 BeV. We assume $\sigma(k) = 100\mu b$ from 400 MeV to 6 BeV. Then for an event rate of 1/sec for each channel we find that

$$\frac{\text{chance coincidence rate}}{\text{event rate}} = 6{\cdot}5\% \qquad \text{Case 1}$$

$$\text{,, \quad ,, \quad ,,} = 64\% \qquad \text{Case 2}$$

$$\text{,, \quad ,, \quad ,,} = 1{\cdot}3 \times 10^{-3}\% \qquad \text{Case 3.}$$

The beam intensities required are about 10^7/sec in all cases.

Note that in the case chosen here, the target is longer than R_{eff} and Case 1 appears more favourable than Case 2. There may, however, be problems about the spread of gamma ray energies due to the shower production.

It is worth noting that for a duty cycle of 2×10^{-4} (SLAC) this experiment could be operated with a count rate of 1/10 sec and a chance to event rate of 30%. An experiment of this type has in fact been proposed (Group D—SLAC). At SLAC, however, the use of a muon beam would be particularly advantageous.

7.6c. *Electron tagging with low momentum-transfer cut-off.* For completeness we note an alternative procedure to make tagging of virtual photons attractive. Muons are attractive because low momentum-transfer events (which predominate in bremsstrahlung) are automatically suppressed and do not appear in the scattered electron counting rate. Hand[7] has suggested a method of accomplishing the same result using electrons. He proposes blocking from the scattered electron detector those electrons scattered at very small angles $\theta < m/E$ and allowing only those with $\theta > 5\ m/E$ to reach the detector. This achieves artificially, but messily, the same suppression of low momentum transfer events as achieved by the use of μ mesons. However, beams of electrons are more easily available and more intense than beams of muons, so that this method has considerable merit.

Two experiments on tagging of *virtual* photons (by inelastic μ scattering) are in active preparation, one at SLAC and one at BNL. At SLAC a μ storage ring has been proposed to make this process very attractive. It is also very obviously going to be a very powerful technique for use at the 200 BeV and 300 BeV accelerators being discussed by the U.S.A. and Europe respectively. We can anticipate 50 BeV muon beams of 10^8-10^9 per second; this will enable electromagnetic interactions to be studied *without* an electron machine of this energy.

7.7. *Time-reversal experiments*

In making the average over spins, etc., in the formula for inelastic scattering, we notice that an interference term between J_T and J_0 can occur if these are complex quantities and the target is polarized.

The elastic scattering $G_M(q^2)$ and $G_E(q^2)$ are real in the space-like region by hermiticity. (See for example the proof in reference 5 on page 7.) For inelastic scattering we need to postulate time-reversal invariance. Thus a search for an interference term becomes a general test for time-reversal invariance. This was pointed out by Christ and Lee[97] in a different notation (that often used for neutrino calculations).

We find an extra term to be added to the curly bracket in the cross-section of [*107*] of the form

$$\pm AS.n\frac{E+E'}{1+\tau'}\cdot\frac{2}{|q|}\cdot[\sigma_0\sigma_T]^{\frac{1}{2}}\cot(\theta/2)$$

$$= \pm 2AS.n[\sigma_0\sigma_T]^{\frac{1}{2}}\frac{\cot(\theta/2)[1+\tau'+\cot^2(\theta/2)]^{\frac{1}{2}}}{1+\tau'} \qquad \ldots[126]$$

where

$$A = \sin\delta\ \frac{|f^-|^2}{|f^+|^2+|f^-|^2} = \frac{|j_x^2|}{|j_x^2|+|j_y^2|}\sin\delta \qquad \ldots[127]$$

and δ is the phase between the longitudinal and transverse photons (= 0 for time-reversal invariance). f^- and f^+ are helicity amplitudes, and we take the + sign in [*126*] for normal parity (*ED*, *EQ* etc.) transitions and the − sign for abnormal parity transitions (*MD*, *MQ*, etc.).

Clearly this will give an asymmetry in scattering from a polarized target. Defining

$$\varepsilon = \frac{\sigma_{\text{up}}-\sigma_{\text{down}}}{\sigma_{\text{up}}+\sigma_{\text{down}}} \qquad \ldots[128]$$

and putting $\sigma_0/\sigma_T = R^2$ and $\cot(\theta/2) = x$, we find

$$\frac{\varepsilon}{A} = \pm\frac{\dfrac{E+E'}{1+\tau'}\dfrac{2}{|q|}Rx}{\left[\dfrac{R^2+1}{1+\tau'}\right]x^2+2} = \pm\frac{\dfrac{2Rx}{1+\tau'}[1+\tau'+x^2]^{\frac{1}{2}}}{\left[\dfrac{R^2+1}{1+\tau'}\right]x^2+2}. \qquad \ldots[129]$$

The problems with the measurement lie with the technology of polarized targets. The only working ones are of lanthanum magnesium double nitrate with 24 waters of crystallization (LMN). The polarized hydrogen is 3% by weight. The asymmetry from this extra term is small (30%) and there are large radiative effects from this high *Z* target. It is estimated[98] that we must search for an asymmetry of 0·8%. It is *not* possible to use a kinematic constraint (by coincidence counting) to reduce the non-hydrogenous background. Moreover, radiation damage of the LMN crystals prevents an adequate intensity of electrons being put onto the crystal. Fortunately there has been recent success in polarizing polystyrene doped with DPPH to make the necessary paramagnetic centres; 30% to 40% polarization has been obtained. The asymmetry one might hope for is now 2% and there is probably little radiation damage.

Of course there are other ways of finding the term in [*126*] or a similar term; we may look for spin of the N^* by the angular distribution of its decay. But these are harder and less definite experimentally.

There are two tremendous attractions to this method of finding the time-reversal violations compared with others proposed by Lee. Firstly, it is fairly general; we can vary M^* and q^2 if no effect is found in one region. Secondly, the apparatus stays still and only the *RF* frequency for the polarized target is changed. There are great advantages over the recent experiments[99] on the reaction

$$\eta^0 \rightarrow \pi^+ + \pi^- + \pi^0$$

where only *one* point is obtained and the symmetry of the experimental situation and errors is not clear.

8

Two Photon Exchange Effects

8.1. *Positron scattering*

In electrodynamics each electron-photon vertex introduces into the amplitude a coefficient $e\gamma_\mu$. (See Figs. 1 and 42.) There are several processes

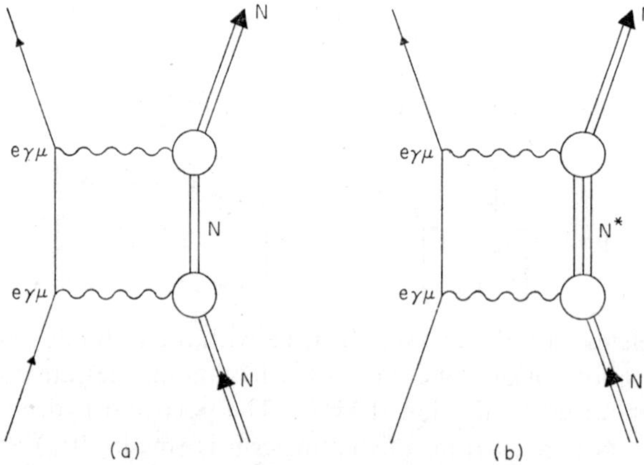

FIG. 42. This shows diagrams for two photon exchange in *ep* scattering (or hyperfine structure).

involving two photon exchange (Table VI). Thus we expect *a priori* the two photon amplitude to be lower than the one photon amplitude by $\alpha = e^2/(\hbar c) = 1/137$. But in the diagram above we note that for two photon exchange, two photons share the momentum transfer and the form factor reduction

TABLE VI

Summary of processes involving two photon exchange
1. Comparison of e^+p and e^-p scattering 2. Polarization of p in ep scattering 3. Nuclear structure term in hydrogen hyperfine structure 4. Nucleon Compton effect (via virtual photon processes)

may be less than for the one photon exchange. So at high momentum transfers two photon exchange effects may be important. The amplitude for the one photon exchange is entirely real; that for the two photon exchange may be complex.

Now if we change from electrons to positrons the amplitude of the first term changes sign and that of the second does not. Therefore, we may write

$$\sigma_{\pm} = |A|^2 = A_{1\gamma}^2 \pm 2A_{1\gamma}\,\mathrm{Re}\,(A_{2\gamma}) + |A_{2\gamma}|^2. \qquad ...[130]$$

Thus

$$R = \frac{\sigma_+ - \sigma_-}{\sigma_+ + \sigma_-} = \frac{2\,\mathrm{Re}\,A_{2\gamma}}{A_{1\gamma}} \qquad ...[131]$$

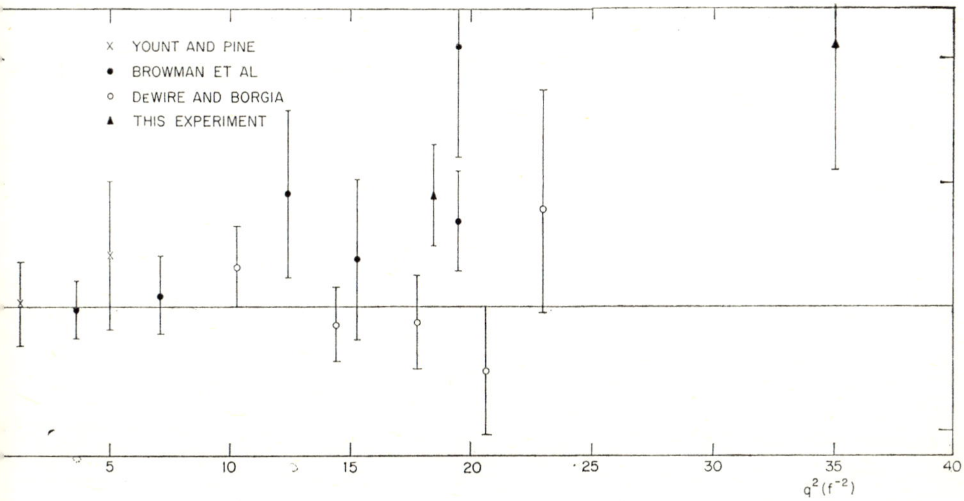

FIG. 43

Experimental data on positron-proton and electron-proton scattering versus q^2.

if $A_{2\gamma} \ll A_{1\gamma}$, and also

$$\tfrac{1}{2}(\sigma_+ + \sigma_-) = A_{1\gamma}^2 \qquad \ldots[132]$$

which can be analyzed as before, *whether or not* two photon effects exist. It is therefore extremely attractive to measure both positron-proton and electron-proton scattering. The problem has been that available positron intensities are small. Typically, positron intensities have been 10^{-4} of electron intensities. Nonetheless, four experiments have been performed.[100, 101, 102, 103] All but the second were at small angles ($\theta \sim 45°$). The data is in Fig. 43.

It seems that there is a significant effect at momentum transfer above $1(\text{BeV/c})^2$. (See Table VII.) Although the results of the one photon exchange analysis given earlier will have to be altered, the major conclusions will be unaffected. These conclusions, we noted earlier, are based almost entirely on the slopes of the form factors at the origin and on the asymptotic limits as $q^2 \to \infty$, both of which are almost unaffected. As SLAC gets on the air, and when CEA gets its new positron injector, we can expect positron scattering data to be standard routine.

TABLE VII

	$q^2(\text{f}^{-2})$	E_{inc}(MeV)	$\theta°$	$100R = \frac{\sigma_+ - \sigma_-}{\sigma_+ + \sigma_-} \times 100$
Stanford	7	710	50°	−0·05 ± 1·5
	12·4	600	100°	−4·5 ± 3·3
	15·3	850	685°	1·9 ± 3·2
	15·9	810	80°	−0·4 ± 4·2
	19·5	850	90°	10·3 ± 4·4
	19·5	850	90°	3·5 ± 2
Cornell	0·40 (BeV/c)²	1200	45°	2·0 ± 2
	0·52 (BeV/c)²	1200	45°	−0·5 ± 1·50
	0·65 (BeV/c)²	1200	45°	−0·5 ± 2
	0·77 (BeV/c)²	1200	45°	0·0 ± 3
Harvard	18		~30°	4 ± 2
	35		~30°	10 ± 5

Another effect can occur with two photon exchange. The two photon exchange amplitude can be spin dependent and have an imaginary part; we can now have a polarization of the outgoing proton (which would vanish for a real part). Two experiments[104, 104A] have attempted to measure the polarization of the forward proton (backward electron) in *ep* scattering by scattering the proton in carbon. The first gives $P = 0{\cdot}040 \pm 0{\cdot}027$ at $q^2 = 15{\cdot}4\,\text{f}^{-2} = 0{\cdot}6(\text{BeV/c})^2$. The second gives $P = 0{\cdot}024 \pm 0{\cdot}050$ at $q^2 = 20\,\text{f}^{-2}$.

We now ask what to expect. There are several calculations. The most elementary is that of Lewis[105] who, in the approximation of a static charge distribution and that magnetic scattering can be neglected, found a number which we represent approximately by

$$R = 0{\cdot}03(q^2/20) \quad (q^2 \text{ in f}^{-2}) \qquad \ldots[133]$$

which is a little smaller than the data. But the theory is overly naïve. It seems to correspond to taking the nucleon pole term in the diagram of Fig. 42a. Iddings and Platzman[111] calculated just this term for the hydrogen hyperfine structure which we discuss later. There are two approaches to improve the theory.

The first is that of Gourdin,[6] which is to consider the different possible angular momentum states of the exchanged multiphoton system, Fig. 44.

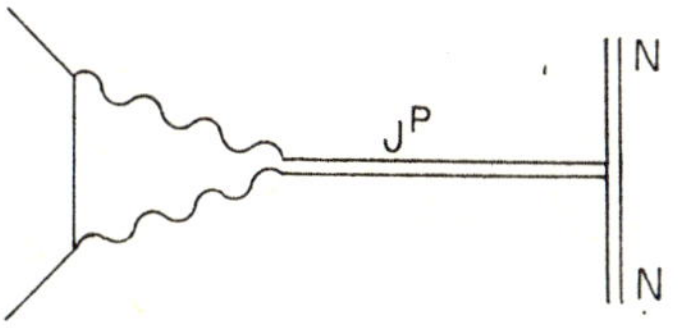

FIG. 44. A diagram for a two photon interaction with exchange of a definite angular momentum.

This is appropriate if our deviation is due to such an exchange. The method is most easily seen by considering the crossed channel

$$e^+ + e^- \rightarrow p + \bar{p}$$

where now we are expanding into partial waves of the $p\bar{p}$ system. It is well known that the same problem arises in elastic pp scattering for example, that for a given angular momentum J of the $p\bar{p}$ system, there exists a maximum complication in the angular distribution.

The cross-section, leaving out trivial multiplicative terms, varies as

$$\sigma \propto \sum_{l=0}^{l=2J} A_l(q^2) P_l(\cos \phi_p). \qquad ...[134]$$

Now we can transform the cos ϕ_p in the $p\bar{p}$ system to the laboratory electron scattering system.

$$\cos^2 \phi_p \rightarrow 1 + \frac{\cot^2(\theta/2)}{1+\tau}. \qquad ...[135]$$

Since charge conjugation for the exchange of one photon changes the amplitude A_p by $(-1)^l$, we *cannot* have a term A_1 for one photon and we obtain only terms A_0 and A_2 and get back to the formula

$$\sigma \propto (C + D \cot^2(\theta/2)) \qquad ...[136]$$

that we had before. (Now A_0 and A_2 are still a *different* pair of form factors from either of the pairs F_1, F_2 or G_E, G_M!)

Now let us consider the other states of the $p\bar{p}$ system up to $J = 1$ and the possible numbers of exchanged photons:

$$\left.\begin{array}{llll} {}^1S_0 & J=0^- & C=+1 & n=2,4\ldots \\ {}^3P_0 & J=0^+ & C=+1 & n=2,4\ldots \\ {}^3S_1 \text{ or } {}^3D_1 & J=1^- & C=-1 & n=1,3\ldots \\ {}^3P_1 & J=1^+ & C=+1 & n=2,4\ldots \\ {}^1P_1 & J=1^+ & C=-1 & n=3,5\ldots \end{array}\right\} \quad \ldots[137]$$

The cross-section arises from interference among these states. We only get an interference between two amplitudes if

$$|J_1 - J_2| \leqq l \leqq J_1 + J_2 \quad \ldots[138]$$

and also if

$$(-1)^l = P_1 P_2. \quad \ldots[139]$$

Now we note that when $q^2 \gg m^2$ we have helicity conservation at the $e\gamma$ vertices. This works in such a way as to annul the $J = 0^+$, $J = 1^-$ interference which would lead to a term in $\cos\phi$.

Thus, neglecting 1^+ exchange, we still get our form of [*136*] but now the terms can no longer be identified with G_E and G_M. *Thus we have a case where the Rosenbluth formula holds exactly but we have two photon exchange* (and a difference between positron and electron scattering).

Next assume we have exchange of states 1^+, $C = +1$, and we have

$$d\sigma \propto \left\{A(q^2)\cot^2(\theta/2) + B(q^2) + C(q^2)\left[1 + \frac{\cot^2(\theta/2)}{1+\tau}\right]^{\frac{1}{2}}\right\} \quad \ldots[140]$$

from the extra term in $\cos\phi$. Can we find this term? We easily see that there is again almost a straight line if $\tan^2(\theta/2) > 1(\theta > \pi/2)$. A term C of the same order of A and B can give a 10% deviation from straight line behaviour at $\theta \sim 10°$. But we already have trouble separating A and B from a straight line plot at high momentum transfer! We cannot find C even when of the same order as A (and producing a large effect in the positron/electron ratio). We can consider terms for $J > 1$. But now we get more terms in $\cot^2(\theta/2)$ and the cross-section becomes too large for unitarity at $\theta = 0$. We have met this before; it is the well known Regge behaviour. But since we are dealing with interactions of order $\alpha = 1/137$, we might expect the higher power of $\cot^2(\theta/2)$ to hold up to an increase of 137 in the cross-section and finally be limited by an iteration of a higher J exchange.‡

Thus we may discuss $J = 2$ exchange, as did Flamm and Kummer.[106] They find of course that there is an interference term

$$\cot^2(\theta/2)\left[1 + \frac{\cot^2(\theta/2)}{1+\tau}\right]^{\frac{1}{2}} \quad \ldots[141]$$

which should show up at small angles. Thus the only hope for a large deviation from a Rosenbluth plot appears to be at small angles.

‡ I am indebted to S. D. Drell for a discussion of this.

We now follow a second approach, that of Drell and Fubini.[107, 108] We no longer break up into partial waves of the exchanged particle. Instead, we consider the Compton scattering of the virtual photons accompanying the electrons by the proton. This uses a dispersion relation, and Drell and Fubini have included terms including the $N^*(1238)$ resonance. Thus they are concerned *not* with a single angular momentum state (and possibly resonance) of the many photon system, but many states, arising from contributions from resonances in the intermediate nucleon system. (See Fig. 42b.) Drell and Fubini did *not* find large effects. For the polarization experiment Guerin[71] finds $p < 0{\cdot}01$ using the isobar model—equivalent to Drell and Fubini's[107] calculation. These two complementary approaches were compared by myself elsewhere.[109]

8.2. *Hydrogen hyperfine structure*

I shall now run quickly through a series of experiments involving two photon exchange effects in the hope that someone can connect them all (Table VI). Firstly consider the hyperfine structure of hydrogen. There exists a fantastically accurate (12 figures) measurement of this quantity.[110] $\nu = 1{,}420{,}405{,}751{\cdot}800 \pm 0{\cdot}028$ cycles per second. The theory is straight forward but complex. The energy difference becomes

$$\Delta E = \frac{32\pi\alpha^2}{3} R_\infty \frac{\mu_p \mu_e}{\mu_0^2} \left(1+\frac{m}{M}\right)^{-3} (1+\tfrac{3}{2}\alpha^2)\, \mathscr{E}\mathscr{R}\mathscr{P}. \qquad \ldots[142]$$

R_∞ is the Rydberg for an electron in a *fixed* state,
μ_e/μ_0 is the magnetic moment of the electron in Bohr magnetons,
μ_p is the magnetic moment of the proton,
$\mathscr{E}$ is a small correction from virtual photons surrounding the electron,
$\mathscr{R}$ is a proton recoil correction, and
$\mathscr{P}$ is the proton shape correction.

The proton shape correction $\mathscr{P}$ is primarily a two photon correction because if a single high energy photon is exchanged between the proton and electron they could not stay bound, but two can be exchanged of opposite momenta. Only high energy photons can give a structure dependent term. The calculations[111, 112, 113] are similar to those of Drell and Fubini and use a dispersion relation to describe the intermediate nuclear state. The major part of the correction

$$\mathscr{P} = (1-23)\times 10^{-6} \qquad \ldots[143]$$

is due to the nucleon term with simple form factors at the vertices.

Of the various numbers, most are known very accurately.[143] R_∞ is known to one part in 10^7,

$$R_\infty = 1093731\ \text{cm}^{-1} + 0{\cdot}1\ \text{ppm}. \qquad \ldots[144]$$

μ_e/μ_0 is known both experimentally and theoretically to this order as we shall see in a moment. Until recently, μ_p was only measured in mineral oil or water and a diamagnetic correction was necessary. Although this correction (due to Ramsey) was believed well known, it is pleasing that a measurement has now been made in atomic hydrogen where this correction is smaller and better known.[142] Fortunately, the early ideas were right. Thus the only uncertainties are those of α and of the proton structure correction $\mathscr{P}$. We can regard the hyperfine structure as a measure of α if we calculate $\mathscr{P}$ or a measure of $\mathscr{P}$ if we know α.

This new measurement of μ_p/μ_e may be understood again by reference to the Breit-Rabi diagram of Fig. 45. We have at low field ($B = 0$) a measure of the hyperfine structure separation just discussed. ($W/h = 1{\cdot}4 \times 10^9$ cps.) This was measured by a hydrogen maser. A similar device, operated at a high field, $B = 3500$ gauss, measures μ_p/μ_e. The field must be kept constant and uniform to about 10^{-7}. This is not enough to obtain maser oscillation, but enough for a measurement.

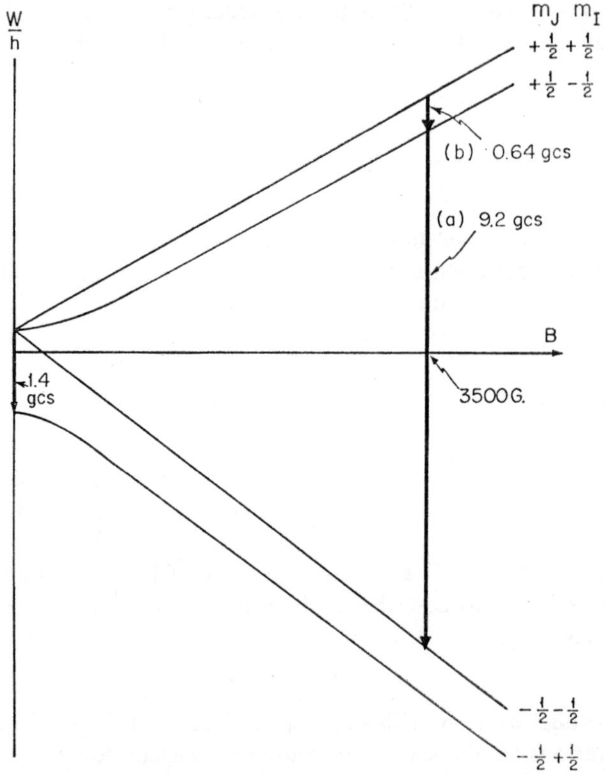

FIG. 45. Breit-Rabi diagram for hydrogen hyperfine structure showing the energy versus magnetic field B.

For $B = 0$ we measure the hyperfine separation. At high field ($B = 3500$ gauss) we can measure μ_p/μ_e.

The transitions

$$[M_J = +\tfrac{1}{2},\ M_I = +\tfrac{1}{2}] \rightarrow [M_J = +\tfrac{1}{2},\ M_I = -\tfrac{1}{2}]$$

and

$$[M_J = +\tfrac{1}{2},\ M_I = -\tfrac{1}{2}] \rightarrow [M_J = +\tfrac{1}{2},\ M_I = -\tfrac{1}{2}]$$

are both measured and correspond to proton spin flip and electron spin flip respectively. The interaction of the proton is primarily with the electron and only secondarily with the external field, so that it must be measured especially well. The authors find for the ratios of the electron and proton g factors in hydrogen

$$g_j(H)/g_p(H) = 658{\cdot}2105\ (0{\cdot}3\ \text{ppm.}). \qquad \ldots[\mathit{145}]$$

This is accurate enough, and in a simple enough atom, to remove the measurement of μ_p/μ_0 as a contender for the discrepancy.

Two measurements are available of α. The main one is the fine structure in deuterium.[123] It is of course logical that the measure of the fine structure constant should come from the process which gives it its name. After minor adjustment we get

$$\alpha^{-1} = 137{\cdot}0388\ (\pm 9\ \text{ppm.}). \qquad \ldots[\mathit{146}]$$

Another measure comes from the hyperfine structure in muonium. This is the same process as we have just discussed for muonium in a hydrogen-like atom.[118] We must now, however, use the muon magnetic moment which is measured to be[141]

$$\mu_\mu/\mu_p = 2{\cdot}18338\ (\pm 13\ \text{ppm.}) \qquad \ldots[\mathit{147}]$$

and there is believed to be no structure of the muon. Thus we can now use the experimental value[115]

$$\Delta\nu\ (\text{expt}) = 4463{\cdot}16 \pm 0{\cdot}05\ \text{Mc/s} \qquad \ldots[\mathit{148}]$$

and we can then *derive* α,

$$\alpha^{-1} = 137{\cdot}0388\ (\pm 9\ \text{ppm.}) \qquad \ldots[\mathit{149}]$$

where the quoted error is believed by many to be a little less conservative than that of [*146*].

Doubt has recently been cast on both of these measures of α. Robiscoe[124] has suggested that a similar measurement, that of the Lamb shift, was too low. If the measurement of deuterium fine structure is low by the same amount, the discrepancy vanishes. But as we see later, this gets us into trouble with quantum electrodynamics. Lamb's measurement is being repeated and the diamagnetic correction for muons in water is believed (by Ruderman[145]) to be incorrectly taken, and therefore [*147*] is in doubt.

Be this as it may, we can use these values of α to derive the theoretical value for the hydrogen hyperfine structure. We find

$$\frac{\delta E_{(\mathrm{exp})}-\delta E_{(\mathrm{theor})}}{\delta E} \simeq (40 \pm 15) \times 10^{-6} \qquad \ldots[150]$$

where the error is in α and includes an average of both measurements of references 123 and 118. This naturally leads to speculation that we made it wrong and $\mathscr{P}$ is really unity.

With this prompting, Nambu[114, 116] has introduced an axial-vector meson in the electron-proton interaction which has $J = 1^+$ and can introduce large two-photon interactions at small angles, as we saw earlier. His fit to the e^+p/e^-p scattering ratio (Table VII and Fig. 43) gives the wrong sign. Nambu[116] also tried a model based on the quark model in which the electron and proton structures never overlap, but the charge and magnetic moment distributions are due to jittering. This has been extended by Drell[117] using a dispersion method. None of these unconventional approaches has been pushed far enough to see what effect it has on other processes.

If we take this discrepancy seriously, we may wish to try to relate Iddings dispersion relation directly to experimental quantities as a check instead of using a model. However, he relates his correction to the scattering of transversely polarized electrons by polarized protons; an experiment which will not soon be performed. There seems to be no disagreement with the computations in spite of two entirely different styles of calculation—Feynman's and Schwinger's (by Iddings[113] and Zemach[112] respectively).

The present indications are that both measures of α may be changed in such a direction as to reduce the discrepancy.

Note added in proof: A recent measurement of hc/e from measuring current jumps in superconducting Josephson junctions excited by microwave radiation,[146] when combined with other measurements of atomic constants,[143] tends to confirm this suggested change in α.

8.3. *Unitarity*

There must clearly be some relation between elastic and inelastic scattering and the size of two photon terms by unitarity. I prefer to see the necessity of such a connection by the following picture. Scattering in the 2nd Born approximation can occur by exciting the nucleon and then de-exciting it; inelastic scattering tells us the probability of this excitation. At high momentum transfers, excitation of the various resonances is comparable to elastic scattering; this is not large and is barely enough to overcome the energy denominator which will appear in the second Born approximation calculation. Thus we see here no good reason for increasing Lewis' calculation much.

The other two photon process is clearly the Compton effect which is now being measured more and more. At forward angles, of course, the forward Compton amplitude is related to the total γ cross-section.

9

QUANTUM ELECTRODYNAMICS EXPERIMENTS

9.1. *General*

No set of talks on electromagnetic interactions is complete without a discussion of experiments which bear on the question of the validity of the theory. It is a remarkably accurate theory and can calculate with great precision fine details—such as the hydrogen hyperfine structure we have discussed —to a precision of 1 in 10^{10} of the energy levels themselves. Perhaps, however, the discrepancy described earlier lies in quantum electrodynamics, but it is hard to interpret a single, integral measurement unless it agrees with theory. The best review of this subject remains that of Drell.[119] We can divide the experiments into two classes: those which check the structure of quantum electrodynamics, including renormalization effects, vacuum polarization, etc. at small momentum transfers and those which check it at large momentum transfers. We tabulate some of these below (see Table VIII).

TABLE VIII

Tests of Quantum Electrodynamics

LOW MOMENTUM TRANSFERS
Hydrogen hyperfine structure
Lamb shift
g-factor of the free electron
Separation between triplet states of positronium
HIGH MOMENTUM TRANSFERS
Electron-proton scattering
Electron-electron scattering by colliding beams
Positron-electron scattering
Production of electron pairs at large angles
g-factor of free muon

9.2. *Lamb shift*

For the separation between the $^2S_{1/2}$ and $^2P_{1/2}$ states in atomic hydrogen, (see Table IX), of the theoretical value, 0·12 mc/s is due to a proton size of 0·8 f. Turning this around, a disagreement of 0·3 mc/s could be due to a deviation at a radius of 2·4 f. We do not know whether to take the discrepancy seriously, but Robiscoe has two separate measurements which agree with each other.

TABLE IX

Lamb Shift

Theory	(Layzer[122])	1057·70 ± 0·15 Mc/s.
Experiment	(Lamb[123])	1057·77 ± 0·10 Mc/s.
Experiment	(Robiscoe[124])	1058·07 ± 0·10 Mc/s.
Theory	(Erickson[122])	1057·64 ± 0·21 Mc/s.
Theory	(M. Soto[122])	1057·50 ± 0·11 Mc/s.

9.3. *g-factor of electron*

For the *g*-factor of the electron (Fig. 46) we have[120, 121]

$$\frac{g-2}{2} = \frac{\alpha}{2\pi} - 0{\cdot}328\left(\frac{\alpha}{\pi}\right)^2 + 0{\cdot}15\left(\frac{\alpha}{\pi}\right)^3 \qquad ...[151]$$

where the third term is an estimate from a dispersion relation.[121] Experiment[125] gives

$$\frac{g-2}{2} = \frac{\alpha}{2\pi} - (0{\cdot}327 \pm 0{\cdot}005)\left(\frac{\alpha}{\pi}\right)^2 \qquad ...[152]$$

which is too poor an accuracy to check the $(\alpha/\pi)^2$ term (or to measure α independently or to verify QED at small distances).

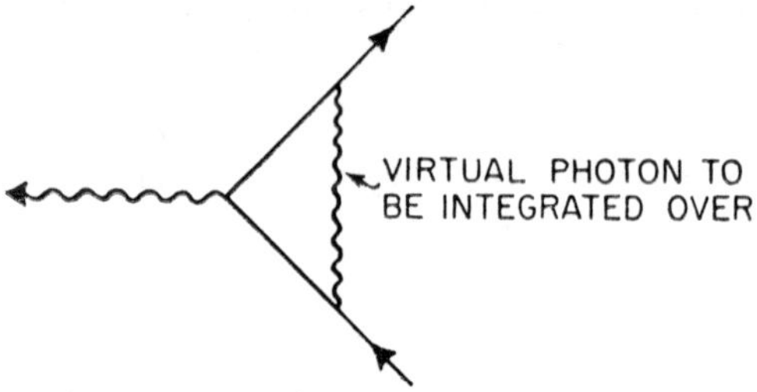

FIG. 46. Principal correction to the magnetic moment of the muon (or electron) to show how the photon (or lepton) propagator enters into the theory.

9.4. *Positronium triplet-singlet separation*

The positronium separation is

Theory[126]	$2{\cdot}0337 \times 10^5$ Mc/s
Experiment[127]	$2{\cdot}0338 \pm 0{\cdot}0004 \times 10^5$ Mc/s.

9.5. *Modification of propagators*

Now we consider the high momentum transfer measurements. We cannot discuss a verification without a phenomenology to describe a deviation. This comes from modifying the propagator. We replace the photon propagator $1/q^2$ by

$$1/q^2 \rightarrow [1/q^2 - 1/(q^2 - \Lambda^2)] = \frac{1}{q^2}\,\frac{1}{1 - q^2/\Lambda^2} \qquad ...[153]$$

with perhaps a similar modification for the electron propagator or the $e\gamma$ vertex in spite of the result, by now well known, that these modifications are not independent and do not always satisfy general field theory requirements.

9.6. *g-factor of free muon*

Neglecting this non-trivial detail, we consider modifying the first order correction to the g-factor of the muon. We replace the virtual photon propagator by that of [*153*] to get

$$\frac{g-2}{2} = \frac{\alpha}{2\pi}\left(1-\tfrac{2}{3}\left(\frac{m}{\Lambda}\right)^2\right) \qquad \ldots[154]$$

where m is the mass of the muon. (Incidentally, we see here that because of its lower mass the electron is less useful in such a study.) The g-factor of the muon is

$$\left(\frac{g-2}{2}\right)_{\text{theory}} = \frac{\alpha}{2\pi}+0{\cdot}76\left(\frac{\alpha}{\pi}\right)^2 \text{ (reference 128)} \qquad \ldots[155]$$

$$\left(\frac{g-2}{2}\right)_{\text{exp}} = 0{\cdot}001162\pm0{\cdot}000005 \text{ (reference 129)}. \ldots[156]$$

From this we derive $1/\Lambda < 0{\cdot}25$ f, $\Lambda > 0{\cdot}8$ BeV/c (for 95% confidence).

9.7. *Electron-electron scattering*

Another recent experiment is with the 300 MeV colliding beams of electrons described earlier.[130] They have 380 events for e^-e^- scattering between 40° and 90°. Fitting these data to the Møller scattering formula we derive for the regulator Λ^2 in the photon propagator:

$$1/\Lambda^2 = +0{\cdot}001\pm0{\cdot}033 \text{ f}^2. \qquad \ldots[157]$$

This puts $1/\Lambda < 0{\cdot}18$ f, $\Lambda > 1{\cdot}1$ BeV/c for one standard deviation or, $\Lambda > 0{\cdot}76$ BeV/c for 95% confidence. It is hard to say more about this beautiful experiment. I have dwelt on the problems of a storage ring, but the experiment itself is simple and clean.

9.8. *Electron-positron scattering*

Another experiment, giving not quite so good a limit, is that of e^+ collision with e^- at rest.[131] One percent measurements have been made of the 180° cross-section at both 300 MeV and 500 MeV and compared with theory with two forms for the radiative correction. The form e^δ is to be preferred theoretically. (See Table X.) This gives $\Lambda > 0{\cdot}4$ BeV/c.

TABLE X

	$\sigma_{th}(1+\delta)$ $\times$ 10^{29} cm²	$\sigma_{th}\exp(\delta)$ $\times$ 10^{29} cm²	σ_{exp} $\times$ 10^{29} cm²	$\frac{\text{experiment}}{\text{theory }(1+\delta)}$	$\frac{\text{experiment}}{\text{theory }(\exp(\delta))}$
500 MeV	0·9222	0·9363	0·9384	1·018±0·011	1·003±0·011
200 MeV	2·303	2·333	2·329	1·011±0·009	0·998±0·009
ratio $\left(\frac{\sigma_{500}}{\sigma_{200}}\right)$	0·4001	0·4013	0·4031	1·007±0·009	1·005±0·009

(Table VI of reference 131)

9.9. *Electron-proton scattering*

Electron-proton scattering can also tell us about the photon propagator. We could consider the discrepancy between the simple theory of the form factor—with vector meson poles alone—and experiment, as a breakdown of quantum electrodynamics. From this we put $\Lambda \simeq 1000$ MeV/c. Thus we see that we must trust QED to a degree far beyond the precision of present experiments to interpret the form factors as we have done. We can turn this around; if we believe the ρ, ω and ϕ poles contribute as much as SU(3) predicts, then we are left with the term $1+q^2/\Lambda^2$ where $\Lambda = 1$ BeV/c. So we have tested QED to this order. This is as well as any direct experiment so far. If we believe Furuichi and Watanabe's calculation[61] too, we are left with 2 BeV/c as a lower limit for a deviation of the photon propagator.

9.10. *Electron-positron pair production*

Lastly, we come to three experiments on wide angle pair production by photons.[132, 133, 134] The theory is straightforward but algebraically complex. The idea, originally due to Feynman and worked out by Drell and collaborators,[135] is to measure pair production at large angles. The theory calculates the two diagrams of Fig. 47.

We note one or two interesting kinematic features of these diagrams. First, when the electron-positron pairs are produced at symmetric angles

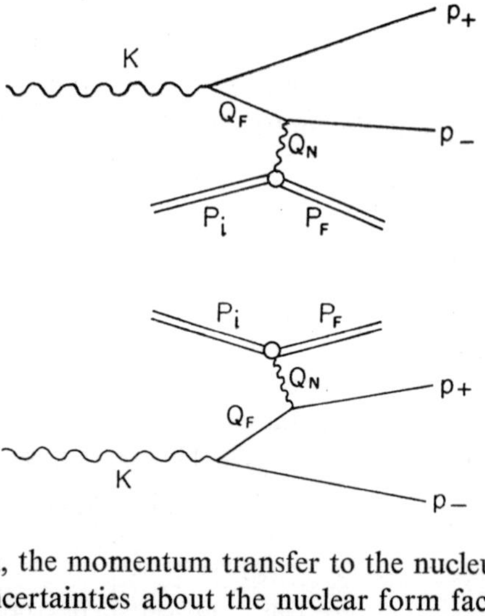

FIG. 47

The Bethe-Heitler diagrams for photoproduction of *e* pairs or μ pairs.

about the incident γ ray direction, the momentum transfer to the nucleus is a minimum and therefore any uncertainties about the nuclear form factors involved are reduced. Secondly, for small angles we can write

$$Q_F^2 = \theta^2 k^2/2 \qquad \ldots[158]$$

but

$$Q_N^2 = \theta^4 k^2/4. \qquad \ldots[159]$$

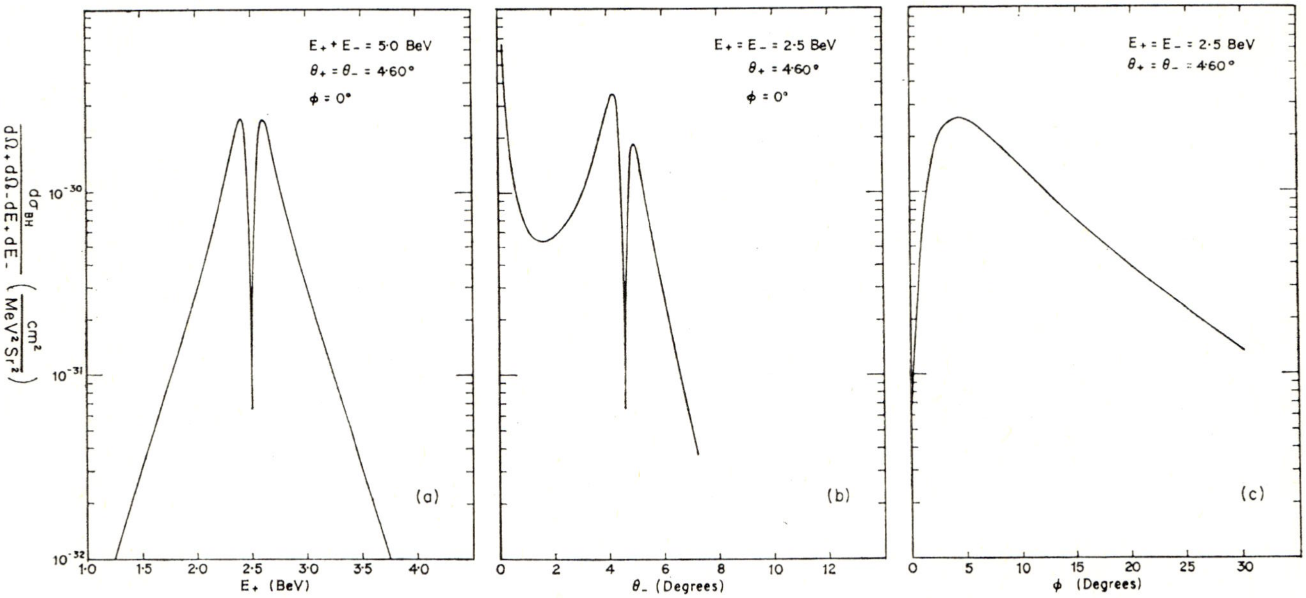

FIG. 48

The Bethe-Heitler cross-section as a function, in turn, of E_+, θ_-, and ϕ, showing the kinematic dip at $\sigma_T = 0$.

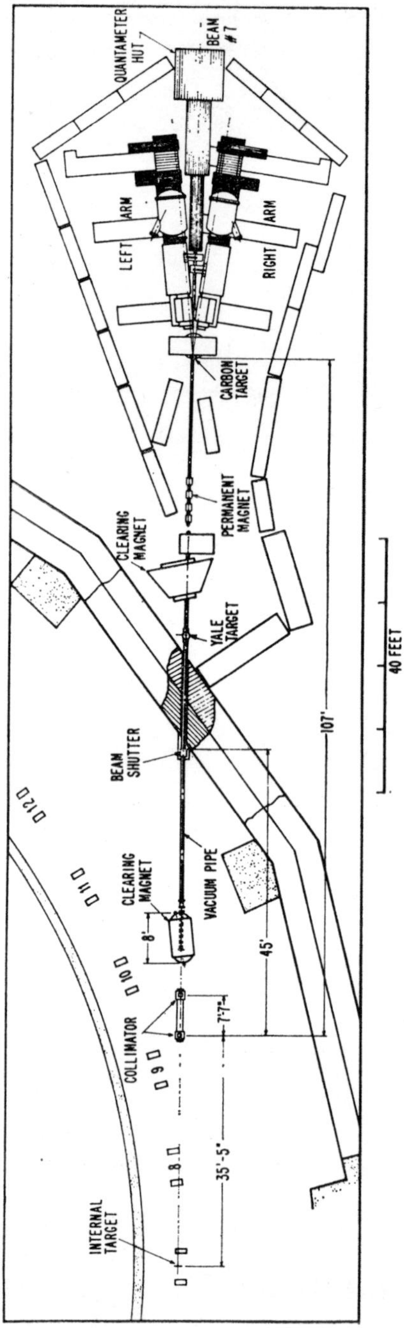

FIG. 49

The apparatus used at CEA to detect the e^+e^- pairs.

Thus the experiment is best performed at large angles. Thirdly, we have a very subtle point; at the minimum the momentum transfer to the nucleus is longitudinal and there is a deep narrow minimum, of order $(\Delta\theta)/\theta \simeq m/k$, in the cross-section (a deeper minimum for electrons than muons). This subtle point may be related to the discrepancies which follow. Some calculations of this are shown in Fig. 48. Fourthly, near this minimum, interfering ampli-

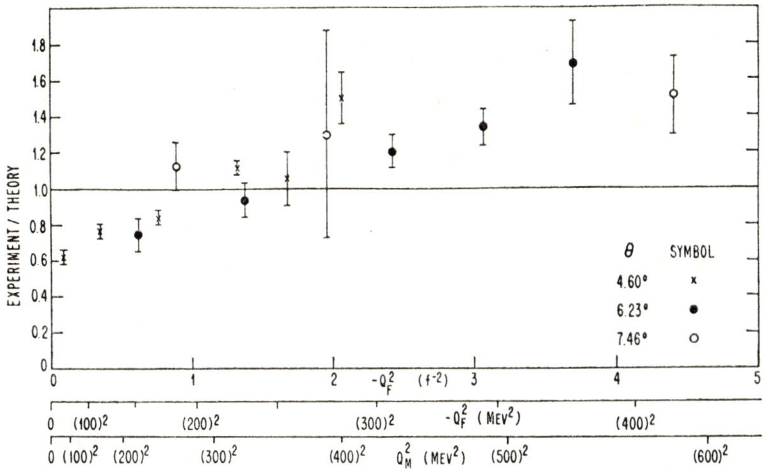

FIG. 50*a*. This figure gives the ratio of the experimental yields to the calculated yields. R is shown as a function Q_f^2 of the virtual electron (and Q_M^2 the invariant mass of the e^+e^- pair).

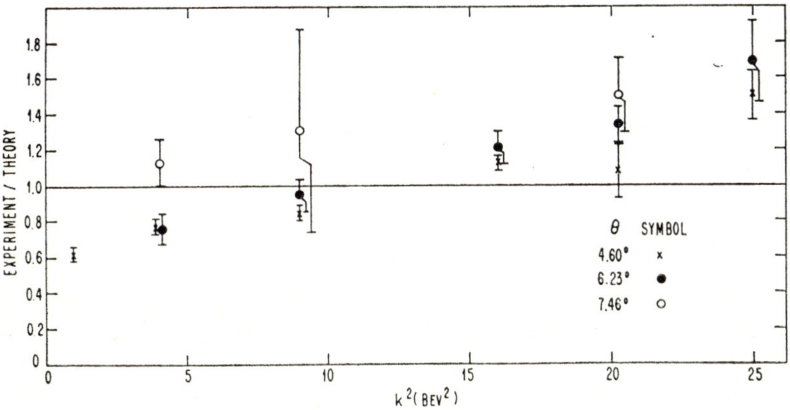

FIG. 50*b*. This shows R as a function of the photon energy $k = E^+ + E^-$. The normalization error is now understood.

tudes change sign and their effect should average to zero if data is taken symmetrically about the minimum.

An experiment on e^+e^- pairs has been carried out at the CEA by Pipkin and collaborators from Harvard.[133] The angular resolution—limited largely by multiple scattering—was not enough to resolve the deep minimum in the theory. Pipkin used two simple electron spectrometers—similar to those I described earlier for electron scattering. His solid angles were defined by the quadrupole magnet faces, but, as we discussed earlier, this should be good for electrons. Electrons are positively separated from pions by a large shower and by a gas Cerenkov threshold counter as in the electron scattering apparatus.

The apparatus is shown in Fig. 49 and the results in Figs. 50a and 50b. The results are found to *disagree* with theory. This disagreement increases with energy or momentum transfer. If we fit by simple relations, we can find the ratio between experiment and theory.

$$R = 0{\cdot}67\,\{(1{\cdot}00 \pm 0{\cdot}04) + Q_F^2/(313 \pm 13\ \text{MeV})^2\} \qquad ...[160]$$

or an alternative fit

$$R = 0{\cdot}62\,\{(1{\cdot}00 \pm 0{\cdot}05) + k^2/(4{\cdot}31 \pm 0{\cdot}17\ \text{BeV})^2\}. \qquad ...[161]$$

There are two distinct aspects of this disagreement, firstly the normalization and secondly the slope as a function of Q_F^2 or k^2. The normalization may be due to counter inefficiencies or monitoring and it is easier to understand how there could be experimental errors in normalization than in the energy dependence. If we assume that there is a normalization error, and Pipkin informs me he now understands this, we are left with a discrepancy—which could be either as a function of Q_F^2 or, a better fit to the data, a function of k^2. If we then replace the electron propagator by the form of [*153*] we get $\Lambda = 313$ MeV, a very low result.

If we assume that [*161*] represents the data better than [*160*], we might expect the deviation to hold down to *low* Q_F values but high k values (in spite of the fact that we could transform this to another frame where k is smaller). Then it is not foolish to search for a deviation in the *total* pair-production cross-section. This has been done by an attenuation measurement to 2% at 4·5 BeV.[136] The result agrees with theory to this accuracy, though the theory was limited by the accuracy with which corrections due to atomic shielding could be applied.

The radiative corrections to the pair-production process have been suggested as a possible answer to the problem. Pairs could be produced at an angle not quite symmetric and then a low energy gamma ray produced to push them into symmetry and thereby fill in the narrow minima of Fig. 48. This has been worked out by Lomon[137] and by Perrin.[138] Although the radiative corrections simulate the form of equation [*161*], they are not large enough to explain the constants. But they suggest that the basic breakdown

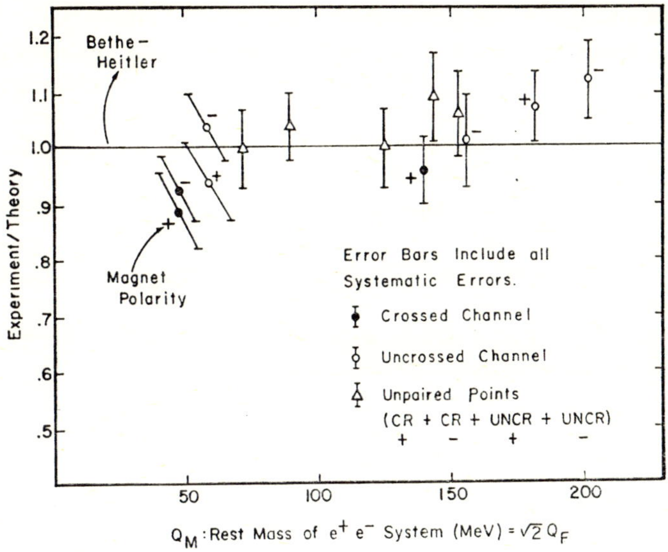

FIG. 51. This shows that ratio *R* of experimental to theoretical yields from the Cornell measurements.

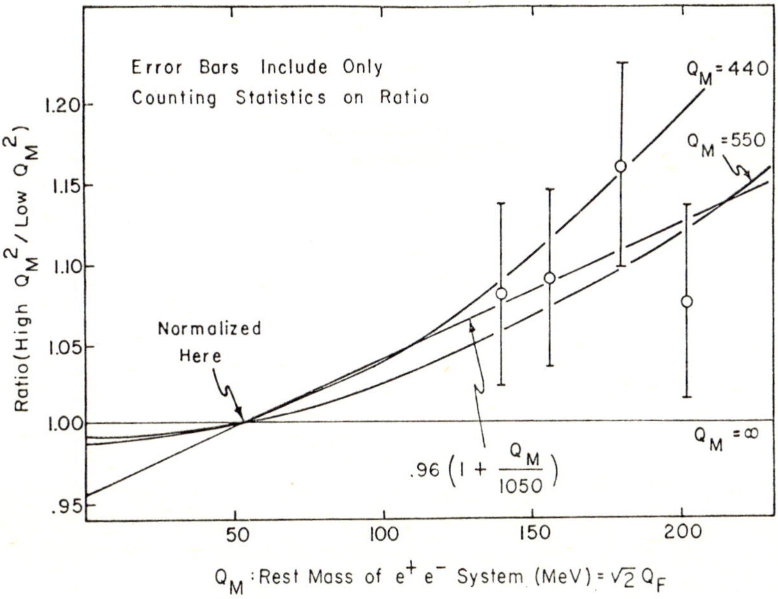

FIG. 52. This shows the ratios *R* from Cornell measurements, normalized to unity low masses.

could be at a distance *smaller* than given by $\Lambda^{-1} = (300\text{Mev/c})^{-1}$, say $\Lambda^{-1} = (2500\text{MeV/c})^{-1}$, but artificially enhanced by the kinematics.

An experiment has also been performed at lower energy and momentum transfer at Cornell.[134] Again, a discrepancy is reported; the measurements were made over only a narrow range of incident photon energies, but a reasonable range of Q_F^2. In Figs 51 and 52 their data are plotted, agreeing with a deviation of the form and magnitude of [*160*] or [*161*] but not with a variation linear in Q_F, which might otherwise fit Pipkin's data. The experiments are being repeated at CEA (two experiments) and DESY; muon bremsstrahlung is being studied at Brookhaven; thus in a year or two one should have a clear picture.

Note added in proof: The DESY experiment disagrees with that of reference 133 and agrees with quantum electrodynamics.

The experiments on the photoproduction of the muon pairs[132] have been described already in so far as they lead to a measure of the branching $\rho^0 \rightarrow \mu^+ + \mu^-$. The muons are identified by range, and their angle measured by counter hodoscopes. This method of separating μ mesons from π mesons only worked well at energies above 2 BeV (photon energies about 4 BeV). Thus the μ pair experiment has not been performed over a large range of photon energies, but the statistical precision is greater than that of the electron pair experiment. The latest results of Fig. 24 show a slightly falling pair cross-section (as Q_F^2 increases and k is kept constant). An equivalent slicing through the data of Fig. 50 gives a rising cross-section if Q_F^2 is kept constant, but only slightly so if we allow a dependence on k. (See [*161*].)

9.11. *Excited lepton states*

Low[139] has postulated an excited electron which decays into a gamma ray and an electron in order to explain these results. In particular he suggests it could appear in a virtual line in Fig. 47. Then we would have a coupling $e'\gamma e$ and we should produce them in the reaction

$$e + p \rightarrow e' + p.$$

Several searches[140] have failed to find them through such a coupling. Of course the particles could exist without any other large coupling except through their charge. They could still be produced copiously in an e^+e^- storage ring.

I close by saying that the cleanest method of studying quantum electrodynamics remains the storage ring—although it studies the photon propagator which may not deviate in the same way as the electron propagator, which is important for pair experiments. But we are intrinsically capable of pushing the limit on Λ to 20 BeV/c, which is high enough to correct electron scattering data.

REFERENCES

1. D. R. Yennie, M. M. Levy and D. G. Ravenhall. 1957. *Rev. Mod. Phys.* **29**, 144.
2. L. N. Hand, D. G. Miller and Richard Wilson. 1963. *Rev. Mod. Phys.* **35**, 335.
3. R. R. Wilson and J. S. Levinger. 1964. *Ann. Rev. Nucl. Sci.* **14**, 135.
4. T. A. Griffy and L. I. Schiff, in *High Energy Physics* (to be published).
5. S. D. Drell and F. Zachariasen. 1961. *Electromagnetic Structure of the Nucleon* (Oxford University Press, Oxford).
6. M. Gourdin. 1966. *Diffusion des Electron de Haute Energie* (Manon et Cie, Paris).
7. L. N. Hand and Richard Wilson. 1963. SLAC Summer Study Report, SLAC 25-II (unpublished).
8. D. R. Yennie, M. M. Levy and D. G. Ravenhall. 1957. *Rev. Mod. Phys.* **29**, 144.
9. D. J. Drickey and L. N. Hand. 1962. *Phys. Rev. Lett.* **95**, 521.
10. T. Janssens, R. Hofstadter, E. B. Hughes and M. R. Yearian. 1966. *Phys. Rev.* **142**, 922.
11. H. Winick. 1965. CEAL-1015 (unpublished).
12. P. Lehmann, R. Taylor and Richard Wilson. 1962. *Phys. Rev.* **126**, 1183.
B. Dudelzak, G. Sauvage and P. Lehmann. 1963. *Nuovo Cim.* **28**, 18.
13. K. W. Chen, J. R. Dunning, Jr., A. A. Cone, N. F. Ramsey, J. K. Walker and Richard Wilson. 1966. *Phys. Rev.* **141**, 1267.
14. J. R. Dunning, K. W. Chen, A. A. Cone, G. Hartwig, N. F. Ramsey, J. K. Walker and Richard Wilson. 1966. *Phys. Rev.* **141**, 1286.
15. L. H. Chan, K. W. Chen, J. R. Dunning, Jr., N. F. Ramsey, J. K. Walker and Richard Wilson. 1966. *Phys. Rev.* **141**, 1298.
16. J. F. Hague, R. F. Jennings and R. E. Rand. 1956. *Nucl. Inst.* **24**, 1956.
17. K. L. Brown and G. W. Tautfest. 1956. *Rev. Sci. Instr.* **27**, 696.
18. Mircea Fotino (private communication).
19. R. R. Wilson. 1957. *Nucl. Inst.* **1**, 101.
20. N. T. Meister and D. R. Yennie. 1963. *Phys. Rev.* **130**, 1210.
21. M. Conversi, T. Massam, Th. Muller and A. Zichichi. 1965. *Nuovo Cim.* **40**, 690.
22. R. Gomez, D. Hartell, A. Mashke, G. Pine, A. Tollestrup and T. Zipf, an AGS experiment (unpublished).
23. E. Fermi and L. Marshall. 1947. *Phys. Rev.* **72**, 1139.
24. V. E. Krohn and G. R. Ringo. 1965. *Phys. Lett.* **18**, 297.
25. E. Melkonian, B. M. Rustad and W. W. Havens. 1959. *Phys. Rev.* **114**, 1571.
26. O. Halpern. 1964. *Phys. Rev.* **133**, B579.
27. D. Hughes, J. A. Harvey, M. D. Goldberger and M. J. Stafue. 1953. *Phys. Rev.* **90**, 497.
28. L. L. Foldy. 1958. *Rev. Mod. Phys.* **30**, 473.
29. L. Durand, III, P. C. DeCelles and R. B. Marr. 1962. *Phys. Rev.* **126**, 1882.
30. M. LeBellac, F. M. Renard and J. Tran. Thanh Van. 1964. *Nuovo Cim.* **33**, 594 and **34**, 450.
31. M. Gourdin, M. LeBellac, F. M. Renard and J. Tran Thanh Van. 1965. *Nuovo Cim.* **37**, 524.
32. F. L. Gross. 1964. *Phys. Rev.* **134**, B405 and **136**, B140.
33. H. F. Jones. 1962. *Nuovo Cim.* **26**, 790.
34. M. Gourdin. 1963. *Nuovo Cim.* **28**, 533; 1964. **32**, 493 and 1965. **35**, 1105.
35. Richard Wilson. 1963. *Nucleon-Nucleon Interaction* (John Wiley, New York).
36. D. J. Drickey and L. N. Hand. 1962. *Phys. Rev. Lett.* **9**, 521. (Up to $q^2 = 1{\cdot}5\ \mathrm{f}^{-2}$.)
37. D. Benaxas, D. Drickey and D. Frèrejacque. 1964. *Phys. Rev. Lett.* **13**, 353. (Up to $q^2 = 2\ \mathrm{f}^{-2}$.) LAL 1155, Orsay Preprint (unpublished).
38. B. Grossetête, D. Drickey and P. Lehmann. 1966. *Phys. Rev.* **141**, 1425. (Up to $q^2 = 2\ \mathrm{f}^{-2}$.)
39. C. D. Buchanan and M. R. Yearian. 1965. *Phys. Rev. Lett.* **15**, 303. (Up to $q^2 = 20\ \mathrm{f}^{-2}$.)
40. J. de Pagter, F. Friedman and H. Kendall, L. P. Van Speybroeck, J. Elias and G. Hartman. 1965. *Proceedings of Hamburg Conference on Electron and Photon Interactions*, II, 50. (Up to $q^2 = 40 f^{-2}$.)
41. N. K. Glendenning and G. Kramer. 1962. *Phys. Rev.* **126**, 2159.

42. H. Feshbach and E. Lomon (private communication).
43. R. J. Adler and S. D. Drell. 1964. *Phys. Rev. Lett.* **13**, 249.
44. B. Grossetête, S. Jullian and P. Lehmann. 1966. *Phys. Rev.* **141**, 1435.
45. E. B. Hughes, T. A. Griffy, M. R. Yearian and R. Hofstadter. 1965. *Phys. Rev.* **139**, B458.
46. C. V. Akerlof, K. Berkelman, G. Rowe and M. Tigner. 1964. *Phys. Rev.* **135**, B810.
47. J. R. Dunning, K. W. Chen, A. A. Cone, G. Hartwig, N. F. Ramsey, J. K. Walker and Richard Wilson. 1966. *Phys. Rev.* **141**, 1286.
48. N. T. Meister and T. A. Griffy. 1964. *Phys. Rev.* **133**, B1032.
49. M. Sugarawa. 1960. *Phys. Rev.* **117**, 614.
50. P. Stein, M. Binkley, R. McAllister, A. Suri and W. Woodward. 1966. *Phys. Rev. Lett.* **16**, 592.
51. L. Durand, III. 1969. *Phys. Rev.* **115**, 1620.
52. M. Le Bellac, F. M. Renard and J. Tran Thanh Van. 1964. *Nuovo Cim.* **33**, 594; 1964. **34**, 450; 1965. **38**, 565 and 1965. **38**, 1688.
53. I. McGee. 1965. Thesis, Yale University.
54. J. Nuttall and M. Whippman. 1963. *Phys. Rev.* **130**, 2495.
54A. B. Bosco, B. Grossetête and P. Quarati. 1966. *Phys. Rev.* **141**, 1441.
54B. B. Bosco, C. Ciochetti and A. Molinari. 1963. *Nuovo Cim.* **28**, 1427.
55. Richard Wilson. 1966. *Springer Tracts* **20** (Springer Verlag).
55A. H. P. Noyes. 1965. Nuclear Physics, **74**, 533.
56. G. F. Chew, R. Karplus, S. Gasiorowicz and F. Zachariasen. 1958. *Phys. Rev.* **110**, 265.
P. Federbush, M. L. Goldberger and S. B. Treiman. 1958. *Phys. Rev.* **112**, 642.
57. M. Gell-Mann, D. Sharp and W. G. Wagner. 1962. *Phys. Rev. Lett.* **8**, 261.
58. M. Gell-Mann. 1962. *Phys. Rev.* **125**, 1067.
M. Gell-Mann and F. Zachariasen. 1961. *Phys. Rev.* **124**, 953.
59. G. Cocho. 1966. *Nuovo Cim.* **42**, 419.
60. T. Massam and A. Zichichi. 1966. *Nuovo Cim.* **43A**, 137 and **44A**, 309.
61. S. Furuichi and K. Watanabe. 1966. *Prog. Theor. Phys.* **35**, 174.
62. J. Bowcock, W. N. Cottingham and D. Lurié. 1960. *Nuovo Cim.* **16**, 918.
63. R. F. Dashen and D. H. Sharp. 1964. *Phys. Rev.* **133**, B1585.
64. G. Cocho, C. Fronsdal, Harun ar-Rashid and R. White. 1966. *Phys. Rev. Lett.* **17**, 275.
65. T. T. Wu and C. N. Yang. 1965. *Phys. Rev.* **137**, B708.
66. R. Serber. 1963. *Phys. Rev. Lett.* **10**, 357.
67. J. K. de Pagter, J. I. Friedman, G. Glass, R. C. Chase, M. Gettner, E. von Goehler, Roy Weinstein and A. Boyarski. 1966. *Phys. Rev. Lett.* **16**, 35.
68. A. Wehmann, E. Engels, Jr., L. N. Hand, C. M. Hoffman, P. G. Innocenti, Richard Wilson, W. A. Blanpied, D. J. Drickey and D. G. Stairs. 1966. *Phys. Rev. Lett.* **17**, 1113, and correction at the International Conference of High Energy Physics, Berkeley, 1966.
69. L. J. Lanzerotti, R. B. Blumenthal, D. C. Ehn, W. L. Faissler, P. M. Joseph, F. M. Pipkin, J. K. Randolph, J. J. Russell, D. G. Stairs and J. Tenenbaum. 1965. *Phys. Rev. Lett.* **15**, 210.
69A. L. W. Jones, D. Harting, H. Middelkoop, B. Zacharov, D. O. Caldwell, E. Bleuler and B. Elsner. 1964. Univ. of Michigan Technical Reports No. 17 and 1966, No. 22 (unpublished).
69B. G. Bellini, E. Fiorini, P. Negri, J. Hennessy, J. Veillet, J. Crussard, J. Ginestet and A. H. Tran. 1964. *Phys. Lett.* **10**, 126.
70. M. Gourdin and P. Salin. 1963. *Nuovo Cim.* **27**, 193.
71. F. Guerin and C. A. Piketty. 1964. *Nuovo Cim.* **32**, 971.
72. N. Cabibbo and R. Gatto. 1961. *Phys. Rev.* **124**, 1577.
73. R. Gatto. 1966. *Springer Tracts* **20** (Springer-Verlag).
74. G. K. O'Neill. 1956. *Phys. Rev.* **102**, 1418 (L).
75. Stanford Summer Study, SLAC 49 (unpublished).
76. F. Amman and D. M. Ritson. 1961. *Brookhaven Accelerator Conference*, p. 262.
77. G. Voss and K. Robinson. 1966. CEAL-1029 (unpublished).

78. W. K. H. PANOFSKY and E. ALLTON. 1959. *Phys. Rev.* **110,** 1155.
79. A. A. CONE, K. W. CHEN, J. R. DUNNING, G. HARTWIG, N. F. RAMSEY, J. K. WALKER and RICHARD WILSON. 1965. *Phys. Rev. Lett.* **14,** 326.
A. A. CONE. 1966. Thesis, Harvard.
80. J. D. BJORKEN and J. D. WALECKA. 1966. *Ann. Phys.* (N.Y.) **38,** 35.
80A. H. JONES. 1965. *Nuovo Cim.* **40,** 1018.
81. L. N. HAND. 1963. *Phys. Rev.* **129,** 1834; Ph.D. Thesis, 1961, Stanford University, Stanford, California (unpublished).
81A. H. LYNCH, J. V. ALLABY and D. M. RITSON. 1965. *Proceedings of Hamburg Conference on Electron and Photon Interactions,* II, 112.
82. J. PEREZ Y JORBA, Orsay Report (unpublished).
83. J. D. BJORKEN. 1963. *Ann. Phys.* (N.Y.) **24,** 201.
83A. See, for example, a good review by W. C. BARBER. 1962. *Ann. Rev. Nucl. Sci.* **12,** 1.
84. N. ZAGURY. 1966. *Phys. Rev.* **145,** 1112.
85. J. ADLER. 1965. *Proceedings of the Argonne Conference on Weak Interactions.*
86. J. P. LOUBATON. 1965. *Nuovo Cim.* **39,** 591.
87. G. CHEW, M. L. GOLDBERGER, F. LOW and Y. NAMBU. 1957. *Phys. Rev.* **106,** 1345.
S. FUBINI, Y. NAMBU and V. WATAGHIN. 1955. *Phys. Rev.* **111,** 329.
88. W. N. COTTINGHAM. 1963. *Ann. Phys.* **25,** 424.
89. R. ALVAREZ, Z. BAR-YAM, W. KERN, D. LUCKEY, L. S. OSBORNE, L. TAZZARI and R. FESSEL. 1964. *Phys. Rev. Lett.* **12,** 707, 710.
90. C. W. AKERLOF, W. W. ASH, K. BERKELMAN and M. TIGNER. 1965. *Phys. Rev. Lett.* **14,** 1036.
91. C. W. AKERLOF, W. W. ASH, K. BERKELMAN and C. A. LICHTENSTEIN. 1966. *Phys. Rev. Lett.* **16,** 147.
92. K. F. WEIZSACKER. 1934. *Physik,* **88,** 612.
93. E. J. WILLIAMS. 1934. *Phys. Rev.* **45,** 729.
E. J. WILLIAMS. 1935. *Dan. Mat. Fys. Medd.* **13,** 4.
94. J. W. WEIL and B. D. MCDANIEL. 1953. *Phys. Rev.* **92,** 391.
95. D. O. CALDWELL, J. P. DOWD, K. HEINLOTH and M. D. ROSSEAU, *Rev. Sci. Instr.* **36,** 283.
96. L. N. HAND. 1964. SLAC 25, II (unpublished).
97. N. CHRIST and T. D. LEE. 1966. *Phys. Rev.* **143,** 1310.
98. J. CHEN. (Private communication and proposal to CEA, Jan. 1966.)
99. CHARLES BALTAY, P. FRANZINI, J. KIM, L. KIRSCH, D. ZANELLO, J. LEE-FRANZINI, R. LOVELEN, J. MCFADYEN and H. YAEGER. 1966. *Phys. Rev. Lett.* **16,** 1224.
100. D. YOUNT and J. PINE. 1962. *Phys. Rev.* **128,** 1842.
101. A. BROWMAN, F. LIU and C. SHAERF. 1965. *Phys. Rev.* **139,** B1079.
102. R. L. ANDERSON, B. BORGIA, G. L. CASSIDAY, J. W. DEWIRE, A. S. ITO and E. C. LOH. 1966. *Phys. Rev. Lett.* **17,** 407.
103. E. ENGELS and L. HAND (private communication).
104. J. C. BIZOT, J. M. BUON, J. LE FRANCOIS, J. PEREZ Y JORBA and P. ROY. 1965. *Phys. Rev.* **140,** B1387.
104A. G. V. DI GIORGIO, E. GANESAUGE, R. GOMEZ, G. GORDONI, I. PENNER, L. SERBASSI, M. L. VINCELLI, E. AMALDI and G. STOPPINI. 1965. *Nuovo Cim.* **39,** 474.
105. R. LEWIS. 1956. *Phys. Rev.* **102,** 537.
106. D. FLAMM and W. KUMMER. 1963. *Nuovo Cim.* **28,** 33.
107. S. D. DRELL and S. FUBINI. 1959. *Phys. Rev.* **113,** 741.
108. S. D. DRELL and M. RUDERMAN. 1957. *Phys. Rev.* **106,** 561.
109. RICHARD WILSON. 1963. SLAC-25, I (unpublished).
110. S. B. CRAMPTON, D. KLEPPNER and N. F. RAMSEY. 1963. *Phys. Rev. Lett.* **11,** 338.
111. C. K. IDDINGS and P. M. PLATZMAN. 1959. *Phys. Rev.* **113,** 192.
112. A. C. ZEMACH. 1956. *Phys. Rev.* **104,** 1771.
113. C. K. IDDINGS. 1965. *Phys. Rev.* **138,** B446.
114. S. FENSTER, R. KOBERLE and Y. NAMBU. 1965. *Phys. Lett.* **19,** 513.
115. J. D. SULLIVAN and S. D. DRELL. 1965. *Phys. Lett.* **19,** 516.
116. S. FENSTER and Y. NAMBU. 1965. *Prog. Theor. Phys. Suppl.*; **Extra Number,** 250.
117. S. D. DRELL. 1966. *Bull. Amer. Phys. Soc.* **21,** 304.
118. W. E. CLELAND, J. M. BAILEY, M. ECKHAUSE, V. W. HUGHES, R. M. MOBLEY, R. PREPOST and J. E. ROTHBERG. 1964. *Phys. Rev. Lett.* **13,** 202, 457(E).

119. S. D. Drell. 1958. *Ann. Phys.* (N.Y.) **4**, 75.
120. C. M. Sommerfield. 1958. *Ann. Phys.* (N.Y.) **5**, 26.
121. S. D. Drell and H. R. Pagels. 1965. *Phys. Rev.* **140**, B397.
122. A. J. Layzer. 1961. *J. Math. Phys.* **2**, 306.
G. W. Erickson and D. R. Yennie. 1965. *Ann. Phys.* (N.Y.) **35**, 271.
M. Soto. 1966. *Phys. Rev. Lett.* **17**, 1153.
123. E. S. Dayhoff, S. Trebwasser and W. E. Lamb. 1953. *Phys. Rev.* **89**, 106.
124. R. T. Robiscoe. 1965. *Phys. Rev.* **138**, B 22.
R. T. Robiscoe. 1966. *Phys. Rev. Lett.* **17**, 69.
125. D. T. Wilkinson and H. R. Crane. 1963. *Phys. Rev.* **130**, 852.
126. R. Karplus and A. Klein. 1952. *Phys. Rev.* **86**, 257.
127. R. Weinstein, M. Deutsch and S. Brown. 1966. *Phys. Rev.* **98**, 223.
128. H. Suura and E. Wickmann. 1957. *Phys. Rev.* **105**, 1930.
129. G. Charpak, F. J. Farley, R. L. Garwin, T. Muller, J. G. Sens and A. Zichichi. 1962. *Phys. Lett.* **1**, 16.
130. W. C. Barber, B. Gittelman, G. K. O'Neill and B. Richter. 1966. *Phys. Rev. Lett.* **16**, 1127.
131. A. Browman, B. Grossetête, and D. Yount. 1966. *Phys. Rev.* **143**, 899.
132. R. Weinstein (private communication and also see reference 67).
133. R. B. Blumenthal, D. C. Ehn, W. L. Faissler, P. M. Joseph, L. J. Lanzerotti, F. M. Pipkin and D. G. Stairs. 1966. *Phys. Rev.* **144**, 1199.
134. R. Talman *et al.* 1966. Invited paper, Washington Meeting A.P.S.
135. S. D. Drell and J. D. Walecka. 1964. *Ann. Phys.* (N.Y.) **28**, 18.
136. J. K. Walker, M. Wong, R. Fessel, R. Little and H. Winick. 1966. *Phys. Rev.* **144**, 1126.
137. E. Lomon, 1966. *Phys. Lett.* **21**, 555.
138. R. Perrin. (Preprint, 1966).
139. F. E. Low. 1965. *Phys. Rev. Lett.* **14**, 238.
140. C. Betourne, H. Nguyen Ngoc, J. Perez y Jorba and J. Tran Thanh Van. 1965. *Phys. Lett.* **17**, 70.
R. Budnitz, J. R. Dunning, M. Goitein, N. F. Ramsey, J. K. Walker and Richard Wilson. 1966. *Phys. Rev.* **141**, 1313.
H. J. Behrend, F. W. Brasse, J. Engler, E. Gaussauge, H. Hutching, S. Galster, G. Hartree and H. Shopper. 1965. *Phys. Rev. Lett.* **15**, 900.
H. Kendall and J. Friedman (private communication).
141. D. P. Hutchinson, J. Menes, A. M. Patlach and G. Shapiro. 1963. *Phys. Rev.* **131**, 1362.
142. Than Myint, D. Kleppner and H. G. Robinson. 1966. *Phys. Rev. Lett.* **17**, 405.
143. E. R. Cohen and J. W. M. DuMond. 1965. *Rev. Mod. Phys.* **37**, 537.
144. R. Dalitz and D. R. Yennie. 1957. *Phys. Rev.* **105**, 1598.
145. M. A. Ruderman. 1966. *Phys. Rev. Lett.* **17**, 794.
146. D. N. Langenberg, W. H. Parker and B. N. Taylor. 1966. *Phys. Rev.* **150**, 186 (and private communication).

4

PRECISION TESTS OF QUANTUM ELECTRODYNAMICS AT LOW ENERGIES

S. D. DRELL

Stanford Linear Accelerator Centre
Stanford University

PREFACE

QUANTUM electrodynamics is a great theory. Some would say it is our only theory in the realm of elementary particle physics, and indeed it has served as the model for all other field theories including the original suggestion of Yukawa for nuclear forces. Others would say that it is no more than a vestige of the dark ages with its atavistic superstructure of local fields and the canonical quantization prescription.

None-the-less, since the great work of Feynman, Schwinger and Tomonaga of 18 years ago that was recognized so deservedly by this year's Nobel Award, it is a theory from which unique finite answers can be derived by well defined calculational techniques of perturbation theory. You merely make the Feynman graph expansion to whatever order of the fine structure constant $\alpha = 1/137$ you require and calculate away, serene in the knowledge that if your strength holds out a number will eventually emerge to compare with any measurable experimental quantity. Herein also lies the limitations of the theory. Is there anything to it beyond the Feynman graph expansion in powers of $\alpha = 1/137$? One has no exact solution so it is not even known whether or not a stable ground state exists. Moreover there are quantities such as charge and mass renormalization constants that can be computed in the theory, if not observed, and which to any finite order of α do not have finite uniquely defined values. Take for example charge renormalization. A bare electron has charge e_0, a parameter introduced in the equations, but a parameter which is not observed when one looks at the electron, say as it scatters in an externally applied electric field. One actually sees it through a cloud of virtual particle-antiparticle pairs—especially electron-positron pairs—that surround it at all times as a result of its interaction with the radiation field. As one penetrates in through this shielding cloud one sees deviations from Coulomb's law:

$$\frac{e^2}{r} \longrightarrow \frac{e^2}{r}\left[1+\int_{4m_e^2}^{\infty} da^2 \rho(a^2) e^{-ar}\right]$$

$$\underset{(r\to 0)}{\longrightarrow} \frac{e^2}{r}\left[1+\int_{4m_e^2}^{\infty} da^2 \rho(a^2)\right] \equiv \frac{e^2}{Z_3}\frac{1}{r} = \frac{e_0^2}{r}.$$

In the above equation $\rho(a^2) \geqq 0$ is the probability for a photon of $(\text{mass})^2 = a^2$ to couple to any physical state of the same mass and quantum numbers and the threshold of the integral is the square of the mass of the lightest state $(e^- + e^+)$ coupling with the radiation field. (To higher order in e we should also include $\gamma \rightarrow 3\gamma$ coupling with threshold at $a^2 = 0$.) We also exclude weak interactions which would give rise to coupling to neutrino pair states via their charge radius and so the range of the shielding cloud is $\approx \dfrac{1}{2m_e^2} \approx 2 \times 10^{-11}$ cm. Whether or not the potential is more singular than $1/r$ is an open question. In perturbation theory $1/Z_3$ is cut-off dependent. If there are any modifications, $\delta\rho$, as may be introduced to avoid a more singular potential, are any experimental consequences implied? Now clearly one cannot change the form of the electromagnetic interaction at large distances such as at atomic dimensions or else the beautiful agreement[1] between theory and experiment in precision atomic spectroscopy as has been tested in the Lamb shift would be lost. For example the quantity known as the vacuum polarization can be altered by less than 2%, i.e.

$$\int_{4m_e^2}^{\infty} \frac{\delta\rho(a^2)}{a^2}\, da^2 < 2\% \text{ of } \int_{4m_e^2}^{\infty} \frac{\rho(a^2)}{a^2}\, da^2$$

since the vacuum polarization contributes −27 Mc/s to the Lamb shift in hydrogen[2, 3, 4] and there is no indicated discrepancy of > 0·5 Mc/s. However what goes on at very large momenta $a \gg 10m_e$; $a^2 \gg 100m_e^2$ will not appear in the Lamb shift.

How can we find out what is going on within this region? This is a general question as to the behaviour of QED at small distances or high momenta and is of course independent of specific questions such as, "is $Z_3 = 0$?".

There are two ways one can get at this. One can do low-energy experiments in addition to the Lamb shift with extreme precision, or experiments of accuracy of a few percent but with very large momentum transfer, probing distances R of the order $R \sim \hbar/q$. Experiments of the latter type need colliding rings if performed on pure electromagnetic systems so that one doesn't waste all of one's energy in the centre-of-mass motion in an electron-electron collision. If one uses nuclear targets to anchor the centre-of-mass in the laboratory system in order to achieve large momentum transfers, one needs a detailed analysis of large momentum transfer Bethe-Heitler pair production and bremmstrahlung processes to be compared with electron scattering from the same nuclear targets and with the same momentum and energy transfers so that the nuclear or nucleon structure can be unambiguously removed in the comparison. Procedures for doing this have been worked out theoretically in great detail and have been applied to experiments which have been reported in the past year. For large angle electron-positron pair production an agreement between theory and experiment is not found. These are very difficult experiments to perform; those at CEA[5] and Cornell[6] involved in these studies are heroes and we anxiously await further confirmation of their results so far obtained. As Dyson remarked recently it is fortunate for quantum electrodynamics that experimental tests are, in fact, being performed and it is indeed very exciting at present that as published they are not in accord with the theory. (*Note added in proof.* New data from DESY reported by S. Ting to the XIII International Conference on High-Energy Physics (Berkeley) agrees well with this theory.)

However in these lectures I want to review and analyze some of the precision measurements at low energy. These include precise $g-2$ measurements of the electron and muon and a very precise hyperfine structure measurement in hydrogen.

Let us start with the hfs in hydrogen—i.e. the splitting of the $1S$ energy levels when the electron and proton couple their spins to form a triplet and singlet state respectively. Transitions between these levels radiate the famous 21 cm line of radio astronomy.

The work on hfs was done in collaboration with Dr. J. D. Sullivan and on the muon $g-2$ value with Dr. J. Trefil and manuscripts prepared in collaboration with them form the basis of the following lectures.

1

The Hyperfine Structure in Hydrogen

1.1. *Introduction*

The hyperfine splitting in hydrogen presents an important and historic confrontation between theory and experiment. Along with positronium the hydrogen atom is the simplest bound state system that is readily available and for which refined and detailed theoretical calculations have been and can be made. It thus affords an important test of relativistic bound state methods. The hyperfine splitting (hfs) in hydrogen is further unique in that it is sensitive to details of proton structure which are usually seen only in high-energy electron scattering experiments. The hfs is therefore an important link between the usually disjoint fields of high energy and precision atomic physics.

At present there exists an apparent discrepancy of approximately 40 ± 20 ppm (parts per million) between experiment and theory for the ground state hyperfine splitting in hydrogen.[1, 7, 8]

The purpose of these lectures is to critically examine the proton structure corrections which go into the theoretical hfs value and to try to assess and improve the accuracy to which they are known. The conclusion of these lectures is that hitherto uncalculated parts of the proton structure corrections may very well account for the magnitude and sign of the apparent discrepancy mentioned above but that they do not seem to be amenable to calculation with the presently available techniques in the theory of strong interactions.

I shall begin by reviewing the theoretical expression for the hfs and the corrections to it with emphasis on the proton polarizability contribution. Next I will construct two non-relativistic models of the proton structure and calculate their contributions to the hfs. Although these models probably bear little resemblance to reality they do contain all the physical effects which are relevant to the hfs problem and thus provide valuable insight into the less transparent and much less complete relativistic calculation of the structure correction. I will also discuss their relation to the recent quark model calculation of Fenster and Nambu[9] as well as to the original analysis of A. Bohr for the polarizability effect in the hfs analysis of deuterium. Next I will discuss the relativistic calculation of the proton structure correction. The work here builds on the recent work of Iddings[10] and the non-relativistic reduction of this work to the earlier discussion by Zemach[11] is displayed. Finally the problems of subtraction constants and contributions from high energy are taken up.

1.2. *Corrections to the hyperfine splitting*

The hyperfine splitting in the ground state of hydrogen has been measured to the incredible precision of 2 parts in 10^{11} by Crampton, Kleppner and Ramsey[7]:

$$\nu = (1420{\cdot}405751800 \pm 28\times 10^{-9})\ \text{Mc/s}. \qquad \ldots[1.1]$$

The limit of precision to which this result can be compared with theory is set by two factors: the uncertainties in measurements of the natural constants and the uncertainties in the calculation of the corrections to the hfs calculations due to proton structure. The purely electrodynamic parts of the calculations have been carried to a higher precision than that to which these factors are known.

The total hyperfine splitting in the H ground state is given by[8]

$$\nu = \tfrac{16}{3}R_\infty c\left(\frac{\mu_p}{\mu_e}\right)\left(\frac{\mu_e}{\mu_0}\right)^2\left(\frac{M}{m+M}\right)^3\alpha^2\Big\{1+\tfrac{3}{2}(Z\alpha)^2-\alpha(Z\alpha)(\tfrac{5}{2}-\ln 2)$$

$$+\frac{\alpha}{\pi}(Z\alpha)^2\left[a_0\ln^2(Z\alpha)+a_1\ln Z\alpha+a_2\right]-\alpha\frac{m}{M}X\Big\} \qquad \ldots[1.2]$$

where: $a_0 = -\frac{8}{3}$; $a_1 = -\frac{37}{35}-\frac{8}{16}+\frac{16}{3}\ln 2$; $a_2 = 18{\cdot}4\pm 5$.

In [*1.2*] m and M denote the electron and proton masses respectively. The natural constants appearing in the first factor are grouped together in the way that they have been most accurately determined.[1] All are known to better than one part in a million except for α which was measured to an accuracy of only 9 ppm in the deuterium fine structure measurements (i.e. $2P_{\frac{3}{2}}-2P_{\frac{1}{2}}$ splitting) by Triebwasser, Dayhoff and Lamb.[2]

In writing the correction terms in the bracket we have distinguished α and $Z\alpha$ as expansion parameters only so that binding and radiative corrections can be readily identified. The first three terms are the Breit binding correction[12] to the electron magnetic moment and the binding corrections (first and second order) to the radiative correction to the electron's electromagnetic interaction as computed by Kroll and Pollack[13] and by Karplus, Klein and Schwinger[14] to order $Z\alpha$ and by Layzer,[15] Zwanziger[16] and Brodsky and Erickson[8] to order $(Z\alpha)^2$. These contributions are now known to better than 1 ppm. The last term of order $\alpha\frac{m}{M}$ represents the effects of finite proton mass and structure, aside from the purely kinematic reduced mass correction, and must be studied if comparisons between equations [*1.1*] and [*1.2*] are to be pursued to the order of 10 ppm.

In order to focus sharply on its significance we write the ratio of [*1.1*] to [*1.2*], using Cohen and DuMond,[1]

$$\frac{\nu_{\text{theory}}}{\nu_{\text{exp}}} = 1-(9\pm 20)\times 10^{-6}-\alpha\frac{m}{M}X \qquad \ldots[1.3]$$

where the uncertainty ± 20 ppm is based on a two standard deviation limit on α in reference 1 or a one standard deviation figure as taken from Cleland *et al.*[17] The question to which we address ourselves is, "does the structure correction disturb the good agreement in [*1.3*] between experiment and theory for a point massive proton?" In studying X we must consider both the ground state average contributions as well as those arising dynamically from the polarizability of the proton. The former have been well studied and can be evaluated quite accurately in terms of the measured electromagnetic form factors of the proton. Referring to the most recent and complete study of Iddings[10] we can write

$$\alpha\frac{m}{M}X = 34\times 10^{-6}+\frac{\Delta\nu^{(p)}}{\nu}. \qquad \ldots[1.4]$$

In [*1.3*] this gives

$$\frac{\nu_{\text{theory}}}{\nu_{\text{exp}}} = 1-(43\pm 20)\times 10^{-6}-\frac{\Delta\nu^{(p)}}{\nu}. \qquad \ldots[1.5]$$

We see then that it is up to the polarizability correction $\Delta\nu^{(p)}/\nu$ to bring the theoretical number back into a clear agreement with observation.

There have been several studies[10, 18, 19] of $\Delta\nu^{(p)}$, all of which have come up with a negligible polarization contribution (less than 1 ppm) and [*1.5*] stands at present as an apparent discrepancy. In analyzing this problem once more we want to shed light on the question as to whether this discrepancy should be viewed as a serious challenge to quantum electrodynamics or whether it is more properly to be viewed as a measure of the inaccuracy in our treatment of the effect of proton structure.

The specific physical idea motivating this study is the same one analyzed within the context of a massive quark model of the proton by Fenster and Nambu.[9] If the orbital electron in the hydrogen atom could completely follow the instantaneous charge and magnetization distribution of the proton then the proton would appear to it as no more than a point. In this case there would be a cancellation of the finite size correction of [*1.5*] that introduced the discrepancy.

There is an analogous effect in deuterium, as first analyzed by A. Bohr.[20] Deuterium is a very weakly bound system and we may think of the proton and neutron as moving very slowly. The electron can then follow the proton —i.e. by virtue of their Coulomb attraction the electron wave function at the nuclear surface can recentre on the instantaneous proton position.[21] In less graphic terms both the deuteron ground state and the electron $1S$ orbit are mutually polarized by mixing in higher excited state amplitudes.

Now the proton is less polarizable than the deuteron, its excited states lying at least $m_\pi = 140$ MeV above the ground state. However, the many resonances contributing to photo and electro production cross-sections are evidence that the proton is nevertheless a highly polarizable structure and there may very well be appreciable corrections to the static average ground state structure as contained in the elastic form factors.

For a proton (or deuteron) with internal dynamical degrees of freedom the Coulomb interaction with the orbital electron has off-diagonal matrix elements to excited proton states. The size of the proton polarizability correction is determined by the magnitude of these off-diagonal matrix elements of the Coulomb excitation as well as by the same off-diagonal matrix elements of the spin dependent magnetic interaction. We therefore expect the electric as well as the magnetic multipole excitations of the proton to play an important role in determining the size of $\Delta\nu^{(p)}$. Previous studies[10, 18, 19] have concentrated exclusively on the magnetic dipole excitations of the 33 resonance in computing negligible contributions to $\Delta\nu^{(p)}$. We extend these considerations to a more complete study of the amplitude for electromagnetic excitation of the proton.

Sizable polarizability contributions are identified in the next section on the basis of a Schrödinger model of the proton structure. When we next turn to a relativistic dispersion analysis we follow the method of Cottingham,[22] as discussed and applied to the hfs problem by Iddings.[10] This approach expresses $\Delta\nu^{(p)}$ in terms of the forward spin-flip Compton scattering of virtual photons, integrated over all photon energies and space-like masses. We will find that one simply does not know enough about this virtual process to make any firm conclusions on the size of $\Delta\nu^{(p)}$. Dispersion theory is at its best when dealing with the interactions of "real" particles in a kinematic region where the absorptive amplitude is dominated by one or by at most a very few resonances. In this problem we do not have this simple circumstance and we find that many small and poorly calculable terms contribute, leading to no firm conclusion with regard to $\Delta\nu^{(p)}$. It is our belief however, derived both from the Schrödinger model considerations as well as the relativistic approach of dispersion theory, that one lacks the information required to infer that the discrepancy in [*1.5*] is at present a serious challenge to quantum electrodynamics. We know too little about proton dynamics.

The recent measurements of the Lamb shift by Robiscoe and collaborators may be interpreted as increasing α by 13 ppm and the hfs by 26 ppm. This would remove the apparent discrepancy in [*1.5*] in the absence of any appreciable polarizability contribution. As this is still an open experimental question, further remarks on this are reserved for the end.

1.3. *Non-relativistic model of proton structure*

In this section the finite size correction to the hyperfine splitting is computed using a non-relativistic model of the proton structure. The purpose of this exercise is not to calculate a quantitatively reliable number but rather to exhibit in a concrete fashion the various physical effects which influence the

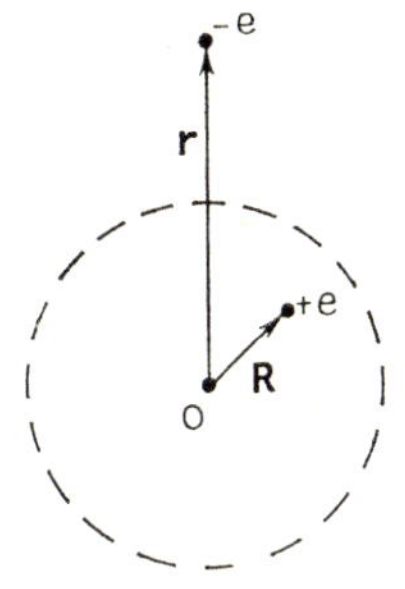

FIG. 1. Model of proton structure with "quarkette" of charge $+e$ and mass μ with coordinate $\boldsymbol{R}$ moving about a fixed centre of mass at the origin. The electron in the hydrogen atom is at $\boldsymbol{r}$.

hfs of a system with structure. In so far as is possible the results are expressed in terms of quantities which are independent of the details of the model and which can be fitted to the observed parameters of the physical proton. In particular we want to gain some insight into the physical parameters governing the size of the polarization correction to the hyperfine splitting in hydrogen.

Our model proton is composed of a particle (called hereafter a quarkette) of charge $+e$ and mass μ which satisfies the Schrödinger equation and is bound to a neutral, infinitely massive centre of force by a non-relativistic potential $V(\boldsymbol{R})$. The infinite mass centre is taken as the origin; $\boldsymbol{R}$ denotes the coordinate of the "quarkette" and $\boldsymbol{r}$ the coordinate of the electron which is bound to the physical proton to form a hydrogen atom (see Fig. 1). Depending on the choice of parity for the quarkette there are two possible cases. A positive parity quarkette is taken to have spin $\frac{1}{2}$ and a point magnetic dipole moment. The proton is the lowest $S_{\frac{1}{2}}^{+}$ bound state of this quarkette. Initially we assume the quarkette to be a point particle; later we consider the possibility that the quarkette itself has finite extension. For negative parity it is most natural to consider the quarkette spinless and take the spin $\frac{1}{2}$ as a degree of freedom of the mass centre. The proton is realised as a $P_{\frac{1}{2}}^{+}$ bound state and the potential $V(\boldsymbol{R})$ must have a spin orbit as well as a central part in order to remove the degeneracy between the $P_{\frac{1}{2}}^{+}$ ground state and a possible $P_{\frac{3}{2}}^{+}$ state. It must also have some property (perhaps a hard core) to prevent a $S_{\frac{1}{2}}^{-}$ state from being the true ground state of the proton system.

In these lectures we treat only the case of the even quarkette parity in detail. Very similar results are found in the odd parity case. We also discuss the influence of different parameters on numerical results from these models as well as their relation to a three massive quark model of the proton as analyzed recently in regard to the hfs by Fenster and Nambu.[9]

The Hamiltonian for the full system is

$$H = H_P(\boldsymbol{R})+H_e'(\boldsymbol{r})+H_c'(\boldsymbol{r}, \boldsymbol{R})+H_M(\boldsymbol{r}, \boldsymbol{R}) \qquad ...[1.6]$$

where H_P is the Hamiltonian of the quarkette including the central potential $V(\boldsymbol{R})$ which binds it to the origin.

$$H_P = \frac{\boldsymbol{P}^2}{2\mu}+V(\boldsymbol{R}). \qquad ...[1.7]$$

μ is the quarkette mass and $\boldsymbol{P}$ is the momentum conjugate to $\boldsymbol{R}$. H_e' is the free Dirac Hamiltonian for the electron which must be treated relativistically as will become clear later

$$H_e' = \boldsymbol{\alpha} \,.\, \boldsymbol{p}+\beta m. \qquad ...[1.8]$$

In [1.8] $\boldsymbol{\alpha}$ and β are the usual Dirac matrices; we will work in the standard representation for them; i.e.

$$\boldsymbol{\alpha} = \begin{pmatrix} 0 & \boldsymbol{\sigma}_e \\ \boldsymbol{\sigma}_e & 0 \end{pmatrix} \quad \text{and} \quad \beta = \begin{pmatrix} 1 & 0 \\ 0 & -1 \end{pmatrix}.$$

H_c' contains the electron-quarkette Coulomb interaction

$$H_c' = -\frac{e^2}{4\pi}\frac{1}{|\,\boldsymbol{r}-\boldsymbol{R}\,|} \qquad ...[1.9]$$

and H_M which is responsible for the hfs describes the interaction of the electron's moment with the magnetic field generated by the proton

$$H_M = +\frac{e}{4\pi}\,\boldsymbol{\alpha} \,.\, \boldsymbol{A}(\boldsymbol{r}). \qquad ...[1.10]$$

The vector potential $A(r)$ seen by the electron is the sum of a moment and a convection current part

$$A(r) = -\frac{e\kappa}{2\mu}\nabla_R \frac{1}{|r-R|}\times\sigma+e\frac{v}{|r-R|} \qquad ...[1.11]$$

where σ is a Pauli matrix for the quarkette and $\frac{\kappa}{\mu} = \frac{\mu_p}{M} = \frac{2{\cdot}79}{M}$ in order to fit the total observed proton magnetic moment. As it stands [*1.11*] lacks a "gauge term" $-\frac{1}{2}e\nabla_r \frac{v\,.\,(r-R)}{|r-R|}$ which should be added to the right-hand side so that the transversality condition $\nabla_r\,.\,A(r) = 0$ is satisfied. However, this "gauge term" makes no contribution to the hfs and we will drop it. Also since v and R do not commute, the convection current term of [*1.11*] should be symmetrized properly. Symmetrization does not alter the hfs so we shall neglect it.[21]

The Hamiltonian [*1.6*] is essentially a Breit Hamiltonian with one of the particles treated as very massive and then given non-relativistic internal structure. It is necessary to keep in mind that when working within the framework of the Breit theory the magnetic term H_M can appear only once in a perturbation calculation. We refer the reader to the discussion of this point in the book of Bethe and Salpeter.[23]

If one could exactly solve the Hamiltonian of [*1.6*], less the magnetic term H_M, for the singlet and triplet ground state wave functions, $\psi_g^s(r, R)$ and $\psi_g^t(r, R)$ respectively, then the hyperfine splitting ν including proton structure effects to all orders would be

$$\nu = \nu_t-\nu_s = \langle\psi_g^t\,|\,H_M\,|\,\psi_g^t\rangle-\langle\psi_g^s\,|\,H_M\,|\,\psi_g^s\rangle. \qquad ...[1.12]$$

In practice it is impossible to construct these exact eigenstates in closed form because of the $r-R$ correlations and we must start an iteration solution with the electron centred at the origin $R = 0$ of the quarkette structure. We rewrite [*1.6*] as

$$H = H_P(R)+H_e(r)+H_c(r, R)+H_M(r, R) \qquad ...[1.13]$$

where

$$H_e(r) = \alpha\,.\,p+\beta m-\frac{e^2}{4\pi}\left(\frac{1}{r}\right) \qquad ...[1.14]$$

and

$$H_c(r, R) = \frac{e^2}{4\pi}\left(\frac{1}{r}-\frac{1}{|r-R|}\right) \qquad ...[1.15]$$

and treat $H_0 = H_P+H_e$ exactly and the remainder H_c+H_M in perturbation theory. The energies and eigenfunctions of H_e and H_P are labelled according to

$$H_e\psi_{n^\pm}(\boldsymbol{r}) = E_{n^\pm}\psi_{n^\pm}(\boldsymbol{r}); \qquad n^\pm = 0, 1, 2, \ldots \qquad \text{...[1.16]}$$

and

$$H_P\Phi_A(\boldsymbol{R}) = W_A\Phi_A(\boldsymbol{R}); \qquad A = 0, 1, 2, \ldots \qquad \text{...[1.17]}$$

where $\psi_{n^\pm}(\boldsymbol{r})$ are four component Dirac spinors for the positive and negative energy Coulomb solutions and the two component quarkette wave functions Φ can be further decomposed into a product of radial, orbital and spin parts

$$\Phi(\boldsymbol{R}) = \frac{\phi(R)}{R}\, Y(\hat{\boldsymbol{R}})\mathrm{X} \qquad \text{...[1.18]}$$

where $Y(\hat{\boldsymbol{R}})$ is a spherical harmonic, X a two component spinor and $\phi(R)$ the radial wave function with normalization

$$\int_0^\infty dR \,|\, \phi(R)\,|^2 = 1. \qquad \text{...[1.19]}$$

The solutions of H_0 are thus simple products

$$\begin{aligned} H_0\psi_{An^\pm}(\boldsymbol{r}, \boldsymbol{R}) &= E_{An^\pm}\psi_{An^\pm}(\boldsymbol{r}, \boldsymbol{R}) \\ \psi_{An^\pm}(\boldsymbol{r}, \boldsymbol{R}) &= \Phi_A(\boldsymbol{R})\psi_{n^\pm}(\boldsymbol{r}) \\ E_{An^\pm} &= W_A + E_{n^\pm}. \end{aligned} \qquad \text{...[1.20]}$$

In particular the zeroth order approximation to the ground state wave function for hydrogen is[24]

$$\begin{aligned} \psi_g &= \Phi_0(R)\psi_0(\boldsymbol{r}) \\ \Phi_0 &= \frac{1}{(4\pi)^{\frac{1}{2}}}\frac{\phi_s(R)}{R}\,\mathrm{X} \end{aligned} \qquad \text{...[1.21]}$$

$\phi_s(R)$ is the S wave ground state radial wave function satisfying [1.17], and

$$\psi_0(\boldsymbol{r}) = \begin{pmatrix} 1 \\ \alpha\dfrac{i\boldsymbol{\sigma}_e\,.\,\boldsymbol{r}}{2r} \end{pmatrix} \chi_e\psi(0)e^{-r/a_0} \qquad \text{...[1.22]}$$

with χ_e the two-component electron spinor, $\psi(0) \equiv \left(\dfrac{\alpha^3 m^3}{\pi}\right)^{\frac{1}{2}}$ and $a_0 = \dfrac{1}{\alpha m}$, the Bohr radius; corrections of order α^2 are dropped in [1.22].

As is well known, in the product spin space of the electron and the proton the hfs reduces to the difference of the triplet and singlet spin expectation value of an operator $\sim \boldsymbol{\sigma}_e\,.\,\boldsymbol{\sigma}$:

$$\langle\boldsymbol{\sigma}_e\,.\,\boldsymbol{\sigma}\rangle_t - \langle\boldsymbol{\sigma}_e\,.\,\boldsymbol{\sigma}\rangle_s = 4\langle\boldsymbol{\sigma}_e\,.\,\boldsymbol{\sigma}\rangle_t. \qquad \text{...[1.23]}$$

Hereafter, we work with the particular triplet state with both the electron and proton spins "up" along the spin axis, and write, suppressing all spin labels and terms,

$$\nu = \nu_t - \nu_s = 4\nu_t. \qquad \ldots[1.24]$$

In first order perturbation theory the only contribution to the hfs comes from the spin part of H_M

$$\begin{aligned}\nu_1 &= 4\langle \psi_g \mid H_M \mid \psi_g \rangle \\ &= -4\alpha \int d^3R \int d^3r \psi_g^\dagger(\mathbf{r}, \mathbf{R}) \frac{\mu_p}{2M} \boldsymbol{\alpha} \cdot \left(\nabla_R \frac{1}{\mid \mathbf{r}-\mathbf{R} \mid} \times \boldsymbol{\sigma} \right) \psi_g(\mathbf{r}, \mathbf{R}).\end{aligned} \qquad \ldots[1.25]$$

The ground state expectation value of the convection current term in [1.11] has no spin dependence and does not contribute to ν in this model. Substituting [1.21] and [1.22] we integrate over the electron coordinates with the aid of the first order identity

$$\int d^3r \psi_0^\dagger(r) \frac{\boldsymbol{\alpha}}{\mid \mathbf{r}-\mathbf{R} \mid} \psi_0(r) = \frac{2\pi}{3} \alpha a_0 \psi^2(0) \langle \boldsymbol{\sigma}_e \rangle \times \mathbf{R} \left(1 - \tfrac{3}{2} \frac{R}{a_0} \right). \qquad \ldots[1.26]$$

This gives in [1.25]

$$\nu_1 = \tfrac{1}{2}\nu_F \left\langle \boldsymbol{\sigma}_e \cdot [(\nabla_R \times \boldsymbol{\sigma}) \times \mathbf{R}] \left(1 - \tfrac{3}{2} \frac{R}{a_0} \right) \right\rangle_0 \qquad \ldots[1.27]$$

where

$$\langle \ldots \rangle_0 \equiv \int d^3R \Phi_0^\dagger(R) \ldots . \Phi_0(R) \qquad \ldots[1.28]$$

denotes the proton ground state expectation value and we are instructed to evaluate the spin operators for a triplet parallel spin up state according to [1.23] and [1.24]. In [1.27]

$$\nu_F = \frac{8\pi\alpha}{3} \frac{\mu_p}{Mm} \psi^2(0) = \tfrac{8}{3}\alpha^4 \mu_p \left(\frac{m^2}{M} \right) \qquad \ldots[1.29]$$

is the point proton (Fermi) hfs energy. Taking the spherical average in the proton ground state gives

$$\nu_1 = \nu_F \left(1 - 2 \frac{\langle R \rangle}{a_0} \right) \qquad \ldots[1.30]$$

where

$$\langle R \rangle \equiv \int_0^\infty dR \mid \phi_s(R) \mid^2 R. \qquad \ldots[1.31]$$

We next consider second order perturbation theory. There are three possibilities: (1) second order in H_c, (2) second order in H_M, and (3) mixed second order in H_c and H_M. The first involves no spin coordinates and so does not contribute to the hfs, and the second must not be computed directly, as noted earlier, when working with the phenomenological Breit Hamiltonian.

In the fully relativistic discussion later on we will see that this second order magnetic dipole interaction is properly included and is only a small correction to the first order result. The mixed second order term occurs with H_c and H_M appearing in either order and gives[21]

$$\nu_2 = -8\,\mathrm{Re}\left\{\sum_{\substack{A,n^+\\(A,n^+)\neq(0,0)}} \frac{\langle 0,0\mid H_M\mid A,n^+\rangle\langle A,n^+\mid H_c\mid 0,0\rangle}{W_A-W_0+E_{n^+}+E_0} - \sum_{A,n^-}\frac{\langle 0,n^-\mid H_M\mid A,0\rangle\langle A,0\mid H_c\mid 0,n^-\rangle}{W_A-W_0-E_{n^-}+E_0}\right\}. \quad ...[1.32]$$

We discuss separately the terms in [1.32] which involve only the ground state of the quarkette. Denoting this contribution by $\nu_2^{(g)}$ we have:

$$\nu_2^{(g)} = -4\frac{\alpha^2\mu_p}{M}\int\frac{d^3R}{4\pi R^2}\mid\phi_s(R)\mid^2\int\frac{d^3R'}{4\pi R'^2}\mid\phi_s(R')\mid^2\int d^3r\int d^3r'\;F(\mathbf{r},\mathbf{r}',\mathbf{R},\mathbf{R}') \quad ...[1.33]$$

with

$$F(\mathbf{r},\mathbf{r}',\mathbf{R},\mathbf{R}') = \mathrm{Re}\left\{\sum_{\substack{n^\pm\\ n^+\neq 0}}\psi_0^\dagger(\mathbf{r})\left[\boldsymbol{\alpha}.\left(\nabla_r\frac{1}{\mid\mathbf{r}-\mathbf{R}\mid}\times\boldsymbol{\sigma}\right)\right]\frac{\psi_{n^\pm}(\mathbf{r})\psi_{n^\pm}^\dagger(\mathbf{r}')}{E_{n^\pm}-E_0}\left(\frac{1}{r'}-\frac{1}{\mid\mathbf{r}'-\mathbf{R}'\mid}\right)\psi_0(\mathbf{r}')\right\}.$$

In arriving at the form of [1.33] we have noted that the quarkette ground state expectation value of $\boldsymbol{\alpha}\,.\,\boldsymbol{v}\,/\mid\mathbf{r}-\mathbf{R}\mid$ is spin independent and so does not contribute to the hfs and in view of the overall real part instruction we have freely taken the complex conjugate of the negative energy sum.

We are interested in contributions to [1.33] that are of no higher than the first power in the nuclear radius and therefore can make a series of approximations that permit us to carry out the sum over the excited electron states $n^\pm$. First we consider the contribution of positive energy electron bound states to the sum over states and take advantage of the spherical symmetry in $\boldsymbol{R}$ to write

$$\int d\Omega_{R'}\left(\frac{1}{r'}-\frac{1}{\mid\mathbf{r}'-\mathbf{R}'\mid}\right) = \theta(R'-r')\left(\frac{1}{r'}-\frac{1}{R'}\right)\int d\Omega_{R'}.$$

Only the bound S-states of the electron contribute in [1.33] and we find

$$\int\psi_n^\dagger(\mathbf{r}')\theta(R'-r)\left(\frac{1}{r'}-\frac{1}{R'}\right)\psi_0(\mathbf{r}')d^3r' \sim \frac{2\pi}{3n^{\frac{3}{2}}}\mid\psi(0)\mid^2 R'^2 \quad ...[1.34]$$

which is of order R^2. The first matrix element in [1.33] can then be evaluated in the $R \to 0$ limit and we obtain a contribution to hfs from the bound state with principal quantum number n that is approximately

$$\nu^{(g)}_{2,\,\text{bound}} \approx \frac{\alpha^2 \mu_p}{M} \mid \psi(0) \mid^2 (\alpha a_0) \frac{1/n^3}{\alpha^2 m \left(1 - \dfrac{1}{n^2}\right)} \frac{R^2}{a_0^3} \approx \left[(R/a_0)^2 \frac{1}{n^3} \frac{1}{1 - \dfrac{1}{n^2}} \right] \nu_F. \qquad \text{...[1.35]}$$

The sum over all $n \neq 1$ converges and we see that the bound states contribute only to order $(R/a_0)^2$ and hence may be dropped. The low lying continuum states of momentum $p \lesssim m$ suffer the same fate and the remaining ones can be treated in plane wave approximation up to higher order corrections in α:

$$\psi_{n^\pm}(\boldsymbol{r}) \to \psi_p^{(\pm)}(\boldsymbol{r}) = e^{i\boldsymbol{p}.\boldsymbol{r}} u^{(\pm)}(\boldsymbol{p}); \quad (u^{\dagger(\pm)}(\boldsymbol{p}), u^{(\pm)}(\boldsymbol{p})) = 1. \qquad \text{...[1.36]}$$

In this approximation we can also replace the ground states wave function by its value at $\boldsymbol{r} = 0$

$$\psi_0(\boldsymbol{r}) \Rightarrow \psi(0)$$

and the relevant matrix elements are tabulated:

$$\int d^3 r' \psi_p^{\dagger(+)}(r') \left(\frac{1}{r'} - \frac{1}{\mid \boldsymbol{r}' - \boldsymbol{R}' \mid} \right) \psi_0(\boldsymbol{r}') = \frac{4\pi}{p^2} \psi(0) (u^{(+)}(\boldsymbol{p}), u(0)) [1 - e^{-i\boldsymbol{p}.\boldsymbol{R}'}] \qquad \text{...[1.37]}$$

$$\int d^3 r \psi_0^\dagger(\boldsymbol{r}) \boldsymbol{\alpha} . \left(\nabla_r \frac{1}{\mid \boldsymbol{r} - \boldsymbol{R} \mid} \times \boldsymbol{\sigma} \right) \psi_p^{(+)}(\boldsymbol{r}) = \frac{4\pi}{p^2} \psi(0) (\boldsymbol{\sigma} \times \nabla_{\boldsymbol{R}} e^{i\boldsymbol{p}.\boldsymbol{R}}) . (u(0), \boldsymbol{\alpha} u^{(+)}(\boldsymbol{p})). \qquad \text{...[1.38]}$$

The negative energy matrix elements differ only in the replacement

$$u^{(+)}(\boldsymbol{p}) \to u^{(-)}(\boldsymbol{p}).$$

Using

$$\sum_{n\pm} = \int \frac{p^2 dp d\Omega_p}{(2\pi)^3}$$

$$\sum_{\text{spins}} u^{(\pm)}(\boldsymbol{p}) (u^{(\pm)}(\boldsymbol{p}))^\dagger = \frac{\pm \boldsymbol{\alpha}.\boldsymbol{p} \pm \beta m + E_p}{2E_p}, \quad E_p = +(p^2 + m_e^2)^{\frac{1}{2}}$$

$$E_0 \approx m,$$

and retaining only spin flip terms

$$\sum_{\text{spins}} (u(0),\, \boldsymbol{\alpha} u^\pm(\boldsymbol{p})) (u^\pm(\boldsymbol{p}), u(0)) \to \pm \frac{i\boldsymbol{p} \times \boldsymbol{\sigma}_e}{2E_p}$$

we find

$$\nu_2^{(g)} = \frac{4\alpha^2\mu_p}{M} \operatorname{Re} \int \frac{d^3R}{4\pi R^2} \,|\, \phi_s(R) \,|^2 \int \frac{d^3R'}{4\pi R'^2} \,|\, \phi_s(R') \,|^2 \,\psi^2(0)\, G(\mathbf{R}, \mathbf{R}') \quad ...[1.39]$$

with

$G(\mathbf{R}, \mathbf{R}') =$

$$\int \frac{p^2 dp d\Omega_p}{8\pi^3} \left(\frac{4\pi}{p^2}\right)^2 (\boldsymbol{\sigma} \times \nabla_R) \cdot (\boldsymbol{\sigma} \times \nabla_R) \left[e^{i\mathbf{p}.\mathbf{R}} - e^{i\mathbf{p}.(\mathbf{R}-\mathbf{R}')}\right] \frac{1}{2E_p} \left(\frac{1}{E_p - m} + \frac{1}{E_p + m}\right).$$

After carrying out the angular integration over $\mathbf{p}$ and taking the gradients in [*1.39*] we have

$$\nu_2^{(g)} = -\nu_F \frac{8}{\pi a_0} \left\langle \int_0^\infty \frac{dp}{p^2} \{j_0(pR) - j_0(p \,|\, \mathbf{R} - \mathbf{R}' \,|)\} \right\rangle_{00} \quad ...[1.40]$$

where j_0 is the spherical Bessel function of zeroth order and the notation

$$\langle \ldots . \rangle_{A0} \equiv \int d^3R \int d^3R' \Phi_0^\dagger(R) \Phi_A^\dagger(R') \{\ldots\} \, \Phi_A(R) \Phi_0(R') \quad ...[1.41]$$

has been introduced. It is convenient to split $\nu_2^{(g)}$ into two finite parts

$$\nu_2^{(g)} = \bar{\nu}_2^{(g)} + \tilde{\nu}_2^{(g)} \quad ...[1.42]$$

where

$$\bar{\nu}_2^{(g)} = +\nu_F \frac{8}{\pi a_0} \left\langle \int_0^\infty \frac{dp}{p^2} \{1 - j_0(pR)\} \right\rangle_{00} \quad ...[1.43]$$

and

$$\tilde{\nu}_2^{(g)} = -\nu_F \frac{8}{\pi a_0} \left\langle \int_0^\infty \frac{dp}{p^2} \{1 - j_0(p \,|\, \mathbf{R} - \mathbf{R}' \,|)\} \right\rangle_{00}. \quad ...[1.44]$$

Using the identity

$$\int_0^\infty \frac{dx}{x^2} (1 - j_0(x)) = \frac{\pi}{4} \quad ...[1.45]$$

we obtain

$$\bar{\nu}_2^{(g)} = +\nu_F \frac{2\langle R \rangle}{a_0} \quad ...[1.46]$$

which cancels the structure term of [*1.30*] and

$$\tilde{\nu}_2^{(g)} = -\nu_F \frac{2\langle |\, \mathbf{R} - \mathbf{R}' \,| \rangle}{a_0} \quad ...[1.47]$$

where

$$\langle |\, \mathbf{R} - \mathbf{R}' \,| \rangle \equiv \int d^3R \frac{|\, \phi_s(R) \,|^2}{4\pi R^2} \int d^3R' \frac{|\, \phi_s(R') \,|^2}{4\pi R'^2} \,|\, \mathbf{R} - \mathbf{R}' \,|. \quad ...[1.48]$$

Adding [1.46] and [1.47] to [1.30] gives the structure correction to hfs due both to the distribution of magnetization in the proton ground state and the alteration of the electron wave function from a point Coulomb solution as a result of the proton structure. The total hfs contribution in this approximation is

$$\nu = \nu_1 + \bar{\nu}_2^{(g)} + \tilde{\nu}_2^{(g)} = \nu_F \left\{ 1 - \frac{2}{a_0} \langle | \boldsymbol{R} - \boldsymbol{R}' | \rangle \right\}. \qquad \ldots[1.49]$$

[1.49] does not yet include any contribution arising from polarization since all excited quarkette states have been omitted in the second order perturbation treatment. For a rigid "baseball" model of the proton—i.e. a proton with the excited state spectrum far removed from the ground state to which it is only negligibly coupled by the perturbation—[1.49] is the correct result for the structure effect on hfs.

A general model-independent derivation of this non-relativistic result was first given in 1956 by Zemach.[11] Specific dependence on a model such as introduced in this section is required in order to formulate the polarizability calculation which is our primary concern and to which we turn now. This involves the terms with $A \neq 0$ in [1.31] corresponding to quarkette excitation. Denoting this contribution by $\nu_2^{(e)}$ we have

$$\nu_2^{(e)} = \frac{4\alpha^2 \mu_p}{M} \sum_{A \neq 0} \int d^3R \int d^3r \int d^3R' \int d^3r' [H^+(\boldsymbol{r}, \boldsymbol{r}', \boldsymbol{R}, \boldsymbol{R}') - H^-(\boldsymbol{r}, \boldsymbol{r}', \boldsymbol{R}, \boldsymbol{R}')] \qquad \ldots[1.50]$$

where

$$H^+(\boldsymbol{r}, \boldsymbol{r}', \boldsymbol{R}, \boldsymbol{R}') =$$

$$\sum_{n^+ \neq 0} \left\{ \frac{\psi_g^\dagger(\boldsymbol{r}, \boldsymbol{R}) \left\{ \boldsymbol{\alpha} . \left(\nabla_r \frac{1}{|\boldsymbol{r} - \boldsymbol{R}|} \times \boldsymbol{\sigma} \right) \right\} \psi_{An^+}(\boldsymbol{r}, \boldsymbol{R}) \psi_{An^+}^\dagger(\boldsymbol{r}', \boldsymbol{R}') \frac{1}{|\boldsymbol{r}' - \boldsymbol{R}'|} \psi_g(\boldsymbol{r}', \boldsymbol{R}')}{W_A - W_0 + E_{n^+} - E_0} \right\}$$

and

$$H^-(\boldsymbol{r}, \boldsymbol{r}', \boldsymbol{R}, \boldsymbol{R}') =$$

$$\sum_{n^-} \left\{ \frac{\psi_{0n^-}^\dagger(\boldsymbol{r}, \boldsymbol{R}) \left\{ \boldsymbol{\alpha} . \left(\nabla_r \frac{1}{|\boldsymbol{r} - \boldsymbol{R}|} \times \boldsymbol{\sigma} \right) \right\} \psi_{A0}(\boldsymbol{r}, \boldsymbol{R}) \psi_{A0}^\dagger(\boldsymbol{r}', \boldsymbol{R}') \frac{1}{|\boldsymbol{r}' - \boldsymbol{R}'|} \psi_{0n^-}(\boldsymbol{r}', \boldsymbol{R}')}{W_A - W_0 - E_{n^-} + E_0} \right\}.$$

Since the $1/r$ term of H_c has no matrix elements to excited quarkette states it has been dropped. Also the convection current term in [1.10] and [1.11] does not contribute to hfs to this order and can be omitted since it contributes only a spin independent energy when inserted in [1.50]. As we found in [1.32], where there was no quarkette excitation, the important contributions to [1.50] come from relativistic intermediate electron energies; in this case $E_n \approx (W_A - W_0) \gg m$. For such energies we may again replace the Dirac Coulomb states $\psi_{n^\pm}$ by free particle wave functions [1.36] with a negligible error $O(\alpha)$. Furthermore, as is apparent in the form of [*1.39*], when the energy denominators are replaced by

$$\frac{1}{E_p \mp m} \xrightarrow[\nu_2^{(g)} \to \nu_2^{(e)}]{} \frac{1}{W_A - W_0 + E_p \mp m} \xrightarrow[p \gg m]{} \frac{1}{W_A - W_0 + p} \qquad \text{...[1.51]}$$

the positive and negative energy sums in [*1.50*] contribute equally to the order of accuracy to which we are working. Thus we need only evaluate the positive energy sum explicitly and double the result to obtain $\nu_2^{(e)}$.

Proceeding in complete analogy with the calculation of $\nu_2^{(g)}$ in [*1.32*] and [*1.39*] we reduce [*1.50*] to

$$\nu_2^{(e)} = -\frac{32\alpha^2 \mu_p \psi^2(0)}{M} \sum_{A \neq 0} \int d^3R \int d^3R' \Phi_0^\dagger(R) \Phi_A^\dagger(R') \Phi_A(R) \Phi_0(R')\, I(\mathbf{R}, \mathbf{R}')$$

with

$$I(\mathbf{R}, \mathbf{R}') = \int_0^\infty \frac{dp}{p^3} \frac{(\sigma \times \nabla_R) . (\sigma_e \times \nabla_R)}{W_A - W_0 + p} \int \frac{d\Omega_p}{4\pi} e^{i\mathbf{p}.(\mathbf{R}-\mathbf{R}')}$$

so that

$$\nu_2^{(e)} = -\nu_F \frac{8}{\pi a_0} \sum_{A \neq 0} \left\langle \int_0^\infty \frac{dp}{p} \frac{1}{(W_A - W_0 + p)} \{1 - j_0(p\,|\mathbf{R} - \mathbf{R}'|)\} \right\rangle_{A0}. \qquad \text{...[1.52]}$$

We may now recognise $\tilde{\nu}_2^{(g)}$ of [*1.44*] as the ground state term, $A = 0$, missing from the sum in [*1.52*] and we can combine them

$$\tilde{\nu}_2^{(g)} + \nu_2^{(e)} = -\nu_F \frac{8}{\pi a_0} \sum_{\text{all } A} \left\langle \int_0^\infty \frac{dp}{p(W_A - W_0 + p)} \{1 - j_0(p\,|\mathbf{R} - \mathbf{R}'|)\} \right\rangle_{A0}. \qquad \text{...[1.53]}$$

Before continuing we can gain some valuable insight into the magnitude of the polarizability correction by considering [*1.53*]. This expression makes it clear that intermediate electron momenta of

$$p \sim \frac{1}{R} \gtrsim 100 \text{ MeV} \gg m$$

are of primary importance—i.e. the hfs is sensitive to the amplitude of finding the electron near the proton, as is well known. The criterion determining whether the polarizability contribution will be important or not can be seen from [*1.53*]. The important region of intermediate electron momenta extends up to $p_{\max} \sim 1/R$. According to the energy denominator in [*1.53*] the contributions from excited states such that $\Delta W_A \lesssim p_{\max} \sim 1/R$ may be of the same approximate magnitude as the ground state term, whereas the contributions from highly excited states with $\Delta W_A R \gg 1$ are reduced.

The physical origin of this criterion is understood by noting that $\Delta W_A R$ is a measure of the quarkette velocity, v_q, as it moves a distance R in time $1/\Delta W_A$. Thus for

$$v_q \sim \Delta W . R < 1 \qquad \text{...[1.54]}$$

the electron moving near the proton surface with the velocity of light ($c = 1$) can follow the proton and recentre its wave function around the instantaneous charge position. In this case the polarizability may be calculated using the recentering theory of A. Bohr[20] and Low,[21] as developed for the deuteron which satisfies [*1.54*], and the correction is sizable. On the other hand for

$$v_q \sim \Delta W.R > 1 \qquad ...[1.55]$$

the electron cannot follow the charge which is moving too rapidly. In this case the proton approaches the hard baseball limit, the polarizability correction becomes unimportant and proton radii or form factors, as measured in elastic electron-proton scattering, completely characterize the correction to the Fermi term in hfs as computed originally by Iddings and Platzman[25] and by Zemach.[11]

These remarks and observations take the following mathematical form. Introducing the integral

$$1-j_0(p\,|\,\mathbf{R}-\mathbf{R}'\,|) = \int_0^{p\,|\,\mathbf{R}-\mathbf{R}'\,|} dz\, j_1(z) \qquad ...[1.56]$$

into [*1.53*] and interchanging orders of integration one has

$$\tilde{v}_2^{(g)}+v_2^{(e)} = -v_F \frac{8}{\pi a_0} \sum_{\text{all } A} \left\langle \left(\int_0^\infty \frac{dz\, j_1(z)}{W_A-W_0} \ln\left(1+\frac{(W_A-W_0)\,|\,\mathbf{R}-\mathbf{R}'\,|}{z}\right)\right)\right\rangle_{A0}. \qquad ...[1.57]$$

The contribution from the proton ground state (i.e. $A = 0$) is explicitly evaluated by observing

$$\lim_{(W_A-W_0)\to 0} \int_0^\infty dz\, j_1(z) \frac{1}{(W_A-W_0)} \ln\left(1+\frac{(W_A-W_0)\,|\,\mathbf{R}-\mathbf{R}'\,|}{z}\right)$$
$$= |\,\mathbf{R}-\mathbf{R}'\,| \int_0^\infty dz \frac{j_1(z)}{z} = \frac{\pi}{4}\,|\,\mathbf{R}-\mathbf{R}'\,| \qquad ...[1.58]$$

which reduces to [*1.49*] as already evaluated. The approximate measure of the excited state contribution is given by the factor

$$\frac{1}{\chi} \ln(1+\chi) \qquad ...[1.59]$$

where $\chi \equiv \overline{\Delta W}.\overline{R}$ is the product of a mean excitation energy of the proton, $\overline{\Delta W}$ and a mean radial size $\overline{R}$. In the limit of a rigid proton $\overline{\Delta W} \to \infty$ and the polarizability correction $\frac{1}{\chi} \ln(1+\chi) \to 0$.

In the opposite limit of a very polarizable structure such that for *all A*, $(W_A - W_0)R \to 0$ we again make the approximations of [*1.58*] and [*1.57*] becomes

$$(\tilde{\nu}_2^{(g)}+\nu_2^{(e)})_{\Delta W\to 0} =$$

$$-\nu_F\frac{2}{a_0}\int d^3R\int d^3R'\sum_{all\,A}\Phi_0^\dagger(R)\Phi_A(R)\Phi_A^\dagger(R')\Phi_0(R')\,|\,R-R'\,|=0$$

...[1.60]

when we make use of completeness of the intermediate state sum

$$\sum_{all\,A}\Phi_A(R)\Phi_A^\dagger(R') = \delta^3(R-R')$$

to apply closure. What is going on in this case is that the electron and proton wave functions have been distorted—i.e. polarized—by their mutual Coulomb attractions so that they follow each other's instantaneous position. There is then no finite size reduction of the hfs in contrast to the rigid proton case in which the electron sees the ground state average charge and magnetization distributions. The recentering theory of A. Bohr[20] applies in the polarizable limit in which case the Born-Oppenheimer approximation is valid for the relevant portion of the electron wave function at the proton (or nuclear) surface. The formal series generated by the recentering transformation of Low[21] is just an expansion in powers of the parameter χ in [1.59] and is appropriate for $\chi < 1$. The present treatment is less elegant in this case but is a convergent perturbation expansion in powers of the fine structure constant α and of $R/a_0 \sim \alpha(mR) < \alpha$ and so can be applied to both the polarizable and the rigid "baseball" limit.

For $\overline{\Delta W}.\bar{R} = \chi \ll 1$ it gives just the same result as found by Low. This is the case appropriate for the deuteron where $\chi \approx$ (2 MeV)/(50 MeV) $< 0{\cdot}1$ and as a result of polarizability, or "recentering", the hfs is increased above the contribution expected for a point structure. This is because the electron, recentred on the proton, sees the full proton moment, whereas the neutron moment, which in the triplet spin state tends to cancel the proton moment, is only partially seen and is thereby less effective—viz.

$$\frac{\nu_{\text{deuteron}}}{\nu_{\text{point}}} = \frac{\mu_p+\mu_n(1-\langle R_d\rangle/a_0)}{\mu_p+\mu_n} = 1+\left|\frac{\mu_n}{\mu_d}\right|\frac{\langle R_d\rangle}{a_0}$$
$$\approx 1+170\times 10^{-6}. \qquad \text{...[1.61]}$$

Model dependent numerical results can also be obtained for an intermediate case with $\chi \sim 1$. Since the meson production threshold gives a lower limit to the excitation spectrum of 140 MeV and the observed proton radius is $R \sim {\cdot}7$ f $\approx 1/300$ MeV neither simple limit of a rigid baseball proton or of a very polarizable one with the electron recentred on the instantaneous charge applies. We may expect then that the actual numerical value of $\Delta\nu^{(p)}$ in equation [1.5] is dependent on details of the proton dynamics.

The simplest approximation for estimating the polarizability contributions from $A \neq 0$ states in [1.57] would be to replace the logarithm by some mean value over the excitation spectrum, as is done with the Bethe logarithm in evaluating Lamb shift. Unfortunately orthogonality and closure may be used very simply to show that in this approximation the contributions vanish so that this approach is inadequate. This shows again that the results depend on details of the excited state spectrum and on no easily estimated simple average parameter alone.

For arbitrary polarizability the hfs is expressed by the sum of [*1.49*] and [*1.57*] with the ground state terms $A = 0$ omitted from the intermediate state sum ($\Delta W_A \equiv W_A - W_0$):

$$\nu = \nu_F\left[1 - \frac{2}{a_0}\langle|\mathbf{R}-\mathbf{R}'|\rangle + \frac{8}{\pi a_0}\int_0^\infty dz\, j_1(z) \sum_{A\neq 0}\left(\frac{1}{\Delta W_A}\left\langle \ln\frac{z}{z+\Delta W_A|\mathbf{R}-\mathbf{R}'|}\right\rangle_{A0}\right)\right]. \qquad \ldots[1.62]$$

For purposes of an estimate we compute the contribution in equation [*1.62*] to the hfs from an excited P state orbital triplet with the radial distribution given by the eigensolutions for a spherical harmonic oscillator well. The wave functions are taken as

$$\left.\begin{aligned}\Phi_0(R) &= \left(\frac{2\lambda}{\pi}\right)^{\frac{3}{4}} \exp(-\lambda R^2)\\ \Phi_1(\mathbf{R}) &= \left(\frac{2\lambda}{\pi}\right)^{\frac{3}{4}} 2(\lambda)^{\frac{1}{2}}\mathbf{R}\exp(-\lambda R^2)\end{aligned}\right\} \qquad \ldots[1.63]$$

where the mean square radius is given in terms of λ by

$$\langle R^2\rangle \equiv \frac{3}{4\lambda}.$$

The ground state contribution in [*1.62*] is (see equation [*1.47*])

$$\begin{aligned}\tilde{\nu}_2^{(g)} &= -\nu_F\frac{2}{a_0}\left(\frac{2\lambda}{\pi}\right)^3\int d^3R\int d^3R' \exp[-2\lambda(R^2+R'^2)]\,|\mathbf{R}-\mathbf{R}'|\\ &= -\nu_F\frac{2}{a_0}\frac{4}{(3\pi)^{\frac{1}{2}}}\langle R^2\rangle^{\frac{1}{2}}. \qquad \ldots[1.64]\end{aligned}$$

The polarizability contribution in [*1.62*] due to the P state triplet can be expressed in a form which permits the angular and one of the radial integrals over the wave functions to be carried out and we find for the contribution from the state $A \neq 0$

$$\left.\begin{aligned}\nu_2^A &= -\nu_F\frac{64\langle R^2\rangle^{\frac{1}{2}}}{\pi(3\pi)^{\frac{1}{2}}a_0}\int_0^\infty d\tau\left[\tfrac{1}{2}\ln\frac{1+\tau^2}{\tau^2} - 1 + \tau\tan^{-1}\frac{1}{\tau}\right]K(\tau),\\ K(\tau) &= \int_0^\infty \rho^3 d\rho\left(\tfrac{3}{2}-\rho^2\right)\exp\{-\rho^2-\rho\tau[W_A - W_0/\lambda^{\frac{1}{2}}]\}.\end{aligned}\right\} \qquad \ldots[1.65]$$

In the limit $(W_A - W_0) \to 0$

$$\lim_{(W_A - W_0) \to 0} \nu_2^A = +\nu_F \frac{2}{a_0} \frac{2}{(3\pi)^{\frac{1}{2}}} \langle R^2 \rangle^{\frac{1}{2}}, \qquad ...[1.66]$$

which cancels one-half of the finite size reduction in [*1.64*]. In the limit $\Delta W_A \langle R^2 \rangle^{\frac{1}{2}} \to 0$ for all A, the sum $\sum_{all\ A} \nu_2^A$ precisely cancels the finite size reduction [*1.64*], as was shown earlier in [*1.60*].

For large values of $(W_A - W_0)\langle R^2 \rangle^{\frac{1}{2}}$ the polarizability contribution decreases roughly as $1/[(W_A - W_0)\langle R^2 \rangle^{\frac{1}{2}}]$. Fixing $\langle R^2 \rangle^{\frac{1}{2}} \approx \cdot 8$ f we have numerically integrated [*1.65*] for a wide range of excitation energies and the ratio $\nu_2^A/\nu_2^{(g)}$ is plotted in Fig. 2 as a function of $(W_A - W_0)$. For excitations up to 1 BeV the correction is still significant. At an excitation energy of $W_A - W_0 = 400$ MeV, for example, this one level contributes $\frac{1}{3}$ as much to the polarizability correction as it does in the completely polarizable limit $(W_A - W_0)R \to 0$ and cancels $\frac{1}{6}$ of the finite size correction to the hfs. Taking into account the fact that the contribution is positive definite for all states $A \neq 0$ it becomes clear that the contributions of several excited states in this model may cancel an appreciable fraction of the finite size reduction in [*1.64*] of the hfs.

Evidently the details of the dynamics of the proton cannot be avoided completely in a quantitative study of the hfs. A similar conclusion follows from the massive three quark model of Fenster and Nambu.[9] They have invoked the recentering idea of Bohr[20] to partially cancel the ground state

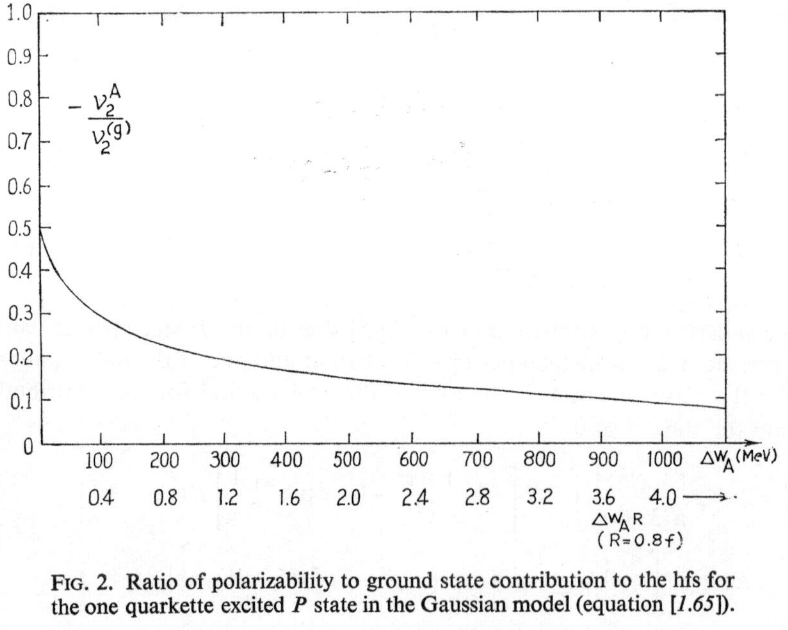

FIG. 2. Ratio of polarizability to ground state contribution to the hfs for the one quarkette excited *P* state in the Gaussian model (equation [*1.65*]).

average form factor effect as expressed in [*1.49*]. Their theory is based on a strict analogy for the proton, constructed of massive slow moving quarks, with a nucleus formed of nucleons. However, there is a fundamental difficulty in taking this analogy literally as proposed by Fenster and Nambu. This arises from the fact that the scale of excitation energies ΔW for the proton is much higher than that for a nucleus. In the deuteron or other weakly bound nuclear systems (with anomalous thresholds) the relevant parameter is $\Delta W\ R < 1$ and the recentering idea of Bohr is applicable. This is not true for a proton for which the minimum ΔW is the pion production threshold, $\approx$ 140 MeV, so that $\Delta W\ R \gtrsim 1$. The Fenster and Nambu prediction is a partial cancellation of the finite size reduction in the hfs reduction based on the recentering assumption which requires $\Delta W\ R < 1$. Whether the proton radius R is attributed to the bound state wave function of a quarkette or to the intrinsic radius of the quarkette itself the simple recentering assumption cannot be applied to the proton as we now show.

Formally the effect of replacing a point by a distributed quarkette is summarized by generalizing [*1.9*] and [*1.11*] to

$$H_c' = -\frac{e^2}{4\pi}\int d^3S\rho_c(\boldsymbol{S}-\boldsymbol{R})\frac{1}{|\boldsymbol{r}-\boldsymbol{S}|} \qquad \ldots[1.67]$$

$$\boldsymbol{A}(\boldsymbol{r}) = -\frac{e\kappa}{2\mu}\int d^3S\rho_m(\boldsymbol{S}-\boldsymbol{R})\,\nabla_S\frac{1}{|\boldsymbol{r}-\boldsymbol{S}|}\times\boldsymbol{\sigma} \qquad \ldots[1.68]$$

where ρ_c and ρ_m are the charge and magnetic densities of the quarkette. As in the point case the structure correction in ν_1 is cancelled by a part, $\bar{\nu}_2^{(g)}$, of the second order correction. The full structure correction is obtained by replacing [*1.53*] by

$$\bar{\nu}_2^{(g)}+\nu_2^{(e)} = -\nu_F\frac{8}{\pi a_0}\sum_{\text{all }A}\Big\langle\int d^3S\rho_m(\boldsymbol{S}-\boldsymbol{R})\int d^3S'\rho_c(\boldsymbol{S}'-\boldsymbol{R}')\,.$$

$$\cdot\int_0^\infty\frac{dp}{p(W_A-W_0+p)}\{1-j_0(p\,|\boldsymbol{S}-\boldsymbol{S}'|)\}\Big\rangle_{A0}. \qquad \ldots[1.69]$$

In the limit of a quarkette of large radius, but with a very polarizable wave function, we may neglect W_A-W_0 in the denominator of [*1.69*], use closure and [*1.45*] to obtain

$$\bar{\nu}_2^{(g)}+\nu_2^{(e)} = -\nu_F\frac{2}{a_0}\int d^3S\int d^3S'\int d^3R\int d^3R'\Phi_0^\dagger(R)\rho_m(\boldsymbol{S}-\boldsymbol{R})\rho_c(\boldsymbol{S}'-\boldsymbol{R}')\,.$$

$$.\ \Phi_0(R')\delta^3(\boldsymbol{R}-\boldsymbol{R}')\,|\boldsymbol{S}-\boldsymbol{S}'| = -\nu_F\frac{2}{a_0}\langle|\boldsymbol{S}-\boldsymbol{S}'|\rangle \qquad \ldots[1.70]$$

where $\langle|\boldsymbol{S}-\boldsymbol{S}'|\rangle = \int d^3Sd^3S'\rho_m(\boldsymbol{S})\rho_c(\boldsymbol{S}')\,|\boldsymbol{S}-\boldsymbol{S}'|.$ ...[*1.71*]

The total hfs is then $$\nu = \nu_F\left(1-\frac{2}{a_0}\langle|\mathbf{S}-\mathbf{S}'|\rangle\right) \qquad ...[1.72]$$

which is the rigid baseball model once more, only now it is the quarkette itself that is the rigid baseball.

The numbers we obtained above may have no quantitative significance whatsoever. However they do carry a warning that simplified approximate inputs in a relativistic dispersion analysis must also be viewed with great caution. At the same time these numbers also provide a strong motivation to perform a dispersion theoretic calculation of the polarizability beyond the ones that have been given and which so far have concentrated on the 33 resonance alone.

1.4. *Relativistic calculation*

Let me turn now to the relativistic calculation of the proton structure correction to the hfs. The general formulation has already been given by Iddings[10] along with the calculation of the proton ground state terms—the analogues of ν_1 and $\nu_2^{(g)}$ of the previous section. As in section 1.3 we focus our attention on the polarizability corrections. Because we work with different invariant amplitudes and make contact with the low energy theorem in a different manner than reference 10 we briefly resketch the development of Iddings.

The contribution to the hfs arising from proton recoil was first calculated by Newcomb and Salpeter[26] and by Arnowitt.[27] Their analysis required a study of the bound state equation for the electron-proton system. If the proton is treated as a point Dirac particle of finite mass but with zero anomalous moment the hfs contribution is

$$-\alpha\frac{m}{M}\mathrm{X} = -\frac{\alpha}{1+\kappa}\frac{m}{M}\left\{\frac{3}{\pi}\ln M/m\right\} = -10 \text{ ppm}.$$

When the proton is also assigned a point Pauli moment of $\kappa = 1{\cdot}79$ nuclear magnetons the additional contribution to the hfs is

$$-\alpha\frac{m}{M}\mathrm{X} = -\frac{\alpha}{1+\kappa}\frac{m}{M}\frac{1}{\pi}\{3(1-\kappa^2/4)\ln M/m+\kappa^2\tfrac{3}{4}(-\tfrac{1}{6}+3\ln\Lambda/M)\}$$
$$\approx -4 \text{ ppm}.$$

A logarithmic cut off $\Lambda \sim 2M$ has been introduced as needed into the square of the Pauli moment interaction and shows a sensitivity of the numerical results to virtual photons of high mass being exchanged between the atomic electron and the proton. This divergence is removed in a natural way when the structure of the proton is taken into account, as first studied by Iddings and Platzman[25] and by Zemach.[11] In particular they show that the effect of structure analogous to the contributions of $(\nu_1-\nu_F)$ and of $\nu_2^{(g)}$ in the Schrödinger models of the preceding section, can be summarized in terms of the electromagnetic form factors of the proton in the interaction current with the electron. In the calculation of these structure corrections to the hfs one can ignore all complications of the bound state problem and treat the electron-proton interaction in second order Born approximation. We saw this to be true in the models of the preceding section where the intermediate electron was described by free particle wave functions and energy eigenvalues. The bound state appears only in the non-relativistic factor $\psi^2(0)$ for the density of the atomic electrons at the origin in [*1.39*].

This formulation in the relativistic calculation was presented by Iddings and Platzman[25] who showed that the complete proton structure contribution to the hfs in hydrogen can be obtained from the two photon exchange amplitude as illustrated in Fig. 3. The "blob" in Fig. 3 is the forward spin-flip Compton scattering amplitude, hereafter denoted by $C_{\mu\nu}$, for virtual

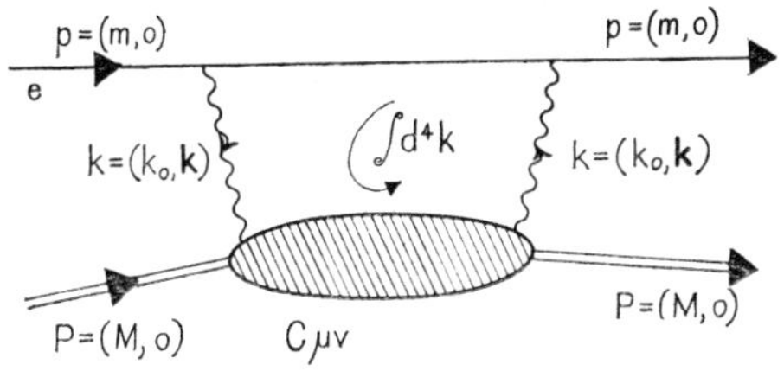

FIG. 3. Two photon exchange amplitude for the proton structure contribution to the hfs in hydrogen. $C_{\mu\nu}$ is the forward spin-flip Compton amplitude for virtual photons of momentum k_μ.

photons from protons at rest. Iddings[10] exploited this result, following the method developed by Cottingham[22] for the neutron-proton mass difference, and rotated the contour of the dk_0 integration in Fig. 3, from the real to the imaginary k_0 axis so that the virtual photon masses that are involved are always space-like ($k^2 \leqq 0$). In this regime of photon masses the Compton amplitude $C_{\mu\nu}$ may be rigorously proved to be analytic in the entire photon energy plane except for the right- and left-hand cuts along the real axis required by unitarity and crossing. Anomalous thresholds are absent. The pole terms in the virtual Compton amplitude contain electromagnetic form factors $F_1(k^2)$ and $F_2(k^2)$ where $k^2 < 0$ is the (mass)^2 of the virtual photon. The structure correction corresponding to $v_1 - v_F$ and $v_2^{(g)}$ in the preceding section comes from the difference between the pole terms calculated with observed values of F_1 and F_2 and with $F_1 = F_2 = 1$ as for a point proton. This difference has been studied by Iddings in detail. The non-pole terms in $C_{\mu\nu}$ contribute to the polarizability correction on which we focus our primary attention.

In place of the invariant amplitudes used by Iddings we decompose $C_{\mu\nu}$ according to

$$C_{\mu\nu}(k^2, \omega) = T_{\mu\nu}^{(1)} H_1(k^2, \omega) + T_{\mu\nu}^{(2)} H_2(k^2, \omega) \qquad \ldots[1.73]$$

where

$$T^{(1)}_{\mu\nu} = \frac{1}{M^3}\{[\gamma_\mu, \gamma_\nu]p^\alpha k_\alpha - [\gamma_\mu, \gamma_\alpha]k^\alpha p_\nu - [\gamma_\nu, \gamma_\alpha]k^\alpha p_\mu\} \qquad ...[1.74]$$

$$T^{(2)}_{\mu\nu} = \frac{p_\beta k^\beta}{M^3}\{[\gamma_\mu, \gamma_\nu]k^2 - [\gamma_\mu, \gamma_\alpha]k^\alpha k_\nu - [\gamma_\nu, \gamma_\alpha]k^\alpha k_\mu\} \qquad ...[1.75]$$

ν is the polarization index for the incident photon line
μ is the polarization index for the outgoing photon line
p_β is the proton four-momentum ($p_\beta = M, 0, 0, 0$)
k_β is the photon four-momentum
$\omega = \dfrac{p_\beta k^\beta}{M}$ and M the proton mass.

This choice of amplitudes is particularly convenient[28]; in the limit of physical Compton scattering ($k^2 = 0$) the amplitude H_2 is completely decoupled and the remaining amplitude H_1 satisfies a sum rule as we shall see below.

With the help of equation (2.2) of reference 10 the two photon exchange contribution to the hfs may be written (after rotation of the k_0 contour of the Feynman loop integration)

$$\frac{\nu_{2\gamma}}{\nu_F} = \frac{2}{a_0}\frac{1}{M(1+\kappa)\pi^3}\int\frac{d^4K}{K^6}\{(2K^2+K_0^2)H_1(-K^2, iK_0) - 3K^2K_0^2H_2(-K^2, iK_0)\} \qquad ...[1.76]$$

where ν_F is the point proton splitting [*1.29*], $\kappa = 1{\cdot}79$ is the proton anomalous moment, and a_0 the Bohr radius ($a_0 = 1/\alpha m$). In [*1.76*] the integration is over a four-dimensional Euclidean space K with $K_0 = ik_0$, $K^2 = -k^2$, and

$$\int d^4K = 4\pi\int_0^\infty K^3 dK \int_0^\pi d\psi \sin^2\psi \qquad ...[1.77]$$

with $K_0 = K\cos\psi$ as in equation (2.14) of reference 10.

Writing dispersion relations in ω for H_1 and H_2, explicitly displaying the contribution of the intermediate one proton pole, one has

$$H_1(k^2, \omega) = \frac{-k^2M^2F_1(k^2)[F_1(k^2)+\kappa F_2(k^2)]}{(k^2)^2 - 4M^2\omega^2} + \frac{1}{\pi}\int_{C(k^2)}^\infty \frac{d\omega'^2}{\omega'^2-\omega^2}\,\mathrm{Im}\,H_1(k^2, \omega') \qquad ...[1.78]$$

$$H_2(k^2, \omega) = \frac{-\kappa M^2F_2(k^2)[F_1(k^2)+\kappa F_2(k^2)]}{(k^2)^2 - 4M^2\omega^2} + \frac{1}{\pi}\int_{C(k^2)}^\infty \frac{d\omega'^2}{\omega'^2-\omega^2}\,\mathrm{Im}\,H_2(k^2, \omega') \qquad ...[1.79]$$

where $F_1(k^2)$ and $F_2(k^2)$ are the proton Dirac and Pauli electromagnetic form factors respectively and the cut starts at the pion-nucleon threshold,

$$C(k^2) = \frac{1}{4M^2}(2M\mu+\mu^2-k^2)^2. \qquad ...[1.80]$$

Throughout this section μ will denote the pion mass. Iddings[10] has further shown through unitarity that Im H_1 and Im H_2 in the space-like region $k^2 < 0$ may be determined experimentally from the cross-section of inelastic scattering of polarized electrons from polarized protons. It will be a long time, however, before such data exist.

In writing [*1.78*] and [*1.79*] we have made the assumption that there need be no subtractions in ω for all $k^2 < 0$. We defer further remarks about this point until the discussion of section 1.5. We must subtract from equations [*1.76*] to [*1.79*] the point proton result, which is obtained by setting $F_1 = F_2 = 1$, Im $H_2 = 0$ and,[29] by the low-energy theorem of Low[30] and Gell-Mann and Goldberger,[31]

$$-\frac{\kappa^2}{4} = \frac{1}{\pi}\int_{C(0)}^{\infty}\frac{d\omega'^2}{\omega'^2}\,\mathrm{Im}\,H_1(0,\omega'). \qquad ...[1.81]$$

We write this difference as

$$\frac{\nu_s}{\nu_F} = S_1+S_2+S_3 \qquad ...[1.82]$$

where

$$S_1 = \frac{1}{a_0M(1+\kappa)\pi^3}\int\frac{d^4K}{K^6}\left\{\frac{2K^2(2K^2+K_0^2)[F_1(F_1+\kappa F_2)-(1+\kappa)]}{4K_0^2+K^4/M^2}\right.$$
$$\left.+\frac{6K^2K_0^2\kappa[F_2(F_1+\kappa F_2)-(1+\kappa)]}{4K_0^2+K^4/M^2}-\frac{(2K^2+K_0^2)\kappa^2[F_2^2-1]}{2}\right\} \qquad ...[1.83]$$

$$S_2 = \frac{1}{a_0M(1+\kappa)\pi}\int_0^{\infty}\frac{dK}{K}\left\{\tfrac{9}{4}\kappa^2(F_2(-K^2))^2\right.$$
$$\left.-\frac{4}{\pi}\int_{C(-K^2)}^{\infty}\frac{d\omega'^2}{\omega'^2}\,\mathrm{Im}\,H_1(-K^2,\omega')(3\theta-2\theta^2-2(2-\theta)[\theta(\theta+1)]^{\frac{1}{2}})\right\}$$
$$...[1.84]$$

$$S_3 = \frac{-12}{a_0M(1+\kappa)\pi^2}$$
$$\cdot\int_0^{\infty}\frac{dK}{K}\int_{C(-K^2)}^{\infty}d\omega'^2\,\mathrm{Im}\,H_2(-K^2,\omega')\{1+2\theta-2[\theta(\theta+1)]^{\frac{1}{2}}\} \qquad ...[1.85]$$

and

$$\theta \equiv \omega'^2/K^2.$$

Note that a non-pole term proportional to κ^2 has been added in S_1 and then subtracted out in S_2. (As a point of reference it might be remarked that S_1 happens to coincide with the Born approximation taken with form factors at the vertices.) The term S_1 has already been evaluated by Iddings[10] and when added to the recoil correction S_R one has

$$S_R+S_1 = -(34{\cdot}5\pm 2) \text{ ppm}. \qquad \text{...[1.86]}$$

Separately S_R and S_1 have logarithmic divergences in the terms proportional to κ^2 but when they are added these divergences cancel provided $F_2(k^2) \to 0$ as $(-k^2) \to \infty$, so that the integral $\int (dK/K)(F_2(-K^2))^2$ converges for $K^2 \to \infty$. The detailed numerical calculations of Iddings[10] show that once the form factors F_1 and F_2 are fitted to their observed r.m.s. radii (and provided $F_2 \to 0$, $(-k^2) \to \infty$) there is surprisingly little freedom ($< \pm 2$ ppm) in the value of S_1.

Most of the contribution to S_1 comes from the first factor in the pole term which is linear in the total moment. The structure correction from this factor contributes $-26{\cdot}6$ ppm using measured form factors whereas the pole term proportional to $\kappa(1+\kappa)$ gives another $-6{\cdot}2$ ppm. Finally the non-pole term is responsible for only $+1{\cdot}3$ ppm.

For a proton without excitations we return to the rigid baseball proton limit of the models in section 1.3. Furthermore we can reduce the result for ν_s to the form found by Zemach[11] for structure corrections in the non-relativistic limit by taking the $M \to \infty$ limit, in which case only [1.83] survives. The last or non-pole term in [1.83], when added to the recoil terms S_R, converges for $K \to \infty$, as remarked above. In the $M \to \infty$ limit it vanishes as a result of the $\dfrac{1}{M}$ factor out in front of the integral. We may do the K_0 integral as a contour integral in the complex K_0-plane. Writing $K_\mu = (K_0, \mathbf{k})$, $k = |\mathbf{k}|$ and $d^4K = d^3k\, dK_0$, we see that poles occur at the roots of $K_0^2+k^2 = \pm 2iM\, K_0$

or

$$K_0 = \pm\frac{ik^2}{2M}\left(1+O\left(\frac{1}{M^2}\right)\right) \qquad \text{...[1.87]}$$

and

$$K_0 = \pm 2iM\left(1+O\left(\frac{1}{M^2}\right)\right). \qquad \text{...[1.88]}$$

In the limit $M \to \infty$ it is only the residue at

$$K_0 = \frac{ik^2}{2M}$$

that survives as we close the contour in the upper half plane and [1.83] reduces to

$$S_1 \xrightarrow[M \to \infty]{} \frac{8}{\pi a_0}\int_0^\infty \frac{dk}{k^2}(F_1(-k^2)G_M(-k^2)-1) \qquad \text{...[1.89]}$$

where

$$G_M(-k^2) \equiv \frac{F_1(-k^2)+\kappa F_2(-k^2)}{1+\kappa} \qquad ...[1.90]$$

is the magnetic form factor of the proton. Introducing the spherical charge and magnetic densities $\rho_c(R)$ and $\rho_M(R)$ respectively,

$$\begin{aligned} F_1(-k^2) &= \int d^3R\,\rho_c(R)\exp(i\boldsymbol{k}.\boldsymbol{R}) \\ G_M(-k^2) &= \int d^3R'\,\rho_M(R')\exp(-i\boldsymbol{k}.\boldsymbol{R}') \end{aligned} \qquad ...[1.91]$$

and using the identity (see [1.45])

$$\int d^3k\left(\frac{1-\exp[i\boldsymbol{k}.(\boldsymbol{R}-\boldsymbol{R}')]}{k^4}\right) = \pi^2\,|\,\boldsymbol{R}-\boldsymbol{R}'\,| \qquad ...[1.92]$$

we obtain

$$S_1 = -\frac{2}{a_0}\langle|\,\boldsymbol{R}-\boldsymbol{R}'\,|\rangle+O\left(\frac{1}{M}\right) \qquad ...[1.93]$$

where

$$\langle|\,\boldsymbol{R}-\boldsymbol{R}'\,|\rangle = \int d^3Rd^3R'\rho_c(R)\rho_M(R')\,|\,\boldsymbol{R}-\boldsymbol{R}'\,|. \qquad ...[1.94]$$

This is just the form found in [1.49].

So far the only attempts at calculations beyond S_1 have been estimates[10,18,19] of the contribution of the $N^*(1238)$ intermediate state to the absorptive amplitude in the dispersion integrals. In view of the dominant role of the 33 resonance in low energy photo-pion processes and its prominent contribution to Compton scattering, which it enhances by roughly one order of magnitude above the pole term at photon energy $\approx$ 300 MeV, it might be expected to contribute significantly to the polarizability correction to hfs. However it was found to play a totally negligible role contributing $\lesssim$ 1 ppm. This result may be understood in terms of the discussions that have been given of the smallness of the two photon exchange contributions to electron-proton scattering at high energies.[32] The arguments given there for the smallness of a narrow resonance's contribution to the dispersive amplitude may be rephrased in language appropriate to the Cottingham formulation. After rotating the k_0 contour to the imaginary axis as in [1.76] the hfs contribution is expressed as an integral over the Compton scattering of a virtual space-like photon. No one energy region is dominant because the energy factors in [1.84] and [1.85] are slowly varying and it is the total area under the photo-pion production curve that contributes to H_1 and H_2, or equivalently to S_2 and S_3. This area exceeds that under any one resonance, as shown in Fig. 4.

Another point emphasized by the models discussed in section 1.3 is that proton states of odd parity, in contrast with even parity ones excited by M1 transitions, play a prominent role in the polarizability contribution. We therefore turn to the continuum and the odd parity resonant excited states of the proton for further corrections to the hfs.

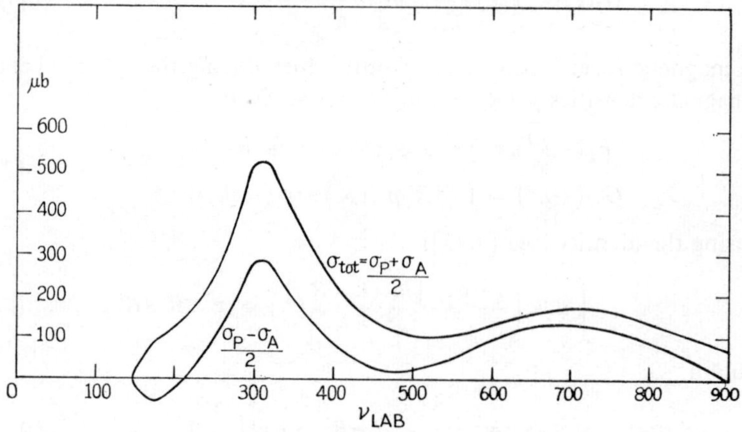

FIG. 4. Photo-pion production cross-section in H according to the model of Gourdin-Salin (reference 34); $\sigma_P(\sigma_A)$ is the cross-section for absorbing a circularly polarized photon with its spin parallel (antiparallel) to the proton spin (taken from reference 29).

We first discuss the term S_2, about which we may hope to get some guide from physical Compton scattering. With the help of the sum rule, [*1.81*], we rewrite S_2 in the convenient ratio form

$$S_2 = \frac{9\kappa^2}{a_0 M(1+\kappa)\pi^2} \int_0^\infty \frac{dK}{K} (F_2(-K^2))^2$$

$$\left\{1 - \left[\int_{C(-K^2)}^\infty \frac{d\omega'^2}{\omega'^2} \operatorname{Im} H_1(-K^2, \omega')\beta\left(\frac{\omega'^2}{K^2}\right)\right]\right.$$

$$\left.\left[(F_2(-K^2))^2 \int_{C(0)}^\infty \frac{d\omega''^2}{\omega''^2} \operatorname{Im} H_1(0, \omega'')\right]^{-1}\right\} \qquad ...[1.95]$$

where

$$\beta(\theta) = \tfrac{4}{9}\{2(2-\theta)[\theta(\theta+1)]^{\frac{1}{2}} - 3\theta + 2\theta^2\}. \qquad ...[1.96]$$

Note that β satisfies

$$0 \leqq \beta(\theta) \leqq 1 \quad \text{for} \quad 0 \leqq \theta \qquad ...[1.97]$$

$$\beta(\theta) \to 1 - \tfrac{5}{18}\frac{1}{\theta} + \ldots \quad \text{as} \quad \theta \to \infty \qquad ...[1.98]$$

$$\beta(\theta) \to \tfrac{4}{9}(4(\theta)^{\frac{1}{2}} - \theta + \ldots) \quad \text{as} \quad \theta \to 0. \qquad ...[1.99]$$

Using [*1.98*] it is immediately clear that in the limit $K \to 0$ the two terms of [*1.95*] cancel and therefore the integrand is well behaved at $K = 0$.

To proceed further we must resort to theoretical models and assumptions.

If one assumes that the dependence of Im H_1 on the photon mass is a simple scale factor, namely

$$\text{Im } H_1(-K^2, \omega) = g(K^2) \text{ Im } H_1(0, \omega), \qquad ...[1.100]$$

then it may be seen from [*1.97*] and [*1.80*], according to which $C(-K^2) \geqq C(0)$, that $S_1 > 0$, provided $g(K^2) \leqq (F_2(-K^2))^2$, and provided Im $H_1(0, \omega)$ is always of the same sign for $\mu < \omega < \infty$. As has been shown[33, 29] Im $H_1(0, \omega)$ is proportional to $(\sigma_P(\omega) - \sigma_A(\omega))$ where $\sigma_P(\omega)[\sigma_A(\omega)]$ is the total cross-section for the absorption of a circularly polarized photon of laboratory energy ω by a proton with its spin parallel [antiparallel] to the photon spin. Using the same fit to $(\sigma_P - \sigma_A)$ as in Fig. 4 we have numerically integrated [*1.95*] for the cases:

(a) $g(K^2) = (F_2(-K^2))^2$
(b) $g(K^2) = 1$ for $0 \leqq K^2 = 1{\cdot}5\,(\text{BeV})^2$; $g(K^2) = 0$ for $K^2 > 1{\cdot}5\,(\text{BeV})^2$
(c) $g(K^2) = 1$ for all K^2.

We approximate $F_2(-K^2)$ by the form

$$F_2(-K^2) = 1/(1+(K/0{\cdot}96M)^2)^2.$$

The results are shown in Table I.

TABLE I
Model Calculations of S_2 in Parts per Million

	(a) $g(K^2) = [1+(K/0{\cdot}96M)^2]^{-4}$	(b) $g(K^2) = \theta(1{\cdot}5\,(\text{BeV})^2 - K^2)$	(c) $g(K^2) = 1$
Sum Rule Fit to $\left(\frac{\sigma_P - \sigma_A}{2}\right)$ Gives $K = 2{\cdot}1$	0·5	−1·5	−1·5
Sum Rule Fit with tail to $\left(\frac{\sigma_P - \sigma_A}{2}\right)$ Gives $K = 1{\cdot}8$	1	−1	−1

Recall that this fit uses low energy S wave parameters, puts in the $N^*(1238)$ with predominantly magnetic dipole excitation as well as the second resonance[34] and takes $(\sigma_P - \sigma_A)$ equal to zero beyond 0·9 BeV.

As a crude test of the effect of assuming that the sum rule [*1.81*] is only apparently saturated by 0·9 BeV—that the integrand in fact is large and oscillating beyond 0·9 BeV—we show also the effect of adding a tail to the above fit

$$\frac{(\sigma_P - \sigma_A)}{2} = -50\,\mu\text{b} \quad 0{\cdot}9\text{ BeV} < \omega < 1{\cdot}8\text{ BeV}$$

which improves the fit to the sum rule. The resulting values for S_2, including this tail, are shown also in Table I. The sign and magnitude of S_2 are dependent on the details of the cut off as well as on the behaviour of $(\sigma_P - \sigma_A)$ which may take positive as well as negative values at high energies. However in all cases the contributions to S_2 amount to no more than 1-2 ppm and thus have little impact on the hfs problem.

We next turn to S_3. Because physical Compton scattering tells nothing about H_2 and there is no sum rule to help normalize the calculation, the magnitude and sign expected for S_3 is even more uncertain than for S_2. Although there are some data on electropion production, the polarized inelastic electron-proton data that is actually required does not at this time exist. We are interested in finding if any terms in Im H_2 are candidates for contributing sizably to S_3 in [*1.85*] or whether a sensitivity of the resulting

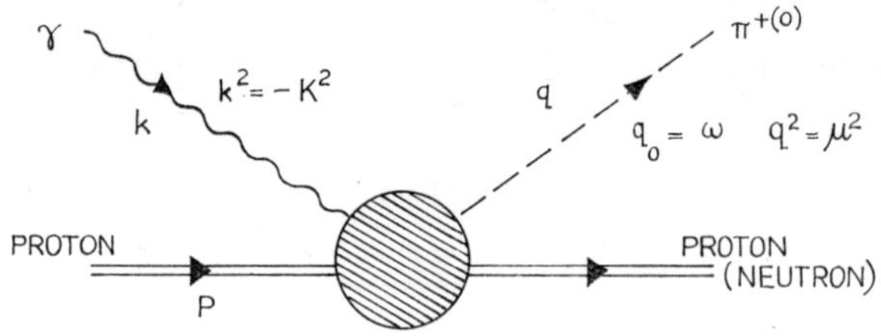

FIG. 5. Kinematics for amplitude of single pion photoproduction by a virtual photon.

value for S_3 to unknown features of the electropion amplitude can be identified.

In the beginning, at least, we are making only a rough search for polarization contributions, so we may use the static model and thereby simplify the kinematics and spinology. This approximation will certainly suffice for order of magnitude calculations. The small nucleon isoscalar anomalous moment will also be neglected; thus $\kappa \equiv \kappa_p \approx -\kappa_n$.

Our method for separating Im H_2 out of the electropion production amplitude is the following. We contract [*1.76*] with polarization vectors $\boldsymbol{e}$ and $\boldsymbol{e}'$ for the incoming and outgoing (virtual) photons respectively and, in the gauge $e_0 = e_0' = 0$, this gives

$$e_j' C^{ji} e_i = \boldsymbol{e}'.\boldsymbol{C}.\boldsymbol{e} = \frac{2i}{M^3}\left[\{\boldsymbol{k}.\boldsymbol{e}'\boldsymbol{\sigma}.(\boldsymbol{k}\times\boldsymbol{e}) - \boldsymbol{k}.\boldsymbol{e}\boldsymbol{\sigma}.(\boldsymbol{k}\times\boldsymbol{e}')(H_1 - M\omega H_2)\}\right.$$
$$\left. - M\omega\,\boldsymbol{\sigma}.(\boldsymbol{e}'\times\boldsymbol{e})(H_1 - K^2 H_2)\right] \qquad \ldots[1.101]$$

where $-K^2 = k_\mu k^\mu < 0$.

Consider first of all the nucleon–one pion intermediate state with momenta as defined in Fig. 5. The photoproduction amplitude $\boldsymbol{A}.\boldsymbol{e}$ that we work with is normalized according to

$$\left(\frac{d\sigma}{d\Omega}\right)_{\text{photoproduction}} = \frac{\alpha f^2 q}{\mu^2 k} \sum_{\substack{\text{spin and} \\ \text{charge states}}} |\boldsymbol{A}.\boldsymbol{e}|^2 \qquad \ldots[1.102]$$

where

$$q = |\boldsymbol{q}|, \quad k = |\boldsymbol{k}| \quad \text{and} \quad f^2 = \frac{\mu^2}{4M^2}\left(\frac{g^2_{\pi NN}}{4\pi}\right) = \cdot 08$$

is the pseudo-vector pion-nucleon coupling constant. Unitarity in the static limit gives

$$\text{Im}\,(e'.\boldsymbol{C}.e) = -f^2\left(\frac{q}{\mu^2}\right)\int \frac{d\Omega_p}{4\pi} \sum_{\substack{\text{spin and}\\ \text{charge states}}} (\boldsymbol{A}.e')^\dagger(\boldsymbol{A}.e). \quad \ldots[1.103]$$

If we take for $\boldsymbol{A}$ the S wave threshold (electric-dipole and longitudinal dipole) production of pions we have

$$(\boldsymbol{A}.\boldsymbol{e})_{\pi^+n} = 2^{\frac{1}{2}}\mathrm{X}^\dagger\left(\boldsymbol{\sigma}.\boldsymbol{e} - \frac{\boldsymbol{\sigma}.\boldsymbol{k}\,\boldsymbol{k}.\boldsymbol{e}}{K^2+2\omega^2}\right)\mathrm{X}$$

$$(\boldsymbol{A}.\boldsymbol{e})_{\pi^0p} = 0 \quad \ldots[1.104]$$

where $\mathrm{X}^\dagger, \mathrm{X}$ are nucleon two component spinors. Using [*1.103*] and [*1.104*] we find

$$\text{Im}\,H_2^S(-K^2,\omega) = -g(K^2)\frac{f^2M^2q}{\mu^2\omega}\frac{1}{(K^2+2\omega^2)}. \quad \ldots[1.105]$$

As a second model we take for $\boldsymbol{A}$ the full relativistic Born approximation

$$(\boldsymbol{A}.\boldsymbol{e})_{\pi^+n} = 2^{\frac{1}{2}}\,\mathrm{X}^\dagger\left(\boldsymbol{\sigma}.\boldsymbol{e} + \frac{\boldsymbol{\sigma}.(\boldsymbol{k}-\boldsymbol{q})(2\boldsymbol{q}-\boldsymbol{k}).\boldsymbol{e}}{K^2+2\omega^2-2\boldsymbol{q}.\boldsymbol{k}} + \frac{i(1+2\kappa)}{2M\omega}\,\boldsymbol{q}.(\boldsymbol{k}\times\boldsymbol{e})\right)\mathrm{X}$$

$$(\boldsymbol{A}.\boldsymbol{e})_{\pi^0p} = -\frac{(1+2\kappa)}{2M\omega}\,\mathrm{X}^\dagger\,(\omega^2\boldsymbol{\sigma}.\boldsymbol{e}+\boldsymbol{\sigma}.[\boldsymbol{q}\times(\boldsymbol{k}\times\boldsymbol{e})])\,\mathrm{X}. \quad \ldots[1.106]$$

We find from [*1.101*] and [*1.103*]

$$\text{Im}\,H_2^B(-K^2,\omega) = g(K^2)\frac{f^2M^2}{\mu^2\omega k}\left[-\tfrac{1}{2}Q_0+\tfrac{3}{2}\frac{q}{k}Q_1-\left(\frac{q}{k}\right)^2 Q_2 + \frac{(1+2\kappa)}{2M\omega}\left\{\tfrac{5}{6}(Q_0-Q_2)-\tfrac{3}{5}\left(\frac{q}{k}\right)(Q_1-Q_3)\right\}\right] \quad \ldots[1.107]$$

where $Q_l = Q_l\left(\dfrac{K^2+2\omega^2}{2kq}\right)$ are the Legendre functions of the second kind. The π^0p amplitude of [*1.106*] makes only a negligible $1/M$ correction to the smallest terms of [*1.107*].

If we take for $\boldsymbol{A}$ the $N^*(1238)$ amplitude with resonant magnetic dipole excitation only

$$(\boldsymbol{A}.\boldsymbol{e})_{\pi^+n} = \frac{2^{\frac{1}{2}}i\mu^2(1+2\kappa)e^{i\delta_{33}}}{12M\,q^3f^2}\sin\delta_{33}\mathrm{X}^\dagger\,\{2\boldsymbol{q}.(\boldsymbol{k}\times\boldsymbol{e})-i\boldsymbol{\sigma}.(\boldsymbol{q}\times(\boldsymbol{k}\times\boldsymbol{e}))\}\mathrm{X}$$

$$(\boldsymbol{A}.\boldsymbol{e})_{\pi^0p} = 2^{\frac{1}{2}}(\boldsymbol{A}.\boldsymbol{e})_{\pi^+n} \quad \ldots[1.108]$$

we find

$$\operatorname{Im} H_2^R(-K^2, \omega) = g(K^2)\,\frac{\mu^2(1+2\kappa)^2 \sin^2\delta_{33}}{48\omega\, q^3 f^2} \qquad \text{...[1.109]}$$

with δ_{33} the pion nucleon phase shift in the 33 channel.

Finally taking for A the Fubini, Nambu, Wataghin[35] approximation to the full static model photoproduction amplitude

$$(\boldsymbol{A}.\boldsymbol{e})_{\pi^+ n} = 2^{\frac{1}{2}}X^\dagger \left\{\boldsymbol{\sigma}.\boldsymbol{e}+\frac{\boldsymbol{\sigma}.(\boldsymbol{k}-\boldsymbol{q})(2\boldsymbol{q}-\boldsymbol{k}).\boldsymbol{e}}{K^2+2\omega^2-2\boldsymbol{k}.\boldsymbol{q}}+\frac{i\mu^2(1+2\kappa)e^{i\delta_{33}}\sin\delta_{33}}{12M\,q^3 f^2}\right.$$

$$\left. .\,[2\boldsymbol{q}.(\boldsymbol{k}\times\boldsymbol{e})-i\boldsymbol{\sigma}.(\boldsymbol{q}\times(\boldsymbol{k}\times\boldsymbol{e}))]\right\} X \qquad \text{...[1.110]}$$

$$(\boldsymbol{A}.\boldsymbol{e})_{\pi^0 p} = \frac{i\mu^2(1+2\kappa)e^{i\delta_{33}}\sin\delta_{33}}{6M^2 q^3 f^2} X^\dagger\, [2\boldsymbol{q}.(\boldsymbol{k}\times\boldsymbol{e})-i\boldsymbol{\sigma}.(\boldsymbol{q}\times(\boldsymbol{k}\times\boldsymbol{e}))]\, X$$

we find

$$\operatorname{Im} H_2^{FNW}(-K^2, \omega) = g(K^2)\left\{\frac{f^2M^2}{\mu^2\omega k}\left[-\tfrac{1}{2}Q_0+\tfrac{3}{2}\left(\frac{q}{k}\right)Q_1-\left(\frac{q}{k}\right)^2 Q_2\right]\right.$$

$$\left. +\frac{\mu^2(1+2\kappa)^2\sin^2\delta_{33}}{48\omega\, q^3 f^2}\right\}. \qquad \text{...[1.111]}$$

In equations [*1.105*], [*1.107*], [*1.109*] and [*1.111*] an *ad hoc* form factor $g(K^2)$ has been added.

TABLE II

Model Calculations of S_3 in Parts per Million

Nucleon-Pion	(a) $g(K^2) = [1+(K/0{\cdot}96M)^2]^{-4}$	(b) $g(K^2) = \theta(1{\cdot}5\ \text{BeV}^2 - K^2)$	(c) $g(K^2) = 1$
S Wave Threshold Electric Dipole Amplitude	+0·9	+2·1	+2·5
Born Amplitude	−0·6	−0·6	−27·0
N^*(1238) Magnetic Dipole Amplitude	−0·3	−0·6	−0·6
Born Amplitude with Final State Enhancement in 33 Channel	−0·1	+0·2	+0·3
N^*(1238)–Pion	−0·01	−0·2	−0·2

Table II gives the results for S_3 for each amplitude taken with three choices of $g(K^2)$:

(a) $g(K^2) = (F_2(-K^2))^2$
(b) $g(K^2) = 1$ for $0 \leqq K^2 \leqq 1{\cdot}5\,(\text{BeV})^2$; $g(K^2) = 0$ for $K^2 > 1{\cdot}5\,(\text{BeV})^2$
(c) $g(K^2) = 1$ for all K^2.

We also make a rough estimate of the $\pi - N^*(1238)$ intermediate state.[34] The momenta are labelled in Fig. 6. The amplitude is

$$
\begin{aligned}
(\boldsymbol{B}.\boldsymbol{e})_{\pi^- N^{*0}} &= X_j^\dagger \left(e_j + \frac{(k-q)_j(2\boldsymbol{q}-\boldsymbol{k}).\boldsymbol{e}}{K^2+2\omega^2-2\boldsymbol{k}.\boldsymbol{q}} \right) X \\
(\boldsymbol{B}.\boldsymbol{e})_{\pi^- N^{*++}} &= (\boldsymbol{B}.\boldsymbol{e})_{\pi^+ N^{*0}} \\
(\boldsymbol{B}.\boldsymbol{e})_{\pi^0 N^{*+}} &= 0,
\end{aligned} \qquad \ldots[1.112]
$$

where X_j is the two component polarization spinor for the N^* satisfying

$$\sum_{\text{spins}} X_j X_i^\dagger = (\delta_{ji} - \tfrac{1}{3}\sigma_j \sigma_i). \qquad \ldots[1.113]$$

We find

$$\text{Im}\, H_2(-K^2, \omega) = g(K^2)\tfrac{1}{3}\frac{f_*^2 M^2}{\mu^2 \omega k}\left[\tfrac{1}{2}Q_0 - \tfrac{3}{2}\left(\frac{q}{k}\right)Q_1 + \left(\frac{q}{k}\right)^2 Q_2\right] \qquad \ldots[1.114]$$

where f_* is the $N\pi N^*$ coupling constant ($f_*^2 \simeq 0{\cdot}39$). Numerical results are shown at the bottom of Table II.

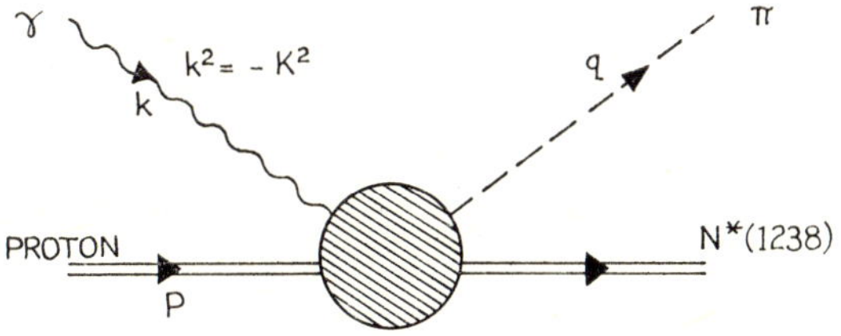

FIG. 6. Kinematics for amplitude of virtual photoproduction of the $N^*(1238)$ plus a pion.

1.5. *Conclusions*

There can be but one conclusion from the varied numbers in these tables: we don't know enough about the amplitude of virtual photon absorption by a proton to *calculate* the polarizability correction to the hfs with any confidence using dispersion theory. Recall from [*1.84*] and [*1.85*] that the input for the structure contribution v_S/v_F is virtual photon absorption by a proton integrated over all energies and space-like masses of photons. The weighting function in the integrals is only slowly convergent at high energies and masses. This is the analogue of our earlier observation in the Schrödinger models that high momentum components of the intermediate electron state corresponding to the electron at the proton surface are important and that numerous excited proton states play an important role in the hfs.

With the aid of the sum rule [*1.81*], based on the low energy theorem and the no subtraction assumption in the dispersion relation for H_1, the S_2 contribution can be tied down to a small value, as shown in Table I. Barring a very large and unanticipated enhancement in Im $H_1(-K^2, \omega)$ for large $-K^2 < 0$ we cannot hope to find any significant contribution to the hfs from this term.

We turn next to S_3 and the entries of Table II. Photo-excitation of the $N^*(1238)$ is the best known and calculable of the amplitudes but plays only a minor role, as noted earlier, contributing < 1 ppm.

The electric dipole production of S wave charged pions has a threshold value fixed by the Kroll-Ruderman theorem for real photons (and in the limit of a massless π^+ being produced). Depending on the form factor appropriate to this amplitude it may contribute up to 2–3 ppm in the direction of removing the apparent discrepancy. However much of its contribution is apparently cancelled when we also include in the pion photoproduction amplitude the meson current contribution in Born approximation and magnetic dipole photoproduction to the 33 resonance state only, as in [*1.111*]. Nevertheless very different and cut-off dependent results are found from the model of [*1.107*] which includes a Pauli coupling to the isovector moment of the proton in Born approximation. This is because the magnetic dipole amplitude continues to grow with increasing photon energy in this model and the integration over Im $H_2(-K^2, \omega)$ acquires very sizable contributions for large ω. Furthermore the numerical results obtained in this way are sensitive both in sign and magnitude to modifications of the Born approximation amplitudes by poorly known phase shifts in the non-33 channels. The last row in Table II shows than an $N^*(1238)$ plus pion state in the photo absorption amplitude is not a promising candidate for a hfs contribution.

During the preparation of these lectures based on analysis and calculations with Dr. J. D. Sullivan at SLAC, a letter arrived from Dr. Francoise Guérin [36] at Orsay describing a dispersion analysis of the polarizability contribution to the hfs along parallel lines to our discussion of this section. She reported a contribution of $\lesssim 1$ ppm from the electric dipole photopion production amplitude as well as a negligible contribution from higher resonances.

We find ourselves then with no clear candidate to add even 10–20 ppm to the hfs. At the same time there is enough sensitivity in the poorly calculable amplitudes so that we can't be sure that such a contribution isn't present. In this our dilemma is similar to that encountered by attempts [37] to use the Cottingham approach to calculate the neutron-proton mass difference. For this problem it is the isovector part of the non spin-flip forward virtual Compton amplitude that must be determined and approximations similar to those we've made above also fail to predict a neutron that is more massive than the proton. As is abundantly clear from the recent rash of analyses, [38] triggered by the work of Dashen and Frautschi, [39] the detailed behaviour of the high-energy and photon mass region is very important.

We might try to achieve further progress by looking into the possibility of a subtracted dispersion relation for H_2 in [*1.85*]. At high energies the K^2 dependence of Im $H_2(-K^2, \omega)$ can only be guessed and as a liberal guess we may ignore form factors altogether, setting

$$\text{Im } H_2(-K^2, \omega') = \text{Im } H_2(0, \omega') \text{ for } \omega' > 2M. \qquad ...[1.115]$$

The orders of integration in [*1.85*] can then be interchanged and we obtain

$$(S_3) \xrightarrow[\text{high energy}]{} -\frac{6}{a_0(1+\kappa)\pi^2}\int_{2M}^{\infty} d\omega' \text{ Im } H_2(0, \omega'). \qquad ...[1.116]$$

Whether or not this integral converges is unknown although it exists for all of the models in Table II. A numerical evaluation based on [*1.105*] gives < 2 ppm to the hfs from [*1.116*].

With the assumption that the dispersion integral in [*1.79*] doesn't exist we make a subtraction writing, in place of [*1.79*],

$$H_2(k^2, \omega) = -\frac{\kappa F_2(F_1+\kappa F_2)M^2}{(k^2)^2-4M^2\omega^2} + G(k^2) \qquad ...[1.117]$$

where we drop the dispersion integral itself as having contributed negligibly to the hfs. Our idea here is to determine G so that in [*1.76*] it adds 20–40 ppm to the hfs; this fixes the integral

$$-\frac{6}{a_0 M(1+\kappa)\pi^3}\int \frac{d^4K}{K^4} K_0^2 G(-K^2) = (20\text{–}40)\times 10^{-6} \qquad ...[1.118]$$

or

$$\int_0^{\infty} dK^2 G(-K^2) = -(30\text{–}60).$$

However an inescapable implication of this assumption is that there is a sizable difference between electron-proton and positron-proton scattering. Since we introduce the subtraction constant in order to increase the attraction between the electron and proton and thereby to increase the hfs it is not surprising that the effect of G is to increase the cross-section for electron-proton scattering above that for positron-proton scattering.[40] This appears to be in conflict with observation[41] and so we abandon the idea of a simple subtraction term. This is not a rigid conclusion for two reasons. First of all one can tailor the K^2 dependence of $G(-K^2)$ so that the hfs is accounted for by the integral [*1.118*] while at the same time the observed cross-sections at large K^2 are not altered. Secondly the scattering is observed at finite angles and so requires, for a complete analysis of the two photon exchange contribution to the electron-positron difference, knowledge of virtual Compton scattering in non forward directions as well as for time-like photon masses. Thus many more amplitudes and parameters are introduced beyond those appearing in the forward spin-flip Compton amplitude for virtual space-like photons. For these reasons it does not appear profitable to explore this approach beyond the discussion of reference 40. The introduction of the subtraction constant $G(k^2)$ in [*1.117*] leads us directly back to the results computed there if we keep only the leading terms for small momentum transfer in the electron (positron)-proton scattering amplitude.

Further counsel against searching for significant contributions to the hfs from high energy parts of the two photon amplitude comes from the current commutator sum rules of Bjorken.[28]

The situation as we view it presently is as follows: a relativistic dispersion approach fails to provide any real insight or quantitative contributions towards a resolution of the hfs problem.

The non-relativistic Schrödinger models show that sizable, albeit model dependent, polarizability contributions may very well be present and it is indeed puzzling as to where they have been lost in the dispersion approach. These models also suggest that one should not expect to find the predominant polarizability contributions to the hfs in the magnetic dipole excitation of the $N^*(1238)$ resonance. The electric excitations are also important.

Both in the hfs as well as the Lamb shift studies[3, 42] the apparent discrepancies between theory and experiment are comparable to the size of the proton recoil current and structure contributions. The possibility cannot be ruled out with certainty that a better understanding and formulation of the contributions due to proton dynamics won't resolve these discrepancies. We cannot claim to understand or to have computed their contributions to the hfs to an accuracy of better than ~ 10 ppm.

A new determination of the fine structure constant from the fine structure[2] in H or D together with the Ruderman[43] analysis of the chemical shift of the muon moment measurements may provide a purely experimental resolution of the hfs problem itself. In presenting the numerical comparisons in [*1.3*] and [*1.5*] we have used the value of the fine structure constant $\alpha^{-1} = 137{\cdot}0388 \pm 9$ ppm obtained from reference 2. The error is a two standard deviation figure. This value of α^{-1} is arrived at by adding the frequencies for the $2P_{\frac{3}{2}} - 2S_{\frac{1}{2}}$ and the $2S_{\frac{1}{2}} - 2P_{\frac{1}{2}}$ transitions in deuterium and comparing them with the theoretical formula for the fine structure splitting $2P_{\frac{3}{2}} - 2P_{\frac{1}{2}}$. If the additional $+0{\cdot}3$ Mc/s found by Robiscoe[3] in H is finally confirmed in D and if the $2P_{\frac{3}{2}} - 2S_{\frac{1}{2}}$ transition does not decrease by a compensating amount there will be an increase in the measured value of α of $+13$ ppm. The theoretical hfs in [*1.2*] is increased by 26 ppm which would bring it into agreement with [*1.1*] to within 20 ppm. This would remove the hfs anomaly but a serious disagreement would remain in the Lamb shift itself between theory and experiment.

The latest published theoretical number on the Lamb shift in H is $1057{\cdot}64 \pm 0{\cdot}21$ Mc/s.[44] This number is further reduced by $0{\cdot}14$ Mc/s to $1057{\cdot}5$ Mc/s by the completion of the fourth order radiative correction by M. Soto (to be published).‡ The experimental result of reference 3 exceeds this by $\sim 0{\cdot}5 \pm 0{\cdot}1$ Mc/s and that of reference 2 by $\sim 0{\cdot}3 \pm 0{\cdot}1$ Mc/s. Since the Lamb shift itself is proportional to $\alpha^3 R$ an increase in α of 70–200 ppm would be required to remove this discrepancy, short of theoretical formula modifications or addenda, and this would disrupt the hfs comparison by (an intolerable) 150–400 ppm.

Clearly then the Lamb shift emerges now as the major challenge to QED. In my view the hfs anomaly recedes to a problem in strong interaction physics to achieve a more reliable calculation of the polarizability contribution. However I feel that it does not stand at this time as a basic challenge to QED whether or not the value of α changes as indicated above. A reanalysis of the finite proton mass and structure contributions to the Lamb shift would be highly desirable.

2

Muon $g-2$ Value

2.1. *Introduction*

The $g-2$ value of the muon is sensitive to electrodynamic behaviour at small distances, or high energies.[45,46,47,48] This is because the muon rest mass determines the scale of momenta contributing to the electromagnetic interactions of the muon; so $g-2$ probes the detailed behaviour of propagators and vertices which transfer hundreds of MeV of momenta or, equivalently, it probes distances as small as $\Delta r \approx \hbar/\Delta q \lesssim 10^{-13}$ cm.

The calculation of the electrodynamic radiative corrections to the muon $g-2$ value has been carried out through second order in the fine structure constant α, with a result[49,50] in accord with the $g-2$ measurements:[45,51,52]

$$\left(\frac{g-2}{2}\right)_{\mu,\,\text{theory}} = \frac{\alpha}{2\pi} + 0{\cdot}7658\left(\frac{\alpha}{\pi}\right)^2 = 1165\times 10^{-6} \qquad \ldots[2.1]$$

$$\left.\begin{aligned}\left(\frac{g-2}{2}\right)_{\mu^+,\,\text{exp}} &= (1162\pm 5)\times 10^{-6}\\ \left(\frac{g-2}{2}\right)_{\mu^-,\,\text{exp}} &= (1165\pm 3)\times 10^{-6}.\end{aligned}\right\} \qquad \ldots[2.2]$$

Limits to various possible forms of electrodynamic modifications set by this comparison have been discussed in reference 45. In particular TCP invariance alone guarantees that the μ^+ and μ^- moments are equal in magnitude.

It is expected that the accuracy of the muon $g-2$ measurement can be and will be carried up to a precision such that contributions of order $(5\text{–}10)\left(\frac{\alpha}{\pi}\right)^3 \approx (5\text{–}10)\times 10^{-8}$ will be seen.[52] At this level of precision one

‡ M. Soto, 1966. *Phys. Rev. Lett.* **17**, 1153.

is not limited in interpreting $g-2$ by the present experimental uncertainty in the value of α itself. α is known to an accuracy of ± 6 parts per million (one standard deviation),[53] and this corresponds to an uncertainty in the numerical value of the Schwinger term of

$$\delta\left(\frac{\alpha}{2\pi}\right) = \pm 0{\cdot}4\left(\frac{\alpha}{\pi}\right)^3 \approx 5\times 10^{-9}. \qquad \text{...[2.3]}$$

However, at this precision one does enter the realm of possible contributions to $g-2$ from both the strong[46, 47] and weak[54] interactions, and we are presented with the very exciting prospect of observing strong interaction effects in a pure electrodynamic system (in this case, a single muon in a weak

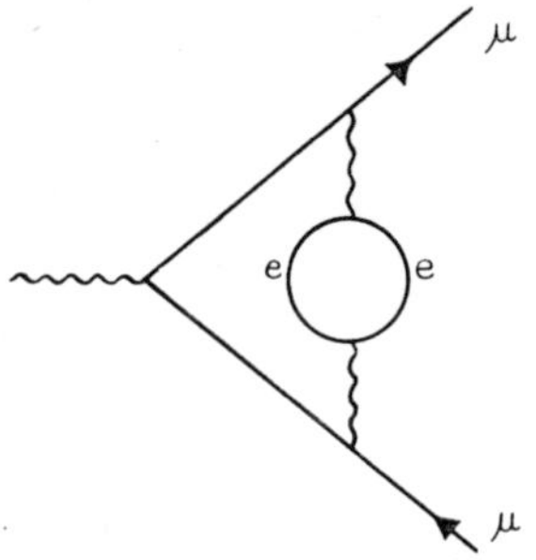

FIG. 7

Electron bubble insertion into the photon propagator of the first order moment diagram.

externally applied magnetic field). In preparation for this we must first compute all purely electrodynamic contributions that may give rise to terms as large as 2–3$(\alpha/\pi)^3$ to $(g-2)/2$ in [2.1].

Although no terms as large as this appeared in the electron $g-2$ calculations to order α^3, we expect to find them in the muon $g-2$ calculation due to the presence of electron-positron bubbles. In particular the difference between the α^2 contribution in [2.1] and the corresponding term, $-0{\cdot}328\ \alpha^2/\pi^2$, in the electron calculation comes from a bubble insertion in the photon propagator, as shown in Fig. 7. It increases the moment[49] by a factor $1+(\alpha/3\pi)(\ln m_\mu^2/m^2)+O(1)$, where m_μ and m are the muon and electron masses respectively. This result is readily understood by studying the electromagnetic vertex as a function of the momentum transfer introduced by the electromagnetic current, with both fermion lines on their mass shells. This is the form of dispersion relation familiar in the electromagnetic form factor studies, viz.

$$\left.\begin{aligned} F_2(q^2) &= \frac{1}{\pi}\int_{4m^2}^{\infty}\frac{d\sigma^2\ \mathrm{Im}\,F_2(\sigma^2)}{\sigma^2-q^2-i\varepsilon} \\ \text{or}\qquad \frac{g-2}{2} &= \frac{1}{\pi}\int_{4m^2}^{\infty}\frac{d\sigma^2}{\sigma^2}\,\mathrm{Im}\,F_2(\sigma^2). \end{aligned}\right\} \qquad \text{...[2.4]}$$

In this approach the absorptive amplitude is obtained by multiplying the current to produce a lepton pair by the Bhabha scattering amplitude in the

$3S_1$ and $3D_1$ states. A perturbation approach reproduces the Schwinger correction and the positive sign of the moment correction corresponds to the fact that the electron and positron attract one another via their Coulomb interaction in the dominant S state interaction. The vacuum polarization contribution due to the electron bubble insertion in the photon propagator in Fig. 1 necessarily enhances the photon propagator and therefore increases the Bhabha amplitude for lepton-antilepton attraction since the spectral function is always greater than or equal to zero.

For the vacuum polarization contribution to the muon $g-2$ value the scales of energy and momentum transfers, and hence of impact parameters of the rescattering of the $\mu^+\mu^-$ pairs in Fig. 7, is determined by the rest masses of the particles, i.e. by the threshold $4m_\mu^2$ in [*2.4*]. The momentum transfer in the scattering, $t = q^2$, is related to the total energy, $s = \sigma^2$, according to

$$t = -\tfrac{1}{2}(s-4m_\mu^2)(1-x), \qquad \ldots[2.5]$$

where x is the cosine of the centre-of-mass scattering angle, which is integrated over in the scattering. The angular momentum and parity selection rules assure us that only the $3S_1$ and $3D_1$ scattering channels contribute (i.e. $J = 1$ and $C = -1$, corresponding to a photon) and so the effective t values are just the energies $s \propto m_\mu^2$ for $x \simeq 0$ in [*2.5*]. The $\mu^+\mu^-$ scattering is thus at small impact parameters, $\approx 1/m_\mu$ and the corresponding vacuum polarization contribution is large. In fact, we may simply extrapolate the Serber-Uehling[55] term for the vacuum polarization contribution for large $|t| \approx \bar{t} = m_\mu^2 \gg m^2$ and enhance the muon $g-2$ calculation by just that amount

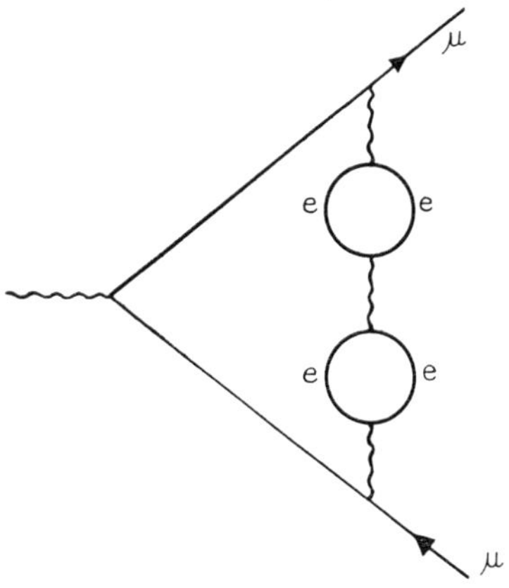

FIG. 8. Insertion of two electron bubbles into the first order diagram.

$$\left[\frac{1}{t}\left(1+\frac{\alpha}{3\pi}\ln\frac{\bar{t}}{m^2}\right)-\frac{1}{t}\right] \approx \frac{1}{t}\left[\frac{\alpha}{3\pi}\left\{\ln\frac{m_\mu^2}{m^2}+O(1)\right\}\right].$$

Similarly an iteration of this bubble as in Fig. 8 will give to leading order in $\ln(m_\mu^2/m^2)$

$$\left(\frac{g-2}{2}\right)_{\text{Fig. 8}} \approx \frac{\alpha}{2\pi}\left[\frac{\alpha}{3\pi}\ln(m_\mu^2/m^2)\right]^2 \approx 6\left(\frac{\alpha}{\pi}\right)^3 \qquad \ldots[2.6]$$

which is of the order of magnitude of the contributions we hope to find. Before discussing the non-electromagnetic contributions to the muon $g-2$

value—or perhaps the failure of the present day theory—all of the purely electromagnetic terms which are enhanced above the electron $g-2$ result by logarithmic factors $\ln^2 (m_\mu^2/m^2)$ as in [2.6] or by $\ln (m_\mu^2/m^2)$ must be computed. The calculation of these terms is our present concern.

2.2. *QED contributions*

To order α^3, there are many graphs which could contribute to $g-2$ for the muon. A direct experimental limit on some contributions can be fixed. For the purposes of this discussion we accept the validity of quantum electrodynamics as applied to the electron $g-2$ value. Both the contribution due to modifications of the vacuum polarization by strongly interacting particles

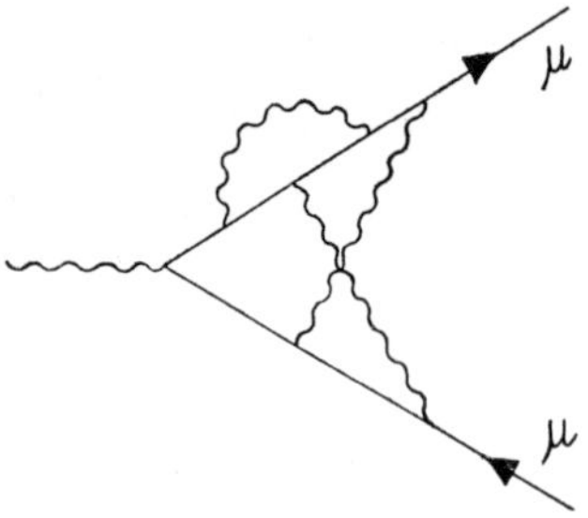

FIG. 9. A typical third order graph with no $\ln (m_\mu^2/m^2)$ terms in $g-2$.

and the weak interaction contributions are proportional to the square of the lepton mass and hence totally negligible for the electron. With these assumptions the beautiful agreement between experiment[56]

$$\left(\frac{g-2}{2}\right)_{e,\,\text{exp}} = \frac{\alpha}{2\pi} - (0{\cdot}327 \pm 0{\cdot}005)\frac{\alpha^2}{\pi^2}$$
$$= \frac{\alpha}{2\pi} - 0{\cdot}328\frac{\alpha^2}{\pi^2} + (0{\cdot}4 \pm 2{\cdot}0)\frac{\alpha^3}{\pi^3} \qquad \ldots[2.7]$$

and theory[57,58]

$$\left(\frac{g-2}{2}\right)_{e,\,\text{theory}} = \frac{\alpha}{2\pi} - 0{\cdot}328\frac{\alpha^2}{\pi^2} + \sim 0{\cdot}15\frac{\alpha^3}{\pi^3} \qquad \ldots[2.8]$$

can be interpreted as bounding the sixth order contributions due to three photon exchange by the maximum value of

$$\left|\left(\frac{g-2}{2}\right)_e^{\alpha^3}\right| \lesssim 2{\cdot}0\left(\frac{\alpha}{\pi}\right)^3. \qquad \ldots[2.9]$$

A tighter limit than [2.9] will emerge if further accuracy in the electron $g-2$ value emerges from measurements now in progress.[59] This means that we can dismiss the class of sixth order contributions to $g-2$ coming from the exchange of three photon lines with all possible time orderings of their

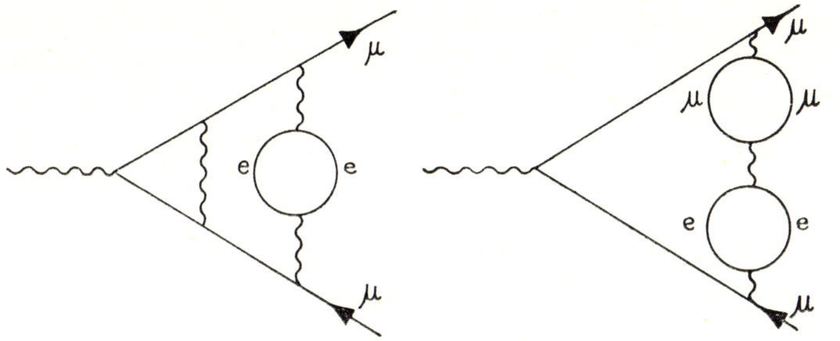

FIG. 10

FIG. 11

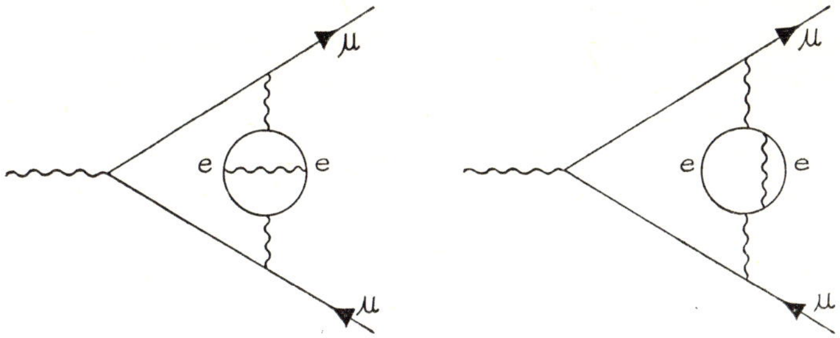

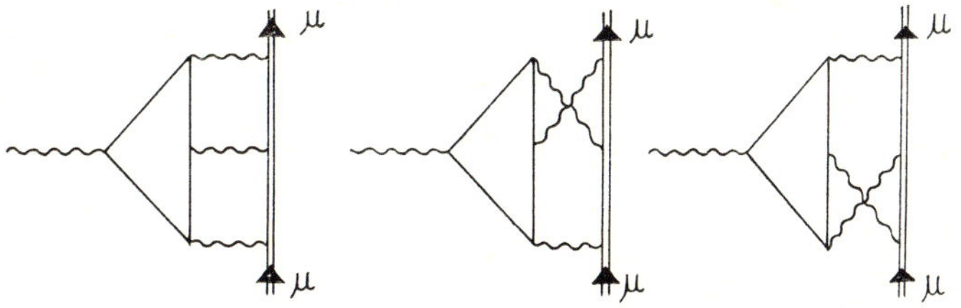

FIG. 12

FIG. 10. Some insertions of electron bubbles into second-order diagrams.

FIG. 11. Insertion of a fourth order electron bubble into the first order diagram.

FIG. 12. Scattering of light by light contribution. The single line represents the electron, the double line represents the muon.

vertices at the muon lines. One such ordering is illustrated in Fig. 9. The $g-2$ contribution of these terms is independent of the lepton mass,‡ and so is the same for the muon and the electron. An approximate calculation of sixth order magnetic moment contributions to the electron has been made[58] using a dispersion theoretic approach with a resulting value

$$\left(\frac{g-2}{2}\right)_e^{\alpha^3} = +0{\cdot}15\left(\frac{\alpha}{\pi}\right)^3. \qquad \ldots[2.10]$$

‡ That is, they contribute a term proportional to $e\hbar/2m_\mu c$ to the magnetic moment only.

This is a full order of magnitude below the limit of accuracy in which we are interested. It is also below the limits of error (equation [*2.3*]) put by uncertainty in the fine structure constant itself.

In search of larger contributions to $g-2$ for the muon we turn to all graphs of order α^3 which contain electron loops and evaluate their singular parts as $m_\mu/m \to \infty$. Thus the only terms we need consider are those which involve the insertion of electron loops into the contributions to the electromagnetic vertex of lower orders in α. The various possibilities are shown in Figs. 8, 10, 11 and 12. Fig. 10 shows two possible ways of inserting second order electron-positron vacuum polarization bubbles into a photon propagator in the α^2 calculation of $g-2$; Figs. 8 and 11 show fourth order insertions into the calculation of order α; Fig. 12 shows the possible scattering of light by light insertions. We shall see that the dominant contribution to $g-2$ for all of these except the last can be obtained simply from the results of the lower order calculation.

To calculate the contribution from the large set of diagrams of the type shown in Fig. 10 we need only recall that the second order electron vacuum polarization insertion into the photon propagator, as in Fig. 7 for the muon $g-2$ calculation, contributes a term

$$\left(\frac{g-2}{2}\right)_{\mu,\ \text{electron bubble}} = \left(\frac{\alpha}{2\pi}\right)\left(\frac{\alpha}{3\pi}\right)\left[\ln\frac{m_\mu^2}{m^2}+O(1)\right] \qquad ...[2.11]$$

that is responsible for the difference between [*2.1*] and [*2.8*] for the muon and electron $g-2$ values through order α^2.

The propagator of a photon, including the electron-positron bubble of Fig. 7, is, for $|q|^2 \gg 4m^2$,

$$D_F(-q^2) = \frac{1}{|q|^2}\left(1+\frac{\alpha}{3\pi}\ln(|q|^2/m^2)\right) \qquad ...[2.12]$$

which gives the Coulomb potential plus the Serber-Uehling correction in momentum space.[55,60] Up to corrections of order unity, we may approximate $|q|^2 \approx m_\mu^2$ in the logarithm in [*2.12*]. The added moment correction in [*2.11*] then emerges directly. The singularity as $m \to 0$ reflects the fact that the Serber-Uehling potential is more singular by a logarithmic factor than the Coulomb interaction, not only in $\boldsymbol{q}$ space but also as $r \to 0$ in coordinate space.

We may similarly extract the logarithmic factors in the α^3 calculation of $g-2$ by making all possible electron loop insertions into the α^2 graphs for $g-2$. Each such insertion multiplies the corresponding amplitude by a factor $+\frac{\alpha}{3\pi}[\ln(m_\mu^2/m^2)+0(1)]$ and there are two possible insertions, one on each photon line, for each graph in the fourth order calculation. It follows then that the sum of all such graphs is a constant multiplier of the Sommerfield-Petermann result to order α^2.

$$\left(\frac{g-2}{2}\right)_{(10)} = (-0{\cdot}328\ \alpha^2/\pi^2)\times 2\times\frac{\alpha}{3\pi}\,[\ln(m_\mu^2/m^2)+O(1)] \quad ...[2.13]$$

where the subscript 10 refers to all diagrams of the type shown in Fig. 10.

The same type of argument can be used in extracting the dominant α^3 contribution from the "double bubble" diagram of Fig. 8. This diagram gives rise both to a $\ln^2(m_\mu^2/m^2)$ and a $\ln(m_\mu^2/m^2)$ contribution that can be identified conveniently by spectralizing the photon propagator. For the iterated bubble the photon propagator is replaced by

$$\frac{1}{q^2}\rightarrow\frac{1}{q^2}\left[q^2\int_0^\infty\frac{\pi(\sigma^2)\,d\sigma^2}{q^2-\sigma^2}\right]^2 \equiv D^{(10)}(q^2) \quad ...[2.14]$$

where $\pi(\sigma^2)$ is the spectral amplitude. It is convenient to rewrite [2.14] in the form

$$D^{(10)}(q^2) = \int_{4m^2}^\infty \pi(\sigma^2)\,d\sigma^2\int_{4m^2}^\infty \pi(\sigma'^2)\,d\sigma'^2\,\frac{1}{\sigma^2-\sigma'^2}\left[\frac{\sigma^2}{q^2-\sigma^2}-\frac{\sigma'^2}{q^2-\sigma'^2}\right]. \quad ...[2.15]$$

If we insert $D^{(10)}(q^2)$ for the simple photon propagator in the ordinary Schwinger term,[60,61] we see that the integral over the photon loop momentum is formally the same, with the photon "mass" $\lambda^2\rightarrow 0$ replaced by σ^2 or σ'^2. We can then use the results of the Schwinger calculation directly to reduce the spin algebra. We find that the contribution from Fig. 8 is

$$\left(\frac{g-2}{2}\right)_{(8)} = \frac{\alpha}{2\pi}m_\mu^2\int_{4m^2}^\infty \pi(\sigma^2)\,d\sigma^2\int_{4m^2}^\infty \pi(\sigma'^2)\,d\sigma'^2\,\frac{1}{\sigma^2-\sigma'^2}\int_0^1 2z(1-z)^2\,dz$$
$$\cdot\left[\frac{\sigma^2}{m_\mu^2(1-z)^2+\sigma^2 z}-\frac{\sigma'^2}{m_\mu^2(1-z)^2+\sigma'^2 z}\right]$$
$$=\frac{\alpha}{2\pi}\int_0^1 2z(1-z)^4\,dz\left[\int_{4m^2}^\infty\frac{\pi(\sigma^2)\,d\sigma^2}{(1-z)^2+\sigma^2 z/m_\mu^2}\right]^2. \quad ...[2.16]$$

To first order in α, the spectral amplitude to produce an electron-positron pair is[60]

$$\pi_{e\bar{e}}(\sigma^2) = \frac{\alpha}{3\pi\sigma^2}[1-4m^2/\sigma^2]^{\frac{1}{2}}(1+2m^2/\sigma^2)\,\theta(1-4m^2/\sigma^2). \quad ...[2.17]$$

Inserting [2.17] into [2.16] we find

$$\left(\frac{g-2}{2}\right)_{(8)} = \frac{\alpha}{2\pi}\left(\frac{\alpha}{3\pi}\right)^2\left\{\ln\frac{m_\mu^2}{m^2}\left[\ln\frac{m_\mu^2}{m^2}-25/3\right]+O(1)\right\}. \quad ...[2.18]$$

The leading contribution from the graph in Fig. 11 can also be found simply by inserting the fourth order vacuum polarization into the photon propagator. This was first computed by Jost and Luttinger[62] and its logarithmically leading term is readily constructed by application of renormalization group arguments.[63, 60] The photon propagator is increased by a factor $\left[1+\left(\frac{\alpha}{2\pi}\right)^2 \ln \frac{|q|^2}{m^2}\right]$ and the corresponding moment contribution is

$$\left(\frac{g-2}{2}\right)_{(11)} = \frac{\alpha}{2\pi}\left(\frac{\alpha}{2\pi}\right)^2 \left[\ln \frac{m_\mu^2}{m^2} + O(1)\right]. \qquad \text{...[2.19]}$$

This brings us finally to the scattering of light by light insertion illustrated in Fig. 12. There are six diagrams representing all possible permutations of the internal photon vertices, but we need only consider the three distinct ones shown in Fig. 12. The remaining three just double the contribution of these. A calculation of these graphs is an extremely arduous task involving three connected loop integrals as well as very lengthy spinor products.‡

To simplify the calculation we take full advantage of the fact that we are interested only in terms that are singular as $m_\mu \to \infty$. We also take advantage of the fact that in calculating the contribution to $g-2$ we need only retain the leading, or first order, term in q as $q^\mu \to 0$. There is as yet no complete proof that the amplitude corresponding to the three graphs in Fig. 12 and linear in q as $q^\mu \to 0$ is finite in the limit $m/m_\mu \to 0$ and although we discuss this calculation in these lectures we have no conclusive result to present.

Ignoring the scattering of light by light insertions for the moment, the total electrodynamic construction of order $\left(\frac{\alpha}{\pi}\right)^3 \ln (m_\mu^2/m^2)$ to the muon $g-2$ is given by the sum of equations [*2.13*], [*2.18*] and [*2.19*] and is just

$$\begin{aligned}\left(\frac{g-2}{2}\right)_\mu^{\alpha^3} &= \frac{\alpha}{2\pi}\left(\frac{\alpha}{3\pi}\right)^2 \ln \frac{m_\mu^2}{m^2}\left[\ln \frac{m_\mu^2}{m^2} - \tfrac{25}{3} + \tfrac{9}{4} - 3{\cdot}936\right] \\ &= \frac{\alpha}{2\pi}\left(\frac{\alpha}{3\pi}\right)^2 \left[0{\cdot}64 \ln \frac{m_\mu^2}{m^2} + O(1)\right] \qquad \text{...[2.20]} \\ &\sim \left(\frac{\alpha}{\pi}\right)^3.\end{aligned}$$

As a result of large cancellation between the $\ln^2 (m_\mu^2/m^2)$ and $\ln (m_\mu^2/m^2)$ we find that the sixth order electromagnetic contributions to $g-2$ due to the electron vacuum polarization insertions are too small to be seen by the planned measurements.

2.3. *Non-QED contributions*

If the intermediate vector boson exists, then the diagram in Fig. 13 would contribute to the muon $g-2$. The amplitude from the graph in Fig. 13 is

‡ The heroic proportion of the scattering of light by light loop alone makes this point clear.[64]

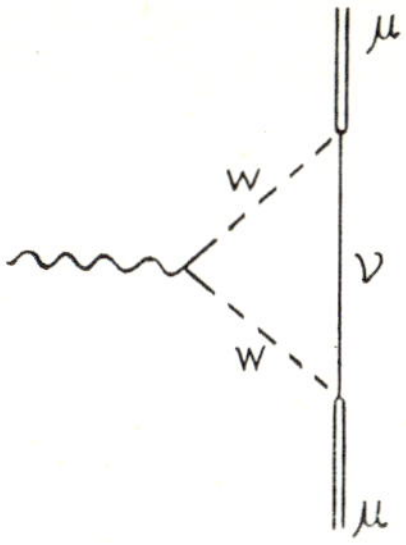

FIG. 13
A possible weak interaction contribution via a W boson.

found to be logarithmically dependent on a cut-off when computed with perturbation theory.[65] With application of the Lee-Yang[66] ξ-limiting procedure for calculating electromagnetic interactions of charged vector mesons, the cut-off dependence is replaced by a logarithm of the fine structure constant. As computed by Segre[54] the weak interaction contribution to the muon magnetic moment is

$$\left(\frac{g-2}{2}\right)_{\mu,\,\text{weak}} = 3\left(\frac{\alpha}{\pi}\right)^3 \qquad \ldots[2.21]$$

if the W meson has no anomalous magnetic moment. Larger contributions are predicted if the W is endowed with an anomalous moment of its own.

The strongly interacting particles will contribute to the muon $g-2$ via their vacuum polarization insertions in the photon propagator as studied extensively by Durand.[46] Denote the spectral amplitude for these insertions by $\pi_s(\sigma^2)$, corresponding to the "blob" in Fig. 14 which represents a strongly interacting system that couples to the photon. This changes the photon propagator to

$$\frac{1}{q^2} \to \int d\sigma^2 \pi_s(\sigma^2)\,\frac{1}{q^2-\sigma^2}$$

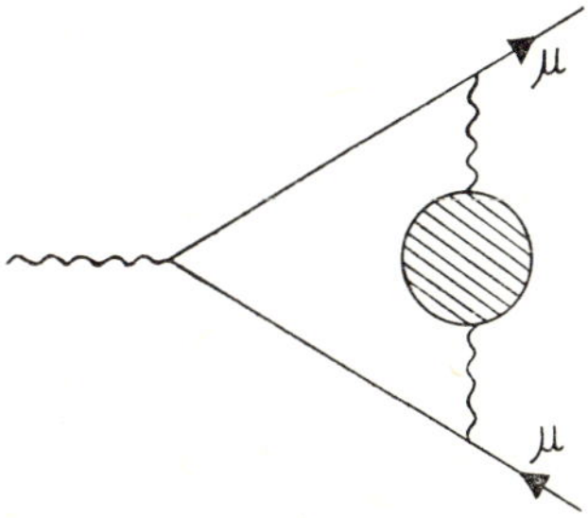

FIG. 14. The effect of strong interactions on the photon vacuum polarization.

and gives in analogy to equations [2.14] to [2.16]

$$\left(\frac{g-2}{2}\right)_s = \frac{\alpha}{2\pi} m_\mu^2 \int_0^\infty \pi_s(\sigma^2)\,d\sigma^2 \int_0^1 \frac{2z(1-z)^2\,dz}{m_\mu^2(1-z)^2+\sigma^2 z}. \qquad \ldots[2.22]$$

If we make the approximation that this blob can be approximated by the exchange of resonances of mass M_i and coupling constants α_i, we can write

$$\pi_s(\sigma^2) = \sum_i \alpha_i \delta(\sigma^2 - M_i^2). \qquad \ldots[2.23]$$

Since all observed vector resonances[67] with quantum numbers in common with the photon are massive, we retain only the leading term in $\left(\frac{m_\mu}{M_i}\right)^2 < 1$ in [2.22] and find

$$\left(\frac{g-2}{2}\right)_s = \frac{\alpha}{2\pi} \sum_i \tfrac{2}{3}\alpha_i \left(\frac{m_\mu^2}{M_i^2}\right). \qquad \ldots[2.24]$$

In particular, if we consider only the effect of the ρ meson, a case which has been extensively studied,[46,47] we find

$$\left(\frac{g-2}{2}\right)_\rho = \frac{\alpha^2}{3\pi}\frac{m_\mu^2}{m_\rho^2} \approx 8\left(\frac{\alpha}{\pi}\right)^3 \qquad \text{...[2.25]}$$

which is well within the range of values of interest to us. Durand[46] has also computed the possible contributions of higher meson states to $\pi_s(\sigma^2)$. Note also that all contributions to $\pi_s(\sigma^2)$ are positive since they represent the square of a matrix element and therefore each intermediate state coupling to the photon increases the contribution to the spectral amplitudes. In particular all $\alpha_i > 0$ in [2.23].

We need not restrict our attention to this approximation, however. Direct, independent experimental information on the photon spectral amplitude

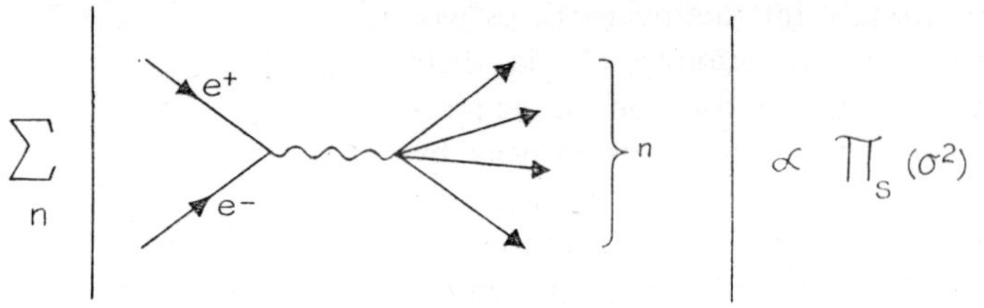

FIG. 15. Electron-positron annihilation to heavy states from which $\pi_s(\sigma^2)$ can be calculated.

$\pi_s(\sigma^2)$ appearing in [2.22] will be obtained from the electron-positron colliding beam experiments now under construction. The total cross section for a colliding electron-positron pair to form all possible states of mass σ^2 is proportional to the square of the amplitude for the electromagnetic current to form these states, as illustrated in Fig. 15, and hence is proportional to $\pi_s(\sigma^2)$. The equation relating the partial cross-section $\sigma_z(2E_{CM})$ to form states (z) of mass $(\sigma^2)^{\frac{1}{2}} = 2E_{CM}$ to the partial spectral amplitude $\pi_z(4E_{CM}^2)$ has been derived by Cabibbo and Gatto[68] and is given by

$$\sigma_z(E_{CM}) = \frac{\pi^2\alpha}{E_{CM}^2}\,\pi_z(4E_{CM}^2). \qquad \text{...[2.26]}$$

In principle, then, the spectral amplitude can be obtained directly from [2.26] and the value of $\pi_s(\sigma^2)$ over the experimentally accessible range can be inserted directly in [2.22]. In this way the major effect of the strong interactions on $g-2$ may be calculated directly from experimental observation. Possible differences between the calculated and observed values would then be attributed to

(i) Contribution to $\pi_s(\sigma^2)$ from massive states not yet accessible to the colliding ring experiments (and not yet "anticipated" by intuition).

(ii) Modifications of QED arising from cut-offs[48,69] in the photon or muon propagator, or in the vertex part.
(iii) The contribution from the W meson.
(iv) Other modifications of the photon propagator, such as the neutrino bubbles, shown in Fig. 16.
(v) The contributions of the scattering of light by light insertions not as yet included.

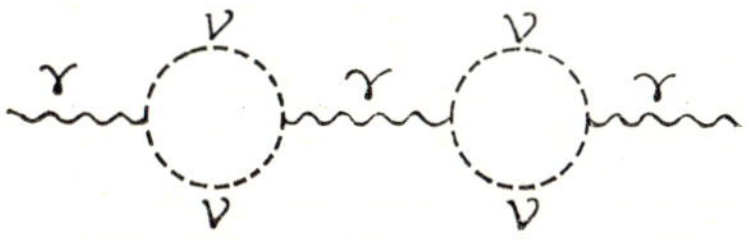

FIG. 16. Possible neutrino effects on the photon vacuum polarization.

As an example of (ii), consider cutting off the photon propagator by letting

$$\frac{1}{K^2} \to \frac{1}{K^2} - \frac{1}{K^2 - \Lambda^2}$$

which, from [2.24], gives[48]

$$\frac{g-2}{2} = \frac{\alpha}{2\pi}\left(1 - \frac{2}{3}\frac{m_\mu^2}{\Lambda^2}\right). \qquad ...[2.27]$$

If no deviations are found to order $5(\alpha/\pi)^3$, [2.27] gives a lower bound $\Lambda \gtrsim 12$ BeV.

The effect of iterated neutrino bubbles, as in Fig. 16, is

$$\left(\frac{g-2}{2}\right)_\nu = \frac{\alpha}{2\pi}\left[\frac{\alpha_\nu}{3\pi}\ln\frac{m_\mu^2}{m_\nu^2} + \left(\frac{\alpha_\nu}{3\pi}\right)^2\left[\ln\frac{m_\mu^2}{m_\nu^2}\right]^2 + \ldots\right]$$

$$= \frac{\alpha}{2\pi}\,\frac{1}{1 - \dfrac{\alpha_\nu}{3\pi}\ln\dfrac{m_\mu^2}{m_\nu^2}} \qquad ...[2.28]$$

where m_ν and α_ν represent the neutrino mass and fine structure constant.

As above, we can use any difference between experiment and theory to bound $\alpha_\nu \ln m_\nu$ (although we note that $m_\nu = 0$ must imply $\alpha_\nu = 0$).

Since all effects listed above may be operating simultaneously the interpretation of any possible deviations between theory and experiment will necessarily be somewhat ambiguous. In particular effect (v) can be computed.

REFERENCES

1. E. R. Cohen and J. W. M. DuMond. 1965. *Rev. Mod. Phys.* **37**, 537.
2. S. Triebwasser, E. S. Dayhoff and W. E. Lamb, Jr. 1953. *Phys. Rev.* **89**, 98.
3. R. T. Robiscoe. 1965. *Phys. Rev.* **138**, A22.
   R. T. Robiscoe. 1966. *Phys. Rev. Lett.* **17**, 69.
   Measurements of the Lamb shift, i.e. the $2S_{\frac{1}{2}}$-$2P_{\frac{1}{2}}$ splitting, in deuterium have also been reported (R. Robiscoe and B. Cosens, *Bull. Amer. Phy. Soc.*, Ser. 11, No. 1, p. 62 (1960).
4. A. J. Layzer. 1961. *J. Math. Phys.* **2**, 292, 308.
   G. W. Erickson and D. R. Yennie. 1965. *Ann. Phys.* (N.Y.) **35**, 271, 447.
5. R. Blumenthal, D. Ehr, W. Faissler, P. Joseph, L. Lanzerotti, F. Pipkin and D. Stairs. 1966. *Phys. Rev.* **144**, 1199.
6. R. Talman. 1966. *Bull. Amer. Phys. Soc.*, Ser. 11, Vol. 11, No. 3, p. 380.
7. S. B. Crampton, D. Kleppner and N. F. Ramsey. 1963. *Phys. Rev. Lett.* **11**, 338.
8. S. J. Brodsky and G. W. Erickson. 1966. *Phys. Rev.* **148**, 26.
9. S. Fenster and Y. Nambu. 1965. *Prog. Theor. Phys. Suppl.* **Extra Number**, 250.
10. C. K. Iddings. 1965. *Phys. Rev.* **138**, B446.
11. A. C. Zemach. 1956. *Phys. Rev.* **104**, 1771.
12. G. Breit. 1930. *Phys. Rev.* **35**, 1447.
13. N. Kroll and F. Pollack. 1951. *Phys. Rev.* **84**, 594; 1952. *Phys. Rev.* **86**, 876.
14. R. Karplus, A. Klein and J. Schwinger. 1951. *Phys. Rev.* **84**, 597.
    R. Karplus and A. Klein. 1952. *Phys. Rev.* **85**, 972.
15. A. J. Layzer. 1964. *Nuovo Cim.* **33**, 1538.
16. D. E. Zwanziger, 1964. *Nuovo Cim.* **34**, 77.
17. W. E. Cleland, J. M. Bailey, M. Eckhause, V. W. Hughes, R. M. Mobley, R. Prepost and J. E. Rothberg. 1964. *Phys. Rev. Lett.* **13**, 202.
18. C. Iddings and P. Platzman. 1959. *Phys. Rev.* **115**, 919.
19. A. Verganelakis and D. Zwanziger. 1965. *Nuovo Cim.* **39**, 613.
20. A. Bohr. 1948. *Phys. Rev.* **73**, 1109.
21. F. Low. 1950. *Phys. Rev.* **77**, 361.
22. W. N. Cottingham. 1963. *Ann. Phys.* (N.Y.) **25**, 424.
23. H. A. Bethe and E. E. Salpeter. 1957. *Quantum Mechanics of One- and Two-Electron Atoms* (Academic Press, New York).
24. J. D. Bjorken and S. D. Drell. 1964. *Relativistic Quantum Mechanics* (McGraw-Hill, New York).
25. C. Iddings and P. Platzman. 1959. *Phys. Rev.* **113**, 192. See also R. Faustov. 1966. *Nucl. Phys.* **75**, 669.
26. W. Newcomb and E. E. Salpeter. 1955. *Phys. Rev.* **97**, 1146.
27. R. Arnowitt. 1953. *Phys. Rev.* **92**, 1002.
28. J. D. Bjorken. 1966. *Phys. Rev.* **148**, 1467.
29. S. D. Drell and A. C. Hearn. 1966. *Phys. Rev. Lett.* **16**, 908.
30. F. Low. 1954. *Phys. Rev.* **96**, 1428.
31. M. Gell-Mann and M. L. Goldberger. 1954. *Phys. Rev.* **96**, 1433.
32. S. D. Drell and F. Zachariasen. 1961. *Electromagnetic Structure of Nucleons*, (Oxford Univ. Press), Chapter 2.
33. M. Gell-Mann, M. L. Goldberger, and W. Thirring. 1954. *Phys. Rev.* **95**, 1612.
34. M. Gourdin and Ph. Salin. 1963. *Nuovo Cim.* **27**, 193.
    Ph. Salin. 1963. *Nuovo Cim.* **28**, 1294.
35. S. Fubini, Y. Nambu and V. Wataghin. 1958. *Phys. Rev.* **111**, 329.
36. F. Guérin (private communication).
37. A. Finn (unpublished).
38. H. Pagels. 1966. *Phys. Rev.* **144**, 1261.
    G. Shaw and D. Wong. *Phys. Rev.*, **147**, 1028.
39. R. Dashen and S. Frautschi. 1964. *Phys. Rev.* **135**, B1190.
    R. Dashen. 1964. *Phys. Rev.* **135**, B1196.
40. S. D. Drell and J. D. Sullivan. 1966. *Phys. Lett.* **19**, 516.

41. A. Browman, F. Liu and C. Schaerf. 1965. *Phys. Rev.* **139**, B1079.
Further experiments on this ratio are in progress at CEA (L. Hand) and Cornell (J. de Witt).
42. E. E. Salpeter. 1952. *Phys. Rev.* **87**, 328; 1953. **89**, 92.
T. Fulton and P. C. Martin. 1954. *Phys. Rev.* **95**, 811.
43. M. A. Ruderman. 1966. *Phys. Rev. Lett.* **17**, 794.
44. G. W. Erickson and D. R. Yennie. 1965. *Ann. Phys.* (N.Y.) **35**, 271, 447.
45. G. Charpak, F. J. M. Farley, R. L. Garwin, T. Muller, J. C. Sens, and A. Zichichi. 1965. *Nuovo Cim.* **37**, 1241.
46. L. Durand, III. 1962. *Phys. Rev.* **128**, 441.
47. G. Bouchiat and L. Michel. 1961. *J. Phys. Radium* **22**, 121.
48. V. B. Berestetski, O. N. Krokhin and A. Khlebnikov. 1956. *Zh. Eksp. i Teoret Fiz.* **30**, 788; (English translation 1958: *Sov. Phys. JETP* **12**, 761.)
49. H. Suura and E. H. Wichmann. 1957. *Phys. Rev.* **105**, 1930.
50. H. H. Elend. 1966. *Phys. Lett.* **20**, 682.
51. G. Schrank, G. R. Henry and R. Swanson. UCRL preprint 16469.
52. F. J. M. Farley, private communication.
F. J. M. Farley *et al.* 1966. Report to XIII International High-Energy Physics Conference (Berkeley) and CERN preprint.
53. E. Cohen and J. DuMond. 1965. *Rev. Mod. Phys.* **37**, 537.
54. G. Segre. 1963. *Phys. Lett.* **7**, 357.
55. E. A. Uehling. 1935. *Phys. Rev.* **48**, 55.
R. Serber. 1935. *Phys. Rev.* **48**, 49.
56. D. T. Wilkinson and H. R. Crane. 1963. *Phys. Rev.* **130**, 852.
57. C. M. Sommerfeld. 1957. *Phys. Rev.* **107**, 328.
A. Petermann. 1957. *Helv. Phys. Acta*, **30**, 407.
58. S. D. Drell and H. R. Pagels. 1965. *Phys Rev.* **140**, B397.
59. W. Fairbank, private communication.
F. C. Witteborn, L. V. Knight and W. M. Fairbank. 1965. *Low Temperature Physics—LT9* (Part B) (Plenum Press, New York), p. 1248.
60. J. D. Bjorken and S. D. Drell. 1965. *Relativistic Quantum Fields* (McGraw-Hill, New York).
61. J. Schwinger. 1948. *Phys. Rev.* **73**, 416L.
62. R. Jost and J. M. Luttinger. 1950. *Helv. Phys. Acta* **23**, 201.
63. M. Gell-Mann and F. Low. 1954. *Phys. Rev.* **95**, 1300.
64. R. Karplus and M. Neuman. 1950. *Phys. Rev.* **80**, 380.
65. G. Feenberg. 1958. *Phys. Rev.* **110**, 1482.
M. Ebel and F. Ernst. 1960. *Nuovo Cim.* **15**, 1731.
N. Byers and F. Zachariasen. 1960. *Nuovo Cim.* **18**, 1289.
66. T. D. Lee and C. N. Yang. 1962. *Phys. Rev.* **128**, 885.
T. D. Lee. 1962. *Phys. Rev.* **128**, 899.
67. A. H. Rosenfeld, A. Barbaro-Galtieri, W. H. Barkas, P. L. Bastien, J. Kirz and M. Roos. 1965. *Rev. Mod. Phys.* **37**, 633.
68. N. Cabibbo and R. Gatto. 1961. *Phys. Rev.* **124**, 1577.
69. B. de Tollis. 1960. *Nuovo Cim.* **16**, 203.

5

THE SYMMETRIES OF THE WEAK INTERACTIONS

P. K. Kabir

Rutherford High Energy Laboratory

1

Introduction

Considerations of symmetry and invariance play an important role in our understanding of natural phenomena. The assumption of a certain regularity or order in Nature is indeed essential to the study of natural philosophy. It is, however, the characteristic feature of physical theory, from which all other sciences derive, that no proposition is considered self-evident and all hypotheses are subject to the scrutiny of experimental test. Any physical theory is a *model*, susceptible to mathematical analysis, which describes phenomena within a certain range of variation of the conditions under which they are observed. The object of physical theory is to abstract from observations on known phenomena the principles to make predictions concerning phenomena as yet unknown. One may go so far as to say that the value of a theory should be judged not at all by how faithfully it reproduces the phenomena already observed but rather by how unambiguously it is able to make predictions outside the domain in which it has proved to furnish a satisfactory description. It is then clear that the projective power of a theory will be the greater as its premises become simpler. What is not evident and, in my opinion, not subject to logical analysis is the following fact. Our entire experience in the study of physics has been that whenever the limit of validity of a particular theory has been reached, by the discovery of phenomena not fully encompassed by that theory, the more general theory which has supplanted it has invariably been in some sense simpler than the previous one. Only vanity would delude us to believe that we could ever guess the laws to predict all possible phenomena (whatever that may mean!). We should therefore not be too surprised when we discover that one of our "universal" laws based on what we consider to be the most aesthetic and logical principles turns out to possess only a limited range of applicability; rather it should encourage us to search for more general principles which we may expect to be even more beautiful and simple.

In the search for the most general basis for the laws governing physical theory, the concept of invariance has always played a central role. The conservation laws which we understand best, governing the conservation of momentum, energy and angular momentum, are best viewed from such a standpoint. Nevertheless, we must bear in mind that even for these, the requirement of invariance is not self-evident. While it may be eminently reasonable to require that the laws governing physical phenomena be unaffected when we move our apparatus from one corner of the room to another, or even from one country to another, there is no logical reason why this should be exactly true. In particular, if we consider displacement by a step comparable in magnitude to the dimensions of the known universe, the premise of invariance becomes much less obvious. For the same reason the special theory of relativity, in spite of its great success, cannot ever be accepted without question and should continue to be tested in wider domains.

It is with such an attitude that we should regard the recent discoveries which challenge our preconceived notions of simplicity and elegance. We may not at the moment have a clear understanding of the phenomena, but, if history is a guide, a simpler picture will surely emerge.

2

The TCP Theorem

The known empirical connection between the spins of particles and the statistics obeyed by them was explained by Pauli as a consequence of relativity: relativistic quantum field theory does not admit of a reasonable interpretation if we use the "wrong" spin-statistics connection. This condition, taken together with the already assumed requirement of invariance under proper orthochronous Lorentz transformations, is sufficient to assure a remarkable theorem due to Lüders and Pauli.[1] A Lagrangian constructed from local field-operators and their finite-order derivatives, which satisfies these conditions, is necessarily also invariant under the operation of "strong reflection" defined by Pauli, which is just the inversion of all co-ordinates

$$x^\lambda \to x^{\lambda\prime} = -x^\lambda \qquad \ldots[1]$$

accompanied by appropriate transformations on the field-operators.‡ In a classical local field theory, invariance under strong reflection follows directly from the fact that the transformation [*1*] can be realized by a rotation in an even-dimensional space.[4] True, Minkowski space is not Euclidean and there is no real transformation actually realizable as a Lorentz transformation

‡ The validity of the theorem had earlier been assumed by Schwinger[2] to make the inverse argument that invariance under strong reflection requires the spin-statistics relation to hold. Explicit reference to Lagrangians is not necessary to establish the equivalent theorem for vacuum expectation values of field-operators,[3] but we prefer to keep the discussion here to the most elementary possible level.

which accomplishes [1]: a time-like vector in the forward light-cone can be rotated into one in the backward cone only if it is allowed to cross the intervening space-like region. Nevertheless, invariance under proper Lorentz transformations is sufficient to guarantee invariance under strong reflection. Consider a Lorentz transformation with velocity βc along the z-axis followed by a 180° rotation around this axis. The transformed co-ordinates are

$$\begin{aligned} x' &= -x, & y' &= -y, \\ z' &= \gamma(z-\beta t) & t' &= \gamma(t-\beta z) \\ &= z \cosh u - t \sinh u, & &= -z \sinh u + t \cosh u \end{aligned} \qquad \ldots[2]$$

or

$$x^{\lambda'} = a^{\lambda}_{\nu} x^{\nu} \qquad \ldots[3]$$

with

$$\begin{aligned} &a^1_1 = a^2_2 = -1, \quad a^3_3 = a^0_0 = \cosh u \\ &a^0_3 = a^3_0 = -\sinh u, \text{ where } \gamma = (1-\beta^2)^{-\frac{1}{2}} \text{ and } \tanh u = \beta \end{aligned} \qquad \ldots[4]$$

all other a's being zero. Now if we have a theory which is invariant under proper Lorentz transformations, the Hamiltonian or energy density, which is required to be the (00) component of the second-rank stress tensor, must be expressed as a corresponding tensor function of the field quantities (which can themselves be classified according to their tensorial transformation properties):

$$\begin{aligned} \mathscr{H}(x) &= T_{00}(x) \\ T_{\alpha\beta}(x) &= T_{\alpha\beta}(\phi^i(x), \phi^j_\lambda(x), \chi_{\lambda\nu}(x) \ldots). \end{aligned} \qquad \ldots[5]$$

Now, under a Lorentz transformation of the form [3], both the L.H.S. and the arguments of the functional appearing on the R.H.S. will undergo corresponding tensor transformations. For any finite u, all finite polynomials in $\cosh u$ and $\sinh u$ can be reduced to cosh and sinh functions, which can be expanded in absolutely convergent series in u. Hence the validity of [5] extends not only to real Lorentz transformations characterized by real values of the parameter u, but to complex values of u as well. Choosing $u = i\pi$, we see from [2] that this results in inversion of all co-ordinates, which leaves the (00) component of a second-rank tensor unchanged. The assumption of a *local* field theory was necessary to limit to finite polynomials the expressions in $\sinh u$ and $\cosh u$ which appear when [5] is subjected to a Lorentz transformation.

For the quantum case, the situation is slightly more involved. The correspondence principle requires that, apart from phase factors, scalar and vector fields transform under strong reflection according to

$$\left.\begin{aligned} \phi(x) &\rightarrow \phi'(x) = \phi(-x) \\ \phi^*(x) &\rightarrow \phi^{*\prime}(x) = \phi^*(-x) \\ \chi_\lambda(x) &\rightarrow \chi'_\lambda(x) = -\chi_\lambda(-x) \\ \chi^*_\lambda(x) &\rightarrow \chi^{*\prime}_\lambda(x) = -\chi^*_\lambda(-x) \end{aligned}\right\} \qquad \ldots[6]$$

with reversal of sign for all odd-rank tensor fields, and no change of sign for even-rank fields. The transformation law for spinor fields cannot be chosen by appeal to classical limits, but invariance of the free-particle equation for the Dirac field

$$\left(i\gamma^\lambda \frac{\partial}{\partial x^\lambda} - m\right)\psi = 0$$

where the Dirac matrices γ^λ are chosen to satisfy

$$\gamma^\lambda\gamma^\mu + \gamma^\mu\gamma^\lambda = 2g^{\lambda\mu}$$

with

$$g^{00} = -g^{11} = -g^{22} = -g^{33}, \text{ and } g^{\lambda\mu} = 0 \text{ for } \lambda \neq \mu,$$

requires

$$\psi(x) \rightarrow \psi'(x) = \gamma_5\psi(-x) \qquad \ldots[7a]$$

and correspondingly,

$$\bar{\psi}(x) \rightarrow \bar{\psi}'(x) = \bar{\psi}(-x)\gamma_5 \qquad \ldots[7b]$$

where

$$\bar{\psi} = \psi^\dagger\gamma^0 \quad \text{and} \quad \gamma_5 = \gamma^0\gamma^1\gamma^2\gamma^3.$$

The Hamiltonian density $\mathscr{H}$ will still be the (00) component of a second-rank tensor, but now it will be an operator constructed from the basic field-operators corresponding to the particles appearing in our theory. We note that if all boson fields are assumed to transform normally under strong reflection, i.e. acquire a sign $(-1)^n$ where n is the rank of the field, then all operators constructed from boson fields, including gradients and higher derivatives of boson field-operators, will also transform normally. Thus all terms in $\mathscr{H}$ which involve only boson operators are invariant under strong reflection [*1*].

On the other hand, tensor covariants formed from the Dirac fields ψ are found to have exactly the opposite of normal behaviour under the transformation [*7a*]-[*7b*]. It is only when we remember that all expressions involving fermion fields must be antisymmetrized with respect to ψ and $\bar{\psi}$ that we have a means for restoring "normal" behaviour for the Dirac covariants, by demanding that strong reflection be accompanied by an inversion of the order of all operator expressions. If such an inversion is always associated with strong reflection, the boson terms are unaffected if they have been symmetrized, but the fermion factors acquire the necessary reversal of sign to make all terms in $\mathscr{H}$ even under strong reflection.

At the same time, the inversion of operator expressions is just what is required to preserve the boson commutation relations

$$[\phi^*(x), \phi(y)] = -i\Delta(x-y), \; [\phi(x), \phi(y)] = [\phi^*(x), \phi^*(y)] = 0$$

$$[\chi_\lambda^*(x), \chi_\mu(y)] = i\left[g_{\lambda\mu} + \frac{\partial_\lambda \partial_\mu}{m^2}\right]\Delta(x-y), \; [\chi_\lambda(x), \chi_\mu(y)] = [\chi_\lambda^*(x), \chi_\mu^*(y)] = 0$$

while the fermion anticommutation relations

$$\{\bar{\psi}(x), \psi(y)\} = i(i\gamma^\mu\partial_\mu + m)\Delta(x-y), \; \{\psi(x), \psi(y)\} = \{\bar{\psi}(x), \bar{\psi}(y)\} = 0$$

remain valid both before and after inversion of the order of factors. Thus, strong reflection, as defined by [*1*], [*6*], [*7*] and the stipulation that all operator expressions be inverted or, what is the same thing, be read from right to left, is a transformation which leaves both the interaction energy density and the commutation rules invariant. The extension of the argument to include spinor fields of spin greater than $\frac{1}{2}$ presents only technical problems and was carried through by Pauli.[1] It only remains to give a physical interpretation of this invariance transformation.

So far we have discussed strong reflection as a symmetry transformation on the operators, which leaves all operator relations invariant if all operator expressions are read backward. To restore the original ordering, it is convenient to consider the hermitian conjugate of the expressions obtained after strong reflection. Since unitarity demands that $\mathscr{H}$ be hermitian, the interaction energy density is still invariant under this transformation which now carries fields into their hermitian conjugates. The field equations and commutation relations are also unaffected. Denoting by Θ the transformation operator which carries out the transformations corresponding to strong reflection followed by hermitian conjugation, we must have

$$\begin{aligned} \Theta\phi(x)\Theta^{-1} &= \phi^*(-x), & \Theta\phi^*(x)\Theta^{-1} &= \phi(-x), \\ \Theta\chi_\lambda(x)\Theta^{-1} &= -\chi_\lambda^*(-x), & \Theta\chi_\lambda^*(x)\Theta^{-1} &= -\chi_\lambda(-x), \\ \Theta\psi(x)\Theta^{-1} &= \bar{\psi}(-x)\gamma_5\gamma^0, & \Theta\bar{\psi}(x)\Theta^{-1} &= \gamma^0\gamma_5\psi(-x). \end{aligned} \qquad \ldots[8]$$

Note that hermitian conjugation includes complex conjugation of c-numbers, therefore the operator Θ is an anti-linear operator. We do not therefore expect any conservation law to result from the invariance under Θ.

The generator of infinitesimal translations P_μ is defined by

$$i\frac{\partial\phi(x)}{\partial x^\mu} = [P_\mu, \phi(x)] \qquad \ldots[9]$$

for all field operators. Transforming this equation by Θ and Θ^{-1},

$$i\frac{\partial}{\partial(-x^{\mu})}\phi^*(-x) = [\Theta P_{\mu}\Theta^{-1}, \phi^*(-x)] \qquad ...[10]$$

making use of [*8*]. Comparing [*9*] and [*10*], we see that $\Theta P_{\mu}\Theta^{-1}$ and P_{μ} have the same commutator with every field operator ϕ and it is not difficult to conclude that they are identical,

$$\Theta P_{\mu}\Theta^{-1} = P_{\mu}. \qquad ...[11]$$

In the same way, the angular momentum tensor is transformed to

$$\Theta J_{\lambda\nu}\Theta^{-1} = -J_{\lambda\nu}. \qquad ...[12]$$

Thus the transformation Θ leaves energies and momenta unchanged, but reverses spins and angular momenta.

If the field ϕ carries a charge q_0, we must have

$$[Q, \phi(x)] = q_0\phi(x), \; [Q, \phi^*(x)] = -q_0\phi^*(x). \qquad ...[13]$$

Transforming the first of these equations by Θ and Θ^{-1}, we have

$$[\Theta Q\Theta^{-1}, \phi^*(-x)] = q_0\phi^*(-x)$$

which, on comparison with the second equation of [*13*] yields

$$\Theta Q\Theta^{-1} = -Q. \qquad ...[14]$$

Thus Θ reverses the signs of all generalized charges. This, combined with [*11*] and [*12*] and the original definition of strong reflection [*1*], shows that Θ may be identified as the combined transformation TCP in which space is inverted, all particles are replaced by their corresponding antiparticles, and all motions are reversed.

2.1. *Consequences of TCP invariance*

The TCP theorem has many interesting and significant implications. From [*11*] and [*12*], we see that to every stable particle there must be a corresponding antiparticle with the same mass and spin, bearing the same generalized charges (with opposite sign). If all generalized charges are zero, particle and antiparticle will in general be indistinguishable and such particles will be described by hermitian fields. From [*11*] and [*12*], we can also write

$$\Theta w^{\kappa}\Theta^{-1} = -w^{\kappa} \qquad ...[15]$$

where w^{κ} is the relativistic spin operator

$$w^{\kappa} = \tfrac{1}{2}\varepsilon^{\kappa\lambda\mu\nu}P_{\lambda}J_{\mu\nu}. \qquad ...[16]$$

[*15*] leads to an interesting conclusion for massless particles. These can have their spin oriented along or opposite to the direction of motion and, contrary

to the case of massive particles, the occurrence of particles with one helicity does not necessarily imply the existence of particles with the opposite helicity. But [*15*] tells us that an antiparticle must exist with the opposite helicity; this conclusion holds whether the particles carry any generalized charges or not.

TCP invariance also relates the interactions of systems of particles to the interaction of the corresponding antiparticles. In terms of S-matrix elements, we must have

$$\langle b \mid S \mid a \rangle = \langle \Theta a \mid S \mid \Theta b \rangle \qquad ...[17]$$

taking account of the anti-linearity of Θ the unitarity of the S-matrix requires that Θ be anti-unitary. $\mid \Theta a \rangle$ represents the state obtained by replacing each particle in $\mid a \rangle$ by its antiparticle and reversing its helicity. We can use [*17*] to compare the decays of particles and antiparticles, assuming that these are caused by interactions which may be treated to lowest order only. We write S as the sum of two parts S_0 and S_1, where S_0 is the S-matrix in the absence of weak interactions and $\langle b \mid S_0 \mid a \rangle = 0$ for all $b \neq a$. The condition that S and S_0 be unitary requires that

$$\langle a \mid S_0{}^{\dagger} \mid a \rangle \langle a \mid S \mid b \rangle + \langle a \mid S^{\dagger} \mid b \rangle \langle b \mid S_0 \mid b \rangle = 0 \qquad ...[18]$$

if we retain terms in S_1 to lowest order only. Choosing $\mid a \rangle$ to be a single particle state and $\mid b \rangle$ to be a scattering eigenstate $\mid \beta \rangle$ of S_0 with eigenphase δ_β, substitution of [*17*] into [*18*] yields

$$\langle \Theta \beta \mid S \mid \Theta a \rangle = -\langle a \mid S^{\dagger} \mid \beta \rangle \, e^{2i\delta_\beta} = -\langle \beta \mid S \mid a \rangle^* \, e^{2i\delta_\beta}.$$

In terms of the T-matrix defined by $S = 1 + iT$,

$$\langle \Theta \beta \mid T \mid \Theta a \rangle = e^{2i\delta_\beta} \langle \beta \mid T \mid a \rangle^*. \qquad ...[19]$$

[*19*] shows that the partial rates are equal for particle and antiparticle to decay into channels which are scattering eigenstates with respect to strong interactions. Therefore this holds for channels in which there is no strong interaction between the outgoing particles. Even when there are strong final state interactions, the partial rate for a particle to decay into a set of channels, so chosen that strong interactions connect any one of these channels only to others within the group, will be necessarily equal to the corresponding partial decay rate for the antiparticle, since all interference terms between different terms of the form [*19*] will drop out in this case.[5] A particular case of this theorem is the requirement that the total decay rates for particle and antiparticle be equal. From [*11*], we have the requirement that the masses be equal, so the equality of mass and lifetime for particle and antiparticle follows from TCP invariance alone[5,6]

$$m_a = m_{\bar{a}}, \quad \tau_a = \tau_{\bar{a}}. \qquad ...[20]$$

The same conclusion may be reached by a more elegant argument given by Lüders and Zumino.[5] The location of the pole in the S-matrix, considered as a function of energy, corresponding to a resonance in a reaction $a \to b$, performed with stable particles in the initial state a and leading to stable particles in the final state b, must (according to [*17*]) be the same as that of a corresponding pole in the inverse reaction performed with antiparticles, $\bar{b} \to \bar{a}$. Unless the channels $a, \bar{a}$ and $b, \bar{b}$ can transform into each other, which is only possible if a and b possess no quantum numbers distinct from the vacuum, the two resonances will be distinct states standing in the relation of particle to antiparticle. Since the position of the pole determines both the mass and lifetime of the resonance, their equality for particle and antiparticle is thus established.

The best comparison of lifetimes is for $\mu^{\pm}$ whose decay rates are found to agree within an error of 0·1%. The $K^0-\bar{K}^0$ system provides an extremely sensitive detector for any $K^0-\bar{K}^0$ mass-difference which we discuss by completing an argument due to Bell.[7] Assuming only that K^0 and $\bar{K}^0$ possess common decay channels, the states undergoing exponential decay will in general be linear superpositions

$$\begin{aligned} |K_1\rangle &= r\,|K^0\rangle + s\,|\bar{K}^0\rangle, \qquad |r|^2+|s|^2 = 1, \\ |K_2\rangle &= p\,|K^0\rangle + q\,|\bar{K}^0\rangle, \qquad |p|^2+|q|^2 = 1. \end{aligned} \qquad \ldots[21]$$

Any complex mass-difference $(M-\bar{M})$ between K^0 and $\bar{K}^0$ is related to the corresponding mass-difference between K_2 and K_1 by

$$M-\bar{M} = (M_2-M_1)(sp+qr)(sp-qr)^{-1}. \qquad \ldots[22]$$

From the low rate of $K_2 \to 2\pi$ relative to $K_1 \to 2\pi$, it can be shown that[7]

$$|\langle K_2 | K_1\rangle| = |pr^*+qs^*| \leqq 0{\cdot}01.$$

Now

$$|sp-qr|^2 = 1-|pr^*+qs^*|^2.$$

Therefore

$$|sp-qr| = 1 \qquad \ldots[23]$$

to an accuracy of better than 1 in 10^4.

Thus [*22*] yields immediately

$$|M-\bar{M}| \leqq 2\,|M_2-M_1| \approx 10^{-14}\,M_{K^0}.$$

An even more restrictive limit can be obtained as follows. Inverting [*21*], we have

$$\begin{aligned} |K^0\rangle &= (sp-qr)^{-1}\,[s\,|K_2\rangle - q\,|K_1\rangle], \\ |\bar{K}^0\rangle &= (sp-qr)^{-1}\,[-r\,|K_2\rangle + p\,|K_1\rangle]. \end{aligned}$$

Now it is known experimentally to an accuracy of a few percent that half of all K^0's produced in strong interactions decay by the K_1 mode, and, to somewhat lower precision, that half decay by the K_2 mode. To the corresponding accuracy, we have therefore

$$|q|^2 \approx |s|^2 \approx |r|^2 \approx |p|^2 \approx \tfrac{1}{2}.$$

Consequently, in view of [*23*], we have

$$\varepsilon = |sp+qr| \ll 1$$

and [*22*] yields

$$|M-\overline{M}| = \varepsilon\,|M_2-M_1| \ll |M_2-M_1|. \qquad \text{...[24]}$$

If the electromagnetic field A_λ is taken to be hermitian (since to date there is no evidence for more than one kind of photon), we have

$$\Theta A_\lambda \Theta^{-1} = -A_\lambda. \qquad \text{...[25]}$$

Then TCP invariance of the electromagnetic interaction requires that the electromagnetic current which is the source of this field satisfies

$$\Theta J_\lambda \Theta^{-1} = -J_\lambda. \qquad \text{...[26]}$$

A necessary consequence of [*26*] is that, apart from obvious reversals of sign, particle and antiparticle have identical electromagnetic moments; this equality extends also to the structure factors associated with these moments. Since magnetic moments can be measured with high precision, a test of TCP invariance would be to compare particle and antiparticle magnetic moments. Note also that the equality of electromagnetic moments holds not only for the familiar E0, M1, E2, etc. moments which have long been the object of study, but also for E1, M2, etc. moments which were formerly believed not to occur in nature, but which should be vigorously sought now. It should be recalled, however, that (according to [*26*]) self-conjugate particles cannot have any static electromagnetic moments; the electromagnetic vertex for such a particle vanishes as a consequence of TCP invariance, whether or not charge-conjugation invariance holds.

TCP invariance is the only symmetry imposed by Lorentz invariance on a local quantum field theory beyond the requirement of invariance under proper Lorentz transformations. Particular systems may, of course, possess further symmetries. For example, for non-interacting particles, it is not difficult to see that one has invariance under the operations of P, C and T separately. It is only proof of the remarkable naïveté of physicists that, until ten years ago, it was considered perfectly natural that these further symmetries should be extended to all interacting systems as well. We now know that this is not true, and experiments have been reported in which the symmetry with respect to P, C and CP have all been found to be less than perfect. What *is* remarkable, and not understood at all, is that there is a large class of phenomena for which these individual operations or combinations of operations are close to being exact symmetries. Thus, even though we know that they are no longer exact symmetry operations for all interactions, the great difference in strength found between different interactions means that there is still considerable practical utility in discussing these transformations.

3

Superselection Rules

Before going on to discuss these, we consider another set of conservation laws, also believed to arise from internal symmetry transformations, which, as far as we know, are exact. These are the conservation laws for baryon number, electric charge, and two kinds of lepton number. Since the stability of ordinary matter depends on their validity, the laws for baryonic and electric charge must hold to a high degree of accuracy, and a figure exceeding 10^{27} years has been quoted as a lower limit for the proton lifetime[8]; the corresponding limit for electrons is 2×10^{21} years.[9] On the other hand, the experimental basis for the lepton conservation laws is very much weaker; all that is known is that interactions which do not respect the laws of lepton conservation are weaker than lepton-conserving interactions by an order of magnitude or more. Since we do not really understand the origin of any of these conservation laws, the conservation laws for leptonic charge are presumed to arise in the same way as the others and, for the moment, are ascribed the same "absolute" validity.

The existence of such exact conservation laws leads to the important concept of superselection rules. The Hilbert space of state-vectors can be divided into sub-spaces A labelled by the values of the absolutely conserved "charges", and no physical operator can have non-vanishing matrix elements between two of these sub-spaces. It is then clear that the relative phase of two vectors lying within distinct sub-spaces is unmeasurable in principle, and it is convenient to define physically realizable states as those whose state-vectors lie entirely within one of these sub-spaces. Any further symmetry transformation which we may wish to introduce will have to satisfy the conditions that every physically realizable state be transformed only into another physically realizable state. That is that each sub-space A transform into a sub-space A' (which may or may not be the same as A) and that the relative phases of the transformation matrices for different sub-spaces (AA'), (BB') be unmeasurable in principle. In this sense an arbitrary isospin transformation, for example one which mixes up neutron and proton states, is not a physically admissible one. The difficulty is only formal, however, since the only isospin transformations we actually consider are those which preserve the charges of physical states. In view of this, the question has been raised whether invariance under a continuous group of isospin or more general (e.g. SU(3)) transformations is necessary, and in particular whether one may not replace this by the requirement of symmetry with respect to a finite group of transformations.[10]

Beside the superselection rules arising from the exactly conserved charges, there is another separating states with integral and half-integral angular momentum. Under a rotation of the co-ordinate system by 2π about any axis, the wave function of a state with integral angular momentum is unaffected while the wave function of a half-integral spin state acquires a change of sign. The relative phase of two such states cannot therefore have any physical significance. All known particles possess baryon and lepton charges such that their sum is even or odd according as the particle is a boson or a fermion. Therefore, the consequences of this superselection rule for these particles are already implied by those discussed earlier but, if particles should be found for which the relation between angular momentum and the sum of baryon and lepton number is relaxed, we should have a new selection rule.

4

Space-Time Symmetries

4.1. *Parity*

Of the geometrical transformations which go beyond TCP invariance, the best studied is the case of space-reflection or parity P. If all interactions were invariant with respect to space-inversion, the operation

$$\boldsymbol{r} \rightarrow -\boldsymbol{r}$$

combined with suitable transformations on the field operators would leave the Hamiltonian invariant. Since the field equations and commutation relations are invariant under space-inversion, the parity operation must be representable as a unitary transformation. For example scalar and Dirac fields will be taken to transform

$$\left.\begin{aligned} \mathscr{P}\phi(\boldsymbol{r}, t)\mathscr{P}^{-1} &= \eta_P\phi(-\boldsymbol{r}, t) \\ \mathscr{P}\phi^*(\boldsymbol{r}, t)\mathscr{P}^{-1} &= \eta_P^*\phi^*(-\boldsymbol{r}, t) \\ \mathscr{P}\psi(\boldsymbol{r}, t)\mathscr{P}^{-1} &= \eta_P'\gamma^0\psi(-\boldsymbol{r}, t) \\ \mathscr{P}\overline{\psi}(\boldsymbol{r}, t)\mathscr{P}^{-1} &= \eta_P'^*\overline{\psi}(-\boldsymbol{r}, t)\gamma^0 \end{aligned}\right\} \qquad \ldots[27]$$

which are the most general local unitary transformations for the field-operators. The factor γ^0 is necessary for the transformation of the Dirac field to assure the invariance of the free-particle equation. The factors η_P will in general be different for different fields; they are called the intrinsic parities and the unitarity of $\mathscr{P}$ restricts them to be phase factors of modulus unity. If the unitary transformation $\mathscr{P}$ leaves the interaction invariant, the scattering operator S will be left invariant under the action of $\mathscr{P}$,

$$\mathscr{P}S\mathscr{P}^{-1} = S \qquad \text{...[28]}$$

whence

$$S\mathscr{P}S^{-1} = \mathscr{P}. \qquad \text{...[29]}$$

Thus reflection-invariance leads to a conservation law of parity

$$P_f = P_i. \qquad \text{...[30]}$$

Two reflections in succession bring us back to the original system, so for any state $|\Psi\rangle$ we must have

$$\mathscr{P}^2 |\Psi\rangle = C_P |\Psi\rangle$$

where the unitarity of $\mathscr{P}$ requires the c-number C_P to have modulus unity. If any superposition of two state-vectors is to transform into the corresponding superposition of the transformed vectors, the factor C_P must be the same for all states $|\Psi\rangle$. For all situations known to occur, we may take advantage of the restriction placed by superselection rules on the measurability of phases to set this factor equal to $+1$, with no loss of generality. Then

$$\mathscr{P}^2 = \mathrm{I}$$

which, taken together with the unitarity of $\mathscr{P}$, tells us that $\mathscr{P}$ is an hermitian operator with eigenvalues ± 1. The intrinsic parity factors η_P can simultaneously all be reduced to ± 1; the sign has an absolute significance only for hermitian fields. For other particles, it is generally dependent on convention, η_P being set equal to $+1$ for one particle in each superselection group. A particular consequence of equations [*27*] is that particle and antiparticle have the same intrinsic parity for bosons and opposite intrinsic parity for fermions. Thus, for example, the even orbital momentum states of positronium have *odd* parity while p, f, h, etc. states have even parity. This results from our assumption that the entire field-operator, i.e. both positive and negative frequency components, transforms in a given manner; a more direct physical interpretation of this simple result would be desirable.

The preceding discussion has been based on the assumption that the interactions were such that a unitary reflection operator $\mathscr{P}$ could be found such that under its action the interaction was left invariant. It is easy to see that this will not always be the case. For example, if the interaction contains a term

$$\bar{\psi}(1+i\gamma_5)\psi\phi$$

there is no choice of the intrinsic parities which will leave this invariant. This is the situation we have to face in the weak interactions, e.g. in nuclear β decay. At the same time, nuclear and electromagnetic interactions are very much stronger than weak interactions and do seem to be reflection-invariant, so that the concept of parity has both meaning and utility for nuclear and electromagnetic processes to a considerable degree of accuracy. Nuclear states in which parity admixtures were sought were found to be parity eigenstates to an accuracy of 1 part in 10^6 or better. Parity selection rules in atomic spectroscopy have been tested to something like 1 part in 10^4.

A consequence of TCP invariance is that if one finds non-invariance with respect to one of the individual transformations P, C or T, one expects non-invariance with respect to at least one of the other "symmetries" as well. In the case of the β decay experiments which demonstrated non-invariance under space inversion, e.g. the detection of a longitudinal polarization for β particles, it was easy to conclude that PT was also not a symmetry of the β interaction. Neither the neutrino nor the electron resulting from β decay has any strong interactions, therefore time-reversal can be taken simply as reversing the motions of all particles, keeping the helicity fixed. But space inversion reverses helicities; therefore the observation of a non-vanishing longitudinal polarization $\langle \boldsymbol{\sigma} . \boldsymbol{p} \rangle$ of electrons from β decay demonstrates that the β decay interaction is invariant neither under P nor under PT. By the TCP theorem we conclude that the existence of longitudinal polarization excludes both P and C invariance. That the weak interactions are indeed not invariant with respect to charge-conjugation was demonstrated more directly by the measurement of the longitudinal polarization of the muons from $\pi^{\pm}$ decay. It was found that whereas the μ^+ from π^+ decay are left-handed (spin opposite to momentum), the μ^- from π^- decay are predominantly right-handed.

An attractive possibility, which seemed to agree with all experiments until the discovery of $K_2^0 \rightarrow \pi^+\pi^-$ decay, was that CP remained an exact symmetry for all interactions while C and P were separately conserved for strong and electromagnetic interactions. A recent experiment reports a violation of C invariance even at the electromagnetic level. Since P invariance has been found to hold to an accuracy of better than 10^{-6} in nuclei, TCP invariance would require T violations in nuclear states with roughly the strength of electromagnetic corrections.

4.2. *Charge-conjugation invariance*

The hypothesis of C invariance, which until recently was believed to hold for both strong and electromagnetic interactions, leads in general to relations between particles and antiparticles and is not easy to check directly. C invariance does not in general lead to a conservation law because, by definition,

$$\mathscr{C} Q \mathscr{C}^{-1} = -Q. \qquad \ldots [31]$$

$\mathscr{C}$ is thus measurable only for $Q = 0$. For such systems and for interactions which are invariant under $\mathscr{C}$, $\mathscr{C}$ will become a measurable operator which is conserved, with eigenvalues ± 1, by arguments similar to those used in the case of $\mathscr{P}$. The electromagnetic field couples directly to the charge, thus C invariance would require

$$\mathscr{C}A_\lambda\mathscr{C}^{-1} = -A_\lambda. \qquad ...[32]$$

This results in the theorem that an odd number of photons may not transform into an even number. For charged fields, charge conjugation may be taken to be simply the interchange of field operators with their conjugates; for Dirac fields, this requires the use of the Majorana representation of the Dirac matrices (in which all the matrices are purely imaginary). The commutation relations are not altered by such an interchange, therefore a unitary transformation $\mathscr{C}$ can be defined such that

$$\begin{aligned} \mathscr{C}\phi(x)\mathscr{C}^{-1} &= \phi^*(x), \qquad \mathscr{C}\phi^*(x)\mathscr{C}^{-1} = \phi(x), \\ \mathscr{C}\psi(x)\mathscr{C}^{-1} &= \psi^\dagger(x), \qquad \mathscr{C}\psi^\dagger(x)\mathscr{C}^{-1} = \psi(x). \end{aligned} \qquad ...[33]$$

Charge conjugation invariance leads in this way to selection rules for positronium and antiproton annihilation and also for decays of neutral particles. For example, the theorem stated above forbids the π^0 to decay into both 2γ and 3γ channels. For states with $Q \neq 0$ the restrictions imposed by C invariance are the same as those due to PT invariance which consist essentially of phase restrictions on different terms which may appear in matrix elements.

A useful extension of charge-conjugation invariance to non-neutral systems which can be connected to their antiparticles through some other internal symmetry transformation is illustrated by the example of G-parity

$$G = \mathscr{C}e^{i\pi I_2}. \qquad ...[34]$$

Non-strange meson states are transformed into themselves under the action of G, consequently for charge-independent interactions which are invariant under charge-conjugation, the G-parity with eigenvalues ± 1 is conserved. An operator similar to G could be defined for strange mesons also. Unfortunately, however, the discrete operation (which is *not* an element of the SU(3) group of transformations) which transforms the members of the meson octet into their conjugate partners, does not seem to be remotely like a symmetry transformation for strong interactions. Thus, such a generalized G operator is not a particularly useful concept.

4.3. *Time-reversal invariance*

We now pass on to consider the consequences of T invariance. Like Θ, $\mathscr{T}$ must be an anti-unitary operator. Thus we do not obtain any conservation laws as a result of T invariance. Writing

$$S = I + iT \qquad ...[35]$$

the unitarity of S requires

$$i(T^\dagger - T) = T^\dagger T. \qquad ...[36]$$

Suppose that we are interested in a transition between states b and a induced by an interaction which may be treated to lowest order. Writing

$$T = T_0 + \varepsilon T_1 \qquad \text{...[37]}$$

and keeping terms in [36] only to the lowest order in ε, we have

$$\begin{aligned} i\langle a \mid (T_1^\dagger - T_1) \mid b\rangle &= \sum_n \langle a \mid T^\dagger \mid n\rangle\langle n \mid T \mid b\rangle \\ &= \langle a \mid T_0^\dagger \mid a\rangle\langle a \mid T_1 \mid b\rangle + \langle a \mid T_1^\dagger \mid b\rangle\langle b \mid T^0 \mid b\rangle \end{aligned} \qquad \text{...[38]}$$

or

$$\frac{\langle a \mid T_1^\dagger \mid b\rangle}{\langle a \mid T_1 \mid b\rangle} = e^{-2i(\delta_a+\delta_b)} \qquad \text{...[39]}$$

taking b and a to be scattering eigenstates with respect to the strong interactions. If time-reversal invariance is valid,

$$\left.\begin{aligned} \langle \mathscr{T}\alpha^{\text{in}} \mid \mathscr{T}\beta^{\text{out}}\rangle &= \langle \alpha^{\text{in}} \mid \beta^{\text{out}}\rangle^* \\ \langle \alpha^{\text{out}} \mid \beta^{\text{in}}\rangle &= \langle \alpha^{\text{in}} \mid \beta^{\text{out}}\rangle^* \end{aligned}\right\} \qquad \text{...[40]}$$

therefore

$$S_{\beta\alpha} = \langle \beta^{\text{in}} \mid \alpha^{\text{out}}\rangle = \langle \alpha^{\text{in}} \mid \beta^{\text{out}}\rangle = S_{\alpha\beta} \qquad \text{...[41]}$$

and

$$\langle a \,|\, T \,|\, b\rangle = \pm \,|\, \langle a \,|\, T \,|\, b\rangle \,|\, e^{i(\delta_a+\delta_b)}. \qquad \text{...[42]}$$

If the scattering phases are negligible, we find that the matrix element for the transition $b \to a$ must be real, i.e. different terms in the transition amplitude must have the same phase. This is equivalent to the usual statement, for interactions which can be treated by perturbation theory, that time-reversal invariance requires that different coupling constants be relatively real. There has so far been no reported instance where time-reversal invariance has directly been demonstrated to break down. For neutron β decay,[11] the vector and axial-vector coupling constants have been found to have a relative phase of $(180 \pm 8)°$, while for $\Lambda^0 \to p\pi^-$ decay,[12] the relative phase of the s and p wave amplitudes has been found to be $(16 \pm 20)°$ which is compatible with the value $(-6{\cdot}5 \pm 0{\cdot}5)°$ to be expected from [42].

An important restriction imposed by time-reversal invariance on electromagnetic interactions is that particles may not have static electric dipole moments. The dipole moment could only be associated with the spin which reverses sign under time-reversal, but since electric fields, which can arise from static charges, do not, an electric dipole interaction would violate time-reversal invariance. Smith, Purcell and Ramsey[13] have found that the electric dipole moment of the neutron cannot exceed $e \times 2.10^{-20}$ cm. After the discovery of parity non-conservation, the failure to detect any dipole moment was generally ascribed to the action of CP or T invariance. In view of the reported CP and C violations, one must instead conclude that the smallness of the dipole moment reflects the high degree of accuracy to which parity is conserved in strong interactions. Considering the importance of the question, it is extremely desirable that the experimental results be refined even further. A limit of $e \times 5.10^{-22}$ cm has been set for the dipole moment of the electron,[14] where the natural unit of length is a thousand times greater, but in this case the relation of the result to the reported CP and C violations is less direct.

A question which may come to mind when discussing non-invariance under time-reversal is whether this could in some way have macroscopic consequences through its influence on the establishment of equilibrium or, even more dramatically, by modifying the second law of thermodynamics. This problem was already considered in classical statistical mechanics by Boltzmann who showed that the validity of the H-theorem, which was his demonstration of the second law, does not depend on the validity of detailed balance. The corresponding discussion‡ for the case of quantum theory was given by Stueckelberg.[15] We shall not reproduce his proof that unitarity alone suffices to assure the validity of the H-theorem, but content ourselves by demonstrating how unitarity assures the possibility of a stable equilibrium condition. The existence of equilibrium requires that the rate of transitions out of any particular state be equal to the total rate of transitions from all other states to the state in question. We shall see that this is a consequence of unitarity alone.§ From

$$S^{\dagger}S = SS^{\dagger} = \mathrm{I}$$

we have, for every state,

$$\sum_n S^{\dagger}_{in} S_{ni} = \sum_n S_{in} S^{\dagger}_{ni}$$

or

$$\sum_n | S_{ni} |^2 = \sum_n | S_{in} |^2.$$

$| S_{ii} |^2$ is common to both sides, so we may subtract it, leaving

$$\sum_{n \neq i} | S_{ni} |^2 = \sum_{n \neq i} | S_{in} |^2$$

which is the previously stated condition for the establishment of equilibrium.

‡ Such a discussion was necessary even when there was no question of giving up T invariance, because T invariance only imposes micro-reversibility $w_{ba} = w_{\bar{a}\bar{b}}$ (where $\bar{b}$ and $\bar{a}$ are the time-reversed states corresponding to b and a) which is not, in general, the same as detailed balance, $w_{ba} = w_{ab}$.

§ Strictly speaking, normality of S, $S^{\dagger}S = SS^{\dagger}$, is all that is required.

5

K^0 AND η DECAYS

5.1. K^0 *decay*

The recent revival of interest in space-time symmetries stems from the important discovery made in 1964 of the $\pi^+\pi^-$ decay mode of long-lived neutral kaons.[16] We briefly review the pertinent details.

If, as was generally assumed after the discovery of parity non-conservation, all interactions were CP invariant, decays of neutral kaons would be most conveniently described in terms of their CP-even and CP-odd components. CP invariance would assure that these components decayed independently and each would transform only into states with the same CP eigenvalue. A $\pi^+\pi^-$ state with zero angular momentum necessarily has CP $= +1$; therefore the observation of both a short-lived and a long-lived $\pi^+\pi^-$ decay mode cannot be reconciled with the existence of a single CP-even K^0 component decaying with a single lifetime. Thus CP invariance must be abandoned.

In what follows, we shall assume TCP invariance, although part of the analysis can be carried out[7,17] without imposing this condition, with only a small increase in complexity. Because of the mass-degeneracy imposed by TCP invariance, K^0 and $\bar{K}^0$ can transform into each other through the weak interactions. Therefore a state prepared originally as K^0 or as $\bar{K}^0$ will in course of time evolve into a superposition of these states. Thus in discussing their decays, we must consider the time-development of a two-component wave function ψ, whose components represent the probability amplitudes for finding the system in the K^0 or $\bar{K}^0$ state respectively. This will be governed by a Schrödinger equation

$$i\frac{d\psi}{d\tau} = \boldsymbol{H}\psi \qquad \ldots[43]$$

where $\boldsymbol{H}$ will be a 2×2 matrix which will *not* be hermitian since we are dealing with a decaying system. TCP invariance requires that the diagonal elements be equal and CP invariance would require that, by a suitable choice of phases, the off-diagonal elements could also be made equal. Assuming TCP invariance, we may write[6,18]

$$\boldsymbol{H} = \begin{bmatrix} \lambda_+ & \lambda_- r^{-1} \\ \lambda_- r & \lambda_+ \end{bmatrix} \qquad \ldots[44]$$

where the parameters appearing in $\boldsymbol{H}$ may be related to physically measurable constants obtained by diagonalizing $\boldsymbol{H}$. The general solution of [*43*], with $\boldsymbol{H}$ given by [*44*], is

$$\psi(\tau) = a\phi_1 e^{-i\lambda_1\tau} + b\phi_2 e^{-i\lambda_2\tau} \qquad \ldots[45]$$

with

$$\lambda_{\pm} = \tfrac{1}{2}(\lambda_1 \pm \lambda_2) \qquad \text{...[46]}$$

and

$$\phi_{1,2} = (1+|r|^2)^{-\frac{1}{2}}\begin{pmatrix}1\\ \pm r\end{pmatrix}. \qquad \text{...[47]}$$

$\phi_{1,2}$ correspond to the particular superpositions of K^0 and $\bar{K}^0$ which undergo purely exponential decay (in the approximation represented by [*43*]). It will be seen that the phase of r depends on the *choice* of relative phase of the K^0 and $\bar{K}^0$ states. Consequently, the separation of [*44*] into its hermitian and anti-hermitian parts also depends on the particular *phase convention* adopted. It is customary to choose the phase so that the amplitude A_0 for $K^0 \to 2\pi$ ($I = 0$) is equal to that for $\bar{K}^0 \to 2\pi$ ($I = 0$) both in magnitude (as required by TCP invariance) and in phase. The labelling of 2π states by isospin ($I = 0$ or 2) rather than by charge indices $\pi^+\pi^-$ or $\pi^0\pi^0$ is preferred because of the possibility of strong final-state interactions between the pions. Then the amplitudes for $\phi_{1,2}$ to decay to $\pi^+\pi^-$ and $\pi^0\pi^0$ channels are given, from [*47*] and [*19*] by[18]

$$\begin{aligned} a_{1,2}(\pi^+\pi^-) &= (1+|r|^2)^{-\frac{1}{2}}[\sqrt{2}A_0(1\pm r)+(A_2\pm A_2^*r)F]/\sqrt{3} \\ a_{1,2}(\pi^0\pi^0) &= (1+|r|^2)^{-\frac{1}{2}}[A_0(1\pm r)-\sqrt{2}(A_2\pm A_2^*r)F]/\sqrt{3} \end{aligned} \qquad \text{...[48]}$$

where A_2 is the amplitude for $K^0 \to 2\pi$ ($I = 2$) and $F = e^{i(\delta_2-\delta_0)}$ takes account of the effect of possible final state interactions. δ_2 and δ_0 are the s-wave π-π scattering phase shifts for $I = 2$ and $I = 0$ respectively, at the c.m. energy equal to the K^0 mass. From the low rate of $K_2 \to 2\pi$ decay relative to $K_1 \to 2\pi$ decay, and the value of the $(\pi^+\pi^-)/(\pi^0\pi^0)$ branching ratio for K_1, one can conclude that r is a number close to unity both in magnitude and phase (with our particular phase convention) so it is convenient to introduce the parameter $\varepsilon = 1-r$. Then it is convenient to approximate the amplitudes for K_2 decay from [*48*] by

$$\left.\begin{aligned} a_2(\pi^+\pi^-) &= A_0[\varepsilon+\delta/\sqrt{2}]/\sqrt{3} \\ a_2(\pi^0\pi^0) &= A_0[\varepsilon-\delta\sqrt{2}]/\sqrt{6} \end{aligned}\right\} \qquad \text{...[49]}$$

where

$$\delta = (A_2 - A_2^* r)F/A_0 \qquad \text{...[50]}$$

is also a small complex parameter. The quantities which are more directly accessible experimentally are the amplitude ratios $\eta = a_2/a_1$, but it is not possible, on the basis of present information, to accurately determine

$$\Delta = (A_2 + A_2^* r)F \approx 2F \,.\, \mathrm{Re}\, A_2. \qquad \text{...[51]}$$

If we *assume* that Δ may be neglected in comparison with A_0 (as would be required by the $|\Delta I| = \frac{1}{2}$ rule), we obtain

$$\left.\begin{aligned}\eta_{+-} &= a_2(\pi^+\pi^-)/a_1(\pi^+\pi^-) = \tfrac{1}{2}[\varepsilon+\delta/\sqrt{2}] \\ \eta_{00} &= a_2(\pi^0\pi^0)/a_1(\pi^0\pi^0) = \tfrac{1}{2}[\varepsilon-\delta\sqrt{2}].\end{aligned}\right\} \quad ...[52]$$

The experimenter's task is thus reduced to the determination of the parameters ε and δ. Information on the phase of η can be obtained from interference between $K_2 \to 2\pi$ and $K_1 \to 2\pi$ decays. If δ can be neglected, η measures ε directly. Now we shall show, following Wu and Yang,[18] that if δ is indeed negligible, the observed rate of $K_2 \to \pi^+\pi^-$ decay not only determines the magnitude of ε but practically fixes its phase as well. From [*44*], we see that for ε small, $\varepsilon\lambda_-$ is the coefficient of the antisymmetric part of $\boldsymbol{H}$.

$$\varepsilon\lambda_- = i\,\mathrm{Im}\,M_{12}+\mathrm{Im}\,\Gamma_{12} \quad ...[53]$$

where $\boldsymbol{M}$ and $i\boldsymbol{\Gamma}$ are the hermitian and anti-hermitian parts of $\boldsymbol{H}$ respectively:

$$\boldsymbol{M} = \tfrac{1}{2}(\boldsymbol{H}+\boldsymbol{H}^\dagger);\ \boldsymbol{\Gamma} = \frac{1}{2i}(\boldsymbol{H}-\boldsymbol{H}^\dagger). \quad ...[54]$$

We can separate $\boldsymbol{\Gamma} = \sum_i \boldsymbol{\Gamma}^i$ into its separate contributions from various channels i. Now our phase convention has been so chosen that the dominant $I = 0$ $\pi\pi$ channel does not contribute at all to $\mathrm{Im}\,\Gamma_{12}$; if the $\Delta S = \Delta Q$ rule holds (see section 6), the leptonic channels cannot contribute to Γ_{12} either. If we neglect the $I = 2$ $\pi\pi$ channel, the only contributions come from 3π channels: for each of the 3π channels, a rigorous upper limit is $|\mathrm{Im}\,\Gamma^i_{12}| \leqq |\Gamma^i_{12}| \leqq [\Gamma(K_1 \to i)\Gamma(K_2 \to i)]^{\frac{1}{2}}$ since each sub-matrix $\boldsymbol{\Gamma}^i$ must be positive definite. Inserting the experimental values, we find that the largest possible value of $\mathrm{Im}\,\Gamma_{12}$ is too small to account for the magnitude of $\varepsilon\lambda_-$ (equation [*53*]) required to fit the $K_2 = \pi^+\pi^-$ decay rate. Even if we relax the $\Delta S = \Delta Q$ rule to the extent allowed by present experimental uncertainty, the maximum possible contribution of leptonic channels to $\mathrm{Im}\,\Gamma_{12}$ is less than a third of $|\varepsilon\lambda_-|$. Thus we conclude that *if* the $I = 2$ channel does not contribute significantly either to [*52*] or to [*53*], the quantity $i\varepsilon\lambda_-$ must be essentially real. Then the phase of ε is given by

$$\phi_\varepsilon = \tan^{-1}\frac{(m_2-m_1)}{\tfrac{1}{2}(\gamma_1-\gamma_2)} \approx \tan^{-1} 2\,\frac{(m_2-m_1)}{\gamma_1}. \quad ...[55]$$

As discussed by several authors,[17,19] the observed CP violation in decays may arise from CP violating interactions of widely different strengths depending on the nature of the hypercharge selection rules obeyed by them.

The simplest and least startling possibility is that the usual weak interaction contains a small CP violating part. If the ratio of $K_2 \to \pi^+\pi^-$ to $K_1 \to \pi^+\pi^-$ decay amplitudes is a representative measure of the relative strengths of the CP violating and CP conserving amplitudes, we would then expect CP violating effects at the level of 0·1%. None of the other experiments performed to test CP or T invariance would have been able to detect effects of this magnitude. Another interesting possibility, suggested by Wolfenstein,[20] is that the CP violating interaction may obey $|\Delta S| = 2$, in which case a very weak interaction, weaker than second-order weak interactions, which mixes K^0 and $\bar{K}^0$ states, would suffice to account for the observed $K_2 \to \pi^+\pi^-$ decay rate. Of course, in that case it would be hopeless to look for CP or P violations other than in K^0 decays. For this theory, the prediction [*55*] for the phase of ε becomes exact and since δ is negligible in this case, the phase of η is predicted exactly. It is interesting that the measurements of the phase of η_{+-} are compatible with this hypothesis.[21]‡

If the CP violating interaction is taken as strangeness-conserving, we arrive at a quite different estimate for its strength. In this case, the CP violating interaction cannot by itself contribute to K^0 decay amplitudes, nor can it give any $K^0-\bar{K}^0$ mixing. The only way in which it can influence K^0 decays is by modifying the phases of decay amplitudes by final-state interaction. The CP violating amplitude is found to be of the order of 10^{-3} of the CP conserving one; if the CP violation arises only through the action of a CP violating interaction obeying $|\Delta S| = 0$, that interaction must be characterized by a dimensionless coupling constant of the order of 10^{-3}, i.e. comparable to the strength of electromagnetism. Since parity conservation is checked to a much higher degree of accuracy in nuclei, such an interaction would have to be primarily a C violating one and, assuming the validity of TCP invariance, also violate T invariance to the same extent. It is somewhat surprising that, until quite recently, there was no experimental information either for or against this hypothesis.

5.2. *η decay*

It was suggested by Friedberg, Schwartz and Lee[22] that η decay, which is believed to occur only through electromagnetic interaction, may serve to test for the existence of a $|\Delta S| = 0$ C violating interaction of the strength which we have discussed. Any asymmetry between the π^+ and the π^- in $\eta \to \pi^+\pi^-\pi^0$ decay would be direct evidence for C non-invariance, assuming that the η is an eigenstate of C, as is only natural if strong and electromagnetic interactions are charge-conjugation invariant.

Since the η has spin zero, the decay configuration is completely specified by the momenta or energies of two of the pions. This is done most conveniently on the Dalitz plot, in which the energies of the three pions for a given $\eta \to \pi^+\pi^-\pi^0$ decay are plotted as the perpendicular distances from the sides of an equilateral triangle, whose height is equal to the η mass. Thus each $\eta \to \pi^+\pi^-\pi^0$ event is represented by a point inside the triangle. If we choose E_{π^0} to be the distance from the base of the triangle, C invariance

‡ *Note added in proof.* Measurements of the rate of $K_2 \to \pi^0\pi^0$ decay (J. M. Galliard *et al.* (1967). *Phys. Rev. Lett.* **18**, 20; J. W. Cronin *et al.* (1967). *Phys. Rev. Lett.* **18**, 25) indicate that $|\eta_{00}|$ is about twice as large as $|\eta_{+-}|$. From [*52*] we see that δ cannot then be negligible compared to ε, so that nothing can be said about the phase of ε in this case.

would require that the population of points be symmetrically distributed with respect to the vertical median.

It is convenient to classify the possible final states according to isospin. A 3π system can have any isospin from 0 to 3, which also serves to classify the values of $\mathscr{C}$ for the neutral charge states. A 3π system necessarily has $G = -1$ so

$$G = \mathscr{C}(-1)^I = -1$$

requires

$$\mathscr{C} = (-1)^{I+1}.$$

Thus $I = 0, 2$ states have $\mathscr{C} = -1$ while $I = 1, 3$ states are even under charge-conjugation. Under the assumption that $\eta \rightarrow 3\pi$ decays occur only through a virtual electromagnetic interaction, which obeys $|\Delta I| = 0, 1$, the 3π state would be uniquely specified to be $I = 1$, to order e^2. Any charge asymmetry would require interference with $I = 0$ or 2 states of opposite $\mathscr{C}$. We now show that the occurrence of such interference also requires the presence of appreciable final state interactions between the pions. From [*19*], we have

$$\langle \Theta\beta \mid T \mid \Theta a \rangle = e^{2i\delta_\beta} \langle \beta \mid T \mid a \rangle^*.$$

If we neglect final state interactions, and choose b to be an eigenstate of charge-conjugation,

$$\langle \Theta b \mid T \mid \Theta\eta \rangle = \mathscr{C}_\eta \mathscr{C}_b \langle b \mid T \mid \eta \rangle = \langle b \mid T \mid \eta \rangle^*. \qquad \text{...[56]}$$

Then the C conserving amplitude is real, while the C violating amplitude is purely imaginary. Therefore any charge asymmetry will be observable only if there is both C violation and strong final state interaction.

An $I = 0$ state of three pions is totally antisymmetric under the exchange of any pair and is therefore invariant under a cyclic permutation of $\pi^+\pi^-\pi^0$. If we take the presumably dominant $I = 1$ state to contain mainly s waves—to zeroth approximation, the $\eta \rightarrow \pi^+\pi^-\pi^0$ Dalitz plot is uniformly populated—interference with an $I = 0$ state would alternate in sign between adjacent 60° sectors of the Dalitz plot. The recently reported results[23] show no effect of this kind but instead a preponderance of energetic π^+, represented by an overpopulation of the corresponding half of the Dalitz plot, of $7{\cdot}2 \pm 2{\cdot}8\%$.‡ This would require interference with $I = 2$ final states.

Another possible test for C-invariance in η decay is to look for

$$\eta \rightarrow \pi^0 + e^+ + e^- \qquad \text{...[57]}$$

decay,[24] which cannot occur via one photon exchange if C invariance holds. The $\eta\pi^0$ electromagnetic vertex can be written in the form

$$\langle \pi^0 \mid j_\lambda \mid \eta \rangle = ef'(q^2)[(m_\eta^2 - m_\pi^2)(p_\eta - p_\pi)_\lambda - q^2(p_\eta + p_\pi)_\lambda] \qquad \text{...[58]}$$

‡ *Note added in proof.* A later experiment with much better statistics (A. M. Cнops *et al.* 1966. *Phys. Lett.* **22**, 546) shows no such effect.

where $f'(q^2)$ is expected to be a slowly varying function of $q^2 = (p_\eta - p_\pi)^2$ over the range $0 < q^2 < (m_\eta - m_\pi)^2$. We replace it by an average value $\frac{1}{6}\langle r^2 \rangle$, and if we neglect the electron mass, [58] can be replaced by

$$\langle \pi^0 \mid j_\lambda \mid \eta \rangle = -\tfrac{1}{6}\langle r^2 \rangle e q^2 (p_\eta + p_\pi)_\lambda \qquad \ldots[59]$$

to calculate the decay [57], which is found to have a rate given by

$$\frac{\Gamma(\eta \to \pi^0 e^+ e^-)}{\Gamma(\eta \to 2\gamma)} \approx 0{\cdot}04\,[\langle r^2 \rangle m_\eta^2]^2 \qquad \ldots[60]$$

which is of the order of unity for $\langle r^2 \rangle = \langle r_p^2 \rangle$. Furthermore, the factor q^2 in [59] cancels the $1/q^2$ from the virtual photon propagator, so the expected mass distribution of (e^+e^-) pairs from [57] is essentially that given by phase space (multiplied by a factor $(1+\cos\theta_{+-})$ to take account of the helicity correlation between e^+ and e^- imposed by the electromagnetic vertex). This favours high mass pairs, as opposed to the low masses which predominantly arise from internal conversion or Dalitz pairs. Thus the process [57] should be clearly recognizable if it occurs, and an upper limit of 6.10^{-3} has been set for the branching ratio [60].[25] The decay [57] is of course allowed through 2-photon exchange even in a C invariant theory, but this will be suppressed relative to the one-photon process by a factor of order $(m_e/m_\eta)^2$ in addition to the factor of α^2 arising from the extra photon. C-even (e^+e^-) states have the opposite helicity correlation to that imposed by the usual γ_λ electromagnetic vertex; therefore their amplitudes must be proportional to m_e, which is the sole agent for breaking the correlation.

The rarity of [57] may be accounted for by assigning suitable isospin and SU(3) transformation properties to the $\eta\pi^0\gamma$ vertex. However, if as some authors suggest, the C violating interaction is purely of electromagnetic origin, the asymmetry in $\eta \to \pi^+\pi^-\pi^0$ decay would seem to require a non-vanishing $\eta\pi^0\gamma$ vertex, unless we are willing to abandon the $|\Delta I| = 0, 1$ transformation property usually ascribed to electromagnetic interactions. One is unfortunately not able to predict exactly how strong the $\eta\pi^0\gamma$ vertex should be, but since the present experimental limit for $\eta \to \pi^0 e^+ e^-$ is many orders of magnitude higher than the rate expected in a C invariant theory, there is considerable room for improvement of the limit.

6

Internal Symmetries

Except for $K_2^0 \to \pi^+\pi^-$ decays, everything we know about weak interactions is consistent with the hypothesis that all weak interactions occur through the action of the current-current coupling interaction density[26]

$$\mathscr{H}(x) = \frac{G}{\sqrt{2}} \mathscr{J}_\lambda^\dagger(x) \mathscr{J}^\lambda(x) + \text{h.c.} \qquad \ldots[61]$$

The positively charged current $\mathscr{J}_\lambda$ contains a hadronic part $\mathscr{J}_\lambda^h$ and a leptonic part L_λ, which we believe we know explicitly

$$L_\lambda = \bar{\nu}\gamma_\lambda(1+\gamma_5)e^- + \bar{\nu}'\gamma_\lambda(1+\gamma_5)\mu^- \qquad \text{...[62]}$$

where the particle symbols represent the corresponding fields, while the corresponding negative current is

$$L_\lambda^\dagger = \bar{e}\gamma_\lambda(1+\gamma_5)\nu + \bar{\mu}\gamma_\lambda(1+\gamma_5)\nu'. \qquad \text{...[63]}$$

Then μ^- decay would be described by the coupling of the second term in [62] to the first term in [63] and μ^+ decay by the other cross-term (which is the hermitian conjugate of the first). All measurements on μ decay agree with the predictions of this theory. [62] and [63] also incorporate an assumption which may be called the principle of $\mu - e$ equivalence, i.e. that the interactions of the muon and its neutrino are identical with those of the electron and its associated neutrino. The strongest evidence in support of this hypothesis is that the branching ratio $(\pi^+ \to e^+\nu)/(\pi^+ \to \mu^+\nu')$ is found to have the value $(1{\cdot}23 \pm 0{\cdot}03) \times 10^{-4}$ exactly as predicted by the theory.[27]

Since the effects of [61] will almost invariably be considered to lowest order only, it may be preferable to regard the current-current coupling [61] as a phenomenological representation of a possibly more complex structure for weak interaction matrix elements

$$\mathscr{H}_{\text{eff}} = 4\pi g^2 \int \mathscr{J}_\lambda^\dagger(x) F^\lambda{}_\sigma(x-y)\mathscr{J}^\sigma(y)\,dy + \text{h.c.} \qquad \text{...[64]}$$

The natural premise underlying [61] and [64] is that the current $\mathscr{J}_\lambda$ is the source of a vector field W_λ which transmits the weak interaction in the same way as the electromagnetic field A_λ is the intermediary field responsible for electromagnetic interaction. If muon decay provides a representative indication of the extent of non-locality represented by the function F, we may conclude that F must be negligible for $|x-y| > 10^{-13}$ cm, from the accuracy with which the electron energy spectrum in μ decay obeys the prediction of the local theory [61]. This sets a lower limit to the possible mass for the quanta of the hypothetical intermediary field W_λ. High energy neutrino experiments, in which real quanta could be produced, set a lower limit $m_W > 2$ GeV,[28] so any possible non-locality would be at considerably smaller distances. The rate of muon decay fixes

$$G = (1{\cdot}02 \pm 0{\cdot}01)\,.\,10^{-5}/M_p^2. \qquad \text{...[65]}$$

The hypothesis of a current-current coupling [61] also makes definite predictions for hitherto unobserved processes, e.g. the reaction

$$\nu + e^- \to \nu + e^- \qquad \text{...[66]}$$

is predicted to have the cross section (which is isotropic in the c.m. system)

$$\sigma(s) = \frac{G^2}{\pi s}(s - m_e^2)^2. \qquad \text{...[67]}$$

Similarly, coupling of the hadronic terms in $\mathscr{J}_\lambda$ to their own hermitian conjugates through [*61*] could give rise to an additional weak interaction between hadrons, whose presence would be recognizable, despite the fact that there are much stronger forces already operating between hadrons, by virtue of the fact that the weak interaction violates parity. Parity admixtures in nuclear states have recently been reported which are of the expected order of magnitude.[29]

For processes induced by [*61*] in which one of the currents is leptonic and the other hadronic, we could, if we wished, write down explicit expressions analogous to [*62*] and [*63*] for the hadronic current, which would phenomenologically describe the structure of the weak matrix-element between various hadron states, for leptonic decays. Clearly, such expressions would be meaningless when both currents participating in [*61*] are hadronic since strong interactions between the hadrons would almost inevitably distort the matrix-elements of the individual currents. Consequently, it is preferable to classify the different components of the hadronic current $\mathscr{J}_\lambda^h$ by their transformation properties with respect to the symmetry operations of the strong interactions. Two other important guiding principles have been the Conserved Vector Current hypothesis and what has come to be called the PCAC (Partial Conservation of Axial-Vector Currents) hypothesis. We shall discuss these later on.

As far as we are aware, baryon number and electric charge are both absolutely conserved, thus the fact that $\mathscr{J}_\lambda$ (as typified by its leptonic terms) carries unit charge and no baryon number, tells us that

$$[B, \mathscr{J}_\lambda^h] = [B, \mathscr{J}_\lambda^{h\dagger}] = 0 \qquad \ldots[68]$$

and

$$[Q, \mathscr{J}_\lambda^h] = \mathscr{J}_\lambda^h; \quad [\mathscr{J}_\lambda^{h\dagger}, Q] = \mathscr{J}_\lambda^{h\dagger}. \qquad \ldots[69]$$

At this point, one could ask whether there could not be interactions analogous to [*61*] involving neutral currents also. It is not possible to answer this question when both currents are hadronic since such a term could not be distinguished from one involving the product of two oppositely charged currents, or for terms which would be hidden under much stronger electromagnetic interactions. However, processes which could arise from the coupling of neutral leptonic currents to strangeness-changing neutral currents, like

$$K_2^0 \to \mu^+\mu^- \qquad \ldots[70]$$

and

$$K^+ \to \pi^+e^+e^- \qquad \ldots[71]$$

although not forbidden by any known selection rule, have not yet been seen and are much rarer than corresponding reactions arising from coupling of $\mathscr{J}_\lambda^h$ and L_λ. For example,[30]

$$\frac{\Gamma(K^+ \to \pi^+e^+e^-)}{\Gamma(K^+ \to \pi^0e^+\nu)} < 2{\cdot}5 \,.\, 10^{-5}. \qquad \ldots[72]$$

Thus it would appear that if neutral currents take part in weak interactions they do not include leptons to any appreciable extent.

Apart from charge and baryon number, hypercharge (or equivalently strangeness) is conserved to a high degree of accuracy (~ 1 part in 10^6) in strong interactions. We can therefore separate $\mathscr{J}^h_\lambda$ into different parts according to their transformation properties with respect to the hypercharge gauge transformation. There must be a part J_λ of the hadronic current, responsible for ordinary nuclear β-decay, which conserves hypercharge,

$$[Y, J_\lambda] = 0 \qquad \text{...[73]}$$

and another S_λ, responsible for hyperon β decays, which changes the hypercharge by one unit

$$[Y, S_\lambda] = S_\lambda. \qquad \text{...[74]}$$

Coupling of S_λ and J_λ is responsible for non-leptonic decay processes while $L^\dagger_\lambda S^\lambda$ gives hyperon β decays. A term S'_λ satisfying

$$[Y, S'_\lambda] = -S'_\lambda \qquad \text{...[75]}$$

would also obey the $|\Delta Y| = 1$ restriction, and direct evidence concerning such terms, which induce so-called $\Delta S = -\Delta Q$ transitions, e.g. $\Sigma^+ \to ne^+\nu$, only limits the coefficient of S'_λ to be not more than about 25% of that of S_λ. However, the coupling of S_λ to S'_λ of such a magnitude through [61] would in general lead to a $|\Delta Y| = 2$ contribution to $\mathscr{H}$ which would be difficult to reconcile with the smallness of the $K^0_1 - K^0_2$ mass-difference.[31] Consequently, S'_λ is probably not present with a relative coefficient much greater than 10^{-6}, at which level terms transforming like S'_λ are expected to be induced by higher order weak interactions anyway. (See Fig. 1.)

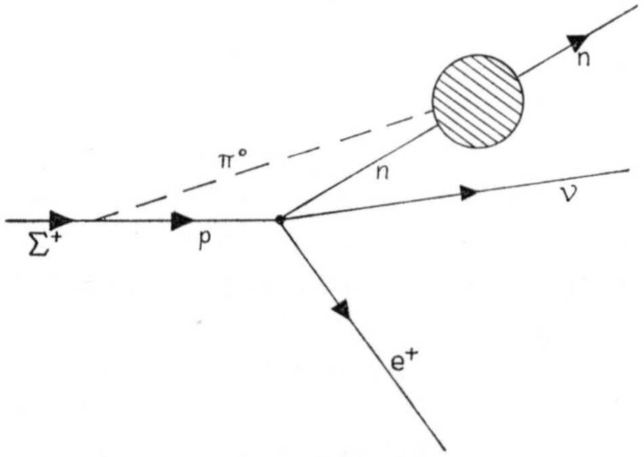

FIG. 1. Example of a $\Delta S = -\Delta Q$ transition induced by higher order weak interaction. The blob represents a strong interaction vertex.

Similar comments apply to possible terms in the hadronic current carrying double hypercharge, i.e. D_λ satisfying

$$[Y, D_\lambda] = \pm 2\, D_\lambda.$$

Direct evidence concerning such terms, e.g. from $\Xi^0 \to pe^- \nu$, is scarce but again a coupling of D_λ (with a coefficient much exceeding 10^{-6}) to J_λ would create an intolerably strong $|\Delta Y| = 2$ interaction. Terms transforming like D_λ are thus expected to appear only at the level of higher-order weak interactions, i.e. with a relative coefficient of the order of 10^{-6} or less.

Parity is conserved to a high degree of accuracy in strong interactions (nuclear levels are believed to be parity eigenstates to an accuracy of 1 part in 10^6) so it is meaningful, to the same accuracy, to separate $\mathscr{J}_\lambda^h$ into its vector and axial-vector parts $\mathscr{V}_\lambda$ and $\mathscr{A}_\lambda$ respectively. The surmise that β decay is correctly described by a V−A interaction led to the hypothesis of the current-current interaction [*61*]. The assumption that $|\Delta Y| = 1$ leptonic decays of hadrons are described by the coupling to lepton currents of vector and axial-vector currents belonging to S_λ is confirmed by

(a) the rarity of $K^+ \to e^+ \nu$ relative to $K^+ \to \mu^+ \nu'$ decay, and

(b) the observed angular correlations in K_{e3}^+ and K_{e3}^0 decays.

The near equality (within 2%) of the μ decay coupling constant [*65*] and the corresponding constant for nuclear Fermi transitions (β decays induced by V_λ, the vector part of J_λ) led to the Conserved Vector Current hypothesis,[26] as a result of which the space-integral of the fourth component of V_λ, the so-called weak charge, does not suffer renormalization under the action of the strong interactions. This implies that V_λ and $V_\lambda^\dagger$ are proportional to the I^+ and I^- components of the total isospin current,‡ guaranteeing the conservation of V_λ and $V_\lambda^\dagger$ by virtue of isospin conservation, apart from small corrections due to electromagnetism. If we also adopt the usual view that the electromagnetic current j_λ consists of an isoscalar part and an isovector part the latter being proportional to the I^3 component of the isospin current, then the three currents

$$\frac{1}{G_\beta^V} V_\lambda, \quad \frac{1}{e} j_\lambda, \quad \frac{1}{G_\beta^V} V_\lambda^\dagger \qquad \ldots[76]$$

transform as the $I^3 = +1, 0, -1$ components of a single isovector. The constant of proportionality to be associated with V_λ and $V_\lambda^\dagger$ depends on whether we regard these currents as the sources of L_λ or of W_λ. In the first case, the factor should be $1/G_\beta^V$ as shown in [*76*], but in the second one should take $1/g$. The isovector current hypothesis for V_λ, $V_\lambda^\dagger$ has the immediate consequence that for weak processes induced by these currents, the matrix-elements may be directly related to corresponding electromagnetic matrix-elements. The hypothesis has been strikingly confirmed by the correct prediction of the rate of pion β decay,[26, 32] $\pi^+ \to \pi^0 e^+ \nu$ and by the confirmation of the "weak magnetism" effect predicted by Gell-Mann,[33] and all other

‡ Certain authors make a distinction between the literal meaning of vector current conservation and this more restrictive assumption which may be called the proportional or isovector current hypothesis. We shall adopt the CVC hypothesis in the second stronger form. Fortunately, it is possible to test the two hypotheses separately.

evidence is in accord with this principle. The assumption that the triplet of currents [76] transform as components of a single isovector also assures that V_λ has a well-defined G-parity:

$$GV_\lambda G^{-1} = V_\lambda. \qquad ...[77]$$

Currents satisfying this property have been named first-class currents by Weinberg, while possible terms V'_λ anticommuting with G are called second-class currents.

We now ask whether the axial-vector parts of J_λ, $J^\dagger_\lambda$ have similarly simple transformation properties. Although this has not been tested as well, all experimental results agree with the simplifying assumption that like V_λ and $V^\dagger_\lambda$, A_λ and $A^\dagger_\lambda$ also transform as components of the same isovector. This, by itself, does not guarantee that A_λ should have a well-defined G-parity, but if we impose time-reversal invariance as well, we have

$$GA_\lambda G^{-1} = -A_\lambda. \qquad ...[78]$$

Conversely, if we demand that A_λ and $A^\dagger_\lambda$ be related by charge symmetry with no further condition, we shall find that A_λ will in general contain a part which violates time-reversal invariance and which is even under G. Cabibbo has suggested that time-reversal violating interactions may be introduced in this way. The terms in the current with abnormal G-transformation property (second-class currents) vanish in the limit of small momentum transfers, consequently there is no strong evidence against the possible existence of such terms. One hopes that high-energy neutrino reactions may be able to settle the question. [77] and [78] may be combined into

$$(G\mathscr{P})\, J_\lambda\, (G\mathscr{P})^{-1} = -J_\lambda \qquad ...[79]$$

as the condition defining first-class currents.

Just as the CVC hypothesis identifies the vector current V_λ with a quantity defined by the strong interactions, the axial-vector A_λ has also been related to a dynamical operator characteristic of strong interactions. This is most simply expressed as the hypothesis of Partial Conservation of the Axial-Vector Current[34] (PCAC):

$$\partial_\lambda A^\lambda = if_\pi m_\pi^2 \phi_\pi. \qquad ...[80]$$

This equation is to be interpreted in the sense that matrix elements of the L.H.S., evaluated between states β, α for which $\langle \beta \mid \phi_\pi \mid \alpha \rangle \neq 0$, may be approximated by those of the R.H.S. for momentum transfers $(p_\beta - p_\alpha)^2$ close to the pion pole. The constant f_π is determined from the rate of π decay which is determined by the matrix-element $\langle 0 \mid \partial_\lambda A^\lambda \mid \pi \rangle$. [80] was originally introduced to account for the success of the famous Goldberger-Treiman relation for π decay, but there are now several additional indications of its validity.

Under isospin transformations, the strangeness-changing current S_λ, must, because of the Gell-Mann-Nishijima relation

$$Q = \tfrac{1}{2}Y + I^3 \qquad \ldots[81]$$

and [74], transform as an object with $I^3 = \frac{1}{2}$. The simplest assumption is that S_λ transforms like an object with $I = \frac{1}{2}$ and leads to the $|\Delta I| = \frac{1}{2}$ rule for strangeness-changing leptonic interactions.

The only reasonably detailed test which is available is from a comparison of K_{l3}^+ and K_{l3}^0 decays,

$$K^+ \to \pi^0 l^+ \nu \qquad \ldots[82a]$$

$$K^0 \to \pi^\pm l^\mp \nu. \qquad \ldots[82b]$$

In general, both $|\Delta T| = \frac{1}{2}$ and $|\Delta I| = \frac{3}{2}$ amplitudes could contribute to these processes; the $|\Delta I| = \frac{1}{2}$ rule stipulates that only the first of these is non-negligible. Then the amplitudes of corresponding processes [82] must have identical structures, e.g.

$$\langle \pi^0(q) \,|\, S_\lambda^\dagger \,|\, K^+ \rangle = \langle \pi^-(q) \,|\, S_\lambda^\dagger \,|\, K^0 \rangle / \sqrt{2}. \qquad \ldots[83]$$

In practice, [82b] is studied through the decay of K_2^0 which, for the present purpose, may be taken to be an equal superposition of K^0 and $\bar{K}^0$ states. By the $\Delta Y = \Delta Q$ rule, only the K^0 component contributes to $\pi^- l^+ \nu$ and $\bar{K}^0$ to $\pi^+ l^- \bar{\nu}$, and TCP invariance requires that the two amplitudes be equal. Thus we expect K_{l3}^+ and $K_2^0 \to \pi l \nu$ to have identical spectra, with the absolute decay rates in the ratio

$$\frac{\Gamma(K_2^0 \to \pi^+ l^- \bar{\nu}) + \Gamma(K_2^0 \to \pi^- l^+ \nu)}{\Gamma(K^+ \to \pi^0 l^+ \nu)} = 2. \qquad \ldots[84]$$

According to Willis,[30] the latest data summed over electron and muon modes gives a ratio 2·12±0·12 while the $K_{\mu 3}/K_{e3}$ branching ratios, which the $|\Delta I| = \frac{1}{2}$ rule would require to be the same for K_2^0 and K^+, are found to be equal within an experimental uncertainty of about 15%. Measurements of the energy spectra in [82] have been reported but unfortunately the results on K_{l3}^0 do not agree with each other, so that a clear test of [83] is not yet possible. The accuracy with which [84] is satisfied provides an upper limit to the relative magnitude of $|\Delta I| = \frac{3}{2}$ amplitudes, which cannot exceed a few percent (unless the $|\Delta I| = \frac{3}{2}$ amplitude is almost exactly 90° out of phase with the $|\Delta I| = \frac{1}{2}$ amplitude). It may also be noted that the $|\Delta I| = \frac{1}{2}$ prediction [84] does not depend on the assumption of $\mu - e$ equivalence.

7

SU(3) Transformation Properties

After the discovery of approximate SU(3) invariance of strong interactions, the question was naturally raised as to whether J_λ and S_λ may not also possess simple SU(3) transformation properties. The hypothesis, first formulated in a complete form by Cabibbo,[35] that the entire hadronic weak current $\mathscr{J}^h_\lambda$ transforms as an octet under SU(3) transformations is a simple and elegant statement which accounts for all known properties of $\mathscr{J}^h_\lambda$. The theory has not yet been verified in all respects, in fact few of its many predictions have been checked in detail. But at the same time, we must remember that it starts with many advantages. It automatically ensures that the charged current $\mathscr{J}^h_\lambda$ comprises only the two parts J_λ and S_λ possessing the properties which we have already discussed. In addition, S_λ and $S^\dagger_\lambda$ are now related by an SU(3) transformation in the same way that J_λ and $J^\dagger_\lambda$ were assumed to be related by an isospin transformation.

If S_λ were coupled to L_λ with the β decay coupling constant, strangeness-changing leptonic decays would occur an order of magnitude faster than they actually do. To account for this fact, Cabibbo set

$$J_\lambda = (j^2_1)_\lambda \cos\theta; \quad S_\lambda = (j^3_1)_\lambda \sin\theta \qquad \text{...[85]}$$

where θ is a new universal constant, and j^2_1 and j^3_1 are octet components which transform like π^+ and K^+ respectively. Universality is redefined to mean that the vector part of $\mathscr{J}^h_\lambda = J_\lambda + S_\lambda$ is coupled to L_λ with the same strength as L_λ is coupled to itself. We shall later re-examine this modified definition of universality.

According to the CVC hypothesis, the vector part of J_λ has already been identified with a component of the isospin current; in SU(3) language, it is a component of the F-spin current which is necessarily conserved by virtue of SU(3) invariance. Since J_λ and S_λ are now postulated to be components of the same octet current, the vector part of S_λ will also be proportional to the corresponding F-spin current, i.e. the entire vector part of $\mathscr{J}^h_\lambda$ will be conserved and the vector charge corresponding to S_λ also will not undergo renormalization by strong interactions in the limit of exact SU(3) invariance. Ademollo and Gatto showed that this absence of renormalization is maintained to lowest order in the SU(3) breaking interaction, so there is some meaning to a discussion of universality.

The parameter θ can be determined from a measurement of the rates for the related processes of K^+ and π^+ β decay. According to the CVC hypothesis, the matrix element for $\pi^+ \to \pi^0 e^+ \nu$ is given by

$$\begin{aligned} \langle \pi^0 | J^\dagger_\lambda | \pi^+ \rangle &= \sqrt{2} G^V_\beta (p_+ + p_0)_\lambda \\ &= \sqrt{2} G \cos\theta (p_+ + p_0)_\lambda \end{aligned} \qquad \text{...[86]}$$

according to Cabibbo's theory [85]. Unfortunately, measurements of this very rare mode of π decay are not yet sufficiently accurate to distinguish $\cos\theta$ from unity. Instead we use the measured rate for $O^{14} \to N^{14}$ β decay which should, according to the CVC hypothesis, be determined by a matrix element of exactly the same form as [86]. This yields

$$\cos\theta = 0{\cdot}975 \pm 0{\cdot}003 \qquad ...[87]$$

where allowance has been made for electromagnetic radiative corrections.[36] In the SU(3) symmetric limit the matrix element for $K^+ \to \pi^0 e^+ \nu$ must be identical, apart from a Clebsch-Gordan coefficient, with that for $\pi^+ \to \pi^0 e^+ \nu$. Thus

$$\langle \pi^0 \mid \mathscr{J}_\lambda \mid K^+ \rangle = \langle \pi^0 \mid S_\lambda \mid K^+ \rangle = G \sin\theta (p_+ + p_0)_\lambda / \sqrt{2}. \qquad ...[88]$$

Actually, the energy-release in K_{e3} decay is appreciable, and measurements on K_{e3}^0 decay show that a form factor $f[(p_+ - p_0)^2]$ should be included on the R.H.S. of [*88*]. In principle, the form factor f could be taken from experiment and the measured rate of K_{e3}^+ decay would thus give us $\sin\theta$. In practice, the spectrum is not yet measured with sufficient accuracy to do this with precision, but Oneda and Sucher have pointed out that quite reasonable form factor variations can depress the value of $\sin^2\theta$ by 10% to 30%. With no form factor corrections,[30]

$$\begin{aligned} \theta_{K_{e3}} &= 0{\cdot}246 \pm 0{\cdot}006 \\ \cos\theta_{K_{e3}} &= 0{\cdot}970 \pm 0{\cdot}001 \end{aligned} \qquad ...[89]$$

so that the two determinations of θ, [*87*] and [*89*], are quite compatible. Comparison of $K^+ \to \mu^+ \nu'$ and π decay rates measures the relative strengths of the axial-vector currents. The matrix-element for $\pi^+ \to \mu^+ \nu'$ has the form

$$\langle 0 \mid \mathscr{J}_\lambda^\dagger \mid \pi^+ \rangle = \langle 0 \mid A_\lambda^\dagger \mid \pi^+ \rangle = f_\pi (p_\pi)_\lambda \qquad ...[90]$$

and similarly,

$$\langle 0 \mid \mathscr{J}_\lambda^\dagger \mid K^+ \rangle = f_K (p_K)_\lambda. \qquad ...[91]$$

But

$$\begin{aligned} \langle 0 \mid \mathscr{J}_\lambda^\dagger \mid \pi^+ \rangle &= \langle 0 \mid (j_1^2)_\lambda^A \mid \pi^+ \rangle \cos\theta \\ \langle 0 \mid \mathscr{J}_\lambda^\dagger \mid K^+ \rangle &= \langle 0 \mid (j_1^3)_\lambda^A \mid K^+ \rangle \sin\theta \end{aligned} \qquad ...[92]$$

and in the limit of SU(3) symmetry,

$$\langle 0 \mid (j_1^2)_\lambda \mid \pi^+ \rangle = \langle 0 \mid (j_1^3)_\lambda \mid K^+ \rangle \qquad ...[93]$$

so that

$$f_K / f_\pi = \tan\theta. \qquad ...[94]$$

In the limit of SU(3) symmetry the inclusion or omission of the "kinematic" factors in [*86*], [*88*], [*90*] and [*91*] would not affect the comparison of the strengths of the strangeness-conserving and the strangeness-changing currents. Since, in fact, the actual masses are quite widely separated, the comparison depends crucially on the insertion of these factors. The prescription advocated by Cabibbo cannot be justified in any logical way from the symmetry approach, but does follow naturally from the assumption of a current algebra (§9).

Translated into decay rates,

$$\tan^2\theta = \frac{m_\pi}{m_K}\left[\frac{1-m_\mu^2/m_\pi^2}{1-m_\mu^2/m_K^2}\right]^2 \frac{\Gamma(K^+\to\mu^+\nu')}{\Gamma(\pi^+\to\mu^+\nu')} \qquad ...[95]$$

whence[30]

$$\theta_{K_{\mu 2}} = 0{\cdot}266 \pm 0{\cdot}0015. \qquad ...[96]$$

This appears to differ significantly from the value of θ[29] obtained for the vector currents but in view of the fact that SU(3) symmetry is far from being exact, the nearness of the two values is remarkable. The small difference quite possibly arises from the SU(3) breaking interaction; the Ademollo-Gatto theorem assures that θ_V is unaffected to lowest order and recent indications using the approach of current algebras suggest that θ_A may also be influenced relatively weakly, which perhaps accounts for the fact that the two empirical angles come out to be so close to each other.

Given the angle θ, all baryonic β decay amplitudes

$$\langle Y|\mathscr{J}_\lambda|X\rangle = \langle Y|(j_1^2)_\lambda\cos\theta + (j_1^3)_\lambda \sin\theta\,|X\rangle \qquad ...[97]$$

will be expressible in terms of two independent amplitudes. This can be seen directly by adopting the matrix representation for the baryon octet B_j^i

$$\boldsymbol{B} = \begin{bmatrix} -\Sigma^0/\sqrt{2}-\Lambda^0/\sqrt{6} & \Sigma^+ & p \\ -\Sigma^- & \Sigma^0/\sqrt{2}-\Lambda^0/\sqrt{6} & n \\ -\Xi^- & \Xi^0 & \sqrt{\tfrac{2}{3}}\Lambda^0 \end{bmatrix}. \qquad ...[98]$$

Correspondingly, the antibaryon octet will be represented by the matrix $\bar{B}_j^i$

$$\bar{\boldsymbol{B}} = \begin{bmatrix} -\bar{\Sigma}^0/\sqrt{2}-\bar{\Lambda}^0/\sqrt{6} & -\bar{\Sigma}^- & -\bar{\Xi}^- \\ \bar{\Sigma}^+ & \bar{\Sigma}^0/\sqrt{2}-\bar{\Lambda}^0/\sqrt{6} & \bar{\Xi}^0 \\ \bar{p} & \bar{n} & \sqrt{\tfrac{2}{3}}\bar{\Lambda}^0 \end{bmatrix}. \qquad ...[99]$$

Then, using the matrix notation j_b^a for the octet currents also, we see that there are two distinct ways in which $\boldsymbol{B}$ and $\bar{\boldsymbol{B}}$ can be combined to form octets, corresponding to the two independent invariant products which can be formed from the three matrices $(\boldsymbol{B})$, $(\bar{\boldsymbol{B}})$ and $(\boldsymbol{j})$:

$$\mathrm{Tr}\,(\bar{\boldsymbol{B}}\boldsymbol{j}\boldsymbol{B}) \quad \text{or} \quad \mathrm{Tr}\,(\bar{\boldsymbol{B}}\boldsymbol{B}\boldsymbol{j}) \qquad ...[100]$$

which it is customary to arrange in combinations which are symmetric and antisymmetric in $\boldsymbol{B}$ and $\bar{\boldsymbol{B}}$ respectively, giving the most general SU(3) invariant as

$$D\,\mathrm{Tr}\,(\bar{\boldsymbol{B}}\{\boldsymbol{B},\boldsymbol{j}\}) + F\,\mathrm{Tr}\,(\bar{\boldsymbol{B}}[\boldsymbol{B},\boldsymbol{j}]). \qquad ...[101]$$

The matrix element $\langle B_j^i\,|\,j_b^a\,|\,B_l^k\rangle$ can then be read off from the coefficient of j_a^b in [101]. Thus all baryon leptonic decay amplitudes are determined by the two amplitudes D, F. We only require the matrix elements of j_1^2 and j_1^3. They are

$$(\bar{B}B)_1^2 = -\bar{\Sigma}^+(\Sigma^0/\sqrt{2}+\Lambda^0/\sqrt{6})-(\bar{\Sigma}^0/\sqrt{2}-\bar{\Lambda}^0/\sqrt{6})\Sigma^- - \bar{\Xi}^0\Xi^-$$

$$(B\bar{B})_1^2 = (\bar{\Sigma}^0/\sqrt{2}+\bar{\Lambda}^0/\sqrt{6})\Sigma^- - \bar{\Sigma}^+(\Sigma^0/\sqrt{2}-\Lambda^0/\sqrt{6})+\bar{p}n$$

$$(\bar{B}B)_1^3 = -\bar{p}(\Sigma^0/\sqrt{2}+\Lambda^0/\sqrt{6})-\bar{n}\Sigma^- - \sqrt{\tfrac{2}{3}}\bar{\Lambda}^0\Xi^-$$

$$(B\bar{B})_1^3 = (\bar{\Sigma}^0/\sqrt{2}+\bar{\Lambda}^0/\sqrt{6})\Xi^- + \bar{\Sigma}^+\Xi^0 + \sqrt{\tfrac{2}{3}}\bar{p}\Lambda^0.$$

Thus, for the decays of interest we have the matrix elements

$$\left.\begin{aligned} \langle p \mid j_1^2 \mid n\rangle &= F+D \\ \langle \Lambda \mid j_1^2 \mid \Sigma^-\rangle &= \langle \Sigma^+ \mid j_2^1 \mid \Lambda\rangle = \sqrt{\tfrac{2}{3}}D \\ \langle p \mid j_1^3 \mid \Lambda\rangle &= \sqrt{\tfrac{3}{2}}(F+D/3) \\ \langle n \mid j_1^3 \mid \Sigma^-\rangle &= F-D \\ \langle \Sigma^+ \mid j_1^3 \mid \Xi^0\rangle &= \sqrt{2}\langle \Sigma^0 \mid j_1^3 \mid \Xi^-\rangle = F+D \\ \langle \Lambda \mid j_1^3 \mid \Xi^-\rangle &= \sqrt{\tfrac{3}{2}}(F-D/3). \end{aligned}\right\} \quad \text{...[102]}$$

Thus all leptonic decay amplitudes are given in terms of the two amplitudes F and D. The general form of the baryonic matrix element contains six terms

$$\langle Y \mid \mathscr{J}_\lambda \mid X\rangle = \bar{u}_Y(p')\{\gamma_\lambda g_1+(p+p')_\lambda g_2+i(p-p')_\lambda g_3$$
$$+i\gamma_5[\gamma_\lambda h_1+(p-p')_\lambda h_2+i(p+p')_\lambda h_3]\}u_X(p) \quad \text{...[103]}$$

where $\bar{u}_Y(p')$ and $u_X(p)$ are free-particle spinors. For a first analysis, we may neglect the induced couplings, which go to zero for vanishing momentum transfer, and retain only the terms in γ_λ and $\gamma_\lambda\gamma_5$. Since the vector current is taken to be conserved, the vector charge should contain F-components only. Thus the relative values of vector matrix elements of S_λ are determined by Clebsch-Gordan coefficients only, just as the matrix elements of V_λ were related to each other by the CVC hypothesis. The relative strength of the strangeness-changing and the strangeness-conserving matrix elements depends on θ. The axial-vector current may contain both F and D contributions; not being conserved, it can develop a D component through the action of strong interactions even if the bare current were pure F. Then

$$\left.\begin{aligned} \langle Y \mid F_\alpha \mid X\rangle &= \frac{G\bar{u}_Y}{\sqrt{2}}(\gamma_\alpha+\gamma_\alpha\gamma_5 f_A)u_X \\ \langle Y \mid D_\alpha \mid X\rangle &= \frac{G\bar{u}_Y}{\sqrt{2}}\gamma_\alpha\gamma_5\, d_A\, u_X \end{aligned}\right\} \quad \text{...[104]}$$

From [102],

$$\langle p \mid (j_1^2)^A \mid n\rangle = d_A+f_A = 1{\cdot}18\pm0{\cdot}02 \quad \text{...[105]}$$

while from measurements on the β decay rate of the Λ,

$$\sqrt{\tfrac{2}{3}}\langle p\,|\,(j_1^3)^A\,|\,\Lambda\rangle = (f_A+d_A/3) = 0{\cdot}68\pm0{\cdot}07 \qquad ...[106]$$

the sign being obtained from the asymmetry of β emission in polarized Λ decay. From these, one can deduce

$$d_A = 0{\cdot}75\pm0{\cdot}10,\quad f_A = 0{\cdot}43\mp0{\cdot}08. \qquad ...[107]$$

A least squares fit to the β decays, $\Lambda \to p$, $\Sigma^- \to n$, $\Sigma^+ \to \Lambda$ and $\Xi^- \to \Lambda$, in terms of the parameters θ, d_A and f_A, by Willis gives

$$\left.\begin{aligned} \theta_B &= 0{\cdot}27\pm0{\cdot}02 \\ d_A/f_A &= 1{\cdot}7\pm0{\cdot}3 \end{aligned}\right\} \qquad ...[108]$$

in close agreement with the parameters we have adopted. The experimental results are shown in Fig. 2.

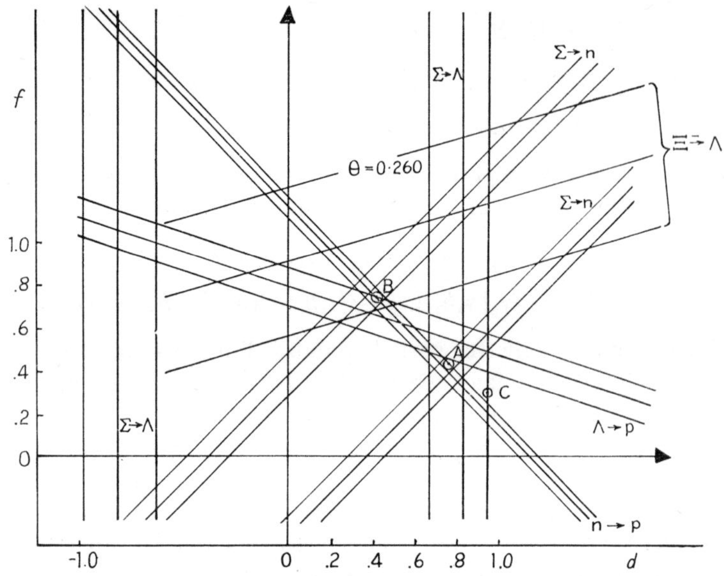

FIG. 2. Linear relations between f and d determined by the axial matrix elements for baryon β decays.

Where only the total decay rate is known, there are two lines corresponding to the two possible signs for the axial matrix element. Cabibbo's theory, which postulates that all amplitudes be determined by a single choice of f and d, requires that all lines intersect in a point. C represents Cabibbo's original choice of parameters, A indicates Willis' best fit.

Once the parameters are known, Cabibbo's theory predicts the matrix elements for all hyperon decays. Thus, for example, for $\Sigma^- \to n$, the g_A/g_V ratio is predicted according to [102] and [107] to be $(d-f) = +0{\cdot}32$ in contrast with the negative values found in the β decays of the neutron and the Λ. A measurement of this ratio is therefore of great interest for Cabibbo's theory. The predicted rate of the β decay $\Xi^- \rightarrow \Lambda$ corresponds to a branching ratio of 5.10^{-4} while the reported value (based on three events, including one doubtful one) is a few times 10^{-3}. A more reliable check on this number would also be interesting inasmuch as it depends only on the assumption that S_λ transforms as an octet, and is therefore independent of the value of θ.

An analysis including some of the additional terms in [*103*] by Brene *et al.*[30a] does not significantly alter the results [*108*]. For the electronic modes, the coefficients g_3 and h_2 in [*103*] may safely be neglected as they contribute terms proportional to m_e; g_3 and h_3 are analogous to Weinberg's second-class currents and vanish in the SU(3) symmetric limit if we demand that all terms in $\mathscr{J}_\lambda$ transform in the same way under CP. Conversely, we can have mixed CP transformation properties for weak currents belonging to a single octet if we admit the presence of second-class terms.

The rather imprecise information that we have about the d/f ratio for the SU(3) invariant strong coupling of the pseudoscalar meson octet, chiefly from a study of the dynamics of the decuplet resonances, yields a range of values which includes the d/f ratio [*108*] found for the weak axial-vector currents. This suggests that the PCAC assumption [*80*] may be generalized for all components of the axial-vector current in the form

$$\partial_\lambda A^i_\lambda = ifm_i^2\phi^i. \qquad ...[109]$$

This is consistent with our assumption relating the $K \rightarrow \mu\nu'$ and $\pi \rightarrow \mu\nu'$ decay amplitudes, and of course goes far beyond it. It is not yet possible to make a severe test of this hypothesis; a test of the generalized Goldberger-Treiman relations implied by [*109*] awaits a determination of both strong and weak coupling constants.

8

Non-Leptonic Decays

According to the hypothesis of current-current interaction, the non-leptonic weak interactions arising from the coupling of two octet currents can only contain $|\Delta Y| = 0, 1$ terms. This should not be surprising since the octet hypothesis for the weak current incorporates the conditions which we have found to be necessary to avoid $|\Delta Y| = 2$. The isospin change for the $|\Delta Y| = 1$ interactions is limited to $|\Delta I| = \frac{1}{2}, \frac{3}{2}$, i.e. there must be no greater isospin contributions. The experimental data suggests that not only is this condition very well satisfied, but almost all results may be accounted for by a $|\Delta I| = \frac{1}{2}$ interaction, with only a small $|\Delta I| = \frac{3}{2}$ admixture occasionally required. There is a considerable body of evidence indicating the dominance of $|\Delta I| = \frac{1}{2}$ amplitudes.

The $|\Delta I| = \frac{1}{2}$ rule for non-leptonic decays was originally proposed by Gell-Mann and Pais to account for the fact that $K^+ \rightarrow 2\pi$ decay proceeds about 500 times more slowly than $K^0 \to 2\pi$. $|\Delta I| = \frac{1}{2}$ forbids $K^+ \to 2\pi$, which must then occur through a violation of the rule. A check on the $|\Delta I| = \frac{1}{2}$ hypothesis is provided by the fact that the 2π state from K^0 decay is then required to be purely $I = 0$, in which all charge projections are equally likely, therefore

$$\frac{\Gamma(K_1^0 \to \pi^+\pi^-)}{\Gamma(K_1^0 \to \pi^0\pi^0)} = 2. \qquad ...[110]$$

The most precise measurements of the charged/neutral branching ratio agree very well with this prediction.

In $\Lambda^0 \to N\pi$ decay, the final state is required to be a pure $I = \frac{1}{2}$ state. Thus the $p\pi^-$ and $n\pi^0$ channels should have identical asymmetry parameters and the absolute rates should be in the ratio 2:1. The asymmetry parameters are indeed found to have the same sign and to agree in magnitude within an accuracy of about 25%. From measurements of polarization and hyperfragment branching ratios, one also knows that in both channels, it is the s wave amplitude which dominates. The latest compilation of results on the branching ratio[37] gives

$$\frac{\Gamma(\Lambda \to p\pi^-)}{\Gamma(\Lambda \to p\pi^-)+\Gamma(\Lambda \to n\pi^0)} = 0{\cdot}675 \pm 0{\cdot}027 \qquad ...[111]$$

in excellent agreement with the theoretical prediction. Thus the Λ decays provide strong support for $|\Delta I| = \frac{1}{2}$ dominance. There are similar predictions for $\Xi \to \Lambda\pi$ decay, where the data is less precise. The asymmetry parameters for $\Xi^- \to \Lambda\pi^-$ and $\Xi^0 \to \Lambda\pi^0$ are found to be in the ratio $1{\cdot}23 \pm 0{\cdot}58$, while the decay rates are reported to be in the ratio $\Gamma_{\Xi^0}/\Gamma_{\Xi^-} = 1{\cdot}66 \pm 0{\cdot}23$ to be compared with the theoretical prediction of 2. For Σ decays, the $|\Delta I| = \frac{1}{2}$ rule allows both $I = \frac{1}{2}$ and $I = \frac{3}{2}$ final states. Since there are three channels, we obtain one relation between the amplitudes

$$\sqrt{2}\mathcal{M}(\Sigma^+ \to p\pi^0) = \mathcal{M}(\Sigma^- \to n\pi^-) - \mathcal{M}(\Sigma^+ \to n\pi^+)$$

which we write in the shorthand notation

$$\sqrt{2}\,\Sigma_0^+ = \Sigma_-^- - \Sigma_+^+. \qquad ...[112]$$

Since this must hold for both s and p wave amplitudes, which time-reversal invariance requires to be relatively real, apart from small phase shifts due to final state interaction, [112] is most conveniently represented as a vector relation in the s-p amplitude space. The absolute rates of the Σ_0^+, Σ_-^- and Σ_0^+ modes are almost exactly equal, within experimental error, so the three vectors in [112] must form a right triangle if the $|\Delta I| = \frac{1}{2}$ relation is to be satisfied. Σ_0^+ is known to have a large asymmetry parameter and therefore comparable s and p wave amplitudes. [112] then requires that one of the amplitudes Σ_+^+ or Σ_-^- contain very little s wave and the other very little p wave. It is known that the asymmetry parameters for both these modes are small, but the dominant amplitude has not been identified in either case.‡ The experimental results are shown in Fig. 3, including only those of the multiple solutions which come close to satisfying [*112*].

Measurements of $K \to 3\pi$ decays are consistent with the $|\Delta I| = \frac{1}{2}$ rule, so we see that all the evidence points to the dominant role of $|\Delta I| = \frac{1}{2}$ amplitudes in the non-leptonic decays.

There are two possible attitudes to the $|\Delta I| = \frac{1}{2}$ rule. One is to view it as an exact property to be satisfied by the non-leptonic weak interaction. It then follows that the weak interaction cannot be simply of the form [*61*], for the product of two charged currents can never yield an exact $|\Delta I| = \frac{1}{2}$ rule. One way of achieving this, suggested by Lee and Yang, is to add to [*61*] a corresponding product of neutral currents, whose $I = \frac{3}{2}$ component exactly cancels the $I = \frac{3}{2}$ part of $J_\lambda^\dagger S_\lambda$. Since any coupling of neutral lepton currents is very much weaker than that of charged leptonic currents, this theory is less universal than one with purely charged currents, and it predicts no more than an exact $|\Delta I| = \frac{1}{2}$ rule for non-leptonic decays. (Should intermediate

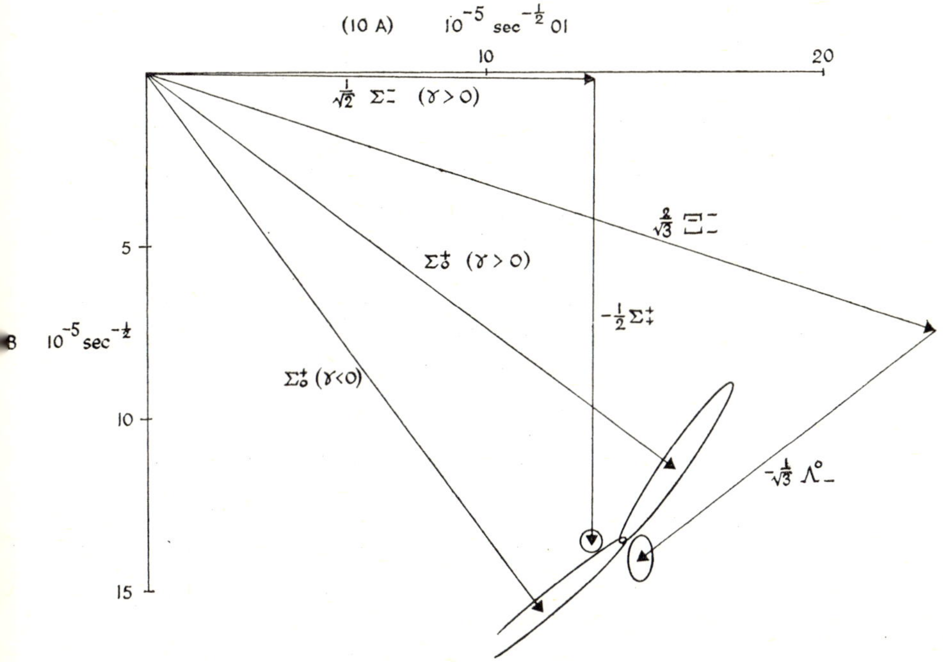

FIG. 3. Comparison of $\frac{1}{\sqrt{2}}(\Sigma_-^- - \Sigma_+^+)$ and $\frac{1}{\sqrt{3}}(2\Xi_-^- - \Lambda_-^0)$ with Σ_0^+, taken from the report by N. Cabibbo at the Berkeley Conference on High-Energy Physics, 1966.

‡ *Note added in proof.* The dominant amplitude for Σ_+^+ has recently been identified as *p* wave (D. Bevley *et al.* 1966. *Phys. Rev. Lett.* **17**, 1071).

bosons $W^{\pm}$ exist, this theory would require the existence of neutral counterparts.)

An alternative view is that the $|\Delta I| = \frac{1}{2}$ rule arises as the result of a "dynamical enhancement" of the $|\Delta I| = \frac{1}{2}$ amplitudes so that $|\Delta I| = \frac{3}{2}$ amplitudes, although present, are overwhelmed by much larger $|\Delta I| = \frac{1}{2}$ terms, whenever these can contribute. Neither of the two approaches leads to any predictions about non-leptonic weak interactions which go beyond those which follow from the $|\Delta I| = \frac{1}{2}$ rule.

The natural extension of the $|\Delta I| = \frac{1}{2}$ rule to SU(3) is to stipulate that the non-leptonic weak interaction also transforms as an octet. Since the product of two identical octets contains components of the **[1]**, $[\mathbf{8}]_s$, and **[27]** representations, this requires that the part of the interaction transforming like a component of **[27]** be negligible compared to the octet part. Note, however, that while the octet hypothesis assures the $|\Delta I| = \frac{1}{2}$ rule, the $|\Delta I| = \frac{1}{2}$ rule does not necessarily require octet dominance.

Unfortunately, even the assumption of octet dominance does not lead to any new verifiable relations beyond those imposed by the $|\Delta I| = \frac{1}{2}$ rule. The number of independent amplitudes, in terms of which the non-leptonic decays must be expressed, is eight, corresponding to the eight independent invariants which can be formed by multiplying the matrices $\boldsymbol{B}$, $\boldsymbol{\bar{B}}$, $\boldsymbol{M}$, and $\boldsymbol{S}$. $\boldsymbol{B}$, $\boldsymbol{\bar{B}}$ represent the initial and final baryon states, $\boldsymbol{M}$ the emitted pion and $\boldsymbol{S}$ an octet "Spurion".

Thus the general form prescribed by octet dominance for the hyperon non-leptonic decay amplitude contains more terms than the number of observable hyperon decay modes! Of course, the $|\Delta I| = \frac{1}{2}$ relations are still obtained.

An additional relation can be obtained by imposing the additional condition of CP invariance. The part of the octet amplitude which is odd under charge-conjugation contains only three independent terms

$$\mathcal{M}_1 \operatorname{Tr}(\boldsymbol{\bar{B}}[\boldsymbol{S},\boldsymbol{M}]\boldsymbol{B}) + \mathcal{M}_2 \operatorname{Tr}(\boldsymbol{\bar{B}}\boldsymbol{B}[\boldsymbol{S},\boldsymbol{M}]) + \mathcal{M}_3[\operatorname{Tr}(\boldsymbol{\bar{B}}\boldsymbol{M})\operatorname{Tr}(\boldsymbol{B}\boldsymbol{S}) - \operatorname{Tr}(\boldsymbol{\bar{B}}\boldsymbol{S})\operatorname{Tr}(\boldsymbol{B}\boldsymbol{M})] \qquad \text{...[113]}$$

where CP invariance restricts the matrix $\boldsymbol{S}$ to be a multiple of λ_6, in the usual (Gell-Mann) notation.[38] Since, according to the $|\Delta I| = \frac{1}{2}$ rule, there are four independent amplitudes governing the processes $\Lambda \to N\pi$, $\Sigma \to N\pi$ and $\Xi \to \Lambda\pi$, we obtain one additional condition relating the parity-violating (s wave) parts of these amplitudes. Even this relation cannot be compared with experiment in an unambiguous manner, because of the approximate nature of SU(3) symmetry. The forms

$$\bar{u}_Y(A + iB\gamma_5)u_X \quad \text{and} \quad (p_X - p_Y)_\lambda \bar{u}_Y \gamma^\lambda (a + ib\gamma_5) u_X$$

are just two different ways of parametrizing the decay amplitude, since they may be transformed into each other by the use of the Dirac equation. However, the statement that the A parameters for different decays obey a certain linear relation is *not* equivalent to having the a's satisfying the same relation. Choosing A as the amplitude to which the additional condition for a C-odd octet interaction is to be applied, one finds

$$2A(\Xi^- \to \Lambda\pi^-) = A(\Lambda \to p\pi^-)+\sqrt{3}A(\Sigma^+ \to p\pi^0) \qquad \ldots[114]$$

a relation first obtained by B. W. Lee and H. Sugawara on the basis of more specific and less justifiable assumptions.

No additional relations are obtained for the parity-conserving (p wave) amplitudes but the Lee-Sugawara relation appears to work as well for the B (p wave) amplitudes as for the A's. (See Fig. 3.) Several theories give the Lee-Sugawara relation, but they cannot avoid making additional predictions which disagree badly with experiment. Perhaps, at this point, the symmetry approach is nearing the limit of its usefulness.

9

Algebra of Vector and Axial-Vector Charges

In the preceding discussions the symmetry operations which were used to classify hadron states and the operations which induce transitions between these states were treated as exact symmetries of the strong interactions. As a matter of fact, however, we know that some of these symmetries, especially the SU(3) invariance of the strong interactions, are only very approximate and therefore such a characterization provides at best a very rough description of the actual phenomena. A particularly interesting and fruitful development over the last year, which promises to remedy this defect and for the first time to give a precise meaning to the notion of approximate symmetry, has been the application of the idea that the weak currents may generate an algebra. Since this is dealt with in detail by Professor Radicati (see accompanying Lecture Notes), we shall not attempt to trace out the full extent to which these ideas have already been developed but rather indicate the principle of the method, and how it may lead to results beyond those accessible to the symmetry approach. In particular, we shall limit ourselves to algebraic relations between integrated fourth-components of currents only, which are subject to less theoretical uncertainty than other more detailed commutation rules which have been used.

We have seen that according to the Conserved Vector Current hypothesis,

$$V_\lambda = V_\lambda^1+iV_\lambda^2, \quad V_\lambda^3, \quad V_\lambda^\dagger = V_\lambda^1-iV_\lambda^2$$

transform as the $I^3 = +1, 0$, and -1 components of an isovector. Then the "charges" (measured in suitable units)

$$F^i(t) = \int V_0^i(\mathbf{x}, t)d^3x \qquad \ldots[115]$$

obey the equal-time commutation relations

$$[F^i(t), F^j(t)] = i\varepsilon_{ijk}F^k(t) \qquad \ldots[116]$$

and by virtue of the fact that $F^i \propto I^i$ are generators of symmetry transformations of the strong interactions, $[F^i, H] = 0$, it follows that the F^i are constants of motion and are conserved, apart from the influence of terms in H which are not invariant under the action of the operators F^i, which in this case would be those arising from electromagnetic interaction. Several years ago, Gell-Mann[39] put forward the very interesting suggestion that relations of the form [*116*] may continue to hold exactly, where the F's are defined by means of [*115*], even when the F^i are not symmetry operators and thus not constant in time, and demonstrated through model examples how this could be the case. In this way, one may be able to define theories of approximate symmetry in a perfectly definite manner.

Fubini and Furlan[40] have shown how the assumption of an algebra can be used to evaluate the effects of symmetry-breaking interactions. The method can be illustrated by considering a specific example. We know from the identification of V_λ with the isospin current that the Fermi constant G_β^V cannot be renormalized by strong (isospin invariant) interactions. But there will be a correction due to electromagnetic violation of isospin invariance, which Fubini and Furlan evaluated in the following manner. From [*116*]

$$[F^+, F^-] = 2I^3. \qquad ...[117]$$

Take the matrix element of [*117*] between physical proton states

$$\langle P(\mathbf{p}') \mid [F^+, F^-] \mid P(\mathbf{p})\rangle = (2\pi)^3 \delta^3(\mathbf{p}-\mathbf{p}'). \qquad ...[118]$$

To evaluate the L.H.S., we insert a complete set of intermediate states and isolate the contribution of the physical neutron state:

$$(2\pi)^3 \delta^3(\mathbf{p}-\mathbf{p}') = (2\pi)^3 r_\beta^2 \delta^3(\mathbf{p}-\mathbf{p}')$$
$$+ \sum_{\alpha \neq N} \langle P \mid F^+ \mid \alpha\rangle\langle\alpha \mid F^- \mid P\rangle - \sum_\beta \langle P \mid F^- \mid \beta\rangle\langle\beta \mid F^+ \mid P\rangle \qquad ...[119]$$

where $r_\beta G_0^V$ is the physically measured (renormalized) coupling constant. Now we recall that the only time-variation of $F^\pm$ arises from the electromagnetic interaction

$$i\dot{F}^\pm(t) = [F^\pm, H] = [F^\pm, H^{\text{em}}] \qquad ...[120]$$

where

$$H^{\text{em}} = \int j_\lambda^{\text{em}} A_\lambda d^3x.$$

If we make use of the fact that the electromagnetic current is the sum of an isoscalar and an isovector part which is just eV_λ^3

$$[F^\pm, H] = \mp e \int V_\lambda^\pm A_\lambda d^3x = \mp H^\pm. \qquad ...[121]$$

Thus

$$\langle a \mid F^\pm \mid b\rangle = \pm e \frac{\langle a \mid H^\pm \mid b\rangle}{E_a - E_b - i\varepsilon} \qquad ...[122]$$

where the $i\varepsilon$ has been inserted to define the R.H.S. in case $E_a = E_b$. Then, from [*119*],

$$\frac{1}{r_\beta^2} = 1 + e^2(A - B) \qquad \ldots[123]$$

where

$$(2\pi)^3 \delta^3(\boldsymbol{p} - \boldsymbol{p}')A = \frac{1}{r_\beta^2} \sum_{\alpha \neq N} \frac{\langle P \mid H^+ \mid \alpha \rangle \langle \alpha \mid H^- \mid P \rangle}{(E_p - E_\alpha - i\varepsilon)(E_p - E_\alpha + i\varepsilon)} \qquad \ldots[124]$$

and a similar expression for B contains the contribution of the last term in [*119*]. We see that the correction is of order e^2, i.e. of second order in the symmetry-violating interaction. To evaluate r_β to this order, we may set $r_\beta = 1$ and identify the currents $V_\lambda^\pm$ appearing in $H^\pm$ with conserved currents in evaluating the correction terms A and B. The matrix elements of $H^\pm$ are just the form factors for the weak vector current V_λ, and by an isospin transformation they may be related to isovector electromagnetic form factors for the nucleon. Using experimental information on either set of quantities, the electromagnetic renormalization of G_β^V could be explicitly calculated. The relevant data are not yet available so a numerical evaluation has not been made, but in principle the electromagnetic correction, of order α, to G_β^V is now known.

The algebra [*116*] may be extended further. The assumption that the axial-vector currents A_λ, $A_\lambda^\dagger$ transform as an isovector means that the "axial charges"

$$G^i(t) = \int A_0^i(\boldsymbol{x}, t)d^3x \qquad \ldots[125]$$

also obey the commutation rules

$$[F^i, G^j] = i\varepsilon_{ijk} G^k \qquad \ldots[126]$$

in a theory in which the strong interactions are exactly isospin invariant. In the same spirit as before, [*126*] will be assumed to hold exactly as far as strong and electromagnetic interactions are concerned, even when the F^i are not constant in time. The algebra defined by [*116*] and [*126*] will be closed under commutation if the G^j satisfy

$$[G^i, G^j] = i\varepsilon_{ijk} F^k \qquad \ldots[127]$$

and models have been constructed in which this is the case. In particular [*116*], [*126*] and [*127*] hold true if the weak currents $V_\lambda^\pm$, $A_\lambda^\pm$ have the simple forms $\bar{\psi}\gamma_\lambda\tau^\pm\psi$ and $i\bar{\psi}\gamma_\lambda\gamma_5\tau^\pm\psi$ respectively, with ψ the bare field-operator of an isospinor spin-$\frac{1}{2}$ particle. If we define

$$K^i = \tfrac{1}{2}(F^i + G^i), \; R^i = \tfrac{1}{2}(F^i - G^i) \qquad \ldots[128]$$

we have

$$[K^i, K^j] = i\varepsilon_{ijk} K^k$$
$$[R^i, R^j] = i\varepsilon_{ijk} R^k \qquad \text{...[129]}$$
$$[K^i, R^j] = 0$$

so we are in fact dealing with two independent groups of transformations, representing a six-parameter SU(2) ⊗ SU(2) algebra.

The obvious generalization of [*116*] and [*126*] and [*127*] to include strangeness changing currents is to adopt Cabibbo's octet hypothesis for F^i and G^i, so that we have commutation relations identical in form to [*116*] and [*126*] with ε_{ijk} replaced by the SU(3) structure constants f_{ijk}. Again, to close the algebra, one needs to postulate the equivalent of [*127*]. Then we can again introduce the "left-handed" and "right-handed" charges K^i and R^i and recover commutation relations analogous to [*129*] for two decoupled F-spins, so we now have a SU(3) ⊗ SU(3) algebra in place of the earlier SU(2) ⊗ SU(2) structure. Note, however, that we can return to the SU(2) ⊗ SU(2) form by making use of the fact that Cabibbo's current

$$\mathscr{J}^h_\lambda = (j^1_\lambda + ij^2_\lambda)\cos\theta + (j^4_\lambda + ij^5_\lambda)\sin\theta \qquad \text{...[130]}$$

can be obtained by a rotation of 2θ in the U-spin space

$$\mathscr{J}^h_\lambda = e^{-2i\theta F^7}\mathscr{J}'_\lambda e^{2i\theta F^7} \qquad \text{...[131]}$$

where

$$\mathscr{J}'_\lambda = j^1_\lambda + ij^2_\lambda \qquad \text{...[132]}$$

so that in a suitable basis system we are back to the SU(2) ⊗ SU(2) algebra.

As pointed out by Gell-Mann, the algebra [*129*] is closely related with the algebra of the weak charges of the leptons. We note that the terms appearing in the leptonic current

$$L_\lambda = 2\bar{\nu}\gamma_\lambda\left(\frac{1+\gamma_5}{2}\right)e + 2\bar{\nu}'\gamma_\lambda\left(\frac{1+\gamma_5}{2}\right)\mu \qquad \text{...[133]}$$

correspond to a leptonic "weak charge" which may be thought of formally as $2(S^1+iS^2)$ where S^1 and S^2 are components of an angular momentum operator in a space in which $\left(\frac{1+\gamma_5}{2}\nu, \frac{1+\gamma_5}{2}e\right)$ form the upper and lower components of a spin-$\frac{1}{2}$ spinor and $\left(\frac{1+\gamma_5}{2}\nu', \frac{1+\gamma_5}{2}\mu\right)$ form another spinor. The electric charge is then S^0+S^3, where S^0 commutes with S^1, S^2, S^3. Thus the weak and electric charges of the leptons generate the algebra of U(1) ⊗ SU(2). Gell-Mann has advanced the view that it is this U(1) ⊗ SU(2) algebra, which is the basic algebra shared by leptons and hadrons, which allows us to define the universality of weak and electromagnetic interactions. This provides the justification for Cabibbo's assumption that it is the current

[*131*] that should have the same coupling strength as the leptonic current. We note also that the definition of universality in this manner requires that the hadronic weak current $\mathscr{J}_\lambda^h$ be reducible to the SU(2) form by a simple transformation of the form [*131*], i.e. one may not introduce different Cabibbo angles for the vector and axial-vector currents. It is only when the weak charges so defined are decomposed, in the case of the hadrons according to the quantum numbers respected by the strong interactions, that the group must be enlarged to SU(3) ⊗ SU(3).

Of course, the validity of the assumed commutation relations can only be tested by comparison with experiment, and there is no *a priori* assurance that the algebra adopted is necessarily the correct one. It may be necessary, assuming that there is an algebra at all, to go to a wider set of commutation relations, incorporating a larger number of operators, before one obtains a group which is closed under commutation. Nevertheless, the SU(3) ⊗ SU(3) algebra is particularly attractive in that it is uniquely selected if the hadronic weak current contains only octet components and if its algebraic properties are the same as those of the weak current of the known leptons, in agreement with the principle of universality.

A significant application, which has given strong support to the current algebra approach, has been the calculation of the axial-vector constant for neutron β-decay. Using one of the commutation relations [*127*],

$$[G^+(t), G^-(t)] = 2I^3 \qquad \ldots[134]$$

Adler and Weisberger[41] were able to fulfil the prediction of Gell-Mann that [*134*] could be used to calculate the axial-vector coupling constant, whose difference from the vector constant must be regarded as a renormalization effect in this theory. Evaluating [*134*] between physical proton states, we have

$$(2\pi)^3\delta^3(\boldsymbol{p}-\boldsymbol{p}') = (2\pi)^3\delta^3(\boldsymbol{p}-\boldsymbol{p}')r_A^2\left(1-\frac{M^2}{E_p^2}\right)+$$
$$\sum_{\alpha \neq N}\langle P(\boldsymbol{p}') \mid G^+ \mid \alpha\rangle\langle\alpha \mid G^- \mid P(\boldsymbol{p})\rangle+\sum_{\beta}\langle P(\boldsymbol{p}') \mid G^- \mid \beta\rangle\langle\beta \mid G^+ \mid P(\boldsymbol{p})\rangle. \qquad \ldots[135]$$

By manipulations similar to those leading to [*108*], we obtain

$$\langle\alpha \mid G^\pm \mid P(\boldsymbol{p})\rangle = \langle\alpha \mid \frac{\int(\partial_\lambda A^{\lambda\pm})d^3x}{E_p-E_\alpha} \mid P(\boldsymbol{p})\rangle$$

where we can now dispense with $i\varepsilon$ since all states α will lie at energies above E_p. Further, we replace the sum over states by an integral over momenta,

$$\sum_{\alpha \neq N} \rightarrow \int\frac{d^3q_\alpha}{(2\pi)^3}\int_{M+m_\pi} d\omega\sum_{\alpha}\delta(\omega-M_\alpha) \qquad \ldots[136]$$

where ω is the total c.m. energy, M the mass of the nucleon and m_π the mass of the pion. The x-integrals give momentum δ-functions and the q_α-integration reduces these to an overall momentum δ-function. Thus [*135*] becomes

$$(2\pi)^3 \delta(\boldsymbol{p}-\boldsymbol{p}') = (2\pi)^3 \delta^3(\boldsymbol{p}-\boldsymbol{p}') r_A^2 \left(1-\frac{M^2}{E_p^2}\right) + (2\pi)^3 \delta^3(\boldsymbol{p}-\boldsymbol{p}')[C^- - C^+] \quad \ldots[137]$$

with

$$C^\pm = \int_{M+m_\pi}^{\infty} d\omega \sum_{\alpha \neq N} \delta(\omega - M_\alpha) \frac{MM_\alpha}{E_p E_\alpha} \frac{|F_\alpha^\pm|^2}{(E_\alpha - E_p)^2}, \quad \ldots[138]$$

or

$$1 = r_A^2 \left(1-\frac{M^2}{E_p^2}\right) + [C^- - C^+] \quad \ldots[139]$$

where now $E_p = \sqrt{M^2+p^2}$, and $E_\alpha = \sqrt{M_\alpha^2+p^2}$, and we have introduced the Lorentz-invariant matrix-elements $F^\pm$ through

$$\langle \alpha(p) \mid \partial_\lambda A^{\lambda\pm} \mid P(p) \rangle = \left(\frac{MM_\alpha}{E_p E_\alpha}\right)^{\frac{1}{2}} F_\lambda^\pm(\omega, q^2) \quad \ldots[140]$$

where

$$q^2 = (\boldsymbol{p}-\boldsymbol{p}_\alpha)^2 = (E_p - E_\alpha)^2 = (\sqrt{M^2+p^2} - \sqrt{M_\alpha^2+p^2})^2. \quad \ldots[141]$$

$F_\alpha^\pm$ will in general also depend on the helicity of the proton state $\mid P(p)\rangle$ but since we shall only be concerned with averages over spin orientations, we shall not display this dependence explicitly, and regard $C^\pm$ as representing the corresponding spin-average quantities.

We see that, in the sum rule [*139*], the relative contribution of the two terms on the R.H.S. depends on the value of p. This should not be surprising since our starting-point was the commutation relation [*134*] which is defined in a given Lorentz frame. The fact that we get an infinity of sum-rules [*139*] dependent on the parameter p only reflects the fact that we start from an infinite family of commutation relations obtained by writing down the relations [*134*] in different Lorentz frames. In the present instance, for example, it is clear that we will get no information whatever concerning r_A if we take $p = 0$. Many previous attempts to relate the value of r_A to the "approximate conservation" of the axial-vector current came to grief at this point. On the other hand, for $p \to \infty$, the first term becomes simply r_A^2 and, as emphasized by Fubini and Furlan, the remaining terms also simplify appreciably. From [*141*], $q^2 \to 0$ as $p \to \infty$, and since the p-dependent factor

$$\frac{1}{E_p E_\alpha} \frac{1}{(E_p - E_\alpha)^2} = \frac{1}{(M_\alpha^2 - M^2)^2} \cdot \frac{(E_p + E_\alpha)}{E_p E_\alpha} \quad \ldots[142]$$

tends to the p-independent limit $4/(M^2-M_\alpha^2)^2$ as $p \to \infty$, we see that $C^\pm$ becomes explicitly p-independent in this limit, as they must since the first term r_A^2 has become independent of p. In this limit, the terms simplify to

$$C^\pm = \int_{M+m_\pi}^{\infty} d\omega \, \frac{4M\omega}{(\omega^2-M^2)^2} \sum_{\alpha \mp N} \delta(\omega - M_\alpha) \, | F_\alpha^\pm(\omega, 0) |^2 \qquad \ldots[143]$$

Now, $F_\alpha^\pm$, which is the matrix-element of the divergence of the axial-vector current, completely determines the cross-section for neutrino reactions for vanishing momentum transfer to the hadrons (which occurs for forward lepton production if we neglect the lepton mass). Thus $C^\pm$ can be evaluated in terms of neutrino cross-sections for this particular configuration of the final lepton. Inserting the appropriate kinematical factors, we obtain Adler's sum rule for r_A in terms of neutrino cross-sections for forward lepton production

$$1 = r_A^2 + \int_{M+m_\pi}^{\infty} \frac{\omega d\omega}{ME_l^2} \frac{1}{G_\beta^{V2}} \left\{ \frac{d^2\sigma_{inel}}{d\Omega_l dE_l} [\bar{\nu}+p \to l^+ + (S=0)] - \frac{d^2\sigma}{d\Omega_l dE_l} [\nu+p \to l^- + (S=0)] \right\} \qquad \ldots[144]$$

where E_l, Ω_l are the lepton energy and solid angle in the laboratory frame. By similar manipulations on the commutator of two strangeness-changing currents, Adler finds the further sum-rules

$$4 = \int \frac{\omega d\omega}{ME_l^2} \cdot \frac{1}{(G \sin\theta)^2} \left\{ \frac{d^2\sigma}{d\Omega_l dE_l} [\bar{\nu}+p \to l^+ + (S=-1)] - \frac{d^2\sigma}{d\Omega_l dE_l} [\nu+p \to l^- + (S=+1)] \right\} \qquad \ldots[145]$$

$$= 2 \int \frac{\omega d\omega}{ME_l^2} \frac{1}{(G \sin\theta)^2} \left\{ \frac{d^2\sigma}{d\Omega_l dE_l} [\bar{\nu}+n \to l^+ + (S=-1)] - \frac{d^2\sigma}{d\Omega_l dE_l} [\nu+n \to l^- + (S=+1)] \right\}. \qquad \ldots[146]$$

Thus, in principle, the algebra assumptions [*116*], [*126*] and [*127*] can be subjected to direct experimental test through the sum-rules [*144*]-[*146*]. Unfortunately, neutrino experiments have not yet progressed to the point where such a test is possible.

Adler and Weisberger therefore devised a more accessible method for evaluating r_A, subject however to an additional assumption. This is to use the PCAC hypothesis in the form

$$\partial_\lambda A^{\lambda\pm}(\mathbf{x}) = if_\pi m_\pi^2 \phi_\pi^\pm(\mathbf{x}). \qquad \ldots[147]$$

Then the matrix elements $F_\alpha^\pm$ appearing in $C^\pm$ [*143*] may be replaced by the matrix elements of the pion field-operator, in which case the correction terms $C^\pm$ can be re-expressed in terms of integrals over total pion cross-sections for a *massless* pion:

$$C^\pm = \frac{4M^2}{\pi g_{\pi N,0}^2} \int \frac{\omega d\omega}{\omega^2 - M^2} \sigma_0^\pm(\omega) \qquad \ldots[148]$$

where the 0 suffixes denote values extrapolated to zero pion mass. Thus, we arrive at the Adler-Weisberger relation for the axial-vector renormalization constant,

$$|r_A| = \left\{1 - \frac{4M^2}{\pi g_{\pi N,0}^2} \int \frac{\omega d\omega}{\omega^2 - M^2} [\sigma_0^+(\omega) - \sigma_0^-(\omega)]\right\}^{-\frac{1}{2}}. \qquad \ldots[149]$$

Inserting the experimental values of the pion cross-sections, Weisberger obtained the value

$$|r_A| = 1{\cdot}16 \qquad \ldots[150]$$

while Adler tried to correct for the extrapolation in pion mass and arrived at the result:

$$|r_A| = 1{\cdot}24. \qquad \ldots[151]$$

Considering the nature of the approximation involved in the PCAC hypothesis, both results should be considered to be in excellent agreement with experiment. Of course, the PCAC assumption could itself be tested by a comparison of [*149*] and [*144*].

As you will see, the current algebra assumption, aided by PCAC, leads to many other promising relations. Although these are usually also of approximate validity, the approximate nature of the results is a consequence of the approximations made in the calculations and not one of principle. This appears to be a more attractive avenue of research by far than the pursuit of approximate symmetries.

REFERENCES

1. G. Lüders. 1954. *Dan. Mat. Fys. Medd.* **28,** No. 5.
   W. Pauli. 1955. *Niels Bohr and the Development of Physics* (Pergamon Press, London).
2. J. Schwinger. 1951. *Phys. Rev.* **82,** 914.
3. R. Jost. 1957. *Helv. Phys. Acta,* **30,** 409.
4. J. S. Bell. 1955. *Proc. Roy. Soc.* **A231,** 479.
5. G. Lüders and B. Zumino. 1957. *Phys. Rev.* **106,** 385.
6. T. D. Lee, R. Oehme and C. N. Yang. 1957. *Phys. Rev.* **106,** 340.
7. J. S. Bell. 1965. *Proceedings of the Oxford International Conference on Elementary Particles.*
8. W. R. Kropp and F. Reines. 1965. *Phys. Rev.* **137,** B710.
9. M. K. Moe and F. Reines. 1965. *Phys. Rev.* **140,** B992.
10. K. M. Case, R. Karplus and C. N. Yang. 1956. *Phys. Rev.* **101,** 874.

11. M. T. BURGY *et al.* 1958. *Phys. Rev.* **110**, 1214.
12. J. W. CRONIN and O. E. OVERSETH. 1963. *Phys. Rev.* **129**, 1795.
13. J. H. SMITH, E. M. PURCELL and N. F. RAMSEY. 1957. *Phys. Rev.* **108**, 120.
14. E. LIPWORTH *et al.* 1966. *Bull. Amer. Phys. Soc.* **11**, 403.
15. E. C. G. STUECKELBERG. 1952. *Helv. Phys. Acta*, **25**, 577.
16. J. H. CHRISTENSON *et al.* 1964. *Phys. Rev. Lett.* **13**, 138.
17. T. D. LEE and L. WOLFENSTEIN. 1965. *Phys. Rev.* **138**, B1490.
18. T. T. WU and C. N. YANG. 1964. *Phys. Rev. Lett.* **13**, 180.
19. L. B. OKUN. 1965. *Sov. J. Nucl. Phys.* **1**, 938.
20. L. WOLFENSTEIN. 1964. *Phys. Rev. Lett.* **13**, 562.
21. C. RUBBIA and J. STEINBERGER. 1966. *Phys. Lett.* **23**, 167.
22. T. D. LEE. 1965. *Phys. Rev.* **139**, B1415.
23. C. BALTAY *et al.* 1966. *Phys. Rev. Lett.* **16**, 1224.
24. J. BERNSTEIN, G. FEINBERG and T. D. LEE. 1965. *Phys. Rev.* **139**, B1650.
25. C. BAGLIN *et al.* 1966. *Phys. Lett.* **22**, 219.
26. R. P. FEYNMAN and M. GELL-MANN. 1958. *Phys. Rev.* **109**, 193.
27. R. FINKELSTEIN and M. RUDERMAN. 1949. *Phys. Rev.* **76**, 458.
H. L. ANDERSON *et al.* 1960. *Phys. Rev.* **119**, 2050.
28. G. BERNARDINI *et al.* 1965. *Nuovo Cim.* **38**, 605.
29. Reviewed by V. TELEGDI. 1965. *Proceedings of the International Conference on Weak Interactions*, Argonne.
30. W. WILLIS. 1965. *ibid.*
30a. N. BRENE, L. VEJE, M. ROOS and C. CRÖNSTRÖM. 1966. *Phys. Rev.* **149**, 1288.
31. L. B. OKUN and B. PONTECORVO. 1957. *Sov. Phys. JETP*, **5**, 1297.
32. P. DE POMMIER *et al.* 1963. *Phys. Rev. Lett.* **5**, 61.
R. BACASTOW *et al.* 1965. *Phys. Rev.* **139**, B407.
33. M. GELL-MANN. 1958. *Phys. Rev.* **111**, 362.
Y. K. LEE, L. W. MO and C. S. WU. 1963. *Phys. Rev. Lett.* **10**, 253.
34. M. GELL-MANN and M. LEVY. 1960. *Nuovo Cim.* **16**, 705.
35. N. CABIBBO. 1963. *Phys. Rev. Lett.* **10**, 531.
36. T. KINOSHITA and A. SIRLIN. 1959. *Phys. Rev.* **113**, 1653.
37. N. SAMIOS. 1965. *Proceedings of the International Conference on Weak Interactions*, Argonne.
38. M. GELL-MANN. 1964. *Phys. Rev. Lett.* **12**, 155.
39. M. GELL-MANN. 1962. *Phys. Rev.* **126**, 1067.
40. S. FUBINI and G. FURLAN. 1965. *Physics*, **1**, 229.
41. S. L. ADLER. 1965. *Phys. Rev. Lett.* **14**, 1051.
W. I. WEISBERGER. 1965. *Phys. Rev. Lett.* **14**, 1047.

6

THE ALGEBRA ASSOCIATED WITH THE VECTOR AND AXIAL-VECTOR CURRENT

L. A. RADICATI

Scuola Normale Superiore, Pisa

SUMMARY

THESE lectures do not cover the whole area which goes under the name of "current algebras". Due to lack of time I have been obliged to leave out a number of topics, unfortunately even some which are very interesting.

After an introductory section on the use of sum rules in atomic and nuclear physics I discuss the $SU(3)$ algebra associated with the time component of the vector current. In section 3 I introduce the $U(6) \otimes U(6)$ algebra generated by the vector and axial-vector currents and the sum rule for the axial-vector current obtained from one of its subalgebras. Section 4 is devoted to some of the consequences which follow from the commutation relations of charges and currents and from those of two currents. Finally, in section 5, I briefly outline the general methods to discuss the consequences of local commutation relations proposed by Fubini and by Gell-Mann and Dashen.

1

HISTORICAL PRELUDE

In atomic spectroscopy a number of important sum rules for the transition amplitudes between multiplets[1] have been derived from the algebras of the orbital and spin angular momentum operators:

$$[L_i, L_j] = i\varepsilon_{ijk}L_k, \qquad \text{...[1.1]}$$

$$[S_i, S_j] = i\varepsilon_{ijk}S_k, \qquad \text{...[1.2]}$$

$$[L_i, S_k] = 0. \qquad \text{...[1.3]}$$

Neither L nor S are rigorous constants of the motion. They commute however with one part of the total Hamiltonian, the part which does not commute giving rise, in most cases, to only small corrections.

The fact that L and S are approximate constants of the motion is not however the important condition in the derivation of the sum rules. Indeed other sum rules are derived from the commutation relations of operators which are not, even approximately, constant. This is the case for the Thomas-Kuhn sum rule for electric dipole transitions,[2] which played an important role in the development of quantum mechanics. We will discuss it briefly here, not only for its historical importance, but also because of its analogy with other sum rules that we shall encounter later.

Let $|\alpha\rangle, |\beta\rangle \ldots.$ be the energy eigenstates of a system and $E_\alpha, E_\beta \ldots.$ the corresponding eigenvalues. No assumption is made about the degeneracy of each level. The amplitude for an electric dipole transition between α and β is

$$F_{\beta\alpha} = e\langle\beta \mid \int d^3x j_k(x) \mid \alpha\rangle \varepsilon_k \qquad \ldots[1.4]$$

where $j_k(x)$ is the electromagnetic current density, e the electron charge and ε_k the photon polarization vector ($k = 1, 2, 3$). $F_{\beta\alpha}$ can be expressed in terms of the electric dipole operator

$$D_k = e\int d^3x x_k j_0(x) \qquad \ldots[1.5]$$

where $j_0(x)$ is the charge density. Since the four-vector $j_\mu(x)$ (j_k, j_0) is divergenceless,‡ the time derivative of D_k is given by

$$\dot{D}_k = e\int d^3x x_k \frac{\partial j_0(x)}{\partial t} = e\int d^3x j_k(x). \qquad \ldots[1.6]$$

The amplitude $F_{\beta\alpha}$ can thus be written

$$F_{\beta\alpha} = i(E_\beta - E_\alpha)\langle\beta \mid D_k \mid \alpha\rangle \varepsilon_k. \qquad \ldots[1.7]$$

It is possible to derive a sum rule for the square of the amplitudes if we know the commutation relations between D_k and $\dot{D}_l$. Let us set

$$[D_k, \dot{D}_l] = iA_{kl}. \qquad \ldots[1.8]$$

We now multiply [1.8] by $\varepsilon_k\varepsilon_l$, take the diagonal matrix element on a state $|\alpha\rangle$ and insert in the L.H.S. a complete set of states

$$\varepsilon_k\varepsilon_l \sum_\gamma \{\langle\alpha \mid D_k \mid \gamma\rangle\langle\gamma \mid \dot{D}_l \mid \alpha\rangle - \langle\alpha \mid \dot{D}_l \mid \gamma\rangle\langle\gamma \mid D_k \mid \alpha\rangle\}$$
$$= 2i \sum_\gamma \frac{|F_{\alpha\gamma}|^2}{E_\gamma - E_\alpha} = i\langle\alpha \mid A_{kl} \mid \alpha\rangle \varepsilon_k\varepsilon_l. \qquad \ldots[1.9]$$

The L.H.S. of [1.9] is related to the total photon cross-section for electric dipole transitions originating from $|\alpha\rangle$:

$$\int d\omega\sigma_\alpha(\omega) = \pi^2 \sum_\gamma \frac{|F_{\alpha\gamma}|^2}{E_\gamma - E_\alpha}. \qquad \ldots[1.10]$$

‡ The metric in space time is such that $A.A = A_\mu A_\mu = A_k A_k - A_0 A_0$ for any four vector A.

To get a sum rule for $\sigma_\alpha(\omega)$ we must therefore know A_{kl}. What we have done so far is quite general and is valid for any system which interacts only through its electric dipole with the electromagnetic field. To proceed further and calculate A_{kl} we must use a definite dynamical model. We choose a non-relativistic model (an approximation valid for atoms and molecules) such that

$$j_k(x) = \sum_{a=1}^{Z} \delta(\mathbf{x}-\mathbf{x}^{(a)})v_k^{(a)} \qquad \text{...[1.11]}$$

and

$$j_0(x) = \sum_{a=1}^{Z} \delta(\mathbf{x}-\mathbf{x}^{(a)}) \qquad \text{...[1.12]}$$

where $v_k^{(a)}$ is the velocity of the a-th particle. With this choice for $j_\mu(x)$ we have

$$D_k = e\sum_{a=1}^{Z} x_k^{(a)} \qquad \text{...[1.13]}$$

and

$$\dot{D}_k = \frac{e}{m}\sum_{a=1}^{Z} p_k^{(a)} \qquad \text{...[1.14]}$$

where $p_k^{(a)} = mv_k^{(a)}$ is the momentum of the a-th particle whose mass is m. We thus get

$$[D_k, \dot{D}_l] = \frac{e^2}{m}\sum_{a,b}^{Z} [x_k^{(a)}, p_l^{(b)}] = i\frac{e^2}{m} Z\delta_{kl}. \qquad \text{...[1.15]}$$

One obtains in this way the Thomas-Kuhn sum rule

$$\sum_\gamma \frac{|F_{\gamma\alpha}|^2}{E_\gamma - E_\alpha} = \frac{e^2}{2m} Z \qquad \text{...[1.16]}$$

where Z is the total number of charged particles in the system. It is interesting to note that it was the requirement that [*1.16*] be satisfied which guided Heisenberg[3] in the choice of the commutation relations between x and p.

The derivation of the sum rule [*1.16*] does not require a knowledge of the potential acting between the Z particles which constitute the system, provided the potential is independent of the momenta. The only essential point in the derivation is the set of canonical commutation relations

$$[x_i, p_j] = i\delta_{ij}. \qquad \text{...[1.17]}$$

The requirement that the interparticle forces should only depend upon the distance is not satisfied in nuclei where exchange forces are present. These can be approximately taken into account and lead to a modification of the sum rule which was given by Bethe and Levinger.[4,5] The result is

$$\int d\omega\sigma(\omega) = \frac{2\pi e^2}{m}\frac{NZ}{A}(1+0{\cdot}8f) \qquad \text{...[1.18]}$$

where N and Z are the number of neutrons and protons, $A = N+Z$, and f is the fraction of exchange (Majorana) forces. In a similar way it is also possible to take approximately into account the effect of other multipoles and the relativistic corrections. We shall however not discuss these generalizations here. The only purpose of this introductory section was to show, on a well-known example, how a sum rule for an observable quantity like the total photon cross-section can be obtained from an algebraic relation between operators.

2

The Algebra Associated with the Hadron Vector Current

2.1. *The SU(3) algebra*

The extension to particle physics of the algebraic methods of atomic spectroscopy depends on the symmetry properties of hadrons with respect to transformations acting on their internal variables. Sum rules for transition amplitudes between hadron states are based on the connection between the currents responsible for the transitions and the elements of the algebra associated with the internal symmetry. In this section we discuss the algebra[6] which obtains from the vector currents responsible for the electromagnetic and for the semileptonic transitions with or without hypercharge conservation.

Let us begin by considering the electromagnetic current density $j_\lambda^{el}(x)$. If we neglect effects due to weak interactions, its matrix elements vanish unless $\Delta T = 0, \pm 1$, and $\Delta Y = 0$. Here T is the isotopic spin and Y the hypercharge. Since $(T = 0 \rightarrow T = 0)$ electromagnetic transitions exist, j_λ splits into two parts

$$j_\lambda^{el} = j_\lambda^3 + \frac{1}{\sqrt{3}} j_\lambda^8. \qquad ...[2.1]$$

j_λ^3 behaves like the third component of an isotopic spin vector and j_λ^8 like an isoscalar. Since j_λ^{el} is conserved the total charge

$$Q^{el} = \int d^3x\, j_0^{el}(x) \qquad ...[2.2]$$

is a constant of the motion. If we disregard weak interactions j_λ^3 and j_λ^8 are separately conserved. We can therefore set

$$T^3 = Q^3 = \int d^3x\, j_0^3(x) \qquad ...[2.3]$$

$$Y = \frac{2}{\sqrt{3}} Q^8 = \frac{2}{\sqrt{3}} \int d^3x\, j_0^8(x) \qquad ...[2.4]$$

where T^3 is the third component of the isotopic spin and Y is the hypercharge. These two quantities are indeed the only quantities conserved by the strong and electromagnetic interactions (besides the geometrical constants of the motion).

We next consider the weak interactions. The lepton-hadron coupling in the $V-A$ theory is described by the Lagrangian

$$\mathscr{L}_{LH} = \frac{G}{\sqrt{2}}\{h_\lambda^+(x)\,l_\lambda^-(x)+h_\lambda^-(x)\,l_\lambda^+(x)\}. \quad ...[2.5]$$

G is the Fermi constant and $l_\lambda^\pm$, $h_\lambda^\pm$ are the lepton and hadron currents.

$$\varepsilon_\lambda[l_\lambda^+(x)]^* = l_\lambda^- = \bar{\mu}\gamma_\lambda(1+\gamma_5)\nu_\mu+\bar{e}\gamma_\lambda(1+\gamma_5)\nu_e \quad ...[2.6]$$

$$\varepsilon_\lambda[h_\lambda^-(x)]^* = h_\lambda^+ = a_V(j_\lambda^1+ij_\lambda^2)+a_A(g_\lambda^1+ig_\lambda^2)+b_V(j_\lambda^4+ij_\lambda^5)+b_A(g_\lambda^4+ig_\lambda^5) \quad ...[2.7]$$

where μ, e, are the muon and electron operators and ν_μ, ν_e the operators for the corresponding neutrinos; $\varepsilon_\lambda = 1$ for $\lambda = 1, 2, 3$; $\varepsilon_4 = -1$; no summation over λ is understood in the L.H.S. of [2.6] and [2.7]. The hadron current h_λ contains four terms: vector currents j_λ and axial currents g_λ, each further divided into a strangeness-conserving (isovector) part $(1\pm i2)$ and a strangeness-changing (isospinor) part $(4\pm i5)$ such that

$$[Y, j_\lambda^1\pm ij_\lambda^2] = 0 \quad ...[2.8]$$

$$[Y, j_\lambda^4\pm ij_\lambda^5] = \pm(j_\lambda^4\pm ij_\lambda^5). \quad ...[2.9]$$

Both the isovector and isospinor parts are of course step operators for the hadron charge.

$$[Q^{el}, j_\lambda^1\pm ij_\lambda^2] = \pm(j_\lambda^1\pm ij_\lambda^2) \quad ...[2.10]$$

$$[Q^{el}, j_\lambda^4\pm ij_\lambda^5] = \pm(j_\lambda^4\pm ij_\lambda^5). \quad ...[2.11]$$

Similar relations hold for the axial currents. a_V, b_V, a_A, b_A are parameters which determine the relative strength of the four kinds of currents. As we shall see later they are not all independent.

If we disregard electromagnetic effects, the isovector current $(j_\lambda^1\pm ij_\lambda^2)$ is conserved. Therefore

$$T^\pm = Q^1\pm iQ^2 = \int d^3x(j_0^1\pm ij_0^2) \quad ...[2.12]$$

are, to this approximation, constants of the motion. $T^\pm = Q^1\pm iQ^2$ are by definition the operators which produce allowed Fermi transitions, defined by the selection rules

$$\left.\begin{array}{l}\Delta J = 0;\ \textit{No parity change}\\ \Delta T = 0;\ \Delta Y = 0;\ \Delta Q^{el} = \pm 1\end{array}\right\} \quad ...[2.13]$$

where J and Q^{el} are the spin and the charge. $T^\pm$ are therefore proportiona to the step operators of the isotopic spin algebra as these are the only con served quantities (apart from electromagnetic corrections) besides T^3 and Y

From the electromagnetic current and the isovector charged currents we have thus constructed the generators of the symmetry group of the medium-strong interactions,

$$SU(2)_T \otimes U(1)_Y.$$

The isospinor current appears in this symmetry scheme in a rather unsymmetrical way. It is known however that hadron states can be grouped according to the representations of a larger symmetry scheme,[7] $SU(3)$, which contains $SU(2)_T \otimes U(1)_Y$ as a subgroup. The Lie algebra of $SU(3)$ has eight elements Q^A ($A = 1 \ldots . 8$) whose Lie product is

$$[Q^A, Q^B] = if^{ABC}Q^C \qquad \ldots[2.14]$$

where f^{ABC} are the structure constants (repeated capital indices are summed from 1 to 8). The basis has been chosen in such a way that Q^1, Q^2 and Q^3 coincide with the elements of the $SU(2)_T$ algebra‡ and $Q^8 = \frac{\sqrt{3}}{2} Y$. With respect to the group $SU(2)_T$ the elements $Q^4 \pm iQ^5$ behave like the components of two (different) spinor operators with $Y = \pm 1$ and $Q^{el} = \pm 1$.

This is exactly the behaviour of the strangeness-changing (isospinor) currents $(j^4_\lambda \pm ij^5_\lambda)$. If in the limit of $SU(3)$ this current is conserved, we can set

$$V^{\pm} = Q^4 \pm iQ^5 = \int d^3x(j^4_0 \pm ij^5_0) \qquad \ldots[2.15]$$

as indeed allowed strangeness-changing transitions satisfy the selection rules

$$\left.\begin{array}{l} \Delta J = 0; \; \textit{No parity change} \\ \Delta T = \frac{1}{2}; \; \Delta Y = \Delta Q^{el} = \pm 1. \end{array}\right\} \qquad \ldots[2.16]$$

It should be remarked however that, whereas the divergence of $j^1_\lambda \pm ij^2_\lambda$ differs from zero only for terms of order e^2, the vanishing of the divergence of $j^4_\lambda \pm ij^5_\lambda$ is much less well satisfied.

The remaining elements Q^6 and Q^7 of the $SU(3)$ algebra, whose matrix elements differ from zero only when

$$\Delta J = 0; \; \Delta T = \tfrac{1}{2}; \; \Delta Y = \pm 1; \; \Delta Q^{el} = 0, \qquad \ldots[2.17]$$

can also be written as integrals over the time components of two neutral vector currents, j^6_λ and j^7_λ. Together with the other six currents they form an octet of vector currents $j^A_\lambda(x)$ such that

$$[Q^A, j^B_\lambda] = if^{ABC}j^C_\lambda. \qquad \ldots[2.18]$$

The currents $j^6_\lambda(x)$ and $j^7_\lambda(x)$ are not coupled to the lepton fields. It remains to be seen whether they are weakly coupled to other hadron currents.

To sum up: from the currents coupled to the electromagnetic and lepton fields we have constructed six operators $Q^1 - Q^5$ and Q^8. To these we adjoin

‡ We use the same symbol for the Lie group and its Lie algebra.

the neutral strangeness-changing operators Q^6 and Q^7 originating from currents which are perhaps only coupled strongly to hadrons. The eight operators so constructed would be independent of time if the corresponding currents were conserved. In this case they are proportional to the generators of $SU(3)$ and satisfy the commutation relations [*2.14*].

None of the eight currents is however rigorously conserved. The non-conservation of $j^4_\lambda, j^5_\lambda, j^6_\lambda, j^7_\lambda$ is primarily due to the mass term and to the medium strong interactions. j^1_λ and j^2_λ are non-conserved due to the presence of electromagnetic interactions. Finally the non-conservation of j^3_λ and j^8_λ depends only on the weak interaction part of the Hamiltonian. This lack of conservation produces a time dependence of the operators

$$Q^A(x_0) = \int d^3x\, j^A_0(\mathbf{x}, x_0)$$

which cannot therefore be rigorously identified with the generators of a symmetry group of the complete Hamiltonian.

We will however assume that at each time x_0 the operators $Q^A(x_0)$ are the basis of an SU(3) algebra, the Lie product of two operators being given by [*2.14*]. We will often drop the variable x_0 but whenever we write a commutator between two non-conserved operators we will imply that they should be taken at the same time.

2.2. *Universality of the weak coupling*

As emphasized by Gell-Mann[6] the commutation relations satisfied by a set of operators lead to definite relations between the coupling constants of the currents associated with them. As an example we discuss briefly Cabibbo's[8] definition of the universality of the weak coupling. This is a relation between the coupling constant in the lepton-lepton Lagrangian

$$\mathscr{L}_{LL} = \frac{G}{\sqrt{2}}\, l^+_\mu l^-_\mu$$

responsible for μ decay, and the coupling $\frac{G}{\sqrt{2}}\, a_V$ and $\frac{G}{\sqrt{2}}\, b_V$ of the isovector and isospinor vector currents to leptons (see equations [*2.5*] and [*2.7*]).

The $SU(3)$ algebra discussed in the previous section admits three sub-algebras which we denote by $SU(2)_T$, $SU(2)_U$, $SU(2)_V$. They are defined as follows:

$$SU(2)_T \text{ commutes with } Y = \frac{2}{\sqrt{3}}\, Q^8$$

$$SU(2)_U \text{ commutes with } Q^{el} = Q^3 + \frac{1}{\sqrt{3}}\, Q^8$$

$$SU(2)_V \text{ commutes with } X = Q^3 - \frac{1}{\sqrt{3}}\, Q^8.$$

The isotopic spin operators belong to $SU(2)_T$; the operators $V^{\pm}$ associated with the strangeness-changing current belong to $SU(2)_V$.

Let us consider unitary transformations on the Q^A, realized by unitary operators $\mathscr{U}$,

$$Q^A \to Q^{A'} = \mathscr{U} Q^A \mathscr{U}^\dagger \qquad ...[2.19]$$

where

$$\mathscr{U} = e^{iG}; \; G = \sum_{L=1}^{8} \alpha_L Q^L. \qquad ...[2.20]$$

The $\mathscr{U}$'s which leave Q^{el} invariant belong to an $SU(2)$ group whose Lie algebra is $SU(2)_U$ (i.e., in equation [2.20] L is restricted to the values $L = 3, 6, 7, 8$). Let $C^{\pm}$ be step operators for Q^{el}

$$[Q^{el}, C^{\pm}] = \pm C^{\pm}. \qquad ...[2.21]$$

For any $\mathscr{U}$ which leaves Q^{el} invariant we have

$$[Q^{el}, C^{\pm\prime}] = \pm C^{\pm\prime} \qquad ...[2.22]$$

where

$$C^{\pm\prime} = \mathscr{U} C^{\pm} \mathscr{U}^\dagger. \qquad ...[2.23]$$

Conversely we will assume that if $C^{\pm}$, $C^{\pm\prime}$ are step operators for Q^{el}, there exists a $\mathscr{U} \in SU(3)$ such that

$$C^{\pm\prime} = \mathscr{U} C^{\pm} \mathscr{U}^\dagger \qquad ...[2.24]$$

$$Q^{el} = \mathscr{U} Q^{el} \mathscr{U}^\dagger \qquad ...[2.25]$$

with $\mathscr{U}\mathscr{U}^\dagger = 1$.

We use this assumption to establish a relation between the coefficients a_V and b_V (equation [2.7]) which determine the relative strength of the isovector and isospinor couplings of the vector current. Let us consider allowed transitions induced by $T^{\pm}$ ($\Delta Y = 0$) and $V^{\pm}$ ($|\Delta Y| = 1$). Both $T^{\pm}$ and $V^{\pm}$ are step operators for Q^{el}. Therefore the operators

$$C^+ = a_V T^+ + b_V V^+, \; C^- = (C^+)^\dagger, \qquad ...[2.26]$$

with a_V and b_V arbitrary complex numbers, are step operators for Q^{el}. According to our assumption there exists a unitary transformation $\mathscr{U}$ such that

$$C^{\pm} = \mathscr{U} T^{\pm} \mathscr{U}^\dagger. \qquad ...[2.27]$$

Let us define an operator C^3 by

$$C^3 = \mathscr{U} T^3 \mathscr{U}^\dagger.$$

Since T^3, $T^{\pm} \in SU(2)_T$, C^3, $C^{\pm}$ are the basis of an $SU(2)$ algebra. This however is only possible if

$$| a_V |^2 + | b_V |^2 = 1, \qquad ...[2.28]$$

as one can verify by direct computation. Equation [*2.28*] is Cabibbo's definition of the universality of the hadron-lepton coupling. As a consequence of [*2.28*] the isovector and isospinor coupling constants are given in terms of the lepton-lepton coupling constant G by

$$G_V^0 = G \cos \theta$$

$$G_V^1 = G \sin \theta$$

where θ is Cabibbo's angle, whose value is, however, not predicted by the theory.

The discrepancy between the coupling constants G derived from μ decay and $G_\beta = G_V^0$ derived from O^{14} is thus seen to be a consequence of the requirement that

$$C^+ = \int d^3x \{a_V(j_0^1 + ij_0^2) + b_V(j_0^4 + ij_0^5)\}, \qquad ...[2.29]$$

together with

$$C^- = (C^+)^\dagger \quad \text{and} \quad C^3 = \tfrac{1}{2}[C^+, C^-], \qquad ...[2.30]$$

are an $SU(2)$ algebra. This ensures that the hadron charge be quantized like the lepton charge.[8,9]

In the $SU(3)$-symmetrical limit the definition of step operators for Q^{el} can only be given modulo a unitary transformation which belongs to $SU(2)_U$. The relative strength of the $\Delta Y = 0$ and $\Delta Y = 1$ vector coupling is thus undefined, the only defined quantity being the sum of the squares of the isovector and isospinor coupling constants. It is this sum that can be compared to G, the lepton-lepton coupling constant.

2.3. *The isospinor vector current*

The conservation of the isovector current is only violated by electromagnetic effects; the isospinor current, on the other hand, is conserved only in the limit of $SU(3)$ symmetry. It is important to know how this lack of conservation affects the results deduced in the symmetric limit. We will show that the assumption of the equal time commutation relations for the Q^A's allows us to estimate the order of magnitude of the symmetry breaking effects.[10]

Neglecting electromagnetic and weak effects we write the hadron Hamiltonian in the form

$$H = H_0 + \lambda H^8 \qquad ...[2.31]$$

where

$$[Q^A, H_0] = 0, \qquad ...[2.32]$$

$$[Q^A, H] = i\lambda f^{A8C} H^C. \qquad ...[2.33]$$

Equation [*2.33*] specifies that the symmetry breaking part of H transforms like the 8-th component of an octet operator H^C. λ is a dimensionless parameter which measures the departure from $SU(3)$ symmetry.‡

H^C is given in terms of the divergence of the vector current

$$\frac{\partial j_\lambda^A}{\partial x_\lambda}(x) = S^A(x) \qquad \text{...[2.34]}$$

by

$$\lambda f^{A8C} H^C = \int d^3x\, S^A(x). \qquad \text{...[2.35]}$$

The matrix elements of each operator Q^A (at time $x_0 = 0$) between physical states with momenta $p \equiv (\boldsymbol{p}, p_0)$, $p' \equiv (\boldsymbol{p}', p_0')$ and spin components s, s' can thus be written as

$$\langle p's' \mid Q^A \mid ps \rangle = (2\pi)^3 \delta(\boldsymbol{p}-\boldsymbol{p}')\delta_{ss'} \frac{\langle p' \mid S^A(0) \mid p \rangle}{i(p_0'-p_0)}. \qquad \text{...[2.36]}$$

If $\mid p\rangle$ and $\mid p'\rangle$ belong to the same $SU(3)$ multiplet both the numerator and the denominator in the R.H.S. of [*2.36*] tend to zero with λ in such a way that the ratio tends to $(2\pi)^3\delta(\boldsymbol{p}-\boldsymbol{p}')$ times the appropriate Clebsch-Gordan coefficient. If $\mid p\rangle$ and $\mid p'\rangle$ belong to different multiplets the R.H.S. vanishes in the symmetric limit.

Equation [*2.36*] relates the matrix element of any Q^A to the divergence of the corresponding vector current. We will use it to evaluate the matrix elements of $V^\pm = Q^4 \pm iQ^5$, which give rise to allowed $\Delta Y = \pm 1$ transitions. Under the group $SU(2)_T$, $V^\pm$ transform like isospinors with $Y = \pm 1$: we can thus write

$$\langle ps \mid V^+ \mid p's' \rangle = (2\pi)^3 \delta(\boldsymbol{p}-\boldsymbol{p}') F(T', Y', T, Y; p) C \delta_{ss'} \qquad \text{...[2.37}$$

where C is a Clebsch-Gordan coefficient and F a renormalization factor which depends on the initial and final isospin and hypercharge and on the (common) momentum. F reduces to 1 in the symmetrical limit. According to our assumption, $V^\pm$, though not constants of the motion, satisfy the equal time commutator (see [*2.14*])

$$[V^+, V^-] = 2V^3 = Q^3 + \sqrt{3}Q^8. \qquad \text{...[2.38]}$$

Let us take the matrix element of [*2.38*] between neutron states. The operators in the R.H.S. are conserved in the approximation defined by [*2.31*] and therefore no renormalization factor appears in the matrix element. We thus have

$$\langle p's' \mid [V^+, V^-] \mid ps \rangle = (2\pi)^3 \delta(\boldsymbol{p}-\boldsymbol{p}')\delta_{ss'}. \qquad \text{...[2.39]}$$

To evaluate the L.H.S. we insert a complete set of states and separate those

‡ From the baryon mass spectrum we estimate $\lambda \simeq \frac{1}{4}$.

which belong to the baryon octet (only Σ^--states contribute). We thus get

$$(2\pi)^3\delta(\mathbf{p}-\mathbf{p}')[F^2(p)+\Delta(p)] = (2\pi)^3\delta(\mathbf{p}-\mathbf{p}'). \qquad ...[2.40]$$

The term F^2 comes from the sum over Σ^- states (the Clebsch-Gordan coefficient is in this case $C_\Sigma = -1$; for simplicity we have suppressed the other labels). The term $\Delta(p)$ arises from the sum over all other states

$$(2\pi)^3\delta(\mathbf{p}-\mathbf{p}')\delta_{ss'}\Delta(p) = \sum_{k\neq\Sigma^-} \langle p \mid V^+ \mid k\rangle\langle k \mid V^- \mid p\rangle - (+\leftrightarrow -)$$

$$= (2\pi)^3\delta(\mathbf{p}-\mathbf{p}')\sum_{k'}\left[\frac{\langle p \mid S^+ \mid k'\rangle\langle k' \mid S^- \mid p\rangle}{(p_0-k_0')^2} - (+\leftrightarrow -)\right]. \qquad ...[2.41]$$

In the last step [*2.36*] has been used; the sum over k' is restricted to states with $\mathbf{k} = \mathbf{p}$. We can thus write

$$1-F^2(p) = \Delta(p) = 0(\lambda^2), \qquad ...[2.42]$$

since by [*2.35*] $S^\pm$ is proportional to λ.

This is the content of the theorem of Ademollo and Gatto[11] which states that the renormalization factor is of second order in the strength of the $SU(3)$ breaking Hamiltonian H^8.

The value of the renormalization factor depends upon p. To study this momentum dependence we write

$$\langle k \mid S^+ \mid p\rangle = \frac{f_k^+(M_k, t)}{(4p_0k_0)^{\frac{1}{2}}}\lambda \qquad ...[2.43]$$

where f_k^+ is an invariant function of the momentum transfer

$$t = -(k-p)^2 = (k_0-p_0)^2 \qquad ...[2.44]$$

and of the mass M_k of the state k. We can thus write

$$\Delta(p)\delta(\mathbf{p}-\mathbf{p}') = \lambda^2\delta(\mathbf{p}-\mathbf{p}')\sum_{k'}\frac{(p_0+k_0)^2}{4p_0k_0}\frac{1}{(M^2-M_k^2)^2}\{\mid f_k^+\mid^2 - \mid f_k^-\mid^2\} \qquad ...[2.45]$$

where we have used the identity

$$(p_0-k_0)^2 = \frac{(p_0^2-k_0^2)^2}{(p_0+k_0)^2} = \frac{(M^2-M_k^2)^2}{(p_0+k_0)^2}. \qquad ...[2.46]$$

For $p_0 \to \infty$, $t \to 0$ and the kinematical factor

$$\frac{(p_0+k_0)^2}{4p_0k_0} \to 1$$

i.e. it tends to its minimal value. If $f_k^\pm(t)$ has a minimum for $t = 0$, the value of F calculated for $p_0 = \infty$ represents the minimum departure from the $SU(3)$ symmetric limit due to the non-conservation of the isospinor current.

We will come back later to the special properties of the reference frame where $p_0 = \infty$.

The actual value of $F(p)$ cannot, however, be evaluated unless one approximates the sum over the intermediate states in $\Delta(p)$ with a few low lying states. It is difficult to assess the validity of such an approximation (which in other cases does not give satisfactory results) so that the calculation of the $SU(3)$-breaking effects on the vector current matrix elements is still an open problem.

3

The Algebra generated by the Vector and Axial-Vector Currents

3.1. *The quark model*

The weak interactions of hadrons involve, besides the vector current $j_\lambda^A(x)$ which we have considered so far, also axial currents g_λ. We know that both strangeness-conserving $(g_\lambda^1 \pm ig_\lambda^2)$ and strangeness-changing $(g_\lambda^4 \pm ig_\lambda^5)$ axial currents must be present. We will assume that besides these axial charged currents there exist four neutral currents and that these eight currents g_λ^A behave like an octet under $SU(3)$, i.e.

$$[Q^A, g_\lambda^B] = if^{ABC} g_\lambda^C. \qquad \ldots[3.1]$$

Let us define

$$R^A(x_0) \equiv R_0^A(x_0) = \int d^3x\, g_0^A(\boldsymbol{x}, x_0). \qquad \ldots[3.2]$$

To find the algebra of the operators $Q^A(x_0)$ and $R^A(x_0)$ we cannot rely on the method used in section 2.1 as there is no realistic approximation in which $g_\lambda^A(x)$ are conserved. Gell-Mann[6] has suggested that this algebra might coincide with the one obtained from the so called quark model.[12] In this model a spinor field ψ (the quark), which transforms like the 3-dimensional representation of $SU(3)$, interacts with a neutral vector boson field. Let us form the bilinear products (current densities)

$$\psi^\dagger(\boldsymbol{x}, x_0)\alpha_\rho\psi(\boldsymbol{x}, x_0) = c_\rho(\boldsymbol{x}, x_0)$$

where α_ρ are 12×12 matrices which satisfy the algebra of $U(12)$ (the index ρ runs from 1 to 144). The equal time commutator of two currents is

$$[c_\rho(\boldsymbol{x}, x_0), c_\sigma(\boldsymbol{x}', x_0)] = ie_{\rho\sigma\tau} c_\tau(\boldsymbol{x}, x_0)\delta(\boldsymbol{x}-\boldsymbol{x}') \qquad \ldots[3.3]$$

where $e_{\rho\sigma\tau}$ are the structure constants of $U(12)$. As a consequence of [3.3] the space integrals of the currents (which we call charges)

$$C_\rho(x_0) = \int d^3x\, c_\rho(\boldsymbol{x}, x_0) \qquad \ldots[3.4]$$

satisfy (at equal times) the algebra of $U(12)$

$$[C_\rho(x_0), C_\sigma(x_0)] = ie_{\rho\sigma\tau} C_\tau(x_0). \qquad \ldots[3.5]$$

It is doubtful whether the whole $U(12)$ algebra [*3.5*] has much physical significance. To discuss its subalgebras we write the 12×12 matrices in the form

$$\alpha_\rho = \Gamma \otimes \lambda$$

where Γ and λ are 4×4 and 3×3 representations of $U(4)$ and $U(3)$ respectively. Amongst the numerous subalgebras we list only those which arise from the vector and axial-vector currents.

(i) $U(6) \otimes U(6)$ with elements

$$Q_a^A(x_0) = \tfrac{1}{2} \int d^3x\, \psi^\dagger \sigma_a \lambda^A \psi \qquad \ldots[3.6]$$

$$R_a^A(x_0) = \tfrac{1}{2} \int d^3x\, \psi^\dagger \gamma_5 \sigma_a \lambda^A \psi \qquad \ldots[3.7]$$

where $A = 0, 1, 2, \ldots 8$; $a = 0, 1, 2, 3$; $\sigma_0 = 1$, and $\sigma_a = i\gamma_4\gamma_5\gamma_a$ for $a = 1, 2, 3$ ($\gamma_\mu = \gamma_\mu^\dagger$; $\gamma_5 = \gamma_5^\dagger$); $\lambda^0 = \sqrt{\tfrac{2}{3}}$; all other λ's defined as in reference 6. One can verify that the equal-time commutators of the Q's and R's are:

$$\left.\begin{aligned} [Q_a^A, Q_b^B] &= i g_{abc}^{ABC} Q_c^C \\ [Q_a^A, R_b^B] &= i g_{abc}^{ABC} R_c^C \\ [R_a^A, R_b^B] &= i g_{abc}^{ABC} Q_c^C \end{aligned}\right\} \qquad \ldots[3.8]$$

where

$$g_{abc}^{ABC} = \varepsilon_{abc} d^{ABC} + \eta_{abc} f^{ABC}.$$

ε_{abc} and f^{ABC} are antisymmetrical tensors such that $\varepsilon_{0ab} = f^{0AB} = 0$. ε_{abc} ($a, b, c \neq 0$) is the Ricci-Levi-Civita tensor and f^{ABC} ($A, B, C \neq 0$) is defined as in reference 6. η_{abc} and d^{ABC} are symmetrical tensors:

$$\eta_{00a} = \delta_{0a};\ \eta_{0ab} = \delta_{ab};\ \eta_{abc} = 0 \text{ for } a, b, c, \neq 0$$

$$d^{00A} = \sqrt{\tfrac{2}{3}}\,\delta^{0A};\ d^{0AB} = \sqrt{\tfrac{2}{3}}\,\delta^{AB}$$

and all other d's as in reference 6. The $U(6) \otimes U(6)$ algebra [*3.8*] is the largest algebra associated with j_λ^A and g_λ^A. The elements Q_a^A are even under space reflection; R_a^A are odd.

(ii) $U(6)$ with elements Q_a^A. It is the largest algebra generated by the vector and axial currents whose generators are all even under space reflection. The representations of the associated $SU(6)$ group correspond to multiplets of particles with the same parity.[13]

(iii) $[U(3) \otimes U(3)]_{\text{chiral}}$ with elements (Q_0^A, R_0^A). It is the subalgebra of (i) generated by the time components of j_λ^A and g_λ^A.[6]

(iv) $[U(3) \otimes U(3)]_{\text{collinear}}$ with elements Q_0^A and

$$Q^A(\hat{p}) = \frac{1}{|\boldsymbol{p}|} p_k Q_k^A \quad (k = 1, 2, 3) \qquad \ldots[3.9]$$

(where p_k is an arbitrary vector), with equal time commutators

$$[Q_0^A, Q_0^B] = if^{ABC} Q_0^C$$
$$[Q_0^A, Q^B(\hat{p})] = if^{ABC} Q^C(\hat{p})$$
$$[Q^A(\hat{p}), Q^B(\hat{p})] = if^{ABC} Q_0^C.$$

The operators $Q^A(\hat{p})$ have even parity like the Q_0^A. They all commute with the generator of a Lorentz transformation in the direction $\hat{p}$

$$R^0(p) = \tfrac{1}{2} \int d^3x \psi^\dagger \gamma_5 (\boldsymbol{\sigma} . \hat{p}) \psi.$$

(v) $U(3)$ with elements $Q_0^A \equiv Q^A$. It is the algebra discussed in section 2.1 with the addition of the element Q_0^0.

The quark model is useful in suggesting the algebraic structure of the operators obtained by integrating the current densities over the whole space. The operators thus obtained are in general time dependent. We will now assume that, *independently of the quark model*, the operators $Q_a^A(x_0)$ and $R_a^A(x_0)$ obtained from the vector and axial-vector densities, namely

$$Q^A(x_0) \equiv Q_0^A(x_0) = \int d^3x j_0^A(\boldsymbol{x}, x_0),$$
$$Q_k^A(x_0) = \int d^3x g_k^A(\boldsymbol{x}, x_0),$$
$$R^A(x_0) \equiv R_0^A(x_0) = \int d^3x g_0^A(\boldsymbol{x}, x_0),$$
$$R_k^A(x_0) = \int d^3x j_k^A(\boldsymbol{x}, x_0),$$

satisfy the equal time commutation relations [*3.8*]. This would be the case if the commutator of the corresponding current densities is of the form [*3.3*]. We do not discuss here the possibility that the R.H.S. of [*3.3*] contains other singular terms besides the δ-function. It has been shown by Schwinger[14] that, even in the simplest cases, terms containing the derivative of $\delta(\boldsymbol{x}-\boldsymbol{x}')$ appear multiplied by infinite constants. These Schwinger terms do not in many cases affect the physical results. A completely different situation would arise if the other singular terms were operators. Such is probably the case for the commutators which involve at least one space component of the current densities. We shall not discuss these cases in the following.

3.2. *The chiral $SU(2) \otimes SU(2)$ generated by the hadron current $h_\lambda^\pm$*

From the vector part of the hadron current $h_\lambda^\pm$ we have constructed the (even) charge raising and lowering operators $C^\pm$ (equation [*2.29*]) which together with their commutator $C^3 = \frac{1}{2}[C^+, C^-]$ form an $SU(2)$ algebra, provided that the coefficients a_V and b_V satisfy Cabibbo's condition [2.28]. Let us consider the axial current

$$a_A(g_\lambda^1 \pm ig_\lambda^2) + b_A(g_\lambda^4 \pm ig_\lambda^5)$$

and denote by $\mathscr{C}^+$ the integral of its time component

$$\mathscr{C}^+ = a_A \int d^3x(g_0^1 + ig_0^2) + b_A \int d^3x(g_0^4 + ig_0^5); \; \mathscr{C}^- = (\mathscr{C}^+)^\dagger, \quad \text{...[3.10]}$$

or, in the notation of section 3.1,

$$\mathscr{C}^+ = a_A(R^1 + iR^2) + b_A(R^4 + iR^5).$$

$\mathscr{C}^\pm$ are the odd (under space reflection) raising and lowering operators for the charge.

Let us require that $C^\pm$, C^3, $\mathscr{C}^\pm$ and $\frac{1}{2}[C^+, \mathscr{C}^-] = \mathscr{C}^3$ generate a chiral $SU(2) \otimes SU(2)$. One verifies easily that this is only possible if

$$a_A = a_V; \quad b_A = b_V \quad \text{...[3.11]}$$

with

$$|a_V|^2 + |b_V|^2 = 1. \quad \text{...[3.12]}$$

i.e., if Cabibbo's angles θ_V and θ_A for the vector and axial currents are equal.

Equations [3.11] and [3.12] define the universality of the weak vector and axial-vector couplings. They are a consequence of the assumption that the integrals of the time components of the vector and axial-vector charged currents, together with their commutators, are an $SU(2) \otimes SU(2)$ algebra.

This is an extension to the odd step operators of the corresponding assumption discussed in section 2.2 to define in an $SU(3)$ invariant way the universality of the vector coupling. Since the charge operator $Q^{el} = Q^3 + \frac{1}{\sqrt{3}} Q^8$ commutes with all the neutral elements of $[U(3) \otimes U(3)]_{\text{chiral}}$, it is plausible to assume that all step operators for Q^{el}, whatever their parity, should be related by a transformation belonging to $[SU(2) \otimes SU(2)]_{\text{chiral}}$, the subgroup of $[SU(3) \otimes SU(3)]_{\text{chiral}}$ generated by its neutral generators.

It should however be stressed that the equality between θ_V and θ_A refers only to the unrenormalized values. Renormalization effects due to the non-conservation of the axial current may well spoil this equality. This seems indeed to be the case.

3.3. *The Adler-Weisberger sum rule*

From the chiral $SU(2) \otimes SU(2)$ algebra whose elements commute with the hypercharge, Adler[15] and Weisberger[16] have independently derived a sum rule relating the ratio $g_A = -G_A/G_V$ of the axial and vector coupling constants to the off-mass-shell pion-proton total cross-section. Several proofs have been given of this sum rule based on seemingly different methods.[17, 18, 19]

Probably the simplest one is based on the method $p \to \infty$ of Fubini and Furlan[10] discussed in section 2.3.

Let us consider the odd parity step operators for the charge

$$\mathscr{T}^{\pm} = R^1(x_0) \pm iR^2(x_0) = \int d^3x(g_0^1(\mathbf{x}, x_0) \pm ig_0^2(\mathbf{x}, x_0)), \quad ...[3.13]$$

which commute with the hypercharge. According to our assumption they satisfy the equal time commutator (see [3.8])

$$[\mathscr{T}^+(0), \mathscr{T}^-(0)] = 2T^3. \quad ...[3.14]$$

The ratio $g_A = -G_A/G_V$ is defined by the matrix element of $\mathscr{T}^+$ between neutron and proton states

$$iG_V\langle ps \mid \mathscr{T}^+(0) \mid p's'\rangle = (2\pi)^3\delta(\mathbf{p}-\mathbf{p}')g_A\bar{u}_P(ps)\gamma_4\gamma_5 u_N(p's') \quad ...[3.15]$$

where u_P and u_N are the proton and neutron wave functions.

Let us take the matrix element of [3.14] between proton states with momenta p, p' and spin components s, s'. The R.H.S. can be immediately computed since T^3 is conserved. We thus have

$$\langle p's' \mid [\mathscr{T}^+, \mathscr{T}^-] \mid ps\rangle = 2(2\pi)^3\delta_{ss'}\delta(\mathbf{p}-\mathbf{p}'). \quad ...[3.16]$$

To evaluate the L.H.S. we insert a complete set of states and separate the neutron contribution. Summing over s, s' we get

$$g_A^2[\delta(\mathbf{p}-\mathbf{p}')g(p) - \delta(\mathbf{p}-\mathbf{p}')\Gamma(p)] = \delta(\mathbf{p}-\mathbf{p}') \quad ...[3.17]$$

where

$$(2\pi)^3\delta(\mathbf{p}-\mathbf{p}')g(p)g_A^2 = \tfrac{1}{2}\sum_{ss'}\sum_{n}\langle p' \mid \mathscr{T}^+ \mid n\rangle\langle n \mid \mathscr{T}^- \mid p\rangle \quad ...[3.18]$$

is the contribution from one-neutron intermediate states and

$$(2\pi)^3 g_A^2\delta(\mathbf{p}-\mathbf{p}')\Gamma(p) = -\tfrac{1}{2}\sum_{ss'}\sum_{k\neq n}\{\langle p' \mid \mathscr{T}^+ \mid k\rangle\langle k \mid \mathscr{T}^- \mid p\rangle - (+\leftrightarrow-)\} \quad ...[3.19]$$

is the contribution from all other states.

With an elementary calculation we get

$$g(p) = 1 - \frac{M^2}{p_0^2} \quad ...[3.20]$$

where M is the nucleon mass and $p_0^2 = M^2 + \mathbf{p}^2$. If we take the limit $p_0 \to \infty$ equation [3.17] can be written

$$1 - \frac{1}{g_A^2} = \lim_{p_0 \to \infty} \Gamma(p). \quad ...[3.21]$$

We notice here an important feature of the system $p_0 \to \infty$: the separation of the sum over intermediate states into a one-neutron contribution and a contribution from other states is independent of $\mathbf{p}$.

The inelastic contribution $\Gamma(p)$ can be related to the divergence of the axial current. Let us put

$$\langle k \mid \mathscr{T}^+ \mid p\rangle = (2\pi)^3 \delta(\boldsymbol{p}-\boldsymbol{k}) \frac{F_k^+(M_k, t)}{i(k_0-p_0)(4p_0k_0)^{\frac{1}{2}}}$$
$$= (2\pi)^3 \delta(\boldsymbol{p}-\boldsymbol{k}) \frac{F_k^+(M_k, t)}{i(M_k^2-M^2)} \cdot \frac{k_0+p_0}{(4p_0k_0)^{\frac{1}{2}}} \qquad ...[3.22]$$

where M_k and k_0 are the mass and energy of the intermediate state k, $t = -(p-k)^2$ and the invariant function F_k^+ is related to the divergence of the axial-vector current

$$\frac{\partial}{\partial x_\lambda}(g_\lambda^1+ig_\lambda^2) = d^+(x) \qquad ...[3.23]$$

by

$$\frac{F_k^+(M_k,t)}{(4p_0k_0)^{\frac{1}{2}}} = \langle k \mid d^+(0) \mid p\rangle. \qquad ...[3.24]$$

(To simplify the notation we have not indicated the dependence of F on the spins of the two states). We now insert [*3.22*] into [*3.19*] and carry out the integration over the momentum $\boldsymbol{k}$ which is implied in Σ_k. We thus get

$$(2\pi)^3 g_A^2 \delta(\boldsymbol{p}-\boldsymbol{p}')\Gamma(p)$$
$$= \tfrac{1}{2}(2\pi)^3\delta(\boldsymbol{p}-\boldsymbol{p}')\sum_{ss'}\int_{(M+m_\pi)}^{\infty} dW \sum_k{}' \delta(W-M_k)\frac{(k_0+p_0)^2}{4p_0k_0(M_k^2-M^2)^2}\{|F_k^+|^2-|F_k^-|^2\} \qquad ...[3.25]$$

where Σ_k' is a sum over all remaining degrees of freedom. Let us now take the limit $p_0 \to \infty$ *and assume it can be exchanged with the integral.* The kinematical factor $(k_0+p_0)^2/4p_0k_0 \to 1$ and $t \to 0$. Under this assumption we can write:

$$1-\frac{1}{g_A^2} = \lim_{p_0 \to \infty} \Gamma(p) = \int \frac{dW}{(W^2-M^2)^2}[A^+(W,0)-A^-(W,0)] \qquad ...[3.26]$$

where

$$A^\pm(W,t) = \tfrac{1}{2}g_A^{-2}\sum_{ss'}\sum_k{}' \delta(W-M_k) \mid F_k^\pm(M_k,t)\mid^2. \qquad ...[3.27]$$

The R.H.S. of [*3.26*] can be directly related to observable quantities by using a theorem due to Adler[20]: The cross-section for emission of a lepton (l) in the forward direction in the scattering of neutrinos (ν_l) by protons (p) depends only on the divergence of the hadron current if terms of order m^2/ν_0^2 are neglected (m is the lepton mass and ν_0 the neutrino energy).

PROOF: Consider the process

$$\nu_l + p \rightarrow l + k \qquad \text{...[3.28]}$$

where k is any final state with $Y = 1$. Let ν, p, l and k be the four-momenta of the particles involved. The cross-section for [3.28] is

$$\sigma = (2\pi)^4 \frac{G^2}{2(\nu . p)} \int \frac{d^3k}{k_0} \frac{d^3l}{l_0} \delta^4(\nu + p - l - k) L_{\alpha\beta}(\nu, l) B_{\alpha\beta}(p, k). \qquad \text{...[3.29]}$$

$L_{\alpha\beta}$ and $B_{\alpha\beta}$ are tensors which depend upon the lepton and baryon variables respectively. If we sum over the lepton spin states we have

$$L_{\alpha\beta} = [-(m^2+q^2)\delta_{\alpha\beta} + q_\alpha q_\beta - P_\alpha P_\beta - \varepsilon_{\alpha\beta\gamma\delta} P_\gamma q_\delta]$$

and

$$B_{\alpha\beta} = \langle k \mid h_\alpha^+ \mid p\rangle \langle p \mid h_\beta^- \mid k\rangle$$

where $q = l - \nu$; $P = l + \nu$ and $h_\alpha^\pm$ is the hadron current.

Let us consider the case when the final lepton momentum $\mathbf{l}$ is parallel to the incident neutrino momentum $\boldsymbol{\nu}$, i.e.:

$$\mathbf{l} = c\boldsymbol{\nu}$$

$$l_0 \simeq c\nu_0 + \frac{m^2}{2c\nu_0}.$$

$L_{\alpha\beta}$ then reduces to

$$L_{\alpha\beta} = -4c\nu_\alpha \nu_\beta + O(m^2/\nu_0^2)$$

and hence

$$L_{\alpha\beta} B_{\alpha\beta} = -4c\nu_\alpha \nu_\beta \langle k \mid h_\alpha^+ \mid p\rangle \langle p \mid h_\beta^- \mid k\rangle.$$

Since

$$k - p = (1-c)\nu + O(m^2/\nu_0^2)$$

we get

$$L_{\alpha\beta} B_{\alpha\beta} = \frac{4c}{1-c} \mid \langle k \mid d^+ \mid p\rangle \mid^2 + O(m^2/\nu_0^2).$$

This concludes the proof of Adler's theorem.

We can therefore relate the quantities $A^\pm(W, 0)$, i.e. essentially the divergence of $g_\mu^\pm$ (see equations [3.23] to [3.27]), to the total cross-sections of neutrinos and antineutrinos on protons. However these cross-sections are not, at the moment, sufficiently well known to obtain in this way a useful sum rule for g_A.

We thus add a third assumption, namely the Partial Conservation of the Axial Current (PCAC).[21] That is

$$d^\pm(x) = \frac{\sqrt{2} M m_\pi^2}{g_r} g_A \phi^\pm(x) \qquad \text{...[3.30]}$$

where $\phi^{\pm}$ is the charged pion field, m_π its mass and g_r the renormalized pion-nucleon coupling constant. $A^{\pm}(W, 0)$ can then be written in the form

$$A^{\pm}(W,0) = \frac{4M^2}{g_r^2} \sum_{ss'} \sum_k{}' \delta(W-M_k)\, p_0 k_0 \mid \langle k \mid J_\pi^{\pm} \mid p \rangle \mid^2$$

where $J_\pi^{\pm}$ are the sources of the pion fields. Using this expression it is possible to relate $A^{\pm}(W, 0)$ to the cross-sections $\sigma^{\pm}$ for scattering of zero mass $\pi^{\pm}$-mesons on protons (for details see reference 17). The result is

$$A^{\pm}(W,0) = \frac{4M^2}{g_r^2 \pi} W(W^2-M^2)\sigma^{\pm}(W).$$

Introducing the variable $\omega = (W^2-M^2)/2M$ we finally obtain the result:

$$1-\frac{1}{g_A^2} = \frac{2M^2}{\pi g_r^2} \int_{m_\pi + m_\pi{}^2/2M}^{\infty} \frac{d\omega}{\omega} [\sigma^+(\omega)-\sigma^-(\omega)]. \qquad ...[3.31]$$

We shall not discuss the numerical evaluation of the integral over the cross-section which involves off-mass-shell contributions. These can be estimated in a reasonably reliable way[15, 16, 17] and lead to the result

$$g_A = 1{\cdot}24$$

which is very close to the experimental value $1{\cdot}18 \pm 0{\cdot}02$[22].

This derivation of the sum rule [*3.31*] rests on three assumptions:

(i) Validity of the commutation relations [*3.14*]

(ii) Validity of PCAC [*3.30*]

(iii) Possibility of exchanging the limit $p \to \infty$ with the integral in [*3.25*].

The last assumption can be shown[17] to be equivalent to requiring that a function $G(\omega)$, which is even in ω, is analytic in the ω-plane, apart from cuts running from $\pm[m_\pi + m_\pi^2/2M]$ to $\pm\infty$, and whose imaginary part is

$$\text{Im}\, G(\omega) = \tfrac{1}{2}[\sigma^+(\omega)-\sigma^-(\omega)],$$

satisfies unsubtracted dispersion relations.

The use of the limit $p \to \infty$ has been essential in this derivation of the sum rule. The importance of this limit is due to the fact that as $p \to \infty$ the momentum transfer t between the initial (final) state and the intermediate state tends to zero *independently of the mass of the intermediate state*. This allows us to relate the divergence of the axial current to the forward lepton scattering,[17, 20] or to the pion-nucleon scattering. Moreover, as we have already remarked, the separation of the sum over intermediate states into its individual terms has, in this frame, an invariant meaning.

3.4. *Approximate saturation of the sum rule*

It is interesting to consider how the experimental value of G_A/G_V can be approximated to if the sum over intermediate states is restricted to a limited number of states. If we neglect all states except the nucleon we get $g_A = 1$.

By including the states contained in the 56-dimensional representation of $SU(6)$ one gets[23]

$$\frac{G_A}{G_V} = -\tfrac{5}{3}$$

which coincides with the result obtained[24] from static $SU(6)$. This is at first sight surprising, as one would not expect the states of $SU(6)$, or of its subgroup $[SU(3) \otimes SU(3)]_{\text{collinear}}$ (see (iv) section 3.1), to produce even approximate saturation of $[SU(3) \otimes SU(3)]_{\text{chiral}}$. The explanation is however quite simple.[25,26] Let us consider two states α, β with the same parity, the same mass, and the same momentum p along the 3rd axis. The matrix element of an odd charge at time $x_0 = 0$, $R^A(0)$, is then

$$\langle \alpha, p \mid R^A(0) \mid \beta, p \rangle = \langle \alpha, 0 \mid \int d^3x [g_0^A(x) + \beta g_3^A(x)] \mid \beta, 0 \rangle \quad ...[3.32]$$

where β is the velocity of the Lorentz transformation which brings the state $\mid p\rangle$ to the rest state, $\mid 0\rangle$. If α and β have the same parity the first term vanishes. Therefore for $p_0 \to \infty$ we get

$$\langle \alpha, p \mid R^A(0) \mid \beta, p \rangle_{p=\infty} = \langle \alpha, 0 \mid \int d^3x \hat{p} \cdot g^A(x) \mid \beta, 0 \rangle = \langle \alpha \mid Q^A(\hat{p}) \mid \beta \rangle \quad ...[3.33]$$

where $Q^A(\hat{p})$ is an element of $[SU(3) \otimes SU(3)]_{\text{collinear}}$. We thus conclude that the matrices of the elements of the collinear $SU(3) \otimes SU(3)$ and those of the chiral $SU(3) \otimes SU(3)$ are related, provided the masses of the two states are equal. This shows why the same results are obtained by saturating the algebra of $SU(6)$, or its subalgebra $[SU(3) \otimes SU(3)]_{\text{collinear}}$, at zero momentum and by using the same set of states to saturate the $[SU(3) \otimes SU(3)]_{\text{chiral}}$ at infinite momentum.

The possibility of improving the $SU(6)$ result by adding other states has been considered by several authors.[26,27] Gatto, Maiani and Preparata[27] take into account the states belonging to the 20-dimensional representation of $SU(6)$ with $L = 1$. In this way they obtain the relation

$$-\frac{G_A}{G_V} = \tfrac{1}{3}\,\frac{D+F}{D-F}$$

where D and F are the coupling constants for symmetrical and antisymmetrical $SU(3)$ couplings. With $G_A/G_V = -1{\cdot}20$ this gives $D/F = 1{\cdot}75$ in good agreement with experiment and close to the $SU(6)$ estimate.[24] However it has been shown by Cabibbo and Ruegg[28] that this agreement is not enough to prove that the choice of '20', $L = 1$, is unique. Other choices would give equally good results.

Let us finally remark that by taking into account the mass difference between baryons and N^* resonances, Adler[29] has obtained the value $G_A/G_V = -1{\cdot}44$. Its departure from the correct value $-1{\cdot}20$ gives a measure, according to Gernstein and Lee,[26] of the breaking of $SU(6)$.

4

Current Density Commutators

4.1. *Low-energy theorems* [30, 31]

Up to now we have considered the commutators between two charges, i.e. between the space integrals of the current densities. If we assume that the current densities have a definite behaviour under the transformations generated by the charges we can use directly the commutators between charges and currents. In this way we obtain relations between form factors instead of relations between their squares.

Let us consider the commutators between the elements of the chiral $SU(3) \otimes SU(3)$ algebra and the vector and axial-vector current densities which are both assumed (equations [*2.18*] and [*3.1*]) to transform according to the adjoint representation of $SU(3)$. Let us assume, in addition, a definite transformation law under $[U(3) \otimes U(3)]_{\text{chiral}}$, namely

$$[Q^A, j^B_\lambda] = [R^A, g^B_\lambda] = if^{ABC} j^C_\lambda \qquad \ldots[4.1]$$

$$[Q^A, g^B_\lambda] = [R^A, j^B_\lambda] = if^{ABC} g^C_\lambda. \qquad \ldots[4.2]$$

If we take the unrenormalized Cabibbo angles for the vector and axial currents to be equal (section 3.2), we can write the hadron current in the form

$$h^+_\lambda = at^+_\lambda + bv^+_\lambda \qquad \ldots[4.3]$$

where

$$|a|^2 + |b|^2 = 1$$

and

$$t^+_\lambda = (j^1_\lambda + ij^2_\lambda) + (g^1_\lambda + ig^2_\lambda) \qquad \ldots[4.4]$$

$$v^+_\lambda = (j^4_\lambda + ij^5_\lambda) + (g^4_\lambda + ig^5_\lambda) \qquad \ldots[4.5]$$

are the isovector (t_λ) and isospinor (v_λ) parts of h_λ.

From [*4.1*] and [*4.2*] we get

$$[R^3, t^\pm_\lambda] = \pm t^\pm_\lambda \qquad \ldots[4.6]$$

$$[R^3, v^\pm_\lambda] = \pm \tfrac{1}{2} v^\pm_\lambda. \qquad \ldots[4.7]$$

The time derivative of R^3 is related by PCAC [*3.30*] to the space integral of the neutral pion field

$$\dot{R}^3 = m_\pi^2 f \int d^3x \, \phi^0(x) \qquad \ldots[4.8]$$

where

$$f = \frac{M}{g_r m_\pi}\left(-\frac{G_A}{G_V}\right) m_\pi = (0{\cdot}60 \pm 0{\cdot}06) m_\pi.$$

By formally integrating [*4.8*] we get

$$R^3(t) = m_\pi^2 f \int d^4x \, \theta(t - x_0) \, \phi^0(x). \qquad \ldots[4.9]$$

The matrix elements of [4.6] and [4.7] are therefore

$$f\langle\beta \mid \int d^4x\,\theta(y_0-x_0)\,m_\pi^2[\phi^0(x), t_\lambda^\pm(y, y_0)] \mid \alpha\rangle = \pm\langle\beta \mid t_\lambda^\pm(y, y_0) \mid \alpha\rangle, \quad \text{...[4.10]}$$

$$f\langle\beta \mid \int d^4x\,\theta(y_0-x_0)\,m_\pi^2[\phi^0(x), v_\lambda^\pm(y, y_0)] \mid \alpha\rangle = \pm\tfrac{1}{2}\langle\beta \mid v_\lambda^\pm(y, y_0) \mid \alpha\rangle. \quad \text{...[4.11]}$$

The L.H.S. are the amplitudes for the transitions $\alpha \to \beta+\pi^0$ when the pion momentum $p \to 0$. Hence

$$f \lim_{p \to 0} (2p_0)^{\frac{1}{2}}\langle\beta+\pi^0 \mid t_\lambda^\pm \mid \alpha\rangle = \pm\langle\beta \mid t_\lambda^\pm \mid \alpha\rangle \quad \text{...[4.12]}$$

$$f \lim_{p \to 0} (2p_0)^{\frac{1}{2}}\langle\beta+\pi^0 \mid v_\lambda^\pm \mid \alpha\rangle = \pm\tfrac{1}{2}\langle\beta \mid v_\lambda^\pm \mid \alpha\rangle. \quad \text{...[4.13]}$$

The parity of the state $\mid \beta+\pi\rangle$ when $p \to 0$ is opposite to that of $\mid \beta\rangle$. Therefore if only the axial part of $t(v)$ contributes to $\alpha \to \beta$, only the vector part contributes to $\alpha \to \beta+\pi$.

By using [4.12] and [4.13] a number of interesting results are obtained:

(i) *K_{l2} and K_{l3} decay.*

Let us take $\mid \alpha\rangle = \mid K^+\rangle$ and $\mid \beta\rangle = \mid 0\rangle$. The two matrix elements in [4.13] are given in terms of three form factors:

$$\sin\theta\,\langle 0 \mid v_\lambda^- \mid K^+\rangle = \frac{1}{(2k_0)^{\frac{1}{2}}} f_K k_\lambda \quad \text{...[4.14]}$$

$$\sin\theta\,\langle \pi^0 \mid v_\lambda^- \mid K^+\rangle = \frac{1}{(4k_0p_0)^{\frac{1}{2}}}\{f_+(q^2)P_\lambda+f_-(q^2)q_\lambda\} \quad \text{...[4.15]}$$

where k is the K^+ four-momentum, $q = k-p$ and $P = k+p$.

From [4.13] we obtain a relation between the axial form factor f_K and the vector form factors f_+ and f_-. The latter are functions of the momentum transfer q^2 and of the pion mass m_π^2. Equation [4.13] gives a condition on their values at the unphysical point $q^2 = -M_K^2$ and $m_\pi^2 = 0$, namely

$$|f_K| = 2f[f_+(-M_K^2)+f_-(-M_K^2)] = 1{\cdot}2m_\pi[f_+(-M_K^2)+f_-(-M_K^2)]. \quad \text{...[4.16]}$$

Experimentally $f_K = (0{\cdot}070 \pm 0{\cdot}001)\,M_K$, where M_K is the K-meson mass. The experimental values for the vector form factors f_+ and f_- are

$$|f_+| = 0{\cdot}16 \pm 0{\cdot}01; \qquad \left|\frac{f_-}{f_+}\right| = 0{\cdot}46 \pm 0{\cdot}27.$$

If we use these values, which means neglecting off-mass-shell corrections arising from the extrapolation to $m_\pi^2 = 0$, we obtain

$$f_K/M_K = 0{\cdot}074 \pm 0{\cdot}014$$

which is very close to the experimental value.

(ii) $\pi^+ \to l^+ + \nu$ *and* $\pi^+ \to \pi^0 + l^+ + \nu$

With the same technique we can compare these two processes by using [*4.12*], which establishes a relation between the axial and vector form factors for $\Delta Y = 0$ processes. If f_π and f_+^0 are the axial and vector form factors ($f_-^0 = 0$ due to the conservation of the $\Delta Y = 0$ current) we obtain

$$f_\pi = f \cdot f_+^0(-m_\pi^2),$$

which is again rather well satisfied ($f_\pi = 0{\cdot}94 m_\pi, f \cdot f_+^0 = 0{\cdot}85 m_\pi$).

The same technique can be applied to derive a relation between the processes K_{l3} and K_{l4}: the results are also in this case rather satisfactory.[30, 32]

The methods discussed in this section have also been extended to non-leptonic processes.[33] The results, though not yet conclusive, seem to indicate that some of the relations attributed to the $\Delta T = \frac{1}{2}$ rule can probably be obtained without introducing explicitly neutral strange currents. Further work is necessary before this result can be considered as established.

4.2. *The axial-vector form factor*[34, 35, 36]

The method discussed in the previous section can be applied to the study of the axial-vector form factor of nucleons. From the set of equations [*4.1*] and [*4.2*] we consider those involving the isovector part of the electromagnetic current $j_\lambda^3(x)$ and the odd operators R^1, R^2, R^3 which commute with the hypercharge

$$[R^a, j_\lambda^3] = i\varepsilon^{a3b} g_\lambda^b \qquad \ldots[4.17]$$

where $a, b = 1, 2, 3$.

Let us take the matrix element of [*4.17*] between nucleon states. The R.H.S. can be written in the form

$$G_V \langle p' \mid g_\lambda^a(0) \mid p \rangle = \tfrac{1}{2}\bar{u}(p')\tau^a[g_A \gamma_\lambda \gamma_5 G(q^2) - \frac{i}{2M} q_\lambda \gamma_5 H(q^2)]u(p) \qquad \ldots[4.18]$$

where τ^a are the isotopic spin 2×2 matrices, $g_A = -G_A/G_V$, $q = p' - p$, and $G(q^2)$, $H(q^2)$ are the two axial-vector form factors ($G(0) = 1$). By using PCAC we can establish a relation between G and H. For this purpose it is convenient to separate from g_λ^a the divergenceless part: we thus rewrite [*4.18*] in the form

$$G_V \langle p' \mid g_\lambda^a \mid p \rangle = \tfrac{1}{2}\left[\bar{u}(p')\tau^a \gamma_5\left(-\gamma_\lambda - \frac{2iMq_\lambda}{q^2}\right)u(p)\right] g_A G(q^2)$$

$$+\frac{i}{2}[\bar{u}(p')\tau^a \gamma_5 u(p)] q_\lambda \left(\frac{-H(q^2)}{2M} + \frac{2Mg_A G(q^2)}{q^2}\right). \qquad \ldots[4.19]$$

One can verify that the first and second terms on the R.H.S. are the matrix elements of the divergenceless part and rotationless part respectively. Indeed

$$G_V i q_\lambda \langle p' \mid g^a_\lambda \mid p\rangle = \frac{1}{2}\bar{u}(p')\tau^a\gamma_5 u(p)\left[-2Mg_A G(q^2)+\frac{q^2H(q^2)}{2M}\right] \qquad ...[4.20]$$

and

$$G_V\left(\varepsilon_\lambda - \frac{q_\lambda(\varepsilon\cdot q)}{q^2}\right)\langle p' \mid g^a_\lambda \mid p\rangle$$
$$= -\tfrac{1}{2}\bar{u}(p')\gamma_5\tau^a\left(\gamma\cdot\varepsilon+\frac{2iM(q\cdot\varepsilon)}{q^2}\right)u(p)g_A G(q^2), \qquad ...[4.21]$$

where ε_λ is an arbitrary unit vector.

Let us assume PCAC and suppose that the matrix element of the pion field is of the form

$$\langle p' \mid \phi^+(0) \mid p\rangle = \frac{g_r}{q^2+m_\pi^2}\cdot\bar{u}\tau^+\gamma_5 u.$$

From [4.20] we then get a relation between $G(q^2)$ and $H(q^2)$, namely

$$-2Mg_A G(q^2)+\frac{q^2H(q^2)}{2M} = \frac{g_r}{q^2+m_\pi^2}\cdot\frac{Mm_\pi^2 g_A}{g_r}, \qquad ...[4.22]$$

that is

$$H(q^2) = \frac{4M^2 g_A}{q^2}\left(G(q^2)+\frac{m_\pi^2}{q^2+m_\pi^2}\right). \qquad ...[4.23]$$

We now consider the matrix element of the commutator in the L.H.S. of [4.17]. With the same argument as used in section 4.1 we write

$$\langle p' \mid [R^a, j^3_\lambda] \mid p\rangle = f\lim_{p_\pi\to 0}(2p_{0\pi})^{\frac{1}{2}}\langle p'+\pi^a \mid j^3_\lambda \mid p\rangle, \qquad ...[4.24]$$

where p_π is the pion momentum. If we multiply [4.24] by $[\varepsilon_\lambda - q_\lambda(\varepsilon\cdot q)/q^2]$ we obtain the amplitude for photoproduction (or electron production) of zero-energy mesons from nucleons. In the limit $p_\pi \to 0$ the amplitude for photomeson production contains only two functions[37] and we can write

$$f\lim_{p_\pi\to 0}(2p_{0\pi})^{\frac{1}{2}}\langle p'+\pi^a \mid j^3_\lambda \mid p\rangle[\varepsilon_\lambda - q_\lambda(q\cdot\varepsilon)/q^2]$$
$$= \tfrac{1}{2}\bar{u}(p')\gamma_5\left\{\left(\gamma\cdot\varepsilon+2i\frac{M(q\cdot\varepsilon)}{q^2}\right)\mathscr{F}_1(q^2)[\tau^a,\tau^3]+\frac{(p+p')\cdot\varepsilon}{2M}\delta_{a3}\mathscr{F}_2(q^2)\right\}u(p). \qquad ...[4.25]$$

The functions $\mathscr{F}_1$ and $\mathscr{F}_2$ can be expressed in terms of the nucleon form factors and integrals over the total photoproduction cross-section. We shall not discuss the general result[34,35] but only quote the approximate solution obtained from a limited number of states.[36] We first remark that from [4.17], [4.21] and [4.25] we get

$$\mathscr{F}_2(q^2) = 0, \quad \mathscr{F}_1(q^2) = g_A G(q^2). \qquad ...[4.26]$$

By separating the nucleon contribution Furlan, Jengo and Remiddi[36] obtained

$$G(q^2) = F_1^V(q^2) + \text{continuum}$$
$$0 = F_2^V(q^2) + \text{continuum}$$

where $F_1^V(q^2)$ and $F_2^V(q^2)$ are the Dirac and Pauli isovector form factors. An approximate evaluation of the continuum, by only taking into account the N^* resonance, yields

$$G(q^2) = F_1^V(q^2) - q^2 C \qquad \text{...[4.27]}$$
$$0 = F_2^V(q^2) + 4q^2 C \qquad \text{...[4.28]}$$

where C is related to the couplings $N^*N\pi$ and $N^*N\gamma$. From [*4.27*] and [*4.28*] one gets

$$G(q^2) = F_1^V(q^2) + \frac{q^2}{4M^2} F_2^V(q^2). \qquad \text{...[4.29]}$$

If we assume that F_1 and F_2 are dominated by the ρ-pole and G by a $J = 1^+$ meson of mass M_A we get

$$M_A \simeq 1{\cdot}1 \text{ GeV}. \qquad \text{...[4.30]}$$

This is close to the mass of the A_1-resonance and to the value obtained from the neutrino experiment ($M_A = (0{\cdot}84 \pm 0{\cdot}3)$ GeV). It should be pointed out however that the errors introduced by the approximate treatment of the continuum contribution are difficult to evaluate. The significance of [*4.30*] is therefore not altogether clear.

4.3. *Sum rule for the isovector charge radii*

In the previous sections we have discussed the commutation relations between two charges, and those between charges and current densities. We now consider the commutators of two current densities under the assumption that they do not contain other singularities besides a δ-function (see section 3.1). In this way it is possible to calculate the commutator of the moments of the currents and derive interesting results between them.

Lee[38] and Gell-Mann and Dashen[39] have used the commutation relations of the space components of the vector current to calculate the commutator between the magnetic moments, defined by

$$\mu_i^A = \tfrac{1}{2}\varepsilon_{ijk} \int d^3x\, x_j j_k^A(x). \qquad \text{...[4.31]}$$

By approximating the sum over the intermediate states with the states of the 56-dimensional representation of $SU(6)$ one gets a relation between the isovector nucleon charge radius and the magnetic moment which is in reasonably good agreement with experiment.

In this section we discuss a different method[25] which only uses the commutator between the time components of the isovector current. Let us assume the following equal time commutator

$$[j_0^+(\boldsymbol{x}, 0), j_0^-(\boldsymbol{x}', 0)] = 2\delta(\boldsymbol{x}-\boldsymbol{x}')j_0^3(\boldsymbol{x}, 0) \qquad \ldots[4.32]$$

where $j_0^{\pm} = (j_0^1 \pm ij_0^2)$ and j_0^3 are the three components of the isovector current. We define the electric dipole moment operators by the relation ($i = 1, 2, 3$)

$$D_i^{\pm} = \int d^3x j_0^{\pm}(x) x_i. \qquad \ldots[4.33]$$

From [4.32] the commutator of the components along the first axis is

$$[D_1^+, D_1^-] = 2 \int d^3x j_0^3(x)(x_1)^2. \qquad \ldots[4.34]$$

Let us take the matrix element of [4.34] between proton states with momentum p along the third axis. In the L.H.S. we insert a complete set of states and separate out the contribution of the neutron intermediate states

$$\begin{aligned}\langle p' \mid [D_1^+, D_1^-] \mid p\rangle &= \sum_n \langle p' \mid D_1^+ \mid n\rangle\langle n \mid D_1^- \mid p\rangle \\ &\quad + \sum_{l \neq n} (\langle p \mid D_1^+ \mid l\rangle\langle l \mid D_1^- \mid p\rangle - (+ \leftrightarrow -)).\end{aligned} \qquad \ldots[4.35]$$

Let us first discuss the one-neutron contribution. The matrix element is of the form

$$\begin{aligned}\langle n \mid D_1^- \mid p\rangle &= \langle p+q \mid D_1^- \mid p\rangle = \\ &= \left(\frac{k_0 M}{(p_0+q_0)p_0}\right)^{\frac{1}{2}} \langle k \mid \int d^3x x_1 \{j_0^-(\boldsymbol{x}, \beta x_3) + \beta j_3^-(\boldsymbol{x}, \beta x_3)\} \mid 0\rangle\end{aligned} \qquad \ldots[4.36]$$

where $q = n-p$ and $k = \Lambda(p)(p+q)$, $\Lambda(p)$ being the Lorentz transformation in the direction of $\boldsymbol{p}$ with velocity $\beta = -p_3/p_0$ ($\boldsymbol{p}$ is supposed to be parallel to the third axis); $\mid 0\rangle$ is the state with zero momentum obtained from $\mid p\rangle$ by means of $\Lambda^{-1}(p)$. By using translation invariance we rewrite [4.36] in the form

$$\begin{aligned}\langle n \mid D_1^- \mid p\rangle &= \left(\frac{k_0 M}{(p_0+q_0)p_0}\right)^{\frac{1}{2}} i(2\pi)^3 \delta(k_1)\delta(k_2)\delta[k_3 - \beta(k_0 - M)] \\ &\quad \times \frac{\partial}{\partial q_1} \langle k \mid [j_0^-(0) + \beta j_3^-(0)] \mid 0\rangle.\end{aligned} \qquad \ldots[4.37]$$

Since $k_1 = q_1$, $k_2 = q_2$, $k_3 = \dfrac{p_0}{M}[p_3+q_3+\beta(q_0+p_0)]$ and $k_0 = M+(p_0q_0 - p_3q_3)/M$, the product of the three δ-functions reduces to $\delta(\boldsymbol{q})\dfrac{p_0}{M}$. Therefore

$$\langle n \mid D_1^- \mid p\rangle = i(2\pi)^3 \delta(\boldsymbol{q}) \left(\frac{k_0}{(p_0+q_0)}\right)^{\frac{1}{2}} \frac{\partial}{\partial q_1} \langle k \mid (j_0^-(0) + \beta j_3^-(0)) \mid 0\rangle. \qquad \ldots[4.38]$$

The matrix element in the R.H.S. of [*4.38*] is of the form

$$u_n^\dagger(0)\,\frac{[p_0+q_0+M+\boldsymbol{\alpha}\,.\,(\boldsymbol{p}+\boldsymbol{q})][p_0+M-\alpha_3 p_3]}{[2M(p_0+q_0+M)2M(p_0+M)]^{\frac{1}{2}}}\,.\,\gamma_4[(-i\gamma_4+\beta\gamma_3)F_1(q^2)$$

$$+\frac{\kappa_p-\kappa_n}{2M}\,.\,(\sigma_{0\nu}+\beta\sigma_{3\nu})q_\nu F_2(q^2)]u_p(0)$$

where $u_n(0)$, $u_p(0)$ are the wave functions for zero momentum of the neutron and proton respectively; κ_p and κ_n are the proton and neutron anomalous magnetic moments and F_1 and F_2 are the Dirac and Pauli isovector form factors.

In the limit $p_3 \to \infty$ equation [*4.38*] becomes

$$\lim_{p_3\to\infty}\langle n \mid D_1^- \mid p\rangle = \frac{\kappa_p-\kappa_n}{2M}(2\pi)^3\delta(\boldsymbol{p}-\boldsymbol{n}')u_n^\dagger\sigma_{31}u_p. \qquad \ldots[4.39]$$

Similarly we can prove that the matrix element of the R.H.S. of [*4.34*] is proportional to $\left(\frac{\partial F_1}{\partial q^2}\right)_{q^2=0}$. Thus from [*4.34*] and [*4.35*] we get

$$-\frac{(\kappa_p-\kappa_n)^2}{8M^2}-\left(\frac{\partial F_1}{\partial q^2}\right)_{q^2=0} = \tfrac{1}{2}\lim_{p_3\to\infty}A(p,p') \qquad \ldots[4.40]$$

where

$$A(p,p') = 4A_1(p,p')-2A_3(p,p') \qquad \ldots[4.41]$$

is the contribution from all intermediate states $\mid l\rangle$ different from the neutron; A_1 and A_3 are the separate contribution from $T = 1/2$ and $3/2$ states. The L.H.S. of [*4.40*] can be rewritten as

$$-\frac{(\mu_p-\mu_n)^2}{8M^2}-\left[\left(\frac{\partial G_E}{\partial q^2}\right)_{q^2=0}-\frac{1}{8M^2}\right] \equiv -\frac{(\mu_p-\mu_n)^2}{8M^2}+\tfrac{1}{6}r^2 \qquad \ldots[4.42]$$

where $\mu_p = 1+\kappa_p$, $\mu_n = \kappa_n$, G_E is the isovector charge form factor, and r^2 is the mean-square charge radius of the proton. To discuss the inelastic contribution we use the identity

$$\langle l \mid D_1^+ \mid p\rangle = (2\pi)^3\delta(\boldsymbol{p}-\boldsymbol{p}')\frac{\langle l \mid j_1^+(0)\mid p\rangle}{i(l_0-p_0)} \qquad \ldots[4.43]$$

where l_0 and p_0 are the energies of the two states. The last matrix element is related by isotopic spin invariance to the matrix element of $\langle l^- \mid j_1^3 \mid p\rangle$ i.e. to the unphysical amplitude for the process $\gamma+p \to l^+(\langle l^- \mid = \langle l \mid T^+)$. Now let $p \to \infty$. Following Adler's derivation of the sum rule for G_A/G_V we can relate $\lim_{p\to\infty} A(p,p')$ to the integral over the total cross-sections for photoproduction by the isovector current j_λ^3 of states with $T = 3/2$, $1/2$ from protons. The final result is

$$\left(\frac{\mu_p-\mu_n}{2M}\right)^2+\frac{1}{2\pi^2\alpha}\int\frac{d\omega}{\omega}\,[2\sigma_1^V(\omega)-\sigma_3^V(\omega)]=\frac{r^2}{3}. \qquad ...[4.44]$$

The derivation of [*4.44*] depends (as in the analogous case of the Adler-Weisberger sum rule) on the possibility of exchanging the limit $p \to \infty$ with the integration over the intermediate energy. This exchange is justified only if $[2\sigma_1(\omega)-\sigma_3(\omega)]$ tends to zero as ω tends to infinity.

To check the plausibility of this convergence assumption let us suppose that the isovector current is dominated by the ρ-resonance. The cross-section for $\gamma+p \to l$ due to the isovector current is in this case proportional to the cross-section for $\rho^0+p \to l$ extrapolated to zero ρ-mass. The latter can be expressed in terms of σ_3^V and σ_1^V. We find:

$$2\sigma_1^V(\gamma+p)-\sigma_3^V(\gamma+p)=b[\sigma(\rho^-p)-\sigma(\rho^+p)] \qquad ...[4.45]$$

where $\sigma(\rho^-p)$ and $\sigma(\rho^+p)$ are the total cross-sections for ρ^- and ρ^+ on protons and b is a coefficient. In this model therefore the vanishing of the denominator is ensured by Pomeranchuk's theorem.

A detailed analysis of the experimental data on photoproduction cross-sections has been carried out by Gilman and Schnitzer[40] who have found

$$\frac{0{\cdot}066}{m_\pi^2}=\frac{r^2}{6}=\frac{0{\cdot}059}{m_\pi^2}+\frac{0{\cdot}004}{m_\pi^2}+\text{contribution from above 800 MeV.} \qquad ...[4.46]$$

The L.H.S. is the value obtained from electron nucleon scattering.[41] The first term in the R.H.S. is the nucleon contribution

$$\frac{(\mu_p-\mu_n)^2}{8M^2}=0{\cdot}059/m_\pi^2$$

and the second comes from $N^*(1239)$, $N^{**}(1520)$ and from E_{0+} multipole. The contribution from above 800 MeV is thus seen to be relatively unimportant.

The commutator [*4.34*] can also be used to evaluate the pion radius.[25] For this we take the matrix element between π^+ states and follow the same procedure used for nucleons (there is of course no one-pion intermediate state contribution in this case). The result is

$$\tfrac{2}{3}r_\pi^2=\frac{1}{2\pi^2\alpha}\int\frac{d\omega}{\omega}\,\{\sigma^V(\gamma+\pi^0\to T=0)+\sigma^V(\gamma+\pi^+\to T=1)$$
$$-\tfrac{5}{4}\sigma^V(\gamma+\pi^0\to T=2)\}. \qquad ...[4.47]$$

Let us isolate the vector meson intermediate state in the integral. The main contribution comes from $\gamma+\pi^0 \to \omega^0$ and can be computed from the rate of the inverse process $\omega^0 \to \pi^0+\gamma$. The contribution from the ϕ meson is probably smaller and we shall neglect it. In this way we get

$$r_\pi^2 = \frac{18}{\alpha}\left(\frac{M_\omega}{M_\omega^2 - m_\pi^2}\right)^3 \Gamma(\omega^0 \to \pi^0 + \gamma). \qquad ...[4.48]$$

Gilman and Schnitzer[40] have estimated the contribution of the continuum and have concluded that the R.H.S. of [*4.14*] should be multiplied by (25/9) $(G_V/G_A)^2 \simeq 2$. With this correction, and using the experimental value $\Gamma(\omega^0 \to \pi^0 + \gamma) = (1{\cdot}3 \pm 0{\cdot}3)$ MeV, they obtain

$$\tfrac{1}{6} r_\pi^2 = (0{\cdot}045 \pm 0{\cdot}010)\, m_\pi^{-2},$$

to be compared with the value deduced from electron pion production[42]

$$[\tfrac{1}{6} r_\pi^2]_{\text{exp}} = (0{\cdot}041 \pm 0{\cdot}020)\, m_\pi^{-2}.$$

5

The Infinite-Parameter Algebra and the Dispersion Method

5.1. *The infinite-parameter algebra associated with the current densities*

In the previous sections we have discussed three types of commutation relations:

(i) Commutators of two charges, a charge being defined as the integral over space of the corresponding current density (see section 3.1). An example of the use of this type of commutation relation is the Adler-Weisberger sum rule.

(ii) Commutators between one charge and one current. This has been used to derive the relation between the vector and axial-vector form factors.

(iii) Commutators between two currents. An example of their use is the sum rule for the isovector charge radius.

Clearly (i) and (ii) follow from (iii) but not vice versa. Addition of gradient terms in the commutation relations of two currents would not alter (i) while modifying the results obtained from (iii).

Only in case (i) has one a Lie algebra: in the other two cases the mathematical structure is far more complicated. Recently Dashen and Gell-Mann[43] have suggested considering (i) and (ii) as special cases of commutators between the elements of an infinite-parameter algebra.

Let us consider a system of current densities $c_\alpha(\boldsymbol{x}, t)$, distinguished by an index α which need not at the moment be specified. The $c_\alpha(\boldsymbol{x}, t)$ are assumed to satisfy the equal time commutator

$$[c_\alpha(\boldsymbol{x}, 0), c_\beta(\boldsymbol{x}', 0)] = i e_{\alpha\beta\gamma} c_\gamma(\boldsymbol{x}, 0)\, \delta(\boldsymbol{x} - \boldsymbol{x}') \qquad ...[5.1]$$

where $e_{\alpha\beta\gamma}$ are the structure constants of a Lie algebra.

The commutator of the Fourier transform of the current densities

$$c_\alpha(\boldsymbol{q}) = \int d^3x\, e^{i\boldsymbol{q}.\boldsymbol{x}} c_\alpha(\boldsymbol{x}, 0) \qquad \ldots[5.2]$$

is given by

$$[c_\alpha(\boldsymbol{q}), c_\beta(\boldsymbol{q}')] = ie_{\alpha\beta\gamma} c_\gamma(\boldsymbol{q}+\boldsymbol{q}'). \qquad \ldots[5.3]$$

The operators $c_\alpha(\boldsymbol{q})$ can thus formally be considered as the basis of an infinite-parameter algebra whose elements are labelled by a discrete parameter, α and a continuous one, $\boldsymbol{q}$. The elements $c_\alpha(\boldsymbol{q} = 0)$ form a finite-dimensional subalgebra, the subalgebra of the charges

$$C_\alpha = \int d^3x c_\alpha(\boldsymbol{x}, 0) = c_\alpha(\boldsymbol{q} = 0). \qquad \ldots[5.4]$$

As an example of this infinite-dimensional algebra let us take $e_{\alpha\beta\gamma}$ to be the structure constants of $SU(2)$ and let us only consider vectors $\boldsymbol{q}$ parallel to one fixed vector $\boldsymbol{a}$. Through the points which represents the elements $c_+(0)$, $c_3(0)$, $c_-(0)$ in the Cartan diagram we draw three lines l_+, l_3, l_- parallel to $\boldsymbol{a}$. (See figure.)

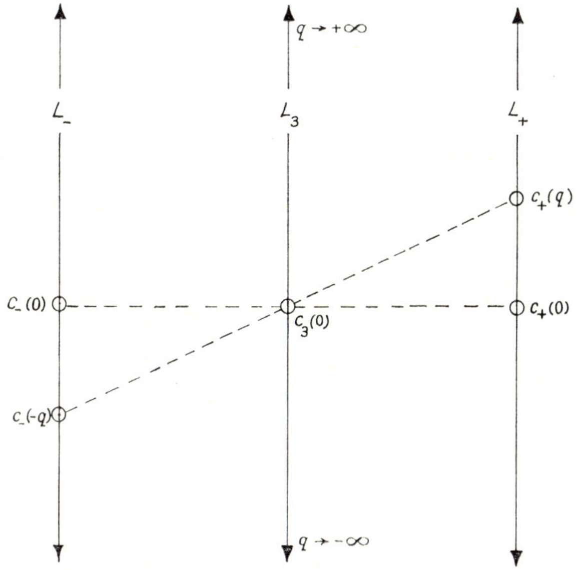

To each point on the lines l_+, l_3, l_-, at distance q from $c_+(0)$, $c_3(0)$, $c_-(0)$ respectively, we associate the elements $c_+(q)$, $c_3(q)$, $c_-(q)$. The following properties are easily verified:

(i) Two points on the same line commute
(ii) The points $c_+(q)$, $c_3(0)$, $c_-(-q)$ belong to a subalgebra denoted by $SU(2)_q$
(iii) Each point on l_+ (l_-) is a raising (lowering) operator for $c_3(0)$.

A representation of this infinite-dimensional algebra, denoted by $[SU(2)]_\infty$, is easily obtained. Let $\frac{1}{2}\sigma_\pm = \frac{1}{2}(\sigma_1 \pm i\sigma_2)$, $\frac{1}{2}\sigma_3$ be the 2×2 unitary representation of $[SU(2)]_{q=0}$. A representation of the algebra with $\boldsymbol{q}$ parallel to $\boldsymbol{a}$ is then

$$\tfrac{1}{2}\sigma_i\, e^{iq.a}.$$

Other representations can be constructed by the usual procedure: take a set of $3N$ matrices $\sigma_i^{(n)}$ such that

$$[\sigma_i^{(n)}, \sigma_j^{(m)}] = 2i\varepsilon_{ijk}\,\sigma_k^{(n)}\,\delta_{nm} \qquad \text{...[5.5]}$$

and a set of vectors $\boldsymbol{a}^{(n)}$; then

$$c_i(\boldsymbol{q}) = \tfrac{1}{2}\sum_{n=1}^{N} \sigma_i^{(n)} \exp\left(iq\,.\,\boldsymbol{a}^{(n)}\right) \qquad \text{...[5.6]}$$

is a representation of the infinite-parameter algebra. Other representations can be obtained from the above by unitary transformations. $[SU(2)]_\infty$ has other representations which are not of the separable form [5.6]. They will not be discussed here.‡

The case of $[SU(3)]_\infty$ can be treated in a similar way. The generalized Cartan diagram is in this case an infinite hexagonal prism whose sides are parallel to a vector $\boldsymbol{a}$. We choose a plane orthogonal to $\boldsymbol{a}$ to represent the Cartan diagram of the $\boldsymbol{q} = 0$ subalgebra

$$[c^A(0), c^B(0)] = if^{ABC}c^C(0)$$

and to fix the origin of the coordinate system on each side. To each point on the edges we can then associate an element of an infinite-parameter algebra defined by the commutator

$$[c^A(\boldsymbol{q}), c^B(\boldsymbol{q}')] = if^{ABC}c^C(\boldsymbol{q}+\boldsymbol{q}'). \qquad \text{...[5.7]}$$

The subalgebras $[SU(3)]_{\boldsymbol{q}}$ are obtained by considering the elements on the planes through one of the symmetry axes of the $\boldsymbol{q} = 0$ plane.

Let us apply this point of view to the commutation relations of the time components of the vector and axial-vector currents,

$$\left.\begin{aligned} [j_0^A(\boldsymbol{x}), j_0^B(\boldsymbol{x}')] &= if^{ABC}j_0^C(\boldsymbol{x})\,\delta(\boldsymbol{x}-\boldsymbol{x}') \\ [j_0^A(\boldsymbol{x}), g_0^B(\boldsymbol{x}')] &= if^{ABC}g_0^C(\boldsymbol{x})\,\delta(\boldsymbol{x}-\boldsymbol{x}') \\ [g_0^A(\boldsymbol{x}), g_0^B(\boldsymbol{x}')] &= if^{ABC}j_0^C(\boldsymbol{x})\,\delta(\boldsymbol{x}-\boldsymbol{x}'). \end{aligned}\right\} \qquad \text{...[5.8]}$$

The Fourier transforms

$$j^A(\boldsymbol{q}) = \int d^3x\, j_0^A(\boldsymbol{x}, 0)\, e^{iq.x} \qquad \text{...[5.9]}$$

$$g^A(\boldsymbol{q}) = \int d^3x\, g_0^A(\boldsymbol{x}, 0)\, e^{iq.x} \qquad \text{...[5.10]}$$

are an infinite-parameter algebra $[SU(3) \otimes SU(3)]_\infty$, the Lie product of two elements being given by

‡ P. De Mottoni, Thesis, Pisa 1966, unpublished.

$$\left.\begin{aligned}[j^A(\boldsymbol{q}), j^B(\boldsymbol{q}')] &= if^{ABC} j^C(\boldsymbol{q}+\boldsymbol{q}') \\ [j^A(\boldsymbol{q}), g^B(\boldsymbol{q}')] &= if^{ABC} g^C(\boldsymbol{q}+\boldsymbol{q}') \\ [g^A(\boldsymbol{q}), g^B(\boldsymbol{q}')] &= if^{ABC} j^C(\boldsymbol{q}+\boldsymbol{q}').\end{aligned}\right\} \qquad \ldots[5.11]$$

The $[SU(3) \otimes SU(3)]_0$ subalgebra of $[SU(3) \otimes SU(3)]_\infty$ coincides with $[SU(3) \otimes SU(3)]_{\text{chiral}}$ discussed in section 3.1. For $q = 0, q' \neq 0$ equations [*5.11*] are equivalent to the condition that j_λ^A and g_λ^A have a definite transformation property under the chiral group $SU(3) \otimes SU(3)$ (see section 3.5). Finally, when both $q, q \neq 0$, we have the commutator of two currents discussed in section 4.1.

Let us consider the matrix element of one element of $[SU(3) \otimes SU(3)]_\infty$, $j(\boldsymbol{q})$ say, between two states with momenta p and k and masses M_p and M_k. Because of the space integration in the definition of $j(\boldsymbol{q})$, the square of the momentum transfer is

$$t = -\boldsymbol{q}^2 + [(\boldsymbol{p}^2 + M_p^2)^{\frac{1}{2}} - ((\boldsymbol{p}-\boldsymbol{q})^2 + M_k^2)^{\frac{1}{2}}]^2.$$

When $\boldsymbol{p} \to \infty$, $t \to -\boldsymbol{q}^2$ (by $\boldsymbol{p} \to \infty$ we mean that one component becomes infinite, the other two remaining finite). In the $\boldsymbol{p} \to \infty$ system the square of the invariant momentum transfer is thus *independent of the masses of two states*. This is a generalization to the matrix elements of $j(\boldsymbol{q})$, $g(\boldsymbol{q})$ of the result already obtained for $j(0)$ and $g(0)$. (See sections 2.3 and 3.3.)

Let us take the matrix elements of equations [*5.11*] between infinite-momentum states α and β and sum over spin states. The R.H.S. are invariant functions of $(q+q')^2$ (if the states are covariantly normalized)

$$\sum_{S_\alpha S_\beta} \langle \beta \mid j^C(\boldsymbol{q}+\boldsymbol{q}') \mid \alpha \rangle = J_{\beta\alpha}^C(q+q')^2 \qquad \ldots[5.12]$$

$$\sum_{S_\alpha S_\beta} \langle \beta \mid g^C(\boldsymbol{q}+\boldsymbol{q}' \mid \alpha \rangle = G_{\beta\alpha}^C(q+q')^2. \qquad \ldots[5.13]$$

The left hand sides are also invariant functions which, after performing the summation over the internal variables of the intermediate, initial, and final states, can only depend upon the vectors q and q'. We thus obtain a system of coupled equations

$$\begin{aligned}\sum_n \{J_{\beta n}^A(q^2) J_{n\alpha}^B(q'^2) + q\,.\,q' L_{\beta n}^A(q^2) L_{n\alpha}^B(q'^2) + \cdots \\ -(A \leftrightarrow B)\} = if^{ABC} J_{\beta\alpha}^C((q+q')^2)\end{aligned} \qquad \ldots[5.14a]$$

$$\begin{aligned}\sum_n \{J_{\beta n}^A(q^2) G_{n\alpha}^B(q'^2) + q\,.\,q' L_{\beta n}^A(q^2) H_{n\alpha}^B(q'^2) + \cdots \\ -(A \leftrightarrow B)\} = if^{ABC} G_{\beta\alpha}^C((q+q')^2)\end{aligned} \qquad \ldots[5.14b]$$

$$\begin{aligned}\sum_n \{G_{\beta n}^A(q^2) G_{n\alpha}^B(q'^2) + q\,.\,q' H_{\beta n}^A(q^2) H_{n\alpha}^B(q'^2) + \cdots \\ -(A \leftrightarrow B)\} = if^{ABC} J_{\beta\alpha}^C((q+q')^2)\end{aligned} \qquad \ldots[5.14c]$$

where the dots stand for terms multiplied by higher powers of $q\,.\,q'$.

In principle, equations [*5.14*] contain all possible relations between the form factors of the vector and axial-vector currents which follow from the constraints imposed by the infinite-parameter algebra defined by [*5.11*].

For example, if we take α and β to be spin zero states with ($T = \frac{1}{2}$, $T_3 = \frac{1}{2}$, $Y = 1$) and ($T = 0$, $Y = 0$) respectively we can obtain from the equation [*5.14b*], with $q' = 0$, the Callan-Treiman relation (section 4.1) between the axial and vector form factors in K_{l2} and K_{l3}. Similarly from [*5.14a*] one obtains, by differentiating with respect to q and q', the sum rule for the isovector charge radius discussed in section 4.3. It should be clear from the preceding discussion that the use of the system $\boldsymbol{p} = \infty$ plays a key role in the deduction. Indeed it is only in this system that one gets automatically invariant results. This should not be surprising in view of the fact that the definition of the time equality for the two operators in a bracket depends upon the reference frame.

5.2. *The dispersion method of Fubini*

In this section we briefly outline an equivalent method proposed by Fubini[44] to discuss all the sum rules connected with a current commutator. Let us consider, for example, the commutator

$$[g_0^+(\boldsymbol{x}, 0)\, g_\nu^-(\boldsymbol{y}, 0)] = 2\delta(\boldsymbol{x}-\boldsymbol{y})\, j_\nu^3(\boldsymbol{x}). \qquad \text{...[5.15]}$$

(Other singular terms in the R.H.S. are probably present: though in some cases they can be taken into account[45] they are here disregarded.)

We now define two tensors:

$$T_{\mu\nu}(\nu, k^2, k'^2, q^2) = \int d^4x e^{ikx}\theta(x_0)\langle p' \mid [g_\mu^+(x), g_\nu^-(0)] \mid p\rangle \qquad \text{...[5.16]}$$

$$t_{\mu\nu}(\nu, k^2, k'^2, q^2) = \int d^4x e^{ikx}\langle p' \mid [g_\mu^+(x), g_\nu^-(0)] p\rangle \qquad \text{...[5.17]}$$

where $\mid p\rangle$ and $\mid p'\rangle$ are (for simplicity) spinless states with equal masses; $q^2 = -(p-p')^2$, $k^2 = -(p+k-p')^2$ and $\nu = \frac{1}{2}(p+p').k$.

$T_{\mu\nu}$ can be considered as the amplitude for the scattering of a virtual vector meson with momentum k by a spin zero meson with momentum p, the final state particles having momenta k' and p' respectively, so that $k+p = k'+p'$. $T_{\mu\nu}$ is the Fourier transform of an advanced causal commutator which vanishes outside the past light cone. It is therefore related to $t_{\mu\nu}$ by a dispersion relation which we shall assume to contain no subtractions. To specify this relation we introduce three independent vectors

$$P_\mu = \tfrac{1}{2}(p+p')_\mu;\; q_\mu = (p'-p)_\mu;\; Q_\mu = \tfrac{1}{2}(k+k')_\mu \qquad \text{....[5.18]}$$

and expand $T_{\mu\nu}$ in terms of them

$$\begin{aligned} T_{\mu\nu} = {} & AP_\mu P_\nu + B_1 P_\mu Q_\nu + B_2 P_\mu q_\nu + B_3 P_\nu Q_\mu + B_4 P_\nu q_\mu + \\ & + C_1 Q_\mu Q_\nu + C_2 Q_\mu q_\nu + C_3 Q_\nu q_\mu + C_4 q_\mu q_\nu + C_5 \delta_{\mu\nu} \end{aligned} \qquad \text{...[5.19]}$$

where A, B_i, C_i are functions of the invariants ν, k^2, k'^2, q^2. A similar expansion can be written for $t_{\mu\nu}$ and we will denote by a, b_i, and c_i the invariant functions which multiply the corresponding terms. The relations between A_1, B_i and C_i and a, b_i, c_i are then

$$\left.\begin{aligned} A(\nu, k^2, k'^2, q^2) &= \frac{1}{2\pi i}\int d\nu' \frac{a(\nu', k^2, k'^2, q^2)}{\nu'-\nu} \\ B_i(\nu, k^2, k'^2, q^2) &= \frac{1}{2\pi i}\int d\nu' \frac{b_i(\nu', k^2, k'^2, q^2)}{\nu'-\nu} \\ C_i(\nu, k^2, k'^2, q^2) &= \frac{1}{2\pi i}\int d\nu' \frac{c_i(\nu', k^2, k'^2, q^2)}{\nu'-\nu}. \end{aligned}\right\} \quad ...[5.20]$$

To use the information contained in the local commutator [*5.15*] we multiply both $T_{\mu\nu}$ and $t_{\mu\nu}$ by k_μ. Using partial integration we get

$$k_\mu T_{\mu\nu} = iU_\nu + i\int d^3x\, e^{ik.x}\langle p' \mid [g_0^+(\mathbf{x}, 0), g_\nu^-(0)] \mid p\rangle \quad ...[5.21]$$

and

$$k_\mu t_{\mu\nu} = iu_\nu \quad ...[5.22]$$

where

$$U_\nu = \int d^4x\, e^{ik.x}\theta(x)\langle p' \mid [\partial_\mu g_\mu^+(x), g_\nu^-(0)] \mid p\rangle \equiv LP_\nu + N_1 Q_\nu + N_2 q_\nu \quad ...[5.23]$$

and

$$u_\nu = \int d^4x\, e^{ik.x}\langle p' \mid [\partial_\mu g_\mu^+(x), g_\nu^-(0)] \mid p\rangle \equiv lP_\nu + n_1 Q_\nu + n_2 q_\nu. \quad ...[5.24]$$

The functions L, N_1, N_2 are related to l, n_1, n_2 by a set of relations similar to [*5.20*]. The second term in the R.H.S. of [*5.21*] contains the local commutator [*5.15*]; we can thus write

$$k_\mu T_{\mu\nu} = iU_\nu - 4iP_\nu F(q^2) \quad ...[5.25]$$

where

$$\langle p' \mid j_\nu^3(0) \mid p\rangle = P_\nu F(q^2). \quad ...[5.26]$$

Using the expansions of $T_{\mu\nu}$, $t_{\mu\nu}$, U_ν and u_ν in terms of P_ν, Q_ν, q_ν ([*5.21*] and [*5.22*]), we get a set of relations between the coefficients of each vector. Equating the coefficients of P_ν we have:

$$A(\nu)\nu + B_3(\nu)(k.Q) + B_4(\nu)(k.q) = iL(\nu) + 2iF(q^2) \quad ...[5.27]$$

$$a(\nu)\nu + b_3(\nu)(k.Q) + b_4(\nu)(k.q) = il(\nu) \quad ...[5.28]$$

where, for simplicity, all other arguments except ν have been suppressed.

With the help of equations [*5.20*], and their analogue for L, we get

$$\frac{1}{2\pi i}\int \frac{d\nu'}{\nu'-\nu}\{a(\nu')\nu + b_3(\nu')(k.Q) + b_4(\nu')(k.q) - il(\nu')\} = +2iF(q^2) \quad ...[5.29]$$

or, (see equation [*5.28*]),

$$\frac{1}{2\pi}\int dv' a(v', k'^2 k^2, q^2) = 2F(q^2). \qquad ...[5.30]$$

This is the more general sum rule for the isovector form factor $F(q^2)$. For $k'^2 = k^2 = q^2 = 0$, equation [*5.30*] corresponds to the analogue of the Adler-Weisberger sum rule. Indeed from [*5.17*] we get

$$k_\mu k'_\nu t_{\mu\nu} = \int d^4x\, e^{ik.x} \langle p' \mid [\partial_\mu g_\mu^+(x), \partial_\nu g_\nu^-(0)] \mid p \rangle.$$

In the limiting case $k^2 = k'^2 = q^2 = 0$, using the expansion corresponding to [*5.19*], one verifies that

$$k_\mu k'_\nu t_{\mu\nu} = \nu^2\, a(\nu).$$

Therefore the function in the integrand of [*5.30*] contains only the commutator of the divergence of the axial current.

The sum rule for the isovector charge radius (section 4.3) can be obtained in a similar way by considering the case $q^2 \neq 0$. For further details we refer to the papers by Fubini quoted in reference 44 and to a set of lectures given by B. Renner[46] at the Rutherford Laboratory in the Summer of 1966.

5.3. *Final remarks*

As pointed out in the introduction, this review of current algebras is far from complete. In particular I would have liked to discuss in some detail the effect of Schwinger terms, the non-renormalization theorems for the axial vector current and the attempts to calculate the renormalization of the Cabibbo angle for the axial current. The dispersion method of Fubini would have deserved a more complete treatment and I strongly advise anyone who wishes to have a complete picture of the subject to study it carefully.

The dispersion method is at the basis of a recent development proposed by de Alfaro, Fubini, Furlan and Rossetti[47] who deduce a number of interesting results from the locality of the commutators. The implications of this method have not yet been fully explored.

Acknowledgment

The text of these lectures has been prepared when I was a Visiting Fellow of Clare Hall, Cambridge. It is a pleasure to thank the President, the Fellows, and in particular, Dr Eden, for their very warm hospitality.

References

1. For a summary of the sum rules of atomic spectroscopy see E. U. Condon and G. H. Shortley. 1935. *The Theory of Atomic Spectra* (Cambridge Univ. Press).
2. L. H. Thomas. 1925. *Naturwiss.* **13**, 627; W. Kuhn, *Zeits. f. Physik*, **33**, 408; see also E. U. Condon and G. H. Shortley quoted in ref. 1 and H. A. Bethe and E. E. Salpeter, *Hand. der Phys.* **35**, 341.

3. W. Heisenberg. 1925. *Zeits. f. Physik*, **33**, 879.
4. H. A. Bethe and J. S. Levinger. 1950. *Phys. Rev.* **78**, 115; see also O. Rojo and J. S. Levinger. 1961. *Phys. Rev.* **123**, 2177.
5. J. S. Levinger. 1960. *Nuclear Photodisintegration* (Oxford Univ. Press).
6. M. Gell-Mann. 1962. *Phys. Rev.* **125**, 1067.
7. See ref. 6 and also Y. Ne'eman. 1961. *Nucl. Phys.* **26**, 222.
8. N. Cabibbo. 1963. *Phys. Rev. Lett.* **10**, 531.
9. M. Gell-Mann. 1964. *Physics*, **1**, 63; see also L. A. Radicati. 1966. Proceedings of the International School "Ettore Majorana". (To be published).
10. S. Fubini and G. Furlan. 1965. *Physics*, **1**, 229.
11. M. Ademollo and R. Gatto. 1965. *Phys. Rev. Lett.* **13**, 264.
12. M. Gell-Mann. 1964. *Phys. Lett.* **8**, 214.
13. F. Gürsey and L. A. Radicati. 1964. *Phys. Rev. Lett.* **13**, 173.
14. J. Schwinger. 1959. *Phys. Rev. Lett.* **3**, 296.
15. S. L. Adler. 1965. *Phys. Rev. Lett.* **14**, 1051.
16. W. I. Weisberger. 1965. *Phys. Rev. Lett.* **14**, 1067; 1966. *Phys. Rev.* **143**, 1302.
17. S. L. Adler. 1965. *Phys. Rev.* **141**, 736.
18. S. Fubini, G. Furlan and C. Rossetti. 1965. *Nuovo Cim.* **40A**, 1171.
19. *Quantum Field Theory Seminar*, unpublished, Princeton 1966.
20. S. L. Adler. 1964. *Phys. Rev.* **135**, B963; see also ref. 17.
21. Y. Nambu. 1960. *Phys. Rev. Lett.* **9**, 380; M. Gell-Mann and M. Levy. 1960. *Nuovo Cim.* **16**, 705.
22. T. D. Lee and C. S. Wu. 1965. *Annual Review of Nuclear Science*, **15**, 381.
23. I. S. Gernstein. 1965. *Phys. Rev. Lett.* **16**, 114.
24. F. Gürsey, A. Pais and L. A. Radicati. 1964. *Phys. Rev. Lett.* **13**, 299.
25. N. Cabibbo and L. A. Radicati. 1966. *Phys. Lett.* **19**, 697.
26. I. S. Gernstein and B. W. Lee. 1966. *Phys. Rev.* **144**, 1142.
27. R. Gatto, L. Maiani and G. Preparata. 1966. *Phys. Rev. Lett.* **16**, 377.
28. N. Cabibbo and H. Ruegg. 1966. *Phys. Lett.* **22**, 85.
29. S. Adler, *Proceedings of the Argonne Conference on Weak Interactions*, 1965.
30. C. G. Callan and S. B. Treiman. 1966. *Phys. Rev. Lett.* **16**, 153.
31. M. Suzuki. 1966. *Phys. Rev. Lett.* **16**, 212; V. S. Mathur, S. Okubo and L. K. Pandit. 1966. *Phys. Rev. Lett.* **16**, 371; V. S. Mathur, L. K. Pandit and R. E. Marshak. 1966. *Phys. Rev. Lett.* **16**, 974.
32. S. Weinberg. 1966. *Phys. Rev. Lett.* **17**, 336.
33. M. Suzuki. 1965. *Phys. Rev. Lett.* **15**, 986; H. Sugawara. 1965. *Phys. Rev. Lett.* **15**, 870, 997 (E). These are the first two papers on the subject; many others have been published since.
34. S. L. Adler, *Proceedings of the Argonne Conference on Weak Interactions*, 1965; S. L. Adler and F. J. Gilman. 1966. *Phys. Rev.* **152**, 1460.
35. Riazuddin and B. W. Lee. 1966. *Phys. Rev.* **146**, 1202.
36. G. Furlan, R. Jengo and E. Remiddi. 1966. *Nuovo Cim.* **44A**, 427.
37. P. Dennery. 1961. *Phys. Rev.* **124**, 2000; see also refs. 35 and 36.
38. B. W. Lee. 1965. *Phys. Rev. Lett.* **14**, 676.
39. R. F. Dashen and M. Gell-Mann. 1965. *Phys. Lett.* **17**, 145.
40. F. J. Gilman and H. J. Schnitzer. 1966. *Phys. Rev.* **150**, 1362.
41. L. H. Chan *et al.* 1966. *Phys. Rev.* **141**, 1298.
42. C. W. Akerloff. 1966. *Phys. Rev. Lett.* **16**, 147.
43. R. F. Dashen and M. Gell-Mann. 1966. *Phys. Rev. Lett.* **17**, 340.
44. S. Fubini. 1966. *Nuovo Cim.* **43**, 475; S. Fubini and G. Segrè. 1966. *Nuovo Cim.* **45**, 641.
45. M. A. Beg. 1966. *Phys. Rev. Lett.* **17**, 333.
46. B. Renner. 1966. *Lectures on Current Algebras*, RHEL/R 126.
47. V. De Alfaro, S. Fubini, G. Furlan and C. Rossetti. 1966. *Phys. Lett.* **21**, 576.

7

PION-NUCLEON PHASE SHIFT ANALYSIS

A. DONNACHIE

CERN—Geneva

1

INTRODUCTION

IN the last few years tremendous strides forward have been made in our knowledge and understanding of the pion-nucleon system. Three years ago, the details of pion-nucleon scattering were known with some uncertainty up to 310 MeV. They are now known in considerable detail to beyond 1310 MeV. This information explosion, which it may fairly be called, had as its basis the development of a realistic dynamical theory of pion-nucleon scattering by partial wave dispersion relations, the advent of a series of experimental results of a quantity and quality hitherto lacking, and, growing out of these, detailed phase shift analysis. It is with this latter topic that these lectures will be concerned.

Beyond recognizing their existence, lack of time and space precludes any further mention of the experiments, although they deserve a better fate than to be mouldering in the references. The theory of partial wave dispersion relations will fare rather better since it plays a crucial role in the most extensive and complete phase shift analysis yet attempted.

Phase shift analysis is a lengthy, painstaking and at times, if the truth be told, a rather tedious business, involving a vast amount of computing. To my knowledge no one has ever dared to estimate the cost per phase, but it must be high. It can have its exciting moments, and this has been particularly true of pion-nucleon phase shift analysis with new resonances popping up, if not every other day, at least with sufficient regularity to make one wonder what is going to happen next. To the outside observer, this unexpected wealth of structure is undoubtedly the most striking and interesting result of the analyses.

Since the nature of the subject is largely a descriptive one, these lectures will follow suit. In preparing lecture notes, it is very tempting to re-derive every formula used, "for completeness". This has not been done, the necessary equations being written down with reference and without proof.

The exceptions to this are some odds and ends which have been gathered into two appendices and consigned to the end, out of harm's way.

2

Kinematics and Formulae

The S-matrix for elastic pion-nucleon scattering can be written as

$$S_{fi} = \delta_{fi} - (2\pi)^4 i\delta^4(p_1+q_1-p_2-q_2)\left\{\frac{m^2}{4E_1E_2\omega_1\omega_2}\right\}^{\frac{1}{2}} \bar{u}_2 T_{fi} u_1 \quad \text{...[1]}$$

where q_1 and q_2 are the four-momenta of the initial and final pions, p_1 and p_2 are the four-momenta of the initial and final nucleons. The nucleon energies E_i and the pion energies ω_i are given by

$$E_i = (m^2+p_i^2)^{\frac{1}{2}},\ \omega_i = (\mu^2+q_i^2)^{\frac{1}{2}},\ i = 1, 2 \quad \text{...[2]}$$

where m and μ are respectively the nucleon and pion masses, and u_1 and u_2 are the Dirac spinors for the initial and final nucleon states.

The scattering amplitude T is of the explicitly Lorentz invariant form[1]

$$T_{fi} = -A_{fi} + i\gamma . QB_{fi} \quad \text{...[3]}$$

where $Q = \frac{1}{2}(q_1+q_2)$, the γ^μ are the Dirac γ matrices and A and B are scalar invariant functions.

The isotopic spin decomposition of the amplitudes A and B is given by[1]

$$A_{\beta\alpha} = \delta_{\beta\alpha}A^{(+)} + \tfrac{1}{2}[\tau_\beta, \tau_\alpha]A^{(-)} \quad \text{...[4a]}$$

$$B_{\beta\alpha} = \delta_{\beta\alpha}B^{(+)} + \tfrac{1}{2}[\tau_\beta, \tau_\alpha]B^{(-)} \quad \text{...[4b]}$$

where α and β are respectively the charge indices of the initial and final pions and the τ_α ($\alpha = 1, 2, 3$) are the usual isotopic spin operators.

The amplitudes $A^{(\pm)}$, $B^{(\pm)}$ are related to the amplitudes $A^{(T)}$, $B^{(T)}$ ($T = \frac{1}{2}, \frac{3}{2}$), the eigenstates of isotopic spin, by

$$A^{(+)} = \tfrac{1}{3}A^{(\frac{1}{2})} + \tfrac{2}{3}A^{(\frac{3}{2})},\quad A^{(-)} = \tfrac{1}{3}A^{(\frac{1}{2})} - \tfrac{1}{3}A^{(\frac{3}{2})} \quad \text{...[5a]}$$

or conversely

$$A^{(\frac{1}{2})} = A^{(+)} + 2A^{(-)},\quad A^{(\frac{3}{2})} = A^{(+)} - A^{(-)} \quad \text{...[5b]}$$

with identical relations holding for the B amplitudes.

The A's and B's are functions of the kinematic invariants s, t, u defined by

$$s = (p_1+q_1)^2,\ t = (q_1-q_2)^2,\ u = (p_1-q_2)^2. \quad \text{...[6]}$$

These are not independent, being related by

$$s+t+u = 2m^2+2\mu^2. \quad \text{...[7]}$$

If we use the subscripts + and − to label the amplitudes for the elastic processes $\pi^+ + p \to \pi^+ + p$ and $\pi^- + p \to \pi^- + p$ respectively, then a convenient expression of crossing symmetry[1,2] is the equality of the matrix elements

$$\langle p_2, q_2 \mid T_+ \mid p_1, q_1 \rangle = \langle p_2, -q_1 \mid T_- \mid p_1, -q_2 \rangle. \qquad ...[8]$$

Under the exchange $q_1 \rightleftharpoons -q_2$, $s \rightleftharpoons u$ and $t \to t$. Since the amplitudes A and B are independent (the term A is independent of nucleon spin, while the B term depends on it via the factor $\gamma . Q$) crossing symmetry can be applied separately to A and $\gamma . QB$. Noting that $(\gamma . Q) \to -(\gamma . Q)$ under crossing, we obtain finally the crossing relation

$$A^{(+)}(s,t,u) = A^{(+)}(u,t,s), \quad A^{(-)}(s,t,u) = -A^{(-)}(u,t,s), \qquad ...[9a]$$

$$B^{(+)}(s,t,u) = -B^{(+)}(u,t,s), \quad B^{(-)}(s,t,u) = B^{(-)}(u,t,s). \qquad ...[9b]$$

We will work invariably in the c.m. system. In this system let W be the total energy, q the magnitude of the pion momentum and θ the scattering angle. Then

$$\left.\begin{aligned} s &= W^2 \\ t &= -2q^2(1-\cos\theta) \\ u &= 2m^2+2\mu^2+2q^2(1-\cos\theta)-W^2. \end{aligned}\right\} \qquad ...[10]$$

Since the ultimate intention is to discuss phase shifts, it is necessary to express the amplitudes A and B in terms of partial wave $\pi - N$ amplitudes. The first step is to relate the transition amplitude [3] to the 2×2 Pauli scattering matrix M.

From [1], the c.m. differential cross-section is

$$\frac{d\sigma}{d\Omega} = \left(\frac{m}{4\pi W}\right)^2 \sum \mid \bar{u}_2 T_{fi} u_1 \mid^2 \qquad ...[11]$$

where Σ implies a sum over final spin states and an average over the initial spin states. In the M form, the c.m. differential cross-section is

$$\frac{d\sigma}{d\Omega} = \sum \mid \langle f \mid M \mid i \rangle \mid^2 \qquad ...[12]$$

where $\mid i\rangle$ and $\mid f\rangle$ are the Pauli spinors for the initial and final nucleon spin states and M is related to T by

$$\langle f \mid M \mid i \rangle = -\frac{m}{4\pi W} \bar{u}_2 T_{fi} u_1. \qquad ...[13]$$

In terms of the $\pi - N$ helicity amplitudes[3] $f_1(\theta), f_2(\theta)$,

$$M = f_1(\theta) + \frac{\boldsymbol{\sigma} . \mathbf{q}_2\, \boldsymbol{\sigma} . \mathbf{q}_1}{q^2} f_2(\theta) \qquad ...[14]$$

where $\boldsymbol{\sigma}$ is the nucleon spin, $\boldsymbol{q}_1$ and $\boldsymbol{q}_2$ the initial and final pion momenta and θ the scattering angle. The helicity amplitudes f_1 and f_2 can now be expressed in terms of A and B, using [*3*], [*13*] and [*14*], yielding

$$f_1 = \frac{(E+m)}{8\pi W}[A+(W-m)B], \qquad \text{...[15a]}$$

$$f_2 = \frac{(E-m)}{8\pi W}[-A+(W+m)B]. \qquad \text{...[15b]}$$

The inverse of [*15*] is

$$\frac{1}{4\pi}A = \frac{W+m}{E+m}f_1 - \frac{W-m}{E-m}f_2, \qquad \text{...[16a]}$$

$$\frac{1}{4\pi}B = \frac{1}{E+m}f_1 + \frac{1}{E-m}f_2. \qquad \text{...[16b]}$$

The helicity amplitudes are given in terms of the partial wave amplitudes $_{l\pm}f$, corresponding to total angular momentum $j = l \pm \frac{1}{2}$ by[4]

$$f_1 = \sum_{l=0}^{\infty}(f_{l+}P'_{l+1}(x) - f_{l-}P'_{l-1}(x)), \qquad \text{...[17a]}$$

$$f_2 = \sum_{l=1}^{\infty}(f_{l-} - f_{l+})P'_l(x) \qquad \text{...[17b]}$$

where $x = \cos\theta$ and $P'_l(x)$ is the first derivative of the Legendre polynomial $P_l(x)$. The partial wave amplitudes $f_{l\pm}$ are expressed in turn in terms of phase shifts $\delta_{l\pm}$ by

$$f_{l\pm} = \frac{e^{2i\delta_{l\pm}}-1}{2iq}. \qquad \text{...[18]}$$

Writing $\delta_{l\pm} = \alpha_{l\pm} + i\beta_{l\pm}$ this becomes

$$f_{l\pm} = \frac{e^{-2\beta_{l\pm}}e^{2i\alpha_{l\pm}}-1}{2iq} = \frac{\eta_{l\pm}e^{2i\alpha_{l\pm}}-1}{2iq} \qquad \text{...[19]}$$

$\alpha_{l\pm}$, $\eta_{l\pm}$ are real, and unitarity requires that $0 \leqq \eta_l \leqq 1$.

In terms of $\xi_J = \eta_J e^{2i\alpha_J}$ the partial wave total, elastic and inelastic cross-sections are given by

$$\sigma^{(J)}_{\text{Total}} = \frac{4\pi}{q}(J+\tfrac{1}{2})\,\text{Im} f_J, \qquad \text{...[20a]}$$

$$\sigma^{(J)}_{\text{Elastic}} = \frac{\pi}{q^2}(J+\tfrac{1}{2})\,|1-\xi_J|^2, \qquad \text{...[20b]}$$

$$\sigma^{(J)}_{\text{Inelastic}} = \frac{\pi}{q^2}(J+\tfrac{1}{2})\{1-|\xi_J|^2\}. \qquad \text{...[20c]}$$

By means of the orthogonality relation

$$\int_{-1}^{1} dx\, P_k'(x)\{P_{l+1}(x)-P_{l-1}(x)\} = \delta_{kl} \qquad \text{...[21]}$$

any partial wave amplitude can be extracted from the helicity amplitudes by the relation

$$f_{l\pm} = \tfrac{1}{2}\int_{-1}^{1} dx\,\{P_l(x)f_1+P_{l\pm 1}(x)f_2\}. \qquad \text{...[22]}$$

An alternative form for the scattering matrix M is in terms of the non-spin-flip and spin-flip amplitudes $f(\theta)$ and $g(\theta)$[5]

$$M = f(\theta)+i\,\frac{\boldsymbol{\sigma}.(\boldsymbol{q}_2\times\boldsymbol{q}_1)}{|\,\boldsymbol{q}_2\times\boldsymbol{q}_1\,|}\,g(\theta). \qquad \text{...[23]}$$

The expansion of $f(\theta)$ and $g(\theta)$ in terms of the scattering amplitudes $f_{l\pm}$ is

$$f(\theta) = \sum_{l=0}^{\infty}[(l+1)f_{l+}+lf_{l-}]\,P_l(\cos\theta) \qquad \text{...[24a]}$$

$$g(\theta) = \sum_{l=1}^{\infty}[f_{l+}-f_{l-}]\,P_l^{(1)}(\cos\theta) \qquad \text{...[24b]}$$

where $P_l^{(1)}(\cos\theta) = \sin\theta(d/d(\cos\theta))P_l(\cos\theta)$.

In terms of $f(\theta)$ and $g(\theta)$ the elastic differential cross-section is given by

$$\frac{d\sigma(\theta)}{d\Omega} = |\,f(\theta)\,|^2+|\,g(\theta)\,|^2 \qquad \text{...[25]}$$

and the polarization $P(\theta)$ of the recoil nucleon by

$$P(\theta) = 2\,\mathrm{Im}\,[f^*(\theta)g(\theta)]\Big/\frac{d\sigma(\theta)}{d\Omega}. \qquad \text{...[26]}$$

The π^+p system, which has a z component of isotopic spin $T_z = \frac{3}{2}$, can exist only in the isotopic spin state $T = \frac{3}{2}$. However, the π^-p system, which has $T_z = -\frac{1}{2}$, is a linear combination of the $T = \frac{1}{2}$ and $T = \frac{3}{2}$ isotopic spin states. The appropriate combinations are

$$A(\pi^+ +p \to \pi^+ +p) = A(T = \tfrac{3}{2}), \qquad \text{...[27a]}$$

$$A(\pi^- +p \to \pi^- +p) = \tfrac{1}{3}[A(T = \tfrac{3}{2})+2A(T = \tfrac{1}{2})], \qquad \text{...[27b]}$$

$$A(\pi^- +p \to \pi^0 +n) = \frac{\sqrt{2}}{3}[A(T = \tfrac{3}{2})-A(T = \tfrac{1}{2})]. \qquad \text{...[27c]}$$

The scattering amplitudes of equations [24] must be extended to take into account the non-relativistic Coulomb effects as well as the first order relativistic Coulomb corrections.[7,8,9] To do this, we introduce the following quantities:

$$\eta = \frac{e^2}{hv} \qquad \text{...[28]}$$

where v is the velocity of the incident pion in the laboratory (η should not be confused with the inelasticity parameter $\eta_{l\pm}$ of [19])

$$B = \frac{\frac{1}{2}(\mu_p\beta_p\beta_\pi)+\frac{1}{4}(2\mu_p-1)\beta_p^2}{1+\beta_\pi\beta_p} \qquad \ldots[29]$$

where μ_p is the magnetic moment of the proton (in nuclear magnetons) and β_p, β_π are respectively the c.m. velocities of the proton and pion, divided by the velocity of light.

The non-relativistic Coulomb phase shift Φ_l of order l is zero for $l = 0$ and is given by

$$\Phi_l = \sum_{n=1}^{l} \tan^{-1}(\eta/n) \qquad \ldots[30]$$

for $l \geqq 1$.

The *total* phase shifts are related to the nuclear phase shifts by

$$\alpha_{l\pm}(+) = \alpha_{l\pm}+\Phi_{l\pm}, \quad \alpha_{l\pm}(-) = \alpha_{l\pm}-\Phi_{l\pm} \qquad \ldots[31]$$

where $\Phi_{l\pm}$ is the complete Coulomb phase shift of order l, consisting of the non-relativistic Coulomb phase shift [30] plus a first order relativistic correction

$$\Phi_{l\pm} = \Phi_l+\Delta\Phi_{l\pm} \qquad \ldots[32]$$

where

$$\Delta\Phi_{0+} \approx \eta[\tfrac{1}{2}(\beta_p\beta_\pi)+\tfrac{1}{4}(2\mu_p-1)\beta_{pl}^2/(1+\beta_\pi\beta_p)], \qquad \ldots[33a]$$

$$\Delta\Phi_{l+} \approx \eta B/(l+1), \quad l \geqq 1, \qquad \ldots[33b]$$

$$\Delta\Phi_{l-} \approx -\eta B/l, \quad l \geqq 1. \qquad \ldots[33c]$$

Using the superscripts $^{(1)}$ and $^{(3)}$ to denote states of isotopic spin $\frac{1}{2}$ and $\frac{3}{2}$ respectively, the Coulomb corrected scattering amplitudes are

(i) $\pi^+ + p \rightarrow \pi^+ + p$

$$f(\theta) = \frac{-\eta}{2q\sin^2(\theta/2)}\exp\{-i\eta\ln[\sin^2(\theta/2)]\} + \sum_{l=0}\left[(l+1)\frac{\eta_{l+}^{(3)}\exp\{2i\alpha_{l+}^{(3)}(+)\}-\exp\{2i\Phi_l\}}{2iq} + l\,\frac{\eta_{l-}^{(3)}\exp\{2i\alpha_{l-}^{(3)}(+)\}-\exp\{2i\Phi_l\}}{2iq}\right]P_l(\cos\theta) \qquad \ldots[34]$$

$$g(\theta) = \frac{\eta B\sin\theta}{2q\sin^2(\theta/2)} + \sum_{l=1}\left[\frac{\eta_{l+}^{(3)}\exp\{2i\alpha_{l+}^{(3)}(+)\}-\eta_{l-}^{(3)}\exp\{2i\alpha_{l-}^{(3)}(+)\}}{2iq} - \frac{\eta B}{q}\,\frac{2l+1}{l(l+1)}\right]P_l^{(1)}(\cos\theta) \qquad \ldots[35]$$

(ii) $\pi^- + p \rightarrow \pi^- + p$

$$f(\theta) = \frac{\eta}{2q\sin^2(\theta/2)} \exp\{i\eta \ln[\sin^2(\theta/2)]\}$$

$$+ \sum_{l=0} \Bigg[(l+1) \frac{\eta_{l+}^{(3)} \exp\{2i\alpha_{l+}^{(3)}(-)\} + 2\eta_{l+}^{(1)} \exp\{2i\alpha_{l+}^{(1)}(-)\}}{6iq}$$

$$+ l \frac{\eta_{l-}^{(3)} \exp\{2i\alpha_{l-}^{(3)}(-)\} + 2\eta_{l-}^{(1)} \exp\{2i\alpha_{l-}^{(1)}(-)\}}{6iq}$$

$$- (2l+1) \frac{\exp\{-2i\Phi_l\}}{2iq} \Bigg] P_l(\cos\theta) \qquad \ldots[36]$$

$$g(\theta) = \frac{\eta B \sin\theta}{2q\sin^2(\theta/2)} + \sum_{l=1} \Bigg[\frac{\eta_{l+}^{(3)} \exp\{2i\alpha_{l+}^{(3)}(-)\} + 2\eta_{l+}^{(1)} \exp\{2i\alpha_{l+}^{(1)}(-)\}}{6iq}$$

$$- \frac{\eta_{l-}^{(3)} \exp\{2i\alpha_{l-}^{(3)}(-)\} + 2\eta_{l-}^{(1)} \exp\{2i\alpha_{l-}^{(1)}(-)\}}{6iq}$$

$$+ \frac{\eta B}{q} \frac{2l+1}{l(l+1)} \Bigg] P_l^{(1)}(\cos\theta). \qquad \ldots[37]$$

Equations [*34*], [*35*], [*36*] and [*37*], together with [*25*] and [*26*], form the basis of the phase shift analysis. The general notation we will use for the partial waves is $L_{2T,\,2J}$ where L is the angular momentum (in the usual spectroscopic notation $S, P, D, F, \ldots$ etc.), T is the isotopic spin and J the total angular momentum.

3

Phase Shift Analysis I

The complexity of the phase shift expansions ([*34*], [*35*], [*36*] and [*37*]) of the scattering amplitude means that there is no simple way of deriving values for the phase shifts and inelasticities from experimental data. The procedure to be followed is to minimize the function

$$M = \sum_i \left\{ \frac{X_i^{(e)} - X_i^{(c)}}{\Delta X_i^{(e)}} \right\}^2 \qquad \ldots[38]$$

where $X_i^{(e)}$ is the quantity X_i as obtained from experiment, $\Delta X_i^{(e)}$ is the experimental error (standard deviation) in $X_i^{(e)}$, $X_i^{(c)}$ is the quantity X_i as calculated for a given set of phase parameters $\alpha_{l\pm}^{(T)}$, $\eta_{l\pm}^{(T)}$ and M is considered as a function of these phase parameters. The quantity M is also called "chi-squared" (χ^2). Having obtained a minimum, errors can be assigned to the phase parameters using the standard error matrix approach. The shape of the M hypersurface near the minimum is examined by evaluating the second partial derivatives of M with respect to each of the phase parameters (p_i, say) used. These partial derivatives form a matrix G defined by

$$G_{ij} = \tfrac{1}{2}(\partial^2 M/\partial p_i \partial p_j). \qquad \ldots[39]$$

The inverse of this matrix yields the error matrix G^{-1} with the properties

$$[(G^{-1})_{ii}]^{\frac{1}{2}} = (\Delta p_i)_{\text{R.M.S.}} \qquad \ldots[40a]$$

and

$$(G^{-1})_{ij} = C_{ij}(\Delta p_i)_{\text{R.M.S.}}(\Delta p_j)_{\text{R.M.S.}} \qquad \ldots[40b]$$

where C_{ij} is the correlation coefficient.

The available experimental data consist of π^+p and π^-p total and elastic cross-sections, elastic differential cross-sections and recoil proton polarizations and π^-p charge exchange differential cross-sections, and are available at a wide variety of angles and energies. This list does not exhaust all the possible experiments, there also being the possibility of measuring charge exchange polarization (i.e. the polarization of the recoil neutron in π^-p charge exchange scattering) and of performing triple scattering experiments. However, these experiments are all ones of considerable difficulty and as yet have not been investigated seriously.

Ideally, it should be possible to obtain a unique set of phase parameters at a given energy by including in the function M all data available at (or very close to) that energy. The number of energies at which there exist sufficient measurements of all the above data to make this theoretically possible is small, and even at such energies the ideal is very seldom realized and the word "unique" must be used with caution. Several things conspire to thwart the hopeful phase shift analyst.

1. Systematic errors can exist in any given measurement. These errors are not included in the standard deviation $X_i^{(e)}$ and frequently it is not possible to modify the latter in a consistent way to take account of these extra errors, even assuming one is aware of their existence. The most common errors of this type arise in absolute normalization, which is a comparatively tractable problem, although occasionally there are discrepancies in the shape of an angular distribution coming from different measurements. This is a much more serious situation.
2. A measured angular distribution can be incomplete in the sense that only a part of the complete angular range has been covered. A common result of this situation is that two (or more) possible phase solutions are obtained, which give very similar predictions in that part of the angular range in which the measurements lie, but which differ considerably in the region where there are no measurements.
3. The errors on an experiment can be sufficiently large to allow different solutions, giving different predictions, to fit the experimental data

equally well in the sense that the chi-squared contribution from the data is approximately the same for each solution.

4. The partial wave expansion must be truncated at some value of l, $l = l_{max}$ say. An interesting anomaly of phase shift analysis is that while it is necessary to include all partial waves up to and including $l = l_{max}$ to obtain a good fit to the data, the highest partial waves are not well determined by the data and the possible experimental errors on them are large. Although in themselves small in magnitude, they can interfere strongly with the lower partial waves causing considerable ambiguity there.

All this makes a conventional energy-by-energy phase shift analysis extremely difficult, posing serious problems of uniqueness and of continuity. This does not mean that such an analysis cannot be performed[10] and although it gives a realistic picture of the large (and consequently dominant) partial waves, the smaller are, at best, only qualitatively correct.

One approach which overcomes many of the above difficulties (although at the expense of introducing others) is that of the energy dependent analysis.[11,12,13] In this approach an energy dependent parametric form is assumed for each partial wave and *all* data at *all* energies are fitted simultaneously with all partial waves up to and including $l = l_{max}$. The advantages of this method are that continuity is built in *ab initio*, systematic errors tend to be ironed out, and the lack of data at some energies and the incompleteness or poor quality of data at others are no longer serious obstacles. The problem of the higher partial waves remains and new problems arise. To make such an analysis numerically tractable, the number of parameters allowed for each partial wave is necessarily small and the choice of parametric form is arbitrary. Consequently, there is the danger of biasing the solution by an unfortunate choice of parametric form, and of ironing out genuine structure along with the systematic singularities. Because of the limited number of parameters, energy dependent fits are invariably statistically poor fits (typically $\chi^2/N_F > 2$, where N_F is the number of degrees of freedom) and it is very difficult to assign realistic errors to the solution.

The parametrization chosen by Roper *et al.*[11] for the scattering amplitudes $A_{l\pm} = qf_{l\pm}$ was a combination of specifically resonant and non-resonant forms

$$A_{l\pm} = \varepsilon_{l\pm} A_{l\pm}\,(\text{resonant}) + A_{l\pm}\,(\text{non-resonant}) \qquad \ldots[41]$$

where

$$A_{l\pm}(\text{resonant}) = \frac{\frac{1}{2}\Gamma_{l\pm}^{(\text{el})}}{(\omega_{l\pm}^{R} - \omega) - \frac{1}{2}i\Gamma_{l\pm}^{(\text{t})}} \qquad \ldots[42]$$

is a resonance amplitude of the Layson form.[14]

$$\Gamma_{l\pm}^{(el)} = \frac{4m}{\omega_{l\pm}^{R}+\omega}(qr_{l\pm})\gamma_{l\pm}^{2}V_{l}(qr_{l\pm}) \qquad \ldots[43]$$

is the resonance elastic width and

$$V_l(qr_{l\pm}) = [(qr_{l\pm})^2\{j_l^2(qr_{l\pm})+n_l^2(qr_{l\pm})\}]^{-1} \qquad \ldots[44]$$

is the barrier penetration factor.

$$\Gamma_{l\pm}^{(t)} = \Gamma_{l\pm}^{(el)}+\Gamma_{l\pm}^{(in)} \qquad \ldots[45]$$

is the resonance total width and

$$\Gamma_{l\pm}^{(in)} = \tilde{\Gamma}_{l\pm}^{(in)}(q-q_I)^{2l+1}. \qquad \ldots[46]$$

q_I is the threshold pion c.m. momentum for one-pion production, $\varepsilon_{l\pm} = 1$ for states in which it was chosen to put a resonance and $\varepsilon_{l\pm} = 0$ for states without a resonance. The non-resonant amplitude

$$A_{l\pm}(\text{non-resonant}) = (\eta_{l\pm}e^{2i\alpha_{l\pm}}-1)/2i \qquad \ldots[47]$$

is the background for resonant states ($\varepsilon_{l\pm} = 1$) or the amplitude for states in which there are no resonances and was parametrized either by

$$\tan\alpha_{l\pm} = q^{2l+1}\sum_{n=0}^{l_{\max}-l}(a_{l\pm})_n q^n \qquad \ldots[48a]$$

or

$$\left.\begin{aligned} \alpha_{l\pm} &= q^{2l+1}\sum_{n=0}^{l_{\max}-l}(a_{l\pm})_n q^n \\ \eta_{l\pm} &= e^{-2\nu_{l\pm}} \end{aligned}\right\} \qquad \ldots[48b]$$

with

$$\left.\begin{aligned} \nu_{l\pm} &= 0 \quad \text{for} \quad q < q_I, \\ \nu_{l\pm} &= (q-q_I)^{2l+1}\sum_{n=0}^{l_{\max}-l}(b_{l\pm})_n(q-q_I)^n \quad \text{for} \quad q \geqq q_I. \end{aligned}\right\} \qquad \ldots[48c]$$

With this parametrization the correct threshold behaviour is guaranteed for the phase shifts and approximated for the absorption parameters.

This parametrization has the merit of comparative simplicity but, since the phase shifts $\alpha_{l\pm}$ and the inelasticities $\eta_{l\pm}$ are parametrized independently, there is no possibility of allowing for correlations between the two. That $\alpha_{l\pm}$ and $\eta_{l\pm}$ *are* correlated as regards their energy dependence is well known, and any changes in $\eta_{l\pm}$ must be reflected in a specific way in $\alpha_{l\pm}$.[15] The parametrization of Bransden *et al.*[12] was chosen to allow for this explicitly.

Their parametrization is based on the dispersion relation for the inverse amplitude $f_{l\pm}^{-1}$ [16]

$$\mathrm{Re}\, f_{l\pm}^{-1}(q) = -\frac{q^2-q_0^2}{\pi}\int_0^{\infty} dq'^2 \frac{q' R_{l\pm}(q')}{(q'^2-q^2)(q'^2-q_0^2)} + \frac{q^2-q_0^2}{\pi}\int_{-\infty}^{-a} dq'^2 \frac{\Delta f_{l\pm}^{-1}(q')}{(q'^2-q^2)(q'^2-q_0^2)} + \sum_{n=0}^{l} \lambda_n/q^{2n} \quad \ldots[49]$$

where

$$a = \mu^2 \frac{(1-\mu^2/4m^2)}{(1+2\mu^2/m^2)} \quad \ldots[50]$$

and the λ_n are (arbitrary) constants. $\Delta f_{l\pm}^{-1}$ is the discontinuity of $-2if_{l\pm}^{-1}$ across the left-hand cut. For those cases in which the amplitude has a zero for a physical value of the energy, an additional pole must be added to the right-hand side of [*49*]. The function $R_{l\pm}$ is given by $\sigma_{l\pm}^{(\mathrm{total})}/\sigma_{l\pm}^{(\mathrm{elastic})}$.

Delta functions for $\Delta f_{l\pm}^{-1}$ were considered to be a reasonable form of parametrization since most of the energies concerned in the analysis were well away from the unphysical region. Thus

$$\Delta f_{l\pm}^{-1} = \sum_{n=1}^{K} (r_{l\pm})_n \delta(q^2+q_n^2) \quad \ldots[51]$$

where $(r_{l\pm})_n$ and q_n^2 are the parametric constants. The value of K differs for different l and in general decreases with increasing l although there is no *a priori* way of choosing it.

$R_{l\pm}$ was parametrized by ratios of polynomials, chosen in such a way as to keep computing time to a minimum and yet remain flexible. In all searches, particular cases of the following general expression were used

$$R_{l\pm}(q) = 1+\theta(q-q_I)\left\{a_{l\pm}\frac{q(q-q_I)}{(1+q^2/A^2)} + b_{l\pm}\frac{q-q_I}{(1+q^2/B^2)}\right\} + \theta(q-q_\eta)\left\{c_{l\pm}\frac{q(q-q_\eta)}{(1+q^2/C^2)} + \delta_{l0} d_{l\pm}\frac{q(q^2-q_\eta^2)^{\frac{1}{2}}}{(1+q^2/D^2)}\right\} \quad \ldots[52]$$

where $a_{l\pm}$, $b_{l\pm}$, $c_{l\pm}$, $d_{l\pm}$, A, B, C, D are constants, q_I is the first inelastic threshold and q_η is the threshold for η production. This form allowed the integral in equation [*49*] to be evaluated analytically.

The numerical technique of finding the best fit in the energy dependent analysis is essentially the same as that of the energy independent analysis except that the summation in equation [*38*] is extended over all energies and the quantities varied to find a minimum of M are the parameters of the phase shifts and inelasticities, not the phase shifts and inelasticities themselves.

Roper *et al.*[11] performed the analysis over three energy ranges: 0–350 MeV, 0–700 MeV and 300–800 MeV with $l_{\max} = 3$ in the first case and $l_{\max} = 3$ or 4 in the two latter, and a series of different starting points were used (32 in the 0–350 MeV analysis, 17 in the 0–700 MeV analysis, 3 in the

300–800 MeV analysis). The best fits in the three cases had χ^2/N_F ratios of 1·86 ($\chi^2 = 1002$, $N_F = 539$, 2·59 ($\chi^2 = 2762$, $N_F = 1067$) and 3·19 ($\chi^2 = 2706$, $N_F = 848$) respectively. As a final check on the consistency of the 0–350 MeV analysis, 0–100 MeV and 0–50 MeV analyses were performed with $l_{\max} = 1$.

The different analyses are consistent in their overlap regions in the behaviour of the largest (and consequently dominant) amplitudes, but there are significant differences in the small amplitudes. From the discussion given above, this latter situation is not unexpected.

Bransden *et al.* performed the analysis in two energy ranges: 300–700 MeV[12] and 750–1100 MeV.[13] In the 300–700 MeV region two different types of solution were found, both with $\chi^2/N_F \approx 2{\cdot}5$ (the differences were not general, being confined to a few partial waves). The solution obtained in the 750–1100 MeV region is a reasonable extrapolation of the better of the two lower energy solutions, although not all amplitudes can be joined on smoothly (again this is a not unexpected result of this type of analysis).

An encouraging feature of the three analyses discussed so far is the consistency in their results for the large amplitudes despite the poor statistical fits of the energy dependent analyses[11,12,13] and the difficulties inherent in the simple energy-by-energy analysis.[10] Although there are certainly quantitative differences, a clear qualitative picture is obtained of the dominant amplitudes in pion-nucleon scattering below 1 GeV. These results will be discussed in more detail in section 7.

There is yet a third approach to phase shift analysis, namely to inject as much theory as possible into the analysis (this is a conspicuous omission from the approaches discussed so far). Although this makes the analysis considerably more complicated, it pays handsome dividends and as we shall see in section 5 it has all the advantages of *both* the energy-by-energy and energy independent analyses, with a few more advantages besides, and has none of the disadvantages. For the moment then we will turn to a brief discussion of the theory to be used and of its validity.

4

Partial Wave Dispersion Relations

The extraction from the Mandelstam representation[17] of the singularities in the complex s-plane of the partial wave amplitudes $f_{l\pm}(s)$, [*18*], is well documented[18] and the resulting singularities are shown in Fig. 1. The dispersion relation for $f_{l\pm}(s)$ is

$$\operatorname{Re} f_{l\pm}(s) = \frac{P}{\pi}\int_{(m+\mu)^2}^{\infty} ds' \frac{\operatorname{Im} f_{l\pm}(s')}{s'-s} + \frac{1}{2\pi i}\int_{\text{(unphysical cuts)}} ds' \frac{\Delta f_{l\pm}(s')}{s'-s} \qquad \ldots[53]$$

where $\Delta f_{l\pm}(s')$ is the discontinuity in $f_{l\pm}(s')$ across the cut at s'. The contribution to [*53*] from the integral along the physical cut $(m+\mu)^2 \leqq s \leqq \infty$

gives the rescattering, and the various unphysical cuts give contributions to [53] which can be regarded as the forces producing the $\pi-N$ scattering. The leading diagrams giving rise to these singularities are shown in Fig. 2 and the different unphysical cuts may be classified as follows.

(*a*) *The short Born cut*: $(m-\mu^2/m)^2 \leqq s \leqq m^2+2\mu^2$.

The discontinuity across this cut is determined completely by the $\pi-N$ coupling constant $f^2 = 0{\cdot}0822 \pm 0{\cdot}0018$.[19,20,21]

(*b*) *The crossed physical cut*: $0 \leqq s \leqq (m-\mu)^2$.

The discontinuity across this cut is given in terms of the imaginary part of the physical $\pi-N$ scattering amplitudes. It is dominated by the contributions from the N^*_{33} resonance (the main contribution) and from the low energy s wave $\pi-N$ amplitudes. Detailed calculation verifies that the contributions from the higher mass $\pi-N$ resonances can be ignored safely. Thus the only information required to evaluate the contributions from this cut is the position and width of the N^*_{33} resonance, and the s wave $\pi-N$ amplitudes up to 100 MeV or so, all of which are well known.[19,20]

(*c*) *The circle cut*: $|s| = m^2-\mu^2$.

The discontinuity across the circle is given in terms of the absorptive parts $\operatorname{Im} f^J_{\pm}(t)$ of the helicity amplitudes for the process $\pi+\pi \rightarrow N+\bar{N}$. J is the angular momentum and t the square of the total energy in this channel. Extended unitarity[22] relates $\operatorname{Im} f^J_{\pm}(t)$ directly to the absorptive part of $\pi-\pi$ scattering at energy $\sqrt{t}$.

Using the notation $s = (m^2-\mu^2) \exp(i\phi)$ for points on the circle, the values of t which can occur at an angle ϕ are

$$4\mu^2 \leqq t \leqq t_{\max}$$

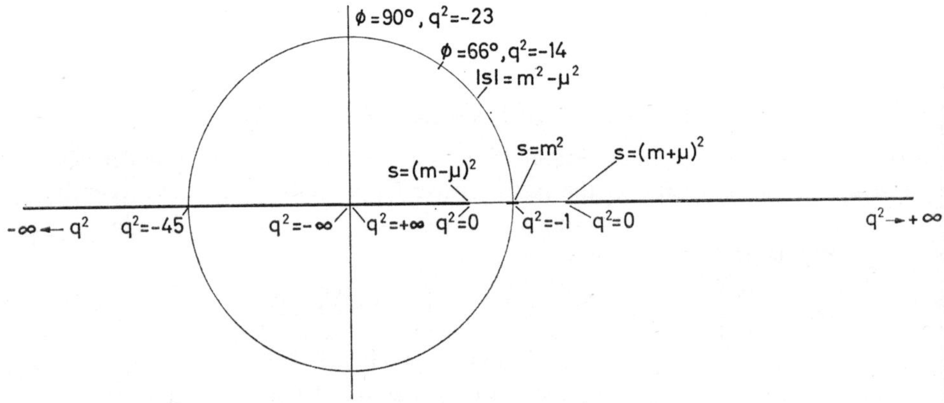

FIG. 1. The singularities of the partial wave amplitudes in the s-plane.

where

$$t_{\max} = 2[(m^2+\mu^2)-(m^2-\mu^2)\cos\phi]$$

and it follows that only low energy $\pi-\pi$ scattering can contribute to $\Delta f_{l\pm}(s)$ for small values of $|\phi|$ and the higher the $\pi-\pi$ scattering energy, the higher is the least value of $|\phi|$ involved. Present methods do not allow us to determine $\Delta f_{l\pm}(s)$ for $|\phi| > 66°$.[23] However, since $|\phi| = 66°$ corresponds to $t_{\max} = 56\mu^2$ and we have no real knowledge of the helicity amplitudes above $t = 50\mu^2$, this is of little practical consequence and the discontinuity across the circle is evaluated only for $0 \leqq |\phi| < 66°$.

There are two distinct contributions to the front of the circle. The first comes from the isospin $T = 0$ helicity amplitudes for $\pi+\pi \rightarrow N+\overline{N}$, which occur for even values of J. The $T = 0$, $J = 2$ state is known to give no significant contribution,[24] so that only the $T = 0$, $J = 0$ (s wave $\pi-\pi$ scattering) need be considered. The parameters of the s wave $\pi-\pi$ contribution to the $\pi-N$ partial wave amplitudes have been

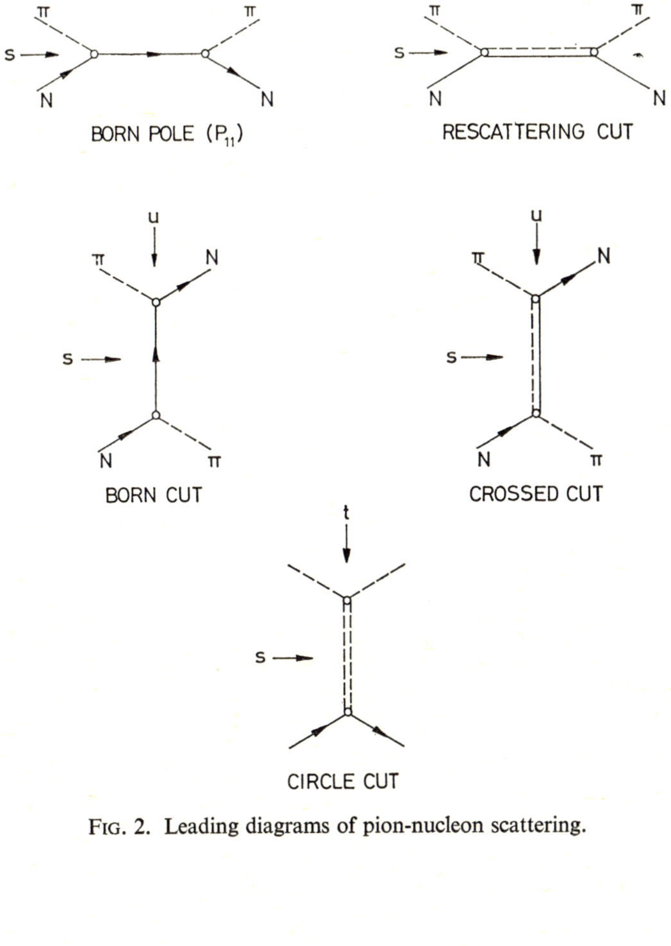

FIG. 2. Leading diagrams of pion-nucleon scattering.

studied in great detail,[25] so for our present purposes we shall simply assume that this contribution is accurately known.

The second contribution comes from the $T = 1, J = 1$ helicity amplitude and can be explained completely by the $T = 1, J = 1$ $\pi-\pi$ resonance, the ρ. Using a narrow resonance approximation, the relevant helicity amplitudes $\mathrm{Im}\, f_{\pm}^{1}(t)$ are given by δ-functions, and the absorptive parts of the invariant $\pi-N$ amplitudes $A^{(-)}$ and $B^{(-)}$ by

$$\mathrm{Im}\, A^{(-)}(s,t) = 12\pi^2 C_2 (s+\tfrac{1}{2}t_R - m^2 - \mu^2)\delta(t-t_R), \qquad \ldots[54a]$$

$$\mathrm{Im}\, B^{(-)}(s,t) = -12\pi^2 (C_1 + 2mC_2)\delta(t-t_R). \qquad \ldots[54b]$$

Here $t_R = m_\rho^2$, where m_ρ is the mass of the ρ, and C_1, C_2 are the $\rho-N$ coupling constants. The gyromagnetic ratio of the nucleon gives[26] $C_2/C_1 = 0{\cdot}27$, leaving one unknown coupling constant. This is readily determined from low energy s wave $\pi-N$ scattering.[25,27]

(*d*) *The left-hand cut*: $-\infty \leqq s \leqq 0$.

All processes so far discussed contribute to this cut also, but in addition there are many other contributions from, for example, the more massive intermediate states in the t channel (4π, 6π, $K\bar{K}$, etc.). Some of these contribute also to the back of the circle and since they are at present quite incalculable, the total contribution from the left-hand cut and the back of the circle is unknown. As we shall see below, they can be classified as "distant singularities" giving rise to "short range forces", in contrast to the other "nearby singularities" which give rise to "long range forces". It will be assumed that the contribution from the distant singularities does not give rise to any structure in the physical region, i.e. it varies slowly with energy and can be adequately represented by a simple parametric form, e.g. one or two simple poles. Note that this assumption says nothing about the *magnitude* of the short range contribution, it merely specifies its *form*. To date all experimental evidence appears to justify this assumption.

Rather than study the amplitude $f_{l\pm}(s)$ it is convenient to introduce $F_{l\pm}(s)$, defined by

$$F_{l\pm}(s) = \frac{f_{l\pm}(s)}{q^{2l}}. \qquad \ldots[55]$$

The relation between s and q^2 is

$$s = \{(m^2+q^2)^{\frac{1}{2}} + (\mu^2+q^2)^{\frac{1}{2}}\}^2 \qquad \ldots[56]$$

and to each value of s there is a unique value of q^2 given by

$$q^2 = \{s-(m-\mu)^2\}\{s-(m+\mu)^2\}/4s. \qquad \ldots[57]$$

Note that q^2 is real on *all* the cuts shown in Fig. 1. On the physical cut

q^2 increases from 0 at $s = (m+\mu)^2$ to ∞ as $s \to \infty$. On the crossed physical cut $q^2 = 0$ at $s = (m-\mu)^2$ and increases to $+\infty$ as $s \to 0$ along this cut, and on the short Born cut $q^2 \simeq -\mu^2$. At the point $s = (m^2-\mu^2)\exp(i\phi)$ on the circle $|s| = m^2-\mu^2$,

$$q^2 = -\{m^2 \sin^2(\phi/2)+\mu^2 \cos^2(\phi/2)\}. \qquad ...[58]$$

Thus $q^2 = -\mu^2$ at $s = m^2-\mu^2$ and as s moves round the circle to $-(m^2-\mu^2)$, q^2 decreases to $-m^2$. Finally, on the cut $-\infty \leqq s \leqq 0$, q^2 has a maximum value of $-m^2$ at $s = -(m^2-\mu^2)$ and then decreases to $-\infty$ both as $s \to 0$ and as $s \to -\infty$.

The well-known threshold behaviour of the phase shift δ_l of the l-th partial wave

$$\delta_l = a_l q^{2l+1}, \qquad ...[59]$$

with a_l a constant as $q \to 0$, implies that $f_{l\pm}(s) = q^{2l}$ near the threshold. It follows from [58] that q^{-2} is a regular function of s except for simple poles at $s = (m+\mu)^2$ and $s = (m-\mu)^2$. Thus, the function $F_{l\pm}(s)$ defined by [55] is bounded at the physical threshold and has a simple branch point there. At the crossed threshold $s = (m-\mu)^2$ the function $F_{l\pm}(s)$ is not bounded in general, the type of singularity there being different from that of $f_{l\pm}(s)$. However, as we shall see, this causes no difficulty.

Thus in place of [53] we shall use the dispersion relation

$$\operatorname{Re} F_{l\pm}(s) = \frac{P}{\pi}\int_{(m+\mu)^2}^{\infty} ds' \frac{\operatorname{Im} F_{l\pm}(s')}{s'-s} + \frac{1}{2\pi i}\int_{\text{(unphysical cuts)}} ds' \frac{\Delta F_{l\pm}(s')}{s'-s} \qquad ...[60]$$

where

$$\Delta F_{l\pm}(s') = (\Delta f_{l\pm}(s'))/(q^2(s'))^l \qquad ...[61]$$

and special care is taken at $s' = (m-\mu)^2$. [60] has considerable advantages over [53] on account of its improved convergence for all $l \geqq 1$. The presence of the factor q^{-2l} in $\Delta F_{l\pm}(s)$ suppresses the contribution to the second integral in [60] from the unknown distant singularities (thereby giving added strength to the assumption about the lack of structure in the contribution from these singularities) and in addition, since q^{-2l} is small at high physical energies, any uncertainties arising from the high-energy portion of the re-scattering integral are very much reduced. Of course, the higher the value of l the greater is the suppression and the smaller one would expect the contribution from the distant singularities to be. This is physically very sensible in that it would not be expected that high partial waves would have any great dependence on the very short range interaction.

One can go further than this. There are two constraints which can be applied which, within the limits of the "no structure" assumption, determine the magnitude and sign of the short range interaction.

(1) *Crossing*

The crossing relations, [*9*], can be used not only to obtain the discontinuity across the crossed cut in terms of the imaginary part of the physical $\pi - N$ scattering amplitudes, but also to obtain the *real part* of the scattering amplitude on the crossed cut in terms of the real parts of the physical $\pi - N$ scattering amplitudes. This crossing relation is of the general form

$$\text{Re}\, F_J(s_C) = \sum_{J'} \int_{\text{some physical } s'} ds'\, K_{JJ'}(s_C, s')\, \text{Re}\, f_{J'}(s') \qquad \ldots[62]$$

where $0 \leqq s_C \leqq (m-\mu)^2$. The summation is over all J' and the integral over some (finite) range of physical energies, the exact limit depending on the particular value of s_C. In practice, of course, the summation over J' must be truncated but, as in evaluating the discontinuity across the cut, the right-hand side of [*62*] is dominated by the N^*_{33} and low energy s waves for values of s_C close to the crossed cut threshold. Further, for $s_C = (m-\mu)^2$, i.e. at the crossed cut threshold, the summation is truncated *exactly* at $l' = l$ (due to the threshold behaviour of the scattering amplitudes the contribution from all partial waves with $l' > l$ vanishes there). The integration also disappears, the right-hand side of [*62*] being evaluated at the physical threshold only. Consequently, at the threshold of the crossed cut the value of the real part of the l-th partial wave is given *exactly* in terms of the physical threshold parameters of the partial waves $l' \leqq l$.

The importance of this is that the real part of the scattering amplitude can be obtained on the crossed cut by a second method, namely evaluation of the dispersion relation [*60*] for $s = s_C$, and the two real parts so obtained should be equal. This crossing constraint is an extremely powerful one to apply in obtaining solutions to partial wave dispersion relations.

(2) *The unitary sum rules*

Suppose, for the convenience of writing (there is no loss of generality in the argument), that we represent the unphysical cuts by the line $-\infty \leqq s \leqq s_1$, so that the amplitude $F_{l\pm}(s)$ satisfies the dispersion relation

$$\text{Re}\, F_{l\pm}(s) = \frac{P}{\pi} \int_{s_0}^{\infty} ds' \frac{\text{Im}\, F_{l\pm}(s')}{s'-s} + \frac{1}{\pi} \int_{-\infty}^{s_1} ds' \frac{\text{Im}\, F_{l\pm}(s')}{s'-s} \qquad \ldots[63]$$

where $s_0 = (m+\mu)^2$ is the physical threshold.

We want to consider the limiting form of [*63*] for infinite physical energies ($s \to +\infty$).

For physical s

$$\operatorname{Re} F_{l\pm}(s) = \frac{\eta_{l\pm}(s)\sin(2\alpha_{l\pm}(s))}{2[q(s)]^{2l+1}}$$

so that

$$|\operatorname{Re} F_{l\pm}(s)| \leqq \tfrac{1}{2}[q(s)]^{-(2l+1)}. \qquad \ldots[64]$$

From the relation [57] between q^2 and s it follows immediately that the magnitude of the left-hand side of [63] is bounded by $2^{2l}/s^{l+\frac{1}{2}}$ for large physical s.

Now look at the right-hand side of [63] as $s \to +\infty$. We require to make the assumption that for $l \geqq 1$, $\operatorname{Im} F_{l\pm}(s)$ decreases faster than $(-s)^{-1-\varepsilon}$, $\varepsilon > 0$, as $s \to -\infty$. This assumption is reasonable, since without it the high-energy behaviour of the dispersion relation must be very pathological. With this assumption,

$$\frac{1}{\pi}\int_{-\infty}^{s_1} ds' \frac{\operatorname{Im} F_{l\pm}(s')}{s'-s} = -\frac{1}{\pi s}\int_{-\infty}^{s_1} ds' \operatorname{Im} F_{l\pm}(s') + O\left(\frac{1}{s^2}\right) \qquad \ldots[65]$$

as $s \to +\infty$.

Since the first term on the right of equation [63] is a principal value integral, it is not so easy to find its asymptotic form. Sufficient conditions[19] for

$$\frac{P}{\pi}\int_{s_0}^{\infty} ds' \frac{\operatorname{Im} F_{l\pm}(s')}{s'-s}$$

to behave like

$$-\frac{1}{\pi s}\int_{s_0}^{\infty} ds' \operatorname{Im} F_{l\pm}(s')$$

as $s \to +\infty$ are that

(i) $$\int^{\infty} ds' \operatorname{Im} F_{l\pm}(s') \text{ exists} \qquad \ldots[66a]$$

and

(ii) given any $\varepsilon > 0$, there exists an S such that

$$\left|\frac{s' \operatorname{Im} F_{l\pm}(s') - s \operatorname{Im} F_{l\pm}(s)}{s'-s}\right| < \frac{\varepsilon}{(ss')^{\frac{1}{2}}} \text{ for all } s' \geqq s > S. \qquad \ldots[66b]$$

Since

$$\operatorname{Im} F_{l\pm}(s) = \frac{1-\eta_{l\pm}(s)\cos(2\alpha_{l\pm}(s))}{2[q(s)]^{2l+1}}$$

condition (i) is satisfied for $l \geqq 1$. However, we can make no definite statement about condition (ii) without detailed knowledge of how $\alpha_{l\pm}$ and $\eta_{l\pm}$ behave for large s, although it should be noted that for simple models of high-energy scattering it is certainly true. What we can say by using [*63*], [*64*] and [*65*] is that

$$\lim_{s \to +\infty} \left[\frac{s}{\pi} P \int_{s_0}^{\infty} ds' \frac{\operatorname{Im} F_{l\pm}(s')}{s'-s} \right] = \frac{1}{\pi} \int_{-\infty}^{s_1} ds' \operatorname{Im} F_{l\pm}(s') \qquad \ldots[67]$$

for $l \geqq 1$, so that the limit on the left-hand side certainly exists, and it is a not unreasonable conjecture that this limit is indeed

$$-\frac{1}{\pi} \int_{s_0}^{\infty} ds' \operatorname{Im} F_{l\pm}(s').$$

Equation [*67*] then gives the unitary sum rule for $l \geqq 1$,

$$\int_{s_0}^{\infty} ds' \operatorname{Im} F_{l\pm}(s') + \int_{-\infty}^{s_1} ds' \operatorname{Im} F_{l\pm}(s') = 0. \qquad \ldots[68]$$

In a similar fashion one can obtain extra sum rules for higher values, e.g. for $l \geqq 2$ one obtains the unitary sum rule

$$\int_{s_0}^{\infty} ds's' \operatorname{Im} F_{l\pm}(s') + \int_{-\infty}^{s_1} ds's' \operatorname{Im} F_{l\pm}(s') = 0 \qquad \ldots[69]$$

and so on.

The consequence of this is that if we are given a solution to the dispersion relation [*63*], then from the values of $\operatorname{Im} F_{l\pm}(s)$ on the physical cut $s_0 \leqq s \leqq \infty$ the unitary sum rule provides a relation governing the values which $\operatorname{Im} F_{l\pm}(s)$ can have on the unphysical cut. As we have seen already, the long and fairly long range parts of the $\pi - N$ interaction are known reasonably well, so that the unitary sum rule gives us information on the short range part of the $\pi - N$ interaction.

A good check on the validity of the sum rule is provided by the N_{33}^* resonance, since the integrand in the physical integral in [*68*] is dominated by the resonance region. The long range forces are, of course, known and representing the short range interaction by a single pole, determined by the sum rule, the resulting P_{33} scattering length is $0{\cdot}219 \pm 0{\cdot}011$ (in units $\hbar = \mu = c = 1$) which is to be compared with the experimental value of $0{\cdot}215 \pm 0{\cdot}005$.[19]

Using this theoretical structure, explicit prediction can be made of the pion-nucleon scattering amplitudes,[28, 29] so long as there is no appreciable inelasticity (the theory is essentially an elastic one). The energy range over which the results are meaningful varies from partial wave to partial wave since usually the higher the partial wave the later the onset of inelasticity, although there are of course exceptions. Simple impact parameter considerations[28] indicate that p waves should be realistic to $\sim$300 MeV, d waves to $\sim$500 MeV, f waves to $\sim$700 MeV and g waves and higher beyond 1 GeV. We will justify these statements *a posteriori*. What justification for the theory can we give *a priori*?

1. The predictions of the theory are in good quantitative agreement with single energy phase shift analyses up to 310 MeV[28] and beyond.[29]
2. The best results of Feld *et al.*[11] in the 0-350 MeV analysis were obtained when the theoretical predictions were used as a starting point, and the experimental values did not differ appreciably from the theoretical ones.
3. General statements can be made about the relative magnitude and signs of the partial wave amplitudes at energies up to 1 GeV or more. These are consistent with experimental evidence.[30]
4. The partial wave amplitudes can be classified into "resonant" and "non-resonant" and upper limits put on the resonance energies of the resonant ones.[31] The resulting quantum numbers are in agreement with and the energies consistent with the known $N^*_{33}(1238)$, $N^*_{13}(1520)$, $N^*_{15}(1670)$ and $N^*_{37}(1920)$. [As we shall see, the forces producing these "old" resonances come from the $\pi-N$ channel itself, although appreciably modified by inelastic effects. In the case of the "new" resonances, the forces come almost completely from channels other than the $\pi-N$ one, and the theory can say nothing about them.]

5

Phase Shift Analysis II

In order to apply the partial wave dispersion relations to a phase shift analysis of the data, a semi-phenomenological technique has been developed whereby the experimental phase shifts and inelasticities in a given partial wave are fitted simultaneously by a parametrization which satisfies the dispersion relation exactly and has the correct long range forces, the short range forces being treated as variables (in effect, they form part of the parametrization) and the whole subjected to the unitary sum rule. The higher angular momentum states, which cannot be expected to be determined by the data but which nevertheless are important (see section 3), are given solely in terms of the theory. Although this does not amount to a complete *a priori* prediction, it imposes strong consistency conditions on the data, enabling decisions to be made between different sets of phase shifts and allowing interpolation from energies with complete experiments to those where they are incomplete. Since errors can be placed on the theoretical predictions, they can be used as "theoretical data" to supplement the inadequate experimental data at these latter energies, and a second (more extensive) fit to the experiments made. The results are then used to re-evaluate the partial wave dispersion relations, and so on until consistency is achieved. By this means, it is possible to obtain all the advantages of the energy dependent fits with none of their attendant disadvantages. Continuity is ensured, systematic errors are smoothed out, the correct energy dependent correlations between the phase shifts $\alpha_{l\pm}$ and the inelasticities $\eta_{l\pm}$ are obtained and a realistic assessment of the errors on $\alpha_{l\pm}$ and $\eta_{l\pm}$ can be made. Further, by fitting one partial wave at a time, it is possible to use a large number of parameters, ensuring that the results come from the dispersion relation itself and not from the particular parametrization, avoiding the danger of bias arising from having too few parameters.

By its very nature, this is a technique which allows the analysis to be extended to ever increasing energy domains and this is what has happened in practice. Starting from a $\pi^{\pm}-p$ analysis between 300–700 MeV[32] the $\pi^{+}-p$ analysis was extended to 1100 MeV[33] and this in turn extended to a $\pi^{\pm}-p$ analysis from 0–1400 MeV.[34] Let us now look at the procedure in some detail.

The starting point is with those energies where $\pi^{+}-p$ and $\pi^{-}-p$ polarization measurements and $\pi^{-}-p$ charge exchange measurements are available, as well as $\pi^{+}-p$ and $\pi^{-}-p$ differential cross-section measurements. The highest partial waves (usually g and h waves) are fixed at the theoretical values and a conventional energy independent phase shift analysis is performed. If necessary, some of the smaller, elastic, amplitudes may be constrained to the theoretical values and used as "theoretical data" to supplement the experimental data. Having found a minimum in M a series of "random searches" is performed. These consist of taking an arbitrary step in parameter space (the arbitrariness being obtained by using a pseudo-random number generator) and carrying out the minimization once again from this new starting point. This procedure, which is repeated many times, is a very necessary one.

There is no guarantee that M has only one minimum. It may have several. Hopefully it has only one deep one, with a statistically significant χ^2, the others being local shallow minima with no real statistical significance. This case presents no serious problem. If the minimization has found a shallow minimum, then the random search procedure takes the parameters away from this region and it is unlikely that this particular "solution" will reappear. On the other hand, if the minimum is a genuinely deep one, then the same solution will be obtained repeatedly. If after a large number of random searches only one such deep minimum has been found, there is some justification for claiming a unique solution. Frequently, however, there exist two (and sometimes more) deep minima, each with approximately the same statistical significance, and further action must be taken. This, and other configurations of parameter space which cause difficulties, will be discussed below.

Having obtained solutions at a number of energies E_i ($i = 1, \ldots, N$) (with corresponding c.m. s values s_i) these are then fitted to the dispersion relations. Denoting the known contribution from the long-range forces by $\mathscr{F}_{l\pm}(s)$ and approximating the short-range contribution by a number of simple poles, the dispersion relation can be written as

$$\text{Re}\, F_{l\pm}(s) = \mathscr{F}_{l\pm}(s)+\frac{P}{\pi}\int_{s_0}^{\infty} ds' \frac{\text{Im}\, F_{l\pm}(s')}{s'-s}+\sum_{j=1}^{n}\frac{r_j}{s-a_j} \quad \ldots[70]$$

and we know (experimentally) $F_{l\pm}(s_i)$, $(i = 1, \ldots, N)$.

To evaluate the dispersion relations, we use a conformal mapping technique, details of which are given in appendix 2. For the moment it is sufficient to know that there exist two sets of functions $g_k(s)$ and $h_k(s)$, $(k = 1, \ldots, \infty)$, with the following properties:

(i) $$g_k(s) = \frac{P}{\pi}\int_{s_0}^{\infty} ds' \frac{h_k(s')}{s'-s}. \quad \ldots[71]$$

(ii) $$\text{Im}\, F_{l\pm}(s) = \sum_{k=1}^{\infty}(a_{l\pm})_k h_k(s), \quad \ldots[72]$$

and this series converges under very general conditions.

(iii) $g_k(s)$ and $h_k(s)$ have the correct threshold and asymptotic behaviour.

The dispersion relation [70] then takes the form

$$\text{Re}\, F_{l\pm}(s) = \mathscr{F}_{l\pm}(s)+\sum_{k=1}^{\infty}(a_{l\pm})_k g_k(s)+\sum_{j=1}^{n}\frac{r_j}{s-a_j}. \quad \ldots[73]$$

Truncating the series at some value $k = K$ and evaluating Re $F_{l\pm}(s)$ and Im $F_{l\pm}(s)$, and their errors, at the points $s = s_i$ $(i = 1, \ldots, N)$ we obtain the equations

$$\text{Im}\, F_{l\pm}(s_i) = \sum_{k=1}^{K}(a_{l\pm})_k h_k(s_i), \quad \ldots[74a]$$

$$[\text{Re}\, F_{l\pm}(s_i)-\mathscr{F}_{l\pm}(s_i)] = \sum_{k=1}^{K}(a_{l\pm})_k g_k(s_i)+\sum_{j=1}^{n}\frac{r_j}{s_i-a_j}, \quad \ldots[74b]$$

for $i = 1, \ldots, N$.

The unknowns $(a_{l\pm})_k$, $(k = 1, \ldots, K)$, and r_j, $(j = 1, \ldots, n)$, can then be varied to obtain a best fit to equations [74]. [The pole positions a_j, $(j = 1, \ldots, n)$, are kept fixed.] The values of K and n cannot be fixed *a priori* and they are automatically adjustable up to specified maximum values $K_{\max}$ and $n_{\max}$. In accordance with the "no-structure" assumption $n_{\max}$ is kept small (≤ 3) but $K_{\max}$ can be larger—in some cases 20 or more. Another variable in the fit is the conformal zero, which gives a considerable measure of control over the behaviour of the fit.

Once a reasonable fit is achieved, we then have a dispersion relation prediction of the amplitudes at *all* energies. This technique is a very powerful one, as the example in Figs. 3a and 3b shows. This is the S_{31} amplitude which is one of the more difficult ones to fit because of the large and rapid changes in α and η above 600 MeV. Even so the dispersion relation *prediction* (the dotted line), based on experiments up to 600 MeV and at the two energies 870 and 990 MeV and the final fit (the solid line) are in remarkably good agreement. In amplitudes which show less structure, the final fits rarely depart significantly from the dispersion relation predictions.

Although the use of the dispersion relations allows the rejection of *most* alternative solutions, it cannot reject them all and above 700 MeV there is at present no unique solution to the analysis. There are three types of non-uniqueness.

1. Two (or more) genuinely different solutions exist. In parameter space this corresponds to having two (or more) distinct deep minima, with statistically meaningful χ^2.
2. Several minor variations exist of the one type of solution. In parameter space this corresponds to there being a number of small local minima grouped close together at the bottom of the deep minimum.
3. Different solutions exist which appear superficially to be type 1, but are in fact a variant of type 2. In this case major differences are confined to only one or two parameters and in parameter space corresponds to there being a flat valley in the direction of each of these parameters, but otherwise the minimum is steep. This type of situation is easily discernible and the extent of the valley can be found readily by tracking across all values of the parameter concerned. For example $\eta(S_{11})$ is one of the frequent culprits and in this case a typical tracking search would be to fix $\eta(S_{11})$ successively at 0, 0·1, 0·2, . . . , 1·0 and minimize with respect to the other parameters for each of those fixed values of $\eta(S_{11})$.

Nothing can be done at present about the type 1 ambiguity. Clarification of this ambiguity will need to wait until the more difficult experiments of charge exchange polarization or of triple scattering are performed.

Types 2 and 3 are however a different matter. Their significance is that the statistical errors obtained in the analysis are not a realistic measure of the true experimental error. The final step is to take all variations of the same solution and combine them, weighted appropriately according to their relative probabilities, into a single composite solution having a realistic experimental error. This is the final output of the phase shift analysis.

6

Resonances

As we shall see later, one of the most interesting results of pion-nucleon phase shift analysis has been the existence of hitherto unsuspected resonances in the pion-nucleon system and in clarifying the structure of the ones which

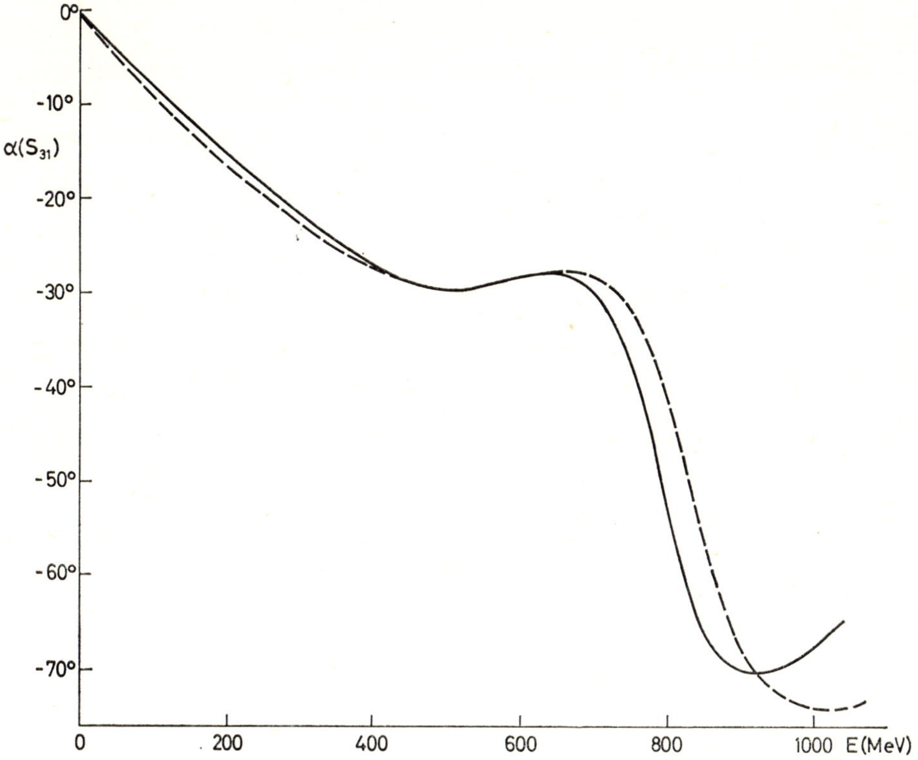

FIG. 3*a*. The S_{31} phase shift plotted as a function of energy. (——— result: - - - - dispersion relation prediction.)

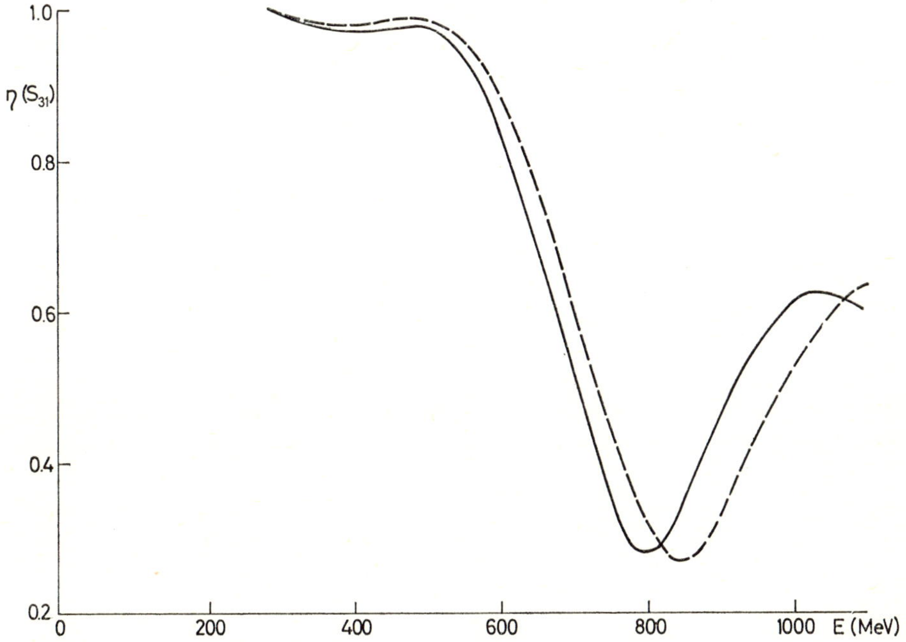

FIG. 3*b*. The S_{31} inelasticity parameter plotted as a function of energy. (——— result: - - - - dispersion relation prediction.)

were known to exist. These new resonances are very different in character from the conventional idea of a resonance, typified for example by the $N_{33}^*(1238)$ resonance, and some explanation of possible resonance structures is necessary.

For all pion-nucleon resonances, apart from the N_{33}^*, the presence of inelastic channels is of considerable importance. In the presence of such channels, the scattering and reaction processes can be described best by the reaction matrix $\boldsymbol{K}$[35,36] which is related to the scattering matrix $\boldsymbol{T}$ by

$$\boldsymbol{T} = \boldsymbol{K}(\boldsymbol{1}-i\boldsymbol{QK})^{-1} \qquad \text{...[75]}$$

where $\boldsymbol{Q}$ is the matrix of available channel momenta.

A resonant state corresponds to the situation when one eigenphase of $\boldsymbol{K}$ increases rapidly through 90°, and in the neighbourhood of the resonance energy the eigenvalue K_R and the eigenphase δ_R are given approximately by

$$K_R = \tan\delta_R = \frac{\gamma}{W_R-W}+\tan\delta_B \qquad \text{...[76]}$$

where δ_B denotes the non-resonant background phase of this eigenstate. This may be described in channel space by specifying the coefficients $C_{R\beta}$ appropriate to channel β, the coefficients being normalized such that $\sum_\beta |C_{R\beta}|^2 = 1$. As emphasized by Adair,[37] in the neighbourhood of the resonance the scattering amplitude has then the form

$$f_J = \frac{1}{2iq}\exp\{2i\delta_\beta\}\frac{\Gamma_1}{\Gamma}\left\{\frac{W^*-W+\frac{1}{2}i\Gamma}{W^*-W-\frac{1}{2}i\Gamma}\right\}+A_J. \qquad \text{...[77]}$$

The Breit-Wigner resonance energy W^* and total width Γ are given by

$$W^* = W_R-\tfrac{1}{2}\gamma\sin 2\delta_B \qquad \text{...[78]}$$

and

$$\Gamma = 2\gamma\cos^2\delta_B \qquad \text{...[79]}$$

and the partial width Γ_1 for the elastic channel by

$$\Gamma_1 = |C_{R1}|^2\,\Gamma. \qquad \text{...[80]}$$

The complex amplitude A_J consists of contributions from all the non-resonant eigenphases and from δ_B, and (hopefully) is slowly varying over the resonance region.

A convenient parameter to introduce is

$$R_J = \left\{\frac{\sigma^{(J)}_{\text{Total}}}{\sigma^{(J)}_{\text{Elastic}}}\right\} \geqq 1 \qquad \text{...[81]}$$

which is given in terms of the phase shift α_J and the inelasticity parameter η_J by

$$R_J = 1+\frac{1-\eta_J^2}{1-2\eta_J\cos 2\alpha_J+\eta_J^2}. \qquad \text{...[82]}$$

The behaviour of R_J as a function of η_J is shown in Fig. 4. For $-1 \leqq \cos 2\alpha_J \leqq 0$, $1 \leqq R_J \leqq 2$, the maximum and minimum values of R_J being attained at the minimum and maximum values of η_J.
For $1 \geqq \eta_J \geqq \cos 2\alpha_J$ (> 0), $1 \leqq R_J \leqq 2$ and for $\cos 2\alpha_J \geqq \eta_J \geqq 0$, $R_J \geqq 2$.

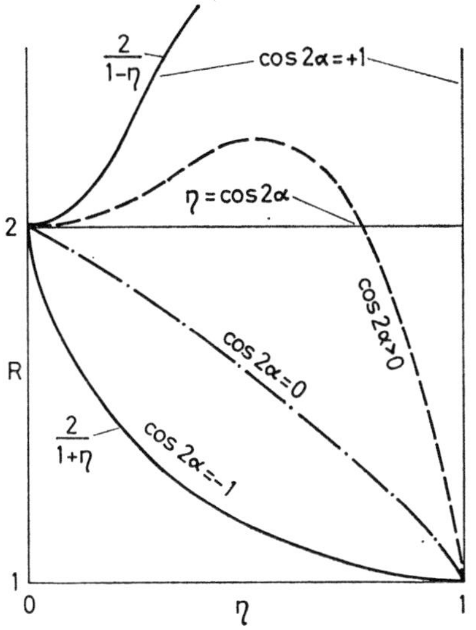

FIG. 4. R as a function of η.

To obtain a prescription for a resonant situation, we now discuss the Adair plot[37] (Argand diagram).

Let $z = 2qf_J(s)$. From the definition, [*19*], of f_J, it follows immediately that

$$|z-i| = \eta_J \qquad \text{...[83]}$$

so that z must lie inside or on the circle with centre i and unit radius, Fig. 5a, the scattering phase being given by $2\alpha_J = \mathrm{A\hat{O}Z}$ and inelasticity by $\eta_J = \mathrm{OZ}$. In terms of R_J the unitarity condition is

$$\{\mathrm{Re}\,(2qf_J)\}^2 + \{\mathrm{Im}\,(2qf_J)\}^2 = 2\,\mathrm{Im}\,(2qf_J)/R_J \qquad \text{...[84]}$$

or,

$$|z-i/R_J| = 1/R_J \qquad \text{...[85]}$$

so that at any given moment z is describing the arc of a circle centre i/R_J and radius $1/R_J$, Fig. 5a.

Equation [*77*] tells us that in the region of a resonance the scattering amplitude is dominated by a pair of Breit-Wigner poles, modified by some background term. Let us consider first the case when the background is zero, when we have the usual multichannel Breit-Wigner description of the scattering amplitude with the various cross-sections given by

$$\sigma^{(J)}_{\text{Total}} = \frac{\pi}{q^2}(J+\tfrac{1}{2})\frac{\Gamma^2}{(W_R-W)^2+\Gamma^2/4} \qquad \ldots[86a]$$

$$\sigma^{(J)}_{\text{Elastic}} = \frac{\pi}{q^2}(J+\tfrac{1}{2})\frac{\Gamma\Gamma_1}{(W_R-W)^2+\Gamma^2/4} \qquad \ldots[86b]$$

$$\sigma^{(J)}_{\text{Inelastic}} = \frac{\pi}{q^2}(J+\tfrac{1}{2})\sum_\beta \frac{\Gamma\Gamma_\beta}{(W_R-W)^2+\Gamma^2/4} \qquad \ldots[86c]$$

with $\Gamma = \Gamma_1 + \sum_\beta \Gamma_\beta$. In this case $\delta_B = 0$ and so from [*78*] and [*79*], $W^* = W_R$ and $\Gamma = 2\gamma$.

In the neighbourhood of the resonance, Γ_1 and Γ_β ($\beta = 2, 3, \ldots$) are approximately constant and so $R_J = \Gamma/\Gamma_1$ will vary slowly with energy with the consequence that the function z will trace out a considerable part of the arc of a circle as the energy W is increased through and beyond the resonance energy W_R. In the purely elastic case, Fig. 5b, the amplitude stays on the circle, starting at the bottom at zero energy and traversing the circle in an

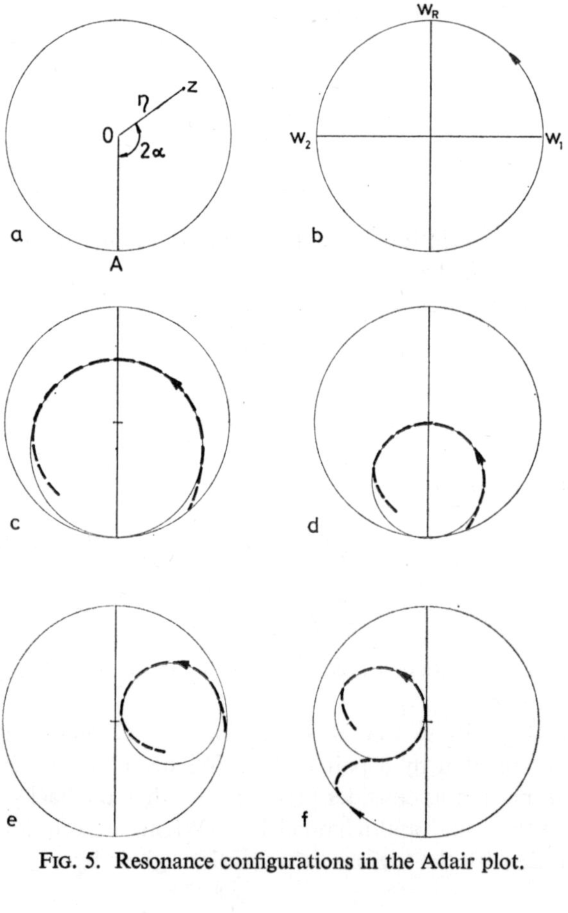

FIG. 5. Resonance configurations in the Adair plot.

anti-clockwise direction. The point $W = W_R$ occurs at the top of the circle ($\alpha_J = 90°$) and the width $\Gamma = W_2 - W_1$, where W_2 and W_1 are the energies at which $\alpha_J = 135°$ and $45°$ respectively.

Two different situations occur in the case of there being some inelasticity.

(i) $1 \leqq R_J \leqq 2$.

In this case, Fig. 5c, which of course includes the purely elastic case ($R_J = 1$), the phase shift α_J passes through 90° at the top of the circle and the resonance width can be defined as above, so long as the departure of R_J from a constant value is small between the upper and lower half widths. Appreciable deviation from a constant causes severe distortion of the arc from the simple circular form, although the resonance position remains at the top of the circle. Outside the resonance region departure of R_J from a constant is certainly to be expected and a complete circle will not be obtained experimentally.

Since $R_J \leqq 2$, $\Gamma_1 > \frac{1}{2}\Gamma$ and this type of resonance can be classified as quasi-elastic.

(ii) $R_J > 2$.

In this class of highly absorptive resonances, the scattering phase shift α_J does not go through 90° at the resonance position, but through 0°, Fig. 5d. The definition of the resonance width is as before, with the same reservations about distortion due to R_J departing from a constant value. In this case $\Gamma_1 < \frac{1}{2}\Gamma$ and so, in the case when there is only one important inelastic channel open, $\Gamma_2 > \frac{1}{2}\Gamma$ and inelastic resonances in one channel are quasi-elastic ones in another channel.

Let us now consider the effect of background, which we have so far ignored. We may rewrite [*77*] as

$$2qf_J = -i\frac{\Gamma_1}{\Gamma}\exp\{2i\delta_\beta\}\exp\{i\phi(W)\} + \tilde{A}_J(W) \qquad \text{...[87]}$$

where

$$\phi(W) = 2\tan^{-1}(\Gamma/2(W_R - W)). \qquad \text{...[88]}$$

If $\tilde{A}_J(W)$ is slowly varying with energy in the resonance region, then [*87*] gives a circle, centre $\tilde{A}_J(W_R)$ and radius $\Gamma_1 \exp\{2i\delta_B\}/\Gamma$, with the phase $\phi(W)$ going from ~ 0 to $\sim 2\pi$ through the resonance. The top of the circle gives the energy W^* but *not* the true resonance energy W_R since δ_B of [*76*] is no longer zero. Unlike the "no-background" cases, which were essentially symmetrical about the vertical diameter of the Adair circle, the effect of the background can move the resonance circle to any part of the diagram. Two typical situations are shown in Figs. 5e and 5f, which are cases of a strong attractive background and a strong repulsive background respectively. Determination of the resonance width is also a hazardous procedure in this case, since not only can there be distortion of the circle due to energy dependence of the radius term, as in the "no-background" cases, but also there can be further distortion due to energy dependence of the background term $\tilde{A}_J(W)$ itself. It is in fact virtually impossible to determine accurately the true parameters of such a resonance from a study of only one of its possible channels.

However, that such resonance circles do occur, and quite noticeably too, can be seen from the Argand plots of the forward pion-nucleon scattering amplitude made by Höhler and Ebel,[38] there being markedly circular paths in the neighbourhood of the $N^*(1238)$ and $N^*(1690)$ resonances, despite the presence of strongly energy-dependent background contributions.

7

RESULTS

The present upper limit in pion-nucleon phase shift analysis is 1·4 GeV, this limit being dictated largely by the lack of sufficient data, although the need to take into account more and more partial waves is beginning to create a barrier which will be difficult to pass. Unfortunately 1·4 GeV is still a long way from the asymptotic region ($\gtrsim$ 3·5 GeV, say) where the scattering can be described reliably in terms of high-energy models, and there is evidence of considerable resonance structure in the intervening energy range. This is clearly of considerable interest, but detailed investigation of this region will be very difficult and almost certainly will be model dependent and not done "exactly" in the sense of the phase shift analyses described here.

To avoid littering the remaining few pages with an overabundance of reference numbers, we will refer to each of the phase shift analyses by the initials of the authors, namely BBSV,[10] RWF,[11] BOM[12,13] and DKLL.[34]

Since the analysis of DKLL is the most extensive yet completed, we quote their results as the basis for discussion (Tables I, II, III and IV) pointing out the similarities and differences in the solutions obtained by BBSV, RWF and BOM. The DKLL results are "semi-final" in the sense that although they represent the final fit to the experimental data they have still to undergo their final theoretical smoothing. Also DKLL have two possible $T = \frac{1}{2}$ solutions between 700 and 1000 MeV. The ambiguity is confined almost entirely to the S_{11} and P_{11} amplitudes. It is hoped that the final application of theory will resolve this ambiguity. [The power of this approach can be gauged from the fact that DKLL found it impossible to obtain a unique $T = \frac{3}{2}$ solution without the aid of the theory and at one point in the analysis they had seven different $T = \frac{1}{2}$ solutions of which the theory has rejected five successfully.]

DKLL considered all partial waves up to, and including, h waves, and the presence of the g and h waves was found to be important (in particular G_{17}). Neither BOM nor BBSV took account of these higher partial waves, truncating the expansion at f waves. Since the g and h waves give a non-negligible contribution, this undoubtedly produces errors, since the other amplitudes distort to try to compensate for these missing contributions.

Further the parametrization of BOM requires that zeros of the amplitudes be inserted at the beginning as poles in the inverse amplitude; they cannot be found in the course of the fit. Since many of the small amplitudes do appear to have zeros (and some more than one) their results for these amplitudes are doubtful. Also their analysis was done in two parts, 300–700 MeV and 750–1050 MeV and there is a distinct discontinuity between the two sets, making it difficult to evaluate the results quantitatively.

$T = \frac{3}{2}$ scattering

S_{31}. This is the most interesting of the $T = \frac{3}{2}$ amplitudes. The results show an inelastic peak at 800 MeV, accompanied by a very rapid drop in the phase shift. When plotted in the Argand diagram, it appears that in the 800 MeV region S_{31} describes part of an anti-clockwise circle, displaced far to the left by the presence of a strongly repulsive background, and it can be interpreted in terms of a resonance of type 5f. BBSV also observe this phenomenon, although the inelastic peak is not so marked as in DKLL ($\eta_{\text{min}} \simeq 0{\cdot}4$ compared with $\eta_{\text{min}} \simeq 0{\cdot}2$) nor is the drop in the phase shift so steep. BOM failed to find this effect, their phase shift and elasticity decreasing slowly together with no marked structure. The RWF analysis does not go to a high enough energy to observe the resonance. At the lower energies (<700 MeV) all solutions are compatible.

P_{31}. In the DKLL solution the phase shift is negative, reaching a minimum of $\sim -25°$ at 1 GeV and then rising, this being accompanied by a steady decrease in η. The BOM and BBSV solutions are roughly similar, the phase shift in the former being less negative ($\delta_{\text{min}} \sim -15°$) and in the latter more negative ($\delta_{\text{min}} \sim -35°$) than that of DKLL, and the BBSV solution is almost elastic. The RWF solution is incompatible with these three, becoming positive at 660 MeV and already being strongly absorptive there ($\eta < 0{\cdot}5$).

P_{33}. This is the well-known N^*_{33} resonance, occurring at an energy of about 200 MeV. The main differences in the solutions are that in BBSV and DKLL the phase shift levels off near 180°, while in RWF and BOM it levels off near 160°. In all solutions there is no appreciable inelasticity below 700 MeV, and thereafter η falls slowly with increasing energy.

D_{33}. The DKLL, BBSV and RWF solutions all find the phase shift to be initially small and positive, then becoming negative in the vicinity of 500 MeV. In the BOM solution the phase shift is small and negative throughout. The BOM solution is essentially elastic, in the DKLL and the BBSV solutions η decreases steadily above 400 MeV.

D_{35}. All solutions find this phase shift to be negative with some inelasticity at higher energies ($\eta \sim 0{\cdot}8$). The BOM and RWF phase shift is more negative ($\sim -10°$) than the DKLL and BBSV phase shift ($\sim -5°$).

TABLE I $T = \frac{3}{2}$ parameters

E_π	S_{31}		P_{31}		P_{33}		D_{33}		D_{35}		F_{35}		F_{37}	
(MeV)	δ	η	δ	η	δ	η	δ	η	δ	η	δ	η	δ	η
24.8	− 3.08 ± 0.14		− 0.37 ± 0.09		1.91 ± 0.08		0.005		− 0.007					
31.0	− 3.86 ± 0.15		− 0.45 ± 0.12		2.39 ± 0.13		0.010		− 0.010					
37.0	− 3.68 ± 0.34		− 0.57 ± 0.17		3.30 ± 0.21		0.013		− 0.016					
41.5	− 4.77 ± 0.33		− 0.79 ± 0.17		3.76 ± 0.25		0.016		− 0.021					
61.0	− 6.34 ± 0.40		− 1.71 ± 0.24		8.71 ± 0.23		0.037		− 0.049		0.002		0.006	
83.0	− 9.81 ± 0.82		− 1.81 ± 0.50		14.35 ± 0.36		0.067		− 0.093		0.005		0.016	
98.0	− 8.49 ± 0.29		− 2.66 ± 0.36		21.12 ± 0.43		0.090		− 0.130 ± 0.011		0.008		0.026	
120.0	− 10.61 ± 0.85		− 2.93 ± 0.44		31.19 ± 0.38		0.125		− 0.19 ± 0.01		0.013		0.045	
150.0	− 12.13 ± 2.00		− 4.71 ± 0.57		53.69 ± 0.57		0.17 ± 0.01		− 0.29 ± 0.01		0.021		0.080	
165.0	− 19.12 ± 3.86		− 4.60 ± 1.27		62.94 ± 2.21		0.20 ± 0.02		− 0.35 ± 0.02		0.024		0.104	
170.0	− 18.03 ± 3.54		− 5.22 ± 1.41		64.79 ± 1.61		0.21 ± 0.02		− 0.36 ± 0.02		0.025		0.112	
176.0	− 17.16 ± 6.59		− 5.27 ± 1.52		71.16 ± 3.86		0.22 ± 0.02		− 0.39 ± 0.02		0.026		0.124	

TABLE I (continued)

E_π (MeV)	S_{31} δ	S_{31} η	P_{31} δ	P_{31} η	P_{33} δ	P_{33} η	D_{33} δ	D_{33} η	D_{35} δ	D_{35} η	F_{35} δ	F_{35} η	F_{37} δ	F_{37} η
200.0	− 8.13 ± 3.11		− 4.89 ± 1.78		96.56 ± 2.23		0.25 ± 0.024		− 0.48 ± 0.02		0.028		0.18	
225.0	− 16.55 ± 0.69		− 4.73 ± 0.94		109.12 ± 0.86		0.29 ± 0.03		− 0.59 ± 0.025		0.028		0.24	
247.5	− 18.05 ± 0.20		− 7.66 ± 0.30		121.02 ± 0.18		0.32 ± 0.04		− 0.69 ± 0.03		0.028		0.30	
270.0	− 18.19 ± 0.84		− 5.56 ± 0.74		125.82 ± 0.74		0.36 ± 0.05	0.997 ± 0.003	− 0.79 ± 0.04		0.027		0.37	
310.0	− 21.59 ± 0.22		− 8.37 ± 0.29		135.64 ± 0.16		0.42 ± 0.05	0.996 ± 0.001	− 1.01 ± 0.04		0.02 ± 0.02		0.52 ± 0.01	
370.0	− 25.29 ± 0.35	0.996 ± 0.004	− 10.51 ± 0.37		144.16 ± 0.19		0.48 ± 0.09	0.983 ± 0.002	− 1.19 ± 0.06		− 0.01 ± 0.02		0.83 ± 0.01	
410.0	− 26.60 ± 0.76	0.982 ± 0.013	− 11.81 ± 0.89	0.971 ± 0.029	148.48 ± 0.51		0.41 ± 0.64	0.986 ± 0.014	− 1.83 ± 0.76		− 0.05 ± 0.03		1.07 ± 0.04	
450.0	− 28.38 ± 0.39	0.984 ± 0.007	− 14.33 ± 0.45	0.999 ± 0.001	153.24 ± 0.24		0.20 ± 0.12	0.959 ± 0.003	− 1.43 ± 0.10		− 0.08 ± 0.04		1.33 ± 0.03	
490.0	− 29.03 ± 0.86	0.940 ± 0.007	− 15.30 ± 2.04	0.985 ± 0.015	157.72 ± 1.00		1.21 ± 1.42	0.955 ± 0.015	− 2.13 ± 1.24		− 0.11 ± 0.03		1.65 ± 0.04	
533.0	− 29.27 ± 1.47	0.951 ± 0.042	− 14.86 ± 3.74	0.943 ± 0.057	161.03 ± 1.99		1.32 ± 2.23	0.939 ± 0.035	− 2.51 ± 1.91		− 0.15 ± 0.10		2.01 ± 0.09	
550.0	− 30.89 ± 0.53	0.905 ± 0.013	− 17.74 ± 0.91	0.999 ± 0.001	162.69 ± 0.69	0.999 ± 0.001	− 0.79 ± 0.12	0.916 ± 0.005	− 1.46 ± 0.14		− 0.12 ± 0.06		2.20 ± 0.05	
581.0	− 28.62 ± 0.48	0.911 ± 0.016	− 20.29 ± 0.63	0.999 ± 0.001	165.41 ± 0.40	0.984 ± 0.012	− 1.24 ± 0.13	0.902 ± 0.006	− 0.87 ± 0.22		− 0.17 ± 0.07		4.14 ± 0.24	

TABLE I (concluded)

E_π (MeV)	S_{31} δ	S_{31} η	P_{31} δ	P_{31} η	P_{33} δ	P_{33} η	D_{33} δ	D_{33} η	D_{35} δ	D_{35} η	F_{35} δ	F_{35} η	F_{37} δ	F_{37} η
600.0	− 28.58 ± 0.65	0.867 ± 0.016	− 20.06 ± 1.11		166.76 ± 0.64	0.980 ± 0.011	− 1.49 ± 0.12	0.896 ± 0.007	− 0.96 ± 0.37		− 0.19 ± 0.08		2.76 ± 0.06	
650.0	− 26.15 ± 1.14	0.673 ± 0.021	− 22.25 ± 0.98		169.69 ± 0.58	0.963 ± 0.019	− 2.36 ± 0.16	0.875 ± 0.007	0.12 ± 0.57		− 0.16 ± 0.10		3.41 ± 0.08	
698.0	− 24.57 ± 4.00	0.500 ± 0.062	− 21.94 ± 2.63	0.913 ± 0.087	171.89 ± 1.58	0.975 ± 0.025	− 3.62 ± 1.67	0.809 ± 0.061	− 1.53 ± 1.84	0.982 ± 0.018	− 0.08 ± 0.29		4.14 ± 0.24	
746.5	− 35.14 ± 5.93	0.315 ± 0.108	− 15.95 ± 6.59	0.866 ± 0.097	170.22 ± 2.53	0.902 ± 0.098	− 0.15 ± 4.40	0.915 ± 0.085	− 3.29 ± 4.79	0.903 ± 0.043	0.39 ± 0.65	0.998 ± 0.002	3.62 ± 1.55	0.988 ± 0.012
795.9	− 52.30 ± 3.24	0.373 ± 0.044	− 17.63 ± 1.45	0.834 ± 0.106	169.73 ± 2.40	0.925 ± 0.057	− 4.59 ± 1.33	0.923 ± 0.043	− 3.01 ± 2.57	0.791 ± 0.019	0.23 ± 0.90	0.985 ± 0.015	6.48 ± 1.62	0.985 ± 0.015
845.4	− 48.12 ± 6.32	0.142 ± 0.023	− 25.94 ± 1.89	0.982 ± 0.018	188.19 ± 1.27	0.995 ± 0.005	− 12.31 ± 1.52	0.544 ± 0.045	− 3.23 ± 1.19	0.991 ± 0.009	0.28 ± 0.80	0.852 ± 0.019	6.86 ± 0.90	0.999 ± 0.001
870.0	− 65.21 ± 1.57	0.458 ± 0.023	− 24.51 ± 0.98	0.864 ± 0.021	178.26 ± 0.62	0.737 ± 0.016	− 6.68 ± 0.27	0.788 ± 0.007	− 3.04 ± 0.46	0.899 ± 0.012	0.51 ± 0.25	0.969 ± 0.005	9.10 ± 0.37	0.988 ± 0.004
899.8	− 62.37 ± 1.13	0.588 ± 0.014	− 22.91 ± 0.55	0.899 ± 0.021	176.03 ± 0.73	0.816 ± 0.019	− 7.97 ± 0.25	0.794 ± 0.006	− 1.02 ± 0.47	0.854 ± 0.009	1.05 ± 0.25	0.949 ± 0.006	8.55 ± 0.34	0.983 ± 0.005
949.4	− 48.44 ± 1.04	0.657 ± 0.011	− 18.63 ± 0.69	0.668 ± 0.015	169.26 ± 0.48	0.905 ± 0.010	− 9.16 ± 0.34	0.569 ± 0.008	− 4.81 ± 0.23	0.990 ± 0.009	− 0.65 ± 0.24	0.833 ± 0.011	13.46 ± 0.26	0.999 ± 0.001
990.0	− 60.27 ± 5.96	0.830 ± 0.106	− 23.59 ± 2.92	0.713 ± 0.112	171.27 ± 2.88	0.857 ± 0.076	− 9.16 ± 2.07	0.826 ± 0.064	− 5.83 ± 2.47	0.764 ± 0.048	3.96 ± 2.08	0.850 ± 0.056	9.53 ± 1.96	0.991 ± 0.009
1048.6	− 69.80 ± 1.38	0.734 ± 0.029	− 21.86 ± 1.43	0.710 ± 0.056	180.00 ± 1.21	0.757 ± 0.038	− 12.39 ± 0.91	0.772 ± 0.019	− 3.73 ± 0.75	0.853 ± 0.022	1.94 ± 0.52	0.830 ± 0.003	12.37 ± 0.55	0.928 ± 0.019
1227.6	− 63.46 ± 2.21	0.367 ± 0.021	− 17.32 ± 2.09	0.609 ± 0.041	172.16 ± 1.38	0.767 ± 0.044	− 16.59 ± 1.33	0.705 ± 0.030	− 6.21 ± 1.97	0.783 ± 0.054	− 6.13 ± 1.22	0.735 ± 0.031	32.38 ± 1.21	0.783 ± 0.040
1311.0	− 60.80 ± 10.05	0.739 ± 0.255	− 4.03 ± 16.63	0.444 ± 0.199	170.17 ± 5.31	0.882 ± 0.106	− 11.88 ± 6.76	0.874 ± 0.126	− 19.33 ± 9.24	0.496 ± 0.131	− 3.92 ± 5.48	0.598 ± 0.085	28.56 ± 2.71	0.563 ± 0.192

F_{35}. This amplitude is not well determined. All solutions agree that the phase shift is very small, BBSV and RWF find it positive, and elastic, BOM find it negative (but tending to zero with increasing energy) and elastic and DKLL find it in turn positive ($<$ 350 MeV), negative ($<$ 700 MeV) and then positive again, with some evidence of increasing inelasticity at the higher energies.

F_{37}. All solutions agree that the phase shift is positive, and increasing with energy, with η decreasing steadily above 750 MeV. This amplitude is the candidate for the resonance observed in $\pi - N$ total cross-sections in the region of 1300 MeV (the old "fourth resonance"). The only solution which extends to this energy is that of DKLL and they find that there is indeed a resonance although it appears to be of the absorptive type 5d rather than the quasi-elastic type 5c, with the phase shift going through zero in the region of 1350 MeV. Thus the resonance energy is probably somewhat higher than that quoted from the total cross-section measurements.

$T = \frac{1}{2}$ *scattering*

S_{11}. DKLL have two possible solutions for this amplitude. In both, the phase shift rises initially to about 50° at 600 MeV and then drops to near zero at 700 MeV, associated with a steep maximum in the inelastic cross-section at about 650 MeV ($\eta < 0{\cdot}2$). Above 700 MeV the solutions diverge; in one case the phase shift goes negative (solution 1) while η rises slowly and in the second case the phase shift rises sharply, passing through 90° at about 820 MeV while η rises steeply and falls again (solution 2). Both results can be interpreted in terms of a resonance, the former of type 5d/5e and the latter a rather deformed type 5c/5e. The BBSV solution is similar to solution 2, although the initial maximum is not reached until 700 MeV, the dip in the phase shift is much less marked and the subsequent rise much slower, the phase shift not reaching 90° until about 950 MeV. The RWF solution (up to 700 MeV) is very similar to the BBSV solution. Because of a large discontinuity, the BOM solution is difficult to assess. Below 700 MeV they have two solutions, one of which is very similar to the DKLL solution 1 and the second of which has a rapidly rising phase shift, reaching 150° by 700 MeV. Their solution between 750 and 1050 MeV is a compromise between DKLL solution 2 and BBSV. All solutions have a resonance interpretation, but not only are the mass and width uncertain, so also is the type. Also a "double resonance" interpretation is not completely excluded.

P_{11}. The ambiguities in S_{11} are reflected in this amplitude. In both DKLL solutions the phase shift has a broad maximum of about 120° between 550 and 600 MeV and then decreases, steeply in solution 1 and more slowly in solution 2, eventually becoming negative in both cases. In both

TABLE II

$T = \frac{1}{2}$ parameters 0–700 MeV

E_π	S_{11}		P_{11}		P_{13}		D_{13}		D_{15}		F_{15}		F_{17}		G_{17}	
(MeV)	δ	η	δ	η	δ	η	δ	η	δ	η	δ	η	δ	η	δ	η
31.0	6.13 ± 0.14		− 0.64 ± 0.02		− 0.14 ± 0.03		0.015		0.012							
41.5	6.58 ± 0.31		− 0.89 ± 0.04		− 0.16 ± 0.04		0.024		0.023							
61.0	6.17 ± 0.47		− 1.26 ± 0.07		− 0.25 ± 0.07		0.064		0.052		0.003					
98.0	7.80 ± 0.30		− 1.54 ± 0.13		− 0.40 ± 0.07		0.17 ± 0.05		0.15 ± 0.03		0.013		0.003			
120.0	7.20 ± 0.36		− 0.96 ± 0.19		− 0.05 ± 0.25		0.29 ± 0.05		0.23 ± 0.05		0.028		0.004			
150.0	9.11 ± 0.22		− 0.92 ± 0.27		− 1.13 ± 0.46		0.59 ± 0.06		0.31 ± 0.07		0.057		0.007			
165.0	8.60 ± 1.00		− 0.22 ± 0.36		− 1.09 ± 0.98		0.69 ± 0.08		0.36 ± 0.09		0.076		0.008			
225.0	9.85 ± 2.19		4.42 ± 0.68	0.999 ± 0.001	− 2.46 ± 0.34		1.60 ± 0.13		0.65 ± 0.15		0.21		0.008			
247.5	12.94 ± 1.95		7.49 ± 0.80	0.993 ± 0.004	− 2.72 ± 0.62		2.01 ± 0.15		0.76 ± 0.14		0.28		0.006			
270.0	14.14 ± 2.09		10.43 ± 0.88	0.986 ± 0.008	− 3.10 ± 0.51		2.52 ± 0.19		0.80 ± 0.15		0.36		0.002			
310.0	16.61 ± 0.59	0.955 ± 0.011	18.82 ± 0.60	0.965 ± 0.008	− 3.80 ± 0.15		3.76 ± 0.16		0.95 ± 0.18		0.72 ± 0.1		0.05 ± 0.06		0.06 ± 0.03	
370.0	19.96 ± 0.85	0.975 ± 0.016	27.12 ± 0.76	0.788 ± 0.020	− 4.24 ± 0.22		6.26 ± 0.19	0.965 ± 0.010	1.40 ± 0.23		1.11 ± 0.14		− 0.04 ± 0.08		0.23 ± 0.03	

TABLE II (concluded)

E_π (MeV)	S_{11} δ	S_{11} η	P_{11} δ	P_{11} η	P_{13} δ	P_{13} η	D_{13} δ	D_{13} η	D_{15} δ	D_{15} η	F_{15} δ	F_{15} η	F_{17} δ	F_{17} η	G_{17} δ	G_{17} η
410.0	20.42 ± 6.93	0.950 ± 0.050	36.74 ± 3.86	0.581 ± 0.194	- 4.62 ± 2.60	0.992 ± 0.008	8.24 ± 1.80	0.926 ± 0.074	0.97 ± 3.20	0.997 ± 0.003	1.57 ± 0.95		- 0.79 ± 1.28		0.57 ± 0.70	
450.0	27.71 ± 1.93	0.862 ± 0.020	49.42 ± 2.42	0.392 ± 0.038	- 5.25 ± 0.64	0.999 ± 0.001	12.85 ± 0.40	0.999 ± 0.001	2.16 ± 0.34	0.999 ± 0.001	2.12 ± 0.28		- 0.01 ± 0.12		0.40 ± 0.09	
490.0	25.56 ± 6.38	0.800 ± 0.200	79.56 ± 15.35	0.290 ± 0.192	- 2.51 ± 3.63	0.960 ± 0.040	20.63 ± 1.91	0.939 ± 0.061	1.77 ± 1.53	0.999 ± 0.001	2.19 ± 0.16		0.25 ± 0.83		1.46 ± 0.95	
533.0	27.47 ± 10.51	0.650 ± 0.350	112.59 ± 48.02	0.304 ± 0.163	- 4.63 ± 6.61	0.937 ± 0.063	32.77 ± 4.69	0.772 ± 0.133	3.18 ± 3.81	0.989 ± 0.011	6.41 ± 4.86		- 0.11 ± 2.57		2.66 ± 2.23	
550.0	41.41 ± 4.38	0.731 ± 0.049	91.56 ± 10.58	0.169 ± 0.055	- 5.79 ± 0.77	0.999 ± 0.001	32.42 ± 1.74	0.633 ± 0.040	4.08 ± 0.42	0.999 ± 0.001	5.65 ± 0.54		0.01 ± 0.19		0.72 ± 0.17	
581.0	44.22 ± 5.35	0.540 ± 0.140	117.45 ± 11.96	0.411 ± 0.154	- 3.88 ± 4.44	0.884 ± 0.116	47.52 ± 8.91	0.390 ± 0.113	5.84 ± 3.18	0.971 ± 0.029	5.05 ± 3.41	0.957 ± 0.043	1.25 ± 1.83		1.82 ± 0.99	
600.0	28.82 ± 3.36	0.523 ± 0.036	120.0 ± 4.35	0.386 ± 0.050	- 4.92 ± 0.71	0.999 ± 0.001	94.75 ± 3.48	0.270 ± 0.037	8.19 ± 0.53	0.991 ± 0.009	6.04 ± 0.73	0.872 ± 0.020	- 0.20 ± 0.23		0.91 ± 0.20	
650.0	10.08 ± 1.17	0.178 ± 0.029	90.49 ± 7.28	0.232 ± 0.028	- 5.93 ± 1.04	0.954 ± 0.024	135.82 ± 1.86	0.532 ± 0.040	10.47 ± 0.74	0.910 ± 0.018	8.59 ± 0.86	0.959 ± 0.023	- 0.20 ± 0.32		1.50 ± 0.29	
698.0	- 1.90 ± 25.92	0.400 ± 0.400	56.86 ± 59.06	0.260 ± 0.260	- 5.71 ± 7.80	0.842 ± 0.158	144.21 ± 10.75	0.693 ± 0.236	11.21 ± 7.39	0.889 ± 0.111	14.58 ± 6.63	0.907 ± 0.093	- 0.85 ± 3.43		5.75 ± 4.54	

TABLE III

$T = \frac{1}{2}$ parameters 700–1400 MeV (solution 1)

E_π	S_{11}		P_{11}		P_{13}		D_{13}		D_{15}		F_{15}		F_{17}		G_{17}	
(MeV)	δ	η	δ	η	δ	η	δ	η	δ	η	δ	η	δ	η	δ	η
746.5	- 36.30	0.300	- 96.34	0.439	- 1.54	0.889	164.38	0.892	15.96	0.691	18.78	0.828	- 0.43		0.96	
	± 49.28	± 0.200	± 15.26	± 0.199	± 3.12	± 0.111	± 7.67	± 0.108	± 3.42	± 0.099	± 2.12	± 0.165	± 2.09		± 2.15	
795.9	0.21	0.278	- 94.31	0.848	7.33	0.951	175.14	0.893	14.98	0.540	28.19	0.611	- 0.77	0.994	2.09	0.999
	± 50.21	± 0.278	± 11.94	± 0.152	± 5.36	± 0.049	± 7.95	± 0.107	± 6.09	± 0.087	± 4.09	± 0.196	± 2.26	± 0.006	± 2.60	± 0.001
845.4	- 92.70	0.500	- 31.64	0.248	3.66	0.871	175.69	0.667	- 4.08	0.291	71.31	0.360	3.86	0.977	7.41	0.972
	± 12.23	± 0.215	± 27.03	± 0.239	± 5.64	± 0.105	± 6.99	± 0.214	± 12.55	± 0.106	± 16.60	± 0.084	± 2.27	± 0.023	± 4.18	± 0.028
870.0	- 83.93	0.550	- 42.94	0.201	5.47	0.916	169.70	0.779	- 18.27	0.285	73.55	0.342	0.35	0.987	8.52	0.999
	± 10.27	± 0.263	± 43.16	± 0.169	± 4.70	± 0.071	± 4.78	± 0.187	± 8.52	± 0.085	± 13.29	± 0.075	± 1.49	± 0.009	± 3.05	± 0.001
899.8	- 58.95	0.746	- 23.57	0.344	9.35	0.686	177.84	0.606	- 18.24	0.566	120.17	0.625	2.17	0.887	11.20	0.932
	± 2.66	± 0.069	± 6.55	± 0.031	± 2.28	± 0.030	± 1.18	± 0.050	± 0.90	± 0.037	± 2.47	± 0.033	± 0.50	± 0.022	± 0.95	± 0.024
949.4	- 63.80	0.528	25.40	0.170	14.71	0.502	186.05	0.660	- 14.58	0.507	136.30	0.733	- 3.25	0.931	6.64	0.998
	± 6.98	± 0.147	± 23.69	± 0.090	± 4.07	± 0.094	± 4.16	± 0.153	± 2.73	± 0.081	± 4.02	± 0.123	± 1.22	± 0.008	± 1.80	± 0.001
990.0	- 63.44	0.603	- 33.84	0.147	2.19	0.537	174.01	0.887	- 19.99	0.601	146.18	0.569	3.37	0.931	3.12	0.890
	± 6.05	± 0.213	± 59.11	± 0.147	± 5.66	± 0.088	± 3.48	± 0.113	± 3.45	± 0.156	± 3.77	± 0.142	± 2.75	± 0.069	± 2.41	± 0.110
1148.0	- 55.56	0.645	- 51.79	0.127	- 6.63	0.656	173.73	0.703	- 12.71	0.795	164.12	0.832	- 0.84	0.901	5.41	0.808
	± 1.13	± 0.226	± 11.51	± 0.042	± 1.19	± 0.038	± 1.30	± 0.025	± 0.95	± 0.030	± 0.95	± 0.025	± 0.55	± 0.022	± 0.53	± 0.022
1227.6	- 84.86	0.193	- 53.68	0.539	- 14.66	0.564	181.22	0.760	- 11.34	0.767	167.86	0.771	1.28	0.789	4.48	0.954
	± 4.99	± 0.019	± 4.27	± 0.075	± 2.89	± 0.047	± 1.28	± 0.073	± 1.67	± 0.068	± 1.59	± 0.040	± 0.86	± 0.044	± 0.78	± 0.031
1311.0	- 92.70	0.308	11.83	0.392	- 16.34	0.524	175.53	0.823	- 13.41	0.657	173.65	0.516	3.45	0.937	5.57	0.886
	± 92.39	± 0.308	± 24.21	± 0.249	± 7.38	± 0.162	± 5.52	± 0.091	± 8.03	± 0.179	± 10.60	± 0.079	± 2.89	± 0.063	± 2.87	± 0.104

TABLE IV

$T = \frac{1}{2}$ parameters 700–1400 MeV (solution 2)

E_π (MeV)	S_{11}		P_{11}		P_{13}		D_{13}		D_{15}		F_{15}		F_{17}		G_{17}	
	δ	η	δ	η	δ	η	δ	η	δ	η	δ	η	δ	η	δ	η
746.5	21.73 ± 3.53	0.600 ± 0.324	83.32 ± 20.60	0.109 ± 0.087	- 5.07 ± 1.52	0.721 ± 0.042	164.38 ± 7.67	0.892 ± 0.108	11.82 ± 1.11	0.815 ± 0.068	18.78 ± 2.12	0.828 ± 0.165	- 1.21 ± 0.76		1.75 ± 1.53	
795.9	39.93 ± 9.93		77.04 ± 7.81	0.666 ± 0.166	- 2.75 ± 4.49	0.949 ± 0.051	170.95 ± 6.11	0.703 ± 0.132	26.45 ± 3.47	0.649 ± 0.151	28.19 ± 4.09	0.611 ± 0.196	- 4.78 ± 2.04		2.65 ± 1.84	
845.4	104.08 ± 11.65	0.826 ± 0.174	61.66 ± 33.40	0.346 ± 0.153	13.20 ± 9.23	0.709 0.093	191.70 ± 6.01	0.952 ± 0.048	- 15.43 ± 6.52	0.389 ± 0.147	97.87 ± 58.01	0.187 ± 0.179	5.93 ± 2.73	0.916 ± 0.084	5.35 ± 3.10	0.876 ± 0.124
870.0	96.07 ± 10.27	0.550 ± 0.263	46.18 ± 24.01	0.339 ± 0.140	8.71 ± 4.70	0.687 ± 0.070	181.47 ± 4.61	0.940 ± 0.060	- 26.85 ± 5.97	0.310 ± 0.139	88.71 ± 23.33	0.175 ± 0.099	2.19 ± 2.69	0.973 ± 0.027	3.95 ± 2.61	0.925 ± 0.075
899.8	109.03 ± 8.85	0.839 ± 0.161	65.56 ± 31.05	0.173 ± 0.125	6.07 ± 9.39	0.527 ± 0.079	187.67 ± 5.78	0.874 ± 0.096	- 20.39 ± 2.34	0.515 ± 0.105	125.58 ± 9.96	0.517 ± 0.112	1.64 ± 1.33	0.916 ± 0.076	6.38 ± 2.97	0.925 ± 0.075
949.4	115.37 ± 4.58	0.748 ± 0.090	- 8.09 ± 15.08	0.218 ± 0.101	11.94 ± 5.83	0.395 ± 0.042	183.49 ± 2.84	0.795 ± 0.096	- 17.27 ± 2.88	0.449 ± 0.050	137.92 ± 2.95	0.666 ± 0.074	- 2.69 ± 1.07	0.931 ± 0.069	6.51 ± 1.53	0.998 ± 0.002
990.0	116.56 ± 6.05	As for solution 1														
1148.0	124.44 ± 1.13	As for solution 1														
1227.6	95.14 ± 4.99	As for solution 1														
1311.0	87.30 ± 92.39	As for solution 1														

the BBSV and RWF solutions, the phase shift passes through 90° at 600 MeV, and continues rising. In the RWF solution it flattens off at about 120°, but in the BBSV it continues up through 180°. In the BOM solution the phase shift rises more slowly, passing through 90° at about 800 MeV. All solutions agree that this amplitude is highly absorptive with $\eta \sim 0{\cdot}15$. A resonance interpretation is again possible, but the details are again obscure.

P_{13}. BBSV, RWF and BOM all find this phase shift small and negative without much inelasticity. In the DKLL solution there is evidence for some structure in the region of 1 GeV, although errors are large. The phase shift, initially negative, becomes positive, rising to $\sim 15°$ before turning and becoming negative again, this being associated with a fairly rapid decrease in η. In the Argand diagram the amplitude describes part of a rather small, squashed anti-clockwise circle (a miniature type 5d). A resonance interpretation is almost certainly too dramatic here, a "woolly cusp" interpretation being more likely.

D_{13}. All solutions have a resonance of type 5c. In the DKLL solution the phase shift passes through 90° close to 600 MeV, in the BOM and BBSV solutions close to 620 MeV and in the RWF solution close to 650 MeV. In all cases η has a deep minimum near this energy ($\eta \sim 0{\cdot}2$). This amplitude is the old "second resonance", and is much narrower than was thought previously.

D_{15}. The DKLL, BBSV and BOM solutions all agree that there is a resonance in this amplitude. The DKLL and BBSV solutions are of type 5d, the BOM of type 5c. Since at the resonance position ($\alpha \simeq 0°$ in BBSV and DKLL, $\alpha \simeq 90°$ in BOM), η is very small ($\simeq 0{\cdot}1$) this difference in interpretation is difficult to make meaningful. The zeros in the real parts of the amplitude occur at different energies: about 850 MeV in BOM and DKLL and about 900 MeV in BBSV.

F_{15}. The DKLL, BBSV and BOM solutions all have a resonance of type 5c at about 900 MeV, or just below. It is fairly absorptive, with $\eta \sim 0{\cdot}3$. This is the old "third resonance", and like D_{13} is much narrower than thought previously.

F_{17}. Like F_{35} this amplitude is badly determined. All that can be said is that it is essentially elastic and the phase shift is very small.

G_{17}. Only DKLL have a solution for this amplitude. They find that the phase shift is surprisingly large for a g wave, rising steadily and reaching $\sim 10°$.

We have discussed in some detail four independent phase shift analyses of pion-nucleon scattering using different techniques. Despite the quantitative differences, a remarkably clear qualitative picture emerges of pion-nucleon scattering below 1·4 GeV. The only possible exceptions to this are the S_{11} and P_{11} amplitudes, and even there the qualitative interpretation of the situation is the same in each case despite the apparently very different results. This difference is in fact due more to the "phase shift and inelasticity" description of the amplitudes than to any fundamental difference, as can be seen if the different solutions are plotted on the Argand diagram, when they all look much more alike. Also it has been shown recently[39] that if the scattering amplitudes are described in terms of the eigenphases and mixing angles of the K matrix then, in certain circumstances, very small changes in these parameters produce very large changes in the phase shift and inelasticities in the elastic scattering channel, i.e., the "phase shift and inelasticity" description is not necessarily the best one to use.

Finally, in Table V we give a composite picture of the resonance structure in pion-nucleon scattering, with possible masses and widths. Further clarification awaits new experiments.

TABLE V

The probable resonance structure of pion-nucleon scattering below 1·4 GeV

Amplitude	Resonance energy (MeV)	Resonance width (MeV)
S_{31}	1690 ± 20	220 ± 20
P_{33}	1240 ± 10	110 ± 10
F_{37}	1920 ± 20	120 ± 30
S_{11} i)	1510 ± 10	100 ± 10
[ii)	1660 ± 60	230 ± 30 (Dubious!)]
P_{11}	1420 ± 20	200 ± ?
D_{13}	1520 ± 10	80 ± 25
D_{15}	1650 ± 20	65 ± 20
F_{15}	1680 ± 10	70 ± 15

APPENDIX 1

THE EVALUATION OF PARTIAL WAVE DISPERSION RELATIONS

The Mandelstam representations for the amplitudes $A^{(T)}$, $B^{(T)}$ are

$$\begin{aligned}B^{(T)}(s,t,u) = {}& \frac{\zeta_1^{(T)}G_r^2}{s-m^2}+\frac{\zeta_2^{(T)}G_r^2}{u-m^2}\\ &+\frac{1}{\pi^2}\int_{(m+\mu)^2}^{\infty} ds' \int_{4\mu^2}^{\infty} dt' \, \frac{B_{12}^{(T)}(s',t')}{(s'-s)(t'-t)}\\ &+\frac{1}{\pi^2}\int_{4\mu^2}^{\infty} dt' \int_{(m+\mu)^2}^{\infty} du' \, \frac{B_{23}^{(T)}(t',u')}{(t'-t)(u'-u)}\\ &+\frac{1}{\pi^2}\int_{(m+\mu)^2}^{\infty} ds' \int_{(m+\mu)^2}^{\infty} du' \, \frac{B_{13}^{(T)}(s',u')}{(s'-s)(u'-u)} \qquad \ldots[A.1]\end{aligned}$$

with $\zeta_1^{(\frac{1}{2})} = -3$, $\zeta_2^{(\frac{1}{2})} = -1$; $\zeta_1^{(\frac{3}{2})} = 0$, $\zeta_2^{(\frac{3}{2})} = 2$, and an identical relation for $A^{(T)}$ with $\zeta_1^{(T)} = \zeta_2^{(T)} = 0$, $T = \frac{1}{2}, \frac{3}{2}$.

By [22] and [15] the projection of $F_{l\pm}^{(T)}(s)$ from the amplitudes $A^{(T)}$ and $B^{(T)}$ is given by

$$F_{l\pm}^{(T)}(s) = \frac{1}{8\pi q^{2l}} \int_{-1}^{1} dx \left\{ P_l(x) \frac{(W+m)^2-\mu^2}{4W^2} [A^{(T)}+(W-m)B^{(T)}] \right.$$
$$\left. + P_{l\pm 1}(x) \frac{(W-m)^2-\mu^2}{4W^2} [-A^{(T)}+(W+m)B^{(T)}] \right\}. \quad \text{...[A.2]}$$

The Born pole

This comes from the term in $(s-m^2)^{-1}$ in the amplitude $B^{(\frac{1}{2})}$. Since there is no $T = \frac{3}{2}$ contribution and since s is independent of x, the projection [A.2] is valid only for F_{1-}, and this term contributes only to the amplitude P_{11}. Evaluating the kinematical factor at $W = m$ and carrying out the integration over x, one obtains the contribution

$$-\frac{6m}{(4m^2-\mu^2)} \cdot \frac{g_r^2}{(s-m^2)} \quad \text{...[A.3]}$$

where $g_r^2 = G_r^2/4\pi$.

The Born cuts

These come from the terms in $(u-m^2)^{-1}$ in the amplitudes $B^{(T)}$ $(T = \frac{1}{2}, \frac{3}{2})$ and are the cuts $-\infty \leqq s \leqq 0$ (the left-hand cut) and $(m-\mu^2/m)^2 \leqq s \leqq (m^2+2\mu^2)$ (the short Born cut). They give the only contribution to this latter cut and it is this contribution we want to evaluate, i.e. we want to find

$$\frac{1}{\pi} \int_{(m-\mu^2/m)^2}^{(m^2+2\mu^2)} ds' \frac{\text{Im } F_l(s')}{s'-s}. \quad \text{...[A.4]}$$

Substituting the pole terms from [A.1] into [A.2] we obtain

$$F_{l\pm}^{(T)} = \frac{\zeta_2^{(T)}}{2q^{2l}} g_r^2 \int_{-1}^{1} dx \left\{ P_l(x) \frac{[(W+m)^2-\mu^2][W-m]}{4W^2} \right.$$
$$\left. + P_{l\pm 1}(x) \frac{[(W-m)^2-\mu^2][W+m]}{4W^2} \right\} \frac{1}{u-m^2}. \quad \text{...[A.5]}$$

In terms of s and x

$$u = -\{[s^2-2s(m^2+\mu^2)-(m^2-\mu^2)^2]+x[s^2-2s(m^2+\mu^2)+(m^2-\mu^2)^2]\}/2s \quad \text{...[A.6]}$$

so that

$$\operatorname{Im} F_{l\pm}^{(T)} = -\frac{s\zeta_2^{(T)}}{q^{2l}} g_r^2 \left\{\frac{[(W+m)^2-\mu^2][W-m]}{4W^2} \operatorname{Im} I_l + \frac{[(W-m)^2-\mu^2][W+m]}{4W^2} \operatorname{Im} I_{l\pm 1}\right\} \quad ...[A.7]$$

where

$$I_l = \int_{-1}^{1} dx \frac{P_l(x)}{a+bx} \quad ...[A.8]$$

with

$$\left.\begin{aligned} a &= s^2-2s\mu^2-(m^2-\mu^2)^2, \\ b &= s^2-2s(m^2+\mu^2)+(m^2-\mu^2)^2. \end{aligned}\right\} \quad ...[A.9]$$

In the region under consideration $|b| \geqq |a|$, so $(a+bx)^{-1}$ has a pole for some x in $-1 \leqq x \leqq 1$ for each value of s in $(m-\mu^2/m)^2 \leqq s \leqq (m^2+2\mu^2)$. Since I_l is a singular integral, we can write

$$I_l = \int_{-1}^{1} dx \frac{P_l(x)}{a+bx} = P\int_{-1}^{1} dx \frac{P_l(x)}{a+bx} \pm i\frac{\pi}{b} P_l(-a/b) \quad ...[A.10]$$

where the $-$ or $+$ sign is to be taken according as to whether the pole is passed from above or below in the x-plane. Letting $s = s+i\varepsilon$ and evaluating x to the first order in ε, we obtain

$$x(s+i\varepsilon) = -a[ab+2i\varepsilon\{(s-\mu^2)b-(s-m^2-\mu^2)a\}]/b^2$$

so that for $\varepsilon > 0$ and for all $s \in (m-\mu^2/m)^2 \leqq s \leqq (m^2+2\mu^2)$, $\operatorname{Im} x < 0$ and we must pass below the pole in the x-plane. Hence

$$\operatorname{Im} I_l = \frac{\pi}{b} P_l(-a/b)$$

and the contribution from the short Born cut is given by

$$-\frac{\zeta_2^{(T)}\cdot g_r^2}{4}\int_{(m-\mu^2/m)^2}^{(m^2+2\mu^2)} ds' \frac{1}{(s'-s)b'q'^{2l}} \{[W'-m][(W'+m)^2-\mu^2]P_l(-a'/b') + [W'+m][(W'-m)^2-\mu^2]P_{l\pm 1}(-a'/b')\} \quad ...[A.11]$$

with

$$\zeta_2^{(\frac{1}{2})} = -1, \quad \zeta_2^{(\frac{3}{2})} = 2$$

and

$$\begin{aligned} a' &= s'^2-2s'\mu^2-(m^2-\mu^2)^2, \\ b' &= s'^2-2s'(m^2+\mu^2)+(m^2-\mu^2)^2. \end{aligned}$$

The crossed cut

We require to evaluate the contribution

$$\frac{1}{\pi}\int_0^{(m-\mu)^2} ds' \, \frac{\operatorname{Im} F_{l\pm}^{(T')}(s')}{s'-s} \qquad \text{...[A.12]}$$

with $\operatorname{Im} F_{l\pm}^{(T')}(s')$ given by the projection

$$\operatorname{Im} F_{l\pm}^{(T')}(s') = \frac{(-1)}{8\pi q'^{2l}} \int_{-1}^{1} dx' \left\{ P_l(x') \frac{[(W'+m)^2-\mu^2]}{4W'^2} [\operatorname{Im} A^{(T')} + (W'-m) \operatorname{Im} B^{(T')}] \right.$$
$$\left. + P_{l\pm 1}(x') \frac{[(W'-m)^2-\mu^2]}{4W'^2} [-\operatorname{Im} A^{(T')} + (W'+m) \operatorname{Im} B^{(T')}] \right\} \qquad \text{...[A.13]}$$

and all quantities are evaluated on the crossed cut.

For the moment, let us denote amplitudes evaluated on the crossed cut by a subscript c, and recalling [9a], we have

$$A_c^{(\frac{1}{2})} = A_c^{(+)} + 2A_c^{(-)} = A^{(+)} - 2A^{(-)} = -\tfrac{1}{3}A^{(\frac{1}{2})} + \tfrac{4}{3}A^{(\frac{3}{2})}$$

and

$$A_c^{(\frac{3}{2})} = A_c^{(+)} - A_c^{(-)} = A^{(+)} + A^{(-)} = \tfrac{2}{3}A^{(\frac{1}{2})} + \tfrac{1}{3}A^{(\frac{3}{2})}.$$

We may combine these results as

$$\begin{bmatrix} A_c^{(\frac{1}{2})} \\ A_c^{(\frac{3}{2})} \end{bmatrix} = \begin{bmatrix} -\frac{1}{3} & \frac{4}{3} \\ \frac{2}{3} & \frac{1}{3} \end{bmatrix} \begin{bmatrix} A^{(\frac{1}{2})} \\ A^{(\frac{3}{2})} \end{bmatrix}$$

or

$$A_c^{(T)} = C_{TT'} A^{(T')} \qquad \text{...[A.14]}$$

where

$$C_{TT'} = \begin{bmatrix} -\frac{1}{3} & \frac{4}{3} \\ \frac{2}{3} & \frac{1}{3} \end{bmatrix} \qquad \text{...[A.15]}$$

is the isospin crossing matrix. By [9b] the corresponding matrix for the B amplitudes is $-C_{TT'}$. Inserting this into [A.13], we obtain

$$\operatorname{Im} F_{l\pm}^{(T')}(s') = \frac{-C_{T'T''}}{8\pi q'^{2l}} \int_{-1}^{1} dx' \left\{ P_l(x') \frac{[(W'+m)^2-\mu^2]}{4W'^2} [\operatorname{Im} A^{(T'')} - (W'-m) \operatorname{Im} B^{(T'')}] \right.$$
$$\left. + P_{l\pm 1}(x') \frac{[(W'-m)^2-\mu^2]}{4W'^2} [-\operatorname{Im} A^{(T'')} - (W'+m) \operatorname{Im} B^{(T'')}] \right\}.$$

Re-expressing the amplitudes $A^{(T'')}$, $B^{(T'')}$ in terms of the amplitudes $F_{l''}(s'')$ via [16] and [17] yields

$$\text{Im}\, F_{l\pm}^{(T')}(s') = \frac{-C_{T'T''}}{8W'^2 q'^{2l}} \int_{-1}^{1} dx' (\mathscr{A}-\mathscr{B}) \qquad \ldots[A.16]$$

where

$$\mathscr{A} = \{[W'+m)^2-\mu^2]\, P_l(x') - [(W'-m)^2-\mu^2]\, P_{l\pm 1}(x')\}$$

$$\left(\frac{W''+m}{E''+m} \sum_{l''=0}^{\infty} [\text{Im}\, F_{l''+}^{(T'')}(s'')\, P'_{l''+1}(x'') - \text{Im}\, F_{l''-}^{(T'')}(s'')\, P'_{l''-1}(x'')](q'')^{2l''}\right.$$

$$\left. +\frac{W''-m}{E''-m} \sum_{l''=1}^{\infty} [\text{Im}\, F_{l''+}^{(T'')}(s'') - \text{Im}\, F_{l''-}^{(T'')}(s'')](q'')^{2l''} P'_{l''}(x'')\right)$$

and

$$\mathscr{B} = \{[W'-m][(W'+m)^2-\mu^2]\, P_l(x') + [W'+m][(W'-m)^2-\mu^2]\, P_{l\pm 1}(x')\}$$

$$\left(\frac{1}{E''+m} \sum_{l''=0}^{\infty} [\text{Im}\, F_{l''+}^{(T'')}(s'')\, P'_{l''+1}(x'') - \text{Im}\, F_{l''-}^{(T'')}(s'')\, P'_{l''-1}(x'')](q'')^{2l''}\right.$$

$$\left. -\frac{1}{E''-m} \sum_{l''=1}^{\infty} [\text{Im}\, F_{l''+}^{(T'')}(s'') - \text{Im}\, F_{l''-}^{(T'')}(s'')]\, P'_{l''}(x'')(q'')^{2l''}\right).$$

The energies s'' and angles x'' which appear on the right-hand side of equation [*A.16*] are real physical values of energy ($s'' \geqq (m+\mu)^2$) and angle ($-1 \leqq x'' \leqq 1$), and they are functions of s' and x'. Since the variable t is unchanged under crossing, the equality

$$t' = t''$$

gives immediately that

$$q'^2(1-x') = q''^2(1-x''). \qquad \ldots[A.17]$$

The crossing relation $s'' = u'$ can be expanded, using [*A.6*], to give

$$s'' = (m^2+\mu^2-\tfrac{1}{2}s')(1+x') + (m^2-\mu^2)^2(1-x')/2s'. \qquad \ldots[A.18]$$

[*A.17*] and [*A.18*] mean of course that the right-hand side of equation [*A.16*] can be expressed entirely as a function of s' and x' and the integration over x' performed.

Alternatively, [*A.17*] and [*A.18*] may be used to express x' as a function of s'' and the x' integration re-cast into an integration over s'' over some (finite) range of energies (the upper and lower limits are controlled by the value of s' occurring in [*A.18*]). In this form [*A.16*] may be written symbolically as

$$\text{Im}\, F_{l\pm}^{(T')}(s') = -C_{T'T''} \sum_{l''} \int_{\substack{\text{Some Physical}\\ \text{Energies}}} ds'' \{K_{ll''}^{+}(s', s'')\, \text{Im}\, F_{l''+}^{(T'')}(s'')$$

$$+K_{ll''}^{-}(s', s'')\, \text{Im}\, F_{l''-}^{(T'')}(s'')\} \qquad \ldots[A.19]$$

where the $K^{\pm}(s', s'')$ are (calculable) kernels, and the contribution from the crossed cut is

$$-\frac{C_{T'T''}}{\pi}\int_0^{(m-\mu)^2}\frac{ds'}{s'-s}\sum_{l''}\int ds''\,\{K^{+}_{ll''}(s', s'')\,\mathrm{Im}\,F^{(T'')}_{l''+}(s'') + K^{-}_{ll''}(s', s'')\,\mathrm{Im}\,F^{(T'')}_{l''-}(s'')\}. \qquad ...[A.20]$$

The only difficulty which occurs in evaluating [*A.20*] is at the crossed threshold $s_1 = (m-\mu)^2$, where there is a divergence for d waves and higher. Evaluating [*A.19*] explicitly in the p wave case[25] shows that

$$\mathrm{Im}\,F_{1\pm}(s'+i0) = \frac{a_{\pm}}{(s_1-s')^{\frac{1}{2}}} + O((s_1-s')^2) \qquad ...[A.21]$$

as s' approaches s_1 along the cut, the constants $a_{\pm}$ depending on the $\pi-N$

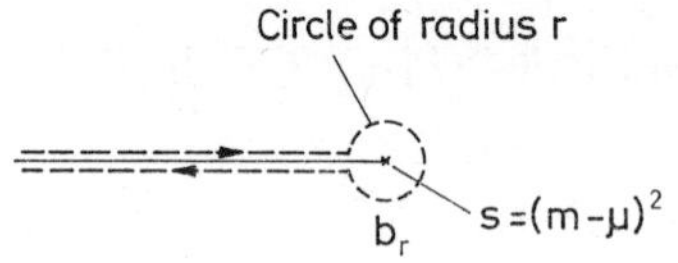

FIG. 6. Contour of integration on the crossed cut.

s wave scattering lengths. Since this is only a square root singularity it gives no trouble.

In the d wave case it is easy to see that

$$\mathrm{Im}\,F_{2\pm}(s'+i0) = \frac{a'_{\pm}}{(s_1-s')^{\frac{3}{2}}} + \frac{b'_{\pm}}{(s_1-s')^{\frac{1}{2}}} + O((s_1-s')^{\frac{1}{2}}) \qquad ...[A.22]$$

as s' approaches s_1 along the cut. As before the constants $a'_{\pm}$ are related to the $\pi-N$ s wave scattering lengths, while $b'_{\pm}$ depend on the behaviour of the $\pi-N$ s waves near the physical threshold. The term in [*A.22*] which contains $(s_1-s')^{-\frac{3}{2}}$ gives rise to a divergence in the integral

$$\frac{1}{\pi}\int_0^{(m-\mu)^2} ds'\,\frac{\mathrm{Im}\,F_{2\pm}(s')}{s'-s}.$$

This divergence is easily removed. In Fig. 6 we show the path of integration around the cut $0 < s' \leqq s_1$. The contribution arising therefrom is

$$\frac{1}{\pi}\int_0^{(s_1-r)} ds'\,\frac{\mathrm{Im}\,F_{2\pm}(s')}{s'-s} + \frac{1}{2\pi i}\int_{C_r} ds'\,\frac{F_{2\pm}(s')}{s'-s} \qquad ...[A.23]$$

where r is a small real number and C_r is the small circle of radius r centred on s_1. Substituting [*A.22*] into the first integral in [*A.23*] gives

$$\frac{1}{\pi}\int_0^{(s_1-r)} ds'\,\frac{\mathrm{Im}\,F_{2\pm}(s')}{s'-s} = \text{(finite term)} + \frac{2}{\pi}\frac{a'_{\pm}}{r^{\frac{1}{2}}}\frac{1}{(s_1-s)} + O(r^{\frac{1}{2}}) \qquad ...[A.24]$$

where the "finite terms" are independent of r.

Substituting [*A.22*] into the second integral in [*A.23*] gives

$$\frac{1}{2\pi i}\int_{C_r} ds' \frac{F_{2\pm}(s')}{s'-s} = -\frac{2}{\pi}\frac{a'_\pm}{r^{\frac{1}{2}}}\frac{1}{s_1-s}+O(r^{\frac{1}{2}}). \qquad ...[A.25]$$

Adding [*A.24*] and [*A.25*] and letting $r \to 0$ gives a finite result.

In practice the integration is done analytically around C_r and along the cut to some value $\bar{s} < s_1 - r$, the remainder being done numerically.

The method can easily be extended to handle the higher order divergences which occur in the higher partial waves.

The circle contribution

The discontinuity in $F_l^{(\pm)}(s)$ across the circle is given by

$$\Delta F_{l\pm}^{(\pm)}(s)$$

$$= \frac{1}{8\pi q^{2l}}\int_{-1}^{1} dx \left\{P_l(x)\frac{(W+m)^2-\mu^2}{4W^2}[\Delta A^{(\pm)}+(W-m)\Delta B^{(\pm)}]\right.$$

$$\left. +P_{l\pm 1}(x)\frac{(W-m)^2-\mu^2}{4W^2}[-\Delta A^{(\pm)}+(W+m)\Delta B^{(\pm)}]\right\}. \qquad ...[A.26]$$

The discontinuity of $A^{(\pm)}$, $B^{(\pm)}$ across the circle (i.e. the difference between the value of $A^{(\pm)}$, $B^{(\pm)}$ just inside the circle and the value just outside) is simply $2i \operatorname{Im} A^{(\pm)}$, $2i \operatorname{Im} B^{(\pm)}$ respectively, where

$$\operatorname{Im} A^{(\pm)} = \sum_{J=0}^{\infty}(J+\tfrac{1}{2})\frac{8\pi}{p_-^2}(iq_3p_-)^J[P_J(\cos\theta_3)\operatorname{Im} f_+^J(t)$$

$$-\frac{m\cos\theta_3}{[J(J+1)]^{\frac{1}{2}}}P'_J(\cos\theta_3)\operatorname{Im} f_-^J(t)] \qquad ...[A.27a]$$

$$\operatorname{Im} B^{(\pm)} = 8\pi\sum_{J=1}^{\infty}\frac{(J+\frac{1}{2})}{[J(J+1)]^{\frac{1}{2}}}(iq_3p_-)^{J-1}P'_J(\cos\theta_3)\operatorname{Im} f_-^J(t) \qquad ...[A.27b]$$

where the $f_\pm^J(t)$ are the helicity amplitudes for the process $\pi+\pi \to N+\bar{N}$. J is even for the $(+)$ case and odd for the $(-)$ case. The momentum of the pion in the process $\pi+\pi \to N+\bar{N}$ is given by

$$q_3^2 = t/4-\mu^2 \qquad ...[A.28]$$

and the scattering angle in $\pi+\pi \to N+\bar{N}$ is given by

$$\cos\theta_3 = \frac{s-p_-^2+q_3^2}{2ip_-q_3} \qquad ...[A.29]$$

where

$$p_-^2 = (m^2-t/4) \qquad ...[A.30]$$

[t, of course, is a function of s and x and consequently so is the right-hand side of [$A.26$]. As in the case of the crossed cut, the integral over x may be converted to an energy integral—this time over t.]

As discussed in section 4, the only important contributions to [$A.27$] come from the $T = 0$, $J = 0$ (s wave $\pi-\pi$) and $T = 1$, $J = 1$ (ρ meson) states, which contribute respectively to the (+) and (−) amplitudes. Information on the former can be obtained from a study of low energy s wave pion-nucleon scattering[25] or from pion-nucleon backward dispersion relations,[40,41] and information on the latter from the same sources and from the isovector form factor of the nucleon.[29] Once these are known, substitution into [$A.27$], and that in turn into [$A.26$], gives the circle contribution

$$\frac{1}{2\pi i}\int ds' \frac{\Delta F_{l\pm}(s')}{(s'-s)}$$

Appendix 2

Conformal Mapping Methods in Evaluating Dispersion Integrals‡

Suppose we require to evaluate the right-hand (principle value) integral in a dispersion relation. For simplicity, let the variable be T, the kinetic energy, so that the integral is from 0 to ∞.

Now make the change of variable to

$$x = \cos\zeta = \frac{T_0 - T}{T_0 + T} \qquad \text{...[A.31]}$$

where T_0 is an arbitrary constant. For obvious reasons $-T_0$ is called the "conformal zero". The conformal mapping [$A.31$] maps the cut 0 to ∞ in the T plane on to the cut -1 to 1 in the x-plane. From [$A.31$] we have that

$$\begin{aligned}(1-x^2)^{\frac{1}{2}} = \sin\zeta &= \frac{2(TT_0)^{\frac{1}{2}}}{T+T_0} \\ &\sim q \quad \text{as} \quad T \to 0 \\ &\sim q^{-1} \quad \text{as} \quad T \to \infty\end{aligned} \qquad \text{...[A.32]}$$

since

$$\begin{aligned}q^2 &= [s-(m+\mu)^2][s-(m-\mu)^2]/4s \\ &= \frac{m^2T(T+4m\mu)}{2mT+(m+\mu)^2}.\end{aligned} \qquad \text{...[A.33]}$$

By unitarity, Im F_l has the behaviour

$$\begin{aligned}\operatorname{Im} F_l &\sim q^{2l+1} \quad \text{as} \quad T \to 0 \\ &\sim q^{-2l-1} \quad \text{as} \quad T \to \infty.\end{aligned}$$

‡ We are indebted to Dr. C. Lovelace for permission to quote freely from his unpublished notes on this subject.

Hence the approximation

$$\operatorname{Im} F_l = \sum_{n=1}^{N} a_n(1-x^2)^{l+\frac{1}{2}} C_{n-1}^{l+1}(x) \qquad \ldots[A.34]$$

where $C_{n-1}^{l+1}(x)$ is a Gegenbauer polynomial,[42] will have the correct behaviour both at threshold and infinity.

The rescattering contribution required is of the form

$$\int_0^{\infty} dT' \frac{\phi(T')}{T'-T} = -\int_{-1}^{1} dx' \left[\frac{1}{x'-x} - \frac{1}{1+x'}\right] \phi(x'). \qquad \ldots[A.35]$$

Therefore the rescattering contribution to the real part will be given by

$$[\operatorname{Re} F_l]_{\text{Rescatt.}} = \sum_{n=1}^{N} a_n[g_n^l(x) - g_n^l(-1)] \qquad \ldots[A.36]$$

where

$$g_n^l(x) = -\frac{P}{\pi}\int_{-1}^{1} dx' \frac{(1-x'^2)^{l+\frac{1}{2}} C_{n-1}^{l+1}(x')}{(x'-x)} \qquad \ldots[A.37]$$

and the subtraction at -1 ensures that the real part vanishes at infinity.

It can be shown[43] that

$$C_{n-1}^{l+1}(x) = \frac{(2l+2)_{n-1}}{(l+\frac{3}{2})_{n-1}} P_{n-1}^{(l+\frac{1}{2}, l+\frac{1}{2})}(x)$$

$$= \frac{\Gamma(2l+n+1)\Gamma(l+\frac{3}{2})}{\Gamma(2l+2)\Gamma(l+n+\frac{1}{2})} P_{n-1}^{(l+\frac{1}{2}, l+\frac{1}{2})}(x). \qquad \ldots[A.38]$$

Therefore[44]

$$2(x^2-1)^{l+\frac{1}{2}} Q_{n-1}^{(l+\frac{1}{2}, l+\frac{1}{2})}(x) = -\int_{-1}^{1} dx' \frac{(1-x'^2)^{l+\frac{1}{2}} P_{n-1}^{(l+\frac{1}{2}, l+\frac{1}{2})}(x')}{x'-x-i\varepsilon} \qquad \ldots[A.39]$$

and we have

$$g_n^l(x) = \frac{1}{\pi}\frac{\Gamma(2l+n+1)\Gamma(l+\frac{3}{2})}{\Gamma(l+n+\frac{1}{2})\Gamma(2l+2)} 2(x^2-1)^{l+\frac{1}{2}} \operatorname{Re} Q_{n-1}^{(l+\frac{1}{2}, l+\frac{1}{2})}(x). \qquad \ldots[A.40]$$

This may be expressed in terms of hypergeometric functions for $-1 \leqq x \leqq 1$[44]

$$2(x^2-1)^{l+\frac{1}{2}} Q_{n-1}^{(l+\frac{1}{2}, l+\frac{1}{2})}(x)$$

$$= \frac{2^{2l+1}\Gamma(l+\frac{1}{2})\Gamma(l+n+\frac{1}{2})}{\Gamma(n+2l+1)} F(n, -n-2l; \tfrac{1}{2}-l; \tfrac{1}{2}-\tfrac{1}{2}x)$$

$$-\frac{\pi(x^2-1)^{l+\frac{1}{2}}}{\sin(\pi[l+\frac{1}{2}])} P_{n-1}^{(l+\frac{1}{2}, l+\frac{1}{2})}(x). \qquad \ldots[A.41]$$

The second term is purely imaginary for $x^2 \leqq 1$ and so in this region (corresponding to $T \geqq 0$) we have

$$g_n^l(x) = \frac{1}{\pi} \frac{\Gamma(l+\frac{1}{2})\Gamma(l+\frac{3}{2})}{\Gamma(2l+2)} 2^{2l+1} F(n, -n-2l; \tfrac{1}{2}-l; \tfrac{1}{2}-\tfrac{1}{2}x)$$

$$= \frac{1}{\sqrt{\pi}} \frac{\Gamma(l+\frac{1}{2})}{l!} F(n, -n-2l; \tfrac{1}{2}-l; \tfrac{1}{2}-\tfrac{1}{2}x). \qquad \text{...[A.42]}$$

The hypergeometric function reduces to a polynomial since the second argument is a negative integer. Convenient transformations to make are[45]

$$F(n, -n-2l; \tfrac{1}{2}-l; \tfrac{1}{2}-\tfrac{1}{2}x) = F\left(\frac{n}{2}, -\frac{n}{2}-l; \tfrac{1}{2}-l; 1-x^2\right) \qquad \text{...[A.43]}$$

and

$$F(n, -n-2l; \tfrac{1}{2}-l; \tfrac{1}{2}-\tfrac{1}{2}x) = xF\left(\frac{n+1}{2}, -\left(\frac{n-1}{2}\right)-l; \tfrac{1}{2}-l; 1-x^2\right). \qquad \text{...[A.44]}$$

The hypergeometric function on the right-hand side of [*A.43*] terminates for n even, and that in [*A.44*] for n odd. Thus we have

$$g_{2m}^l(x) = \binom{l-\frac{1}{2}}{l} F(m, -m-l; \tfrac{1}{2}-l; 1-x^2), \qquad \text{...[A.45]}$$

$$g_{2m+1}^l(x) = \binom{l-\frac{1}{2}}{l} x F(m+1, -m-l; \tfrac{1}{2}-l; 1-x^2) \qquad \text{...[A.46]}$$

and these series terminate more quickly than that in [*A.42*]. It follows from [*A.45*] and [*A.46*] that

$$g_n^l(-1) = (-1)^n \binom{l-\frac{1}{2}}{l}. \qquad \text{...[A.47]}$$

Thus given $\operatorname{Im} F_l$ by [*A.34*], then [*A.36*], [*A.45*], [*A.46*] and [*A.47*] together determine the rescattering.

REFERENCES

1. G. F. Chew, M. L. Goldberger, F. E. Low and Y. Nambu. 1957. *Phys. Rev.* **106**, 1337.
2. G. F. Chew. 1961. In *Dispersion Relations*, Ed. G. R. Screaton (Oliver and Boyd, Edinburgh).
3. M. Jacob and G. C. Wick. 1959. *Ann. Phys.* **7**, 404.
4. M. L. Goldberger. 1960. In *Dispersion Relations and Elementary Particles*, Ed. C. De Witt (Wiley, New York).
5. J. Ashkin. 1959. *Nuovo Cim. Supp.* **14**, 221.
6. H. A. Bethe and F. de Hoffmann. 1955. *Mesons and Fields*, vol. II (Row, Peterson and Company).
7. H. P. Stapp, T. J. Ypsilantis and N. Metropolis. 1957. *Phys. Rev.* **105**, 302.
8. F. T. Solmitz. 1954. *Phys. Rev.* **94**, 1799.

9. J. Foote, O. Chamberlain, E. Rogers and H. Steiner. 1961. *Phys. Rev.* **122**, 959.
10. P. Bareyre, C. Brickman, A. V. Stirling and G. Villet. 1965. *Phys. Lett.* **18**, 342.
11. L. D. Roper, R. M. Wright and B. T. Feld. 1965. *Phys. Rev.* **138**, B190.
12. B. H. Bransden, P. J. O'Donnell and R. G. Moorhouse. 1965. *Phys. Rev.* **139**, B1566.
13. B. H. Bransden, P. J. O'Donnell and R. G. Moorhouse. 1965. *Phys. Lett.* **19**, 420.
14. W. M. Layson. 1961. *Nuovo Cim.* **20**, 1207.
15. J. S. Ball and W. R. Frazer. 1961. *Phys. Rev. Lett.* **7**, 204.
16. J. W. Moffat. 1961. *Phys. Rev.* **121**, 926.
17. S. Mandelstam. 1958. *Phys. Rev.* **112**, 1344 and 1959; *Phys. Rev.* **115**, 1741 and 1752.
18. J. Hamilton and T. D. Spearman. 1961. *Ann. Phys.* **12**, 172.
19. J. Hamilton and W. S. Woolcock. 1963. *Rev. Mod. Phys.* **35**, 737.
20. J. Hamilton. 1964. In *Strong Interactions and High Energy Physics*, Ed. R. G. Moorhouse (Oliver and Boyd, Edinburgh).
21. W. S. Woolcock and V. K. Samaranayake. 1965. *Phys. Rev. Lett.* **15**, 936.
22. S. Mandelstam. 1960. *Phys. Rev. Lett.* **4**, 84.
23. W. R. Frazer and J. R. Fulco. 1960. *Phys. Rev.* **119**, 1420.
24. G. C. Oades. 1963. *Phys. Rev.* **132**, 1277.
25. J. Hamilton, P. Menotti, G. C. Oades and L. L. J. Vick. 1962. *Phys. Rev.* **128**, 1881.
26. J. Hamilton, T. D. Spearman and W. S. Woolcock. 1962. *Ann. Phys.* **17**, 1.
27. A. Donnachie and J. Hamilton. 1964. *Phys. Rev.* **133**, B1053.
28. A. Donnachie, J. Hamilton and A. T. Lea. 1964. *Phys. Rev.* **135**, B515.
29. A. Donnachie and J. Hamilton. 1965. *Phys. Rev.* **138**, B678.
30. P. Auvil and C. Lovelace. 1964. *Nuovo Cim.* **33**, 473.
31. A. Donnachie and J. Hamilton. 1965. *Ann. Phys.* **31**, 410.
32. P. Auvil, A. Donnachie, A. T. Lea and C. Lovelace. 1964. *Phys. Lett.* **12**, 76.
33. A. Donnachie, A. T. Lea and C. Lovelace. 1965. *Phys. Lett.* **19**, 146.
34. A. Donnachie, R. G. Kirsopp, A. T. Lea and C. Lovelace. 1966. "πp Phase Shifts up to 1311 MeV", in preparation.
35. R. H. Dalitz. 1961. *Rev. Mod. Phys.* **33**, 471.
36. R. H. Dalitz. 1964. *Ann. Rev. Nucl. Sci.* **13**, 339.
37. R. K. Adair. 1959. *Phys. Rev.* **113**, 338.
38. G. Höhler and G. Ebel. 1963. *Nucl. Phys.* **48**, 470.
39. H. Goldberg. 1966. *Phys. Rev.* **151**, 1186.
40. D. Atkinson. 1962. *Phys. Rev.* **128**, 1908.
41. C. Lovelace, R. M. Heinz and A. Donnachie. 1966. *Phys. Lett.* **22**, 332.
42. A. Erdelyi, Ed. 1953. *Higher Transcendental Functions*, vol. II (McGraw-Hill).
43. ibid., p. 174 and p. 396.
44. ibid., p. 171.
45. ibid., p. 112.

8

QUARK MODELS

EUAN J. SQUIRES
University of Durham

1

INTRODUCTION

MODELS of strongly interacting particles and resonances (hadrons), in which they are supposed to be made out of quarks in an analogous way to that in which nuclei are made of nucleons, have become very fashionable in the last two years. Since there is no evidence that quarks exist, the motivation for work on quark models lies in their appealing simplicity, in the fact that they offer some hope of understanding the success of the higher symmetry groups like $SU(6)$—without getting involved in the morass of combining the Lorentz group with internal symmetry groups!—and, finally, in the surprising agreement which in some cases exists between experiment and quark models treated in the most naïve way.

In these lectures we first look at the direct evidence for quarks (section 3) and ask whether they should be regarded as "elementary" (section 4), then we discuss in detail one of the basic assumptions of all quark models, namely that the amount of admixture of many quark (and anti-quark) channels—some of which have much lower thresholds than the simplest channels—is relatively small. This topic is treated by means of a simplified dynamical model in sections 5, 6, and 7. A large part of the work involved in these sections was done by P. J. S. Watson of the University of Durham.

Sections 8 and 9 discuss two possible extreme types of quark model, and section 10 provides a review of the types of predictions one can make from quark models, and a comparison of some of these predictions with experiment.

2

UNITARY TRIPLETS

The unitary symmetry classification of hadrons became relevant to physics when Gell-Mann[1] and Ne'eman[2] had the idea of putting the simplest known baryons and mesons into the 8-dimensional representation of $SU(3)$, rather than into lower dimensional representations as, for example, in the Sakata model.[3] The resulting symmetry scheme has had considerable success in providing a phenomenological framework for describing experimental data on elementary particles. It is, however, still conceivable that particles transforming as the lowest dimensional representation of $SU(3)$ might exist. Gell-Mann[4] and Zweig[5] independently suggested the existence of the $SU(3)$ triplet of spin $\frac{1}{2}$ particles, each with baryon number $\frac{1}{3}$. Following Gell-Mann[4] we refer to these particles as *quarks*. An $SU(3)$ triplet contains an isotopic doublet, which we denote by n', p', and an isotopic singlet, Λ'. Using the Gell-Mann–Nishijima rule we find the quantum numbers for these particles as given in Table I. We shall confine our discussion in these lectures to this type of quark, although other schemes, involving, for example, triplets with only integral charges have been suggested.[6]

TABLE I
The Quantum Numbers of Quarks

	Spin	Baryon Number	T	T_3	Q	S
p'	$\frac{1}{2}$	$\frac{1}{3}$	$\frac{1}{2}$	$\frac{1}{2}$	$\frac{2}{3}$	0
n'	$\frac{1}{2}$	$\frac{1}{3}$	$\frac{1}{2}$	$-\frac{1}{2}$	$-\frac{1}{3}$	0
Λ'	$\frac{1}{2}$	$\frac{1}{3}$	0	0	$-\frac{1}{3}$	-1

We now note that, at least formally, we can regard baryons as being made up of three quarks, and mesons as being made up of a quark and an antiquark. In fact in terms of the dimensions of $SU(3)$ representations we have:

For quarks $\quad \mathbf{3} \otimes \mathbf{3} \otimes \mathbf{3} = \mathbf{10} \oplus \mathbf{8} \oplus \mathbf{8} \oplus \mathbf{1}.$ ...[*2.1*]

For quark-antiquark $\quad \mathbf{3} \otimes \overline{\mathbf{3}} = \mathbf{8} \oplus \mathbf{1}.$ ...[*2.2*]

In exact $SU(3)$ the representations on the right are the only states that could exist with 3 quarks or a quark-antiquark pair. We see that all particles which have so far been fitted into $SU(3)$ representations can be accommodated. The same remark applies to the $SU(6)$ classification.[7] Here the quark belongs to the 6 and the corresponding results are

$$\mathbf{6} \otimes \mathbf{6} \otimes \mathbf{6} = \mathbf{70} \oplus \mathbf{70} \oplus \mathbf{56} \oplus \mathbf{20} \qquad \text{...[2.3]}$$

$$\mathbf{6} \otimes \overline{\mathbf{6}} = \mathbf{35} \oplus \mathbf{1}. \qquad \text{...[2.4]}$$

An important distinction now arises. We can regard the "quark structure" of elementary particles as merely a formal device for doing group theory (i.e. by tensor methods), or we can consider the possibility that it is "physically useful" to regard hadrons as being many quark or quark-antiquark bound states. Later we shall try to be more precise about the meaning of the term "physically useful"; here an illustration will suffice: it is clearly useful to regard a hydrogen atom as being a bound state of an electron and a proton (e.g. one can then attempt to calculate electron-hydrogen scattering), however it is not useful to regard a neutron as a bound state of an electron, proton and neutrino, or a ρ-meson as a bound state of a triton and an anti-triton—even though in the latter case strong interaction quantum numbers are not violated. In these lectures, we shall not be concerned with doing group theory by quark methods, but will consider the second possibility.

As we shall see (section 10) the quark models of hadrons seem to produce some excellent agreement with experiment, and this has encouraged the belief that quarks exist; we shall first, however, briefly consider the possibility of more direct evidence.

3

Do Quarks Exist?

According to the simpler types of quark model, the quarks have charges which are not integral multiples of the electron charge. Since non-integral charges have not (so far) been observed it is necessary, if we are to believe in quarks, to explain their non-appearance. It should be noted that, given charge conservation, at least one non-integral charged particle should be stable. The simplest way of explaining the lack of observed quarks is to assume that they are very massive. If this is the case then the most likely sources of quarks would be in high energy reactions produced by accelerated particles, or in cosmic-rays.

Several groups have searched for quarks in accelerator experiments, but without success.[8,9,10,11,12] Since quarks must be produced in pairs (from charge conservation) a centre of mass energy of at least $2M_q$ is required for their production, so present machines are not likely to see quarks with a mass much above 4 BeV. However, if there exist quarks with mass below 4 BeV, then they must have unreasonably low production cross-sections to explain their non-appearance.

Domokos and Fulton[13] have made a statistical model calculation of the expected flux of quarks in cosmic rays as a function of the quark mass. They assume that the only source of quarks is the primary proton radiation. The predicted counting rate of quarks is as follows:

M_q(BeV)	2·5	4·8	6·2	7·4	8·7
Counting rate: one quark per	sec	hour	week	year	century

Although the assumptions involved in this calculation are certainly not wholly reliable, it suggests that only a narrow range of possible quark masses are susceptible to cosmic ray experiments.

Chilton, Horn and Jabbur[14] obtain very different values for the expected flux of quarks. They use the peripheral model, with ρ-exchange, and include some absorptive corrections. The absolute values here have little meaning since we have no knowledge of the $\rho q\bar{q}$ coupling. However, there is a much smaller dependence on the quark mass than in the statistical model calculations. Other results are obtained by Adair and Price.[15]

Some cosmic ray experiments have already been performed[16,17] but so far they have not obtained any results which go beyond the negative results of machine experiments. However, further experiments are in progress.

Domokos and Fulton[13] estimate on the basis of the above figures that if quarks of mass 5 BeV have been accumulating on the earth throughout its lifetime, then the density in the oceans will be about 1 quark per 10^{20} water molecules. This makes it extremely improbable that any low-energy experiment (of the Milikan type) would observe a quark. Some refinements of the Milikan experiment, however, are being made in order to enlarge the probability.[18,19,20] The idea here is to perform the Milikan experiment on larger bodies than those used in the "oil-drop" type experiments. Since it is not then possible to hold such bodies in position by electrostatic forces, they are magnetized and held by suitable magnetic fields. These experiments are in progress.

Chupka, Schiffer and Stevens[21] have looked for fractionally charged particles by collecting them from vapours passing through electric fields. They studied material from meteorites, air and water. For the latter, for example, they found a quark concentration less than 3×10^{-24} quarks per nucleon, which would suggest a quark mass of greater than about 7 BeV on Domokos and Fulton's predicted flux, and 50 BeV on the basis of Adair's estimate. If one could be sure that there is no mechanism whereby quarks could be "lost" from the oceans during times of the order of 10^9 years, then this experiment would suggest that the chances of finding quarks in cosmic rays are poor.

To summarize the present situation we can say that so far there is no direct evidence for the existence of quarks. This implies strongly that quarks of mass 4 BeV or less do not exist. On the other hand, if quarks with mass greater than about 8 BeV exist, then even experiments at present in progress are unlikely to see them.

4

Are Quarks Fundamental?

In a quark model hadrons are taken to be formed from quarks and antiquarks. It is thus possible to regard the quarks as playing some fundamental role in the world and as being the objects from which all other strongly interacting particles are derived. However, one is not forced to adopt this view, and it is possible to extend the "nuclear democracy" idea of Chew to include quarks.

To illustrate this we recall that, according to the reciprocal bootstrap idea,[22,23] the existence of the nucleon and '33' resonance is a consequence of the existence of the pion, i.e. they make themselves in a self-consistent way. Wazed[24] has shown that the quarks can be formed as quark-pion bound states in an analogous way. He considers quark-pion scattering, which in $SU(3)$ leads to

$$\mathbf{3} \otimes \mathbf{8} = \mathbf{3} \oplus \bar{\mathbf{6}} \oplus \mathbf{15}.$$

From the $SU(3)$ crossing matrix and the spin crossing matrix (which is identical to that which occurs in the N, N* case), it follows that the exchange of a quark produces attraction in the p wave $SU(3)$—**15** with total angular momentum $\frac{3}{2}$. The exchange of this state (which we refer to as q^*) in turn produces strong attraction in the triplet (**3**) $SU(3)$ state with total angular momentum $\frac{1}{2}$. Thus there is again the possibility of a reciprocal bootstrap. Numerical calculations yield self-consistent masses of 2·4 and 5·3 GeV for the q and q^* respectively.

One cannot regard the numerical values as having much meaning in view of the approximation involved. In particular, we have ignored completely the effect of meson exchange. According to M. Nieto[25] this spoils the above calculation, and no self-consistent bootstrap is possible. However, this assumes the $\rho q\bar{q}$ coupling equals the $\rho N\bar{N}$ coupling, which may be false. In general there is no objection in principle to the idea that quarks are self-consistently bootstrapped particles and, for the purpose of the dynamical calculations that we shall consider below, it is quite irrelevant whether we regard the quarks as "fundamental" or not.

5

Are Quark Models Reasonable?

We now assume that quarks do exist and turn to the problem of making hadrons out of them. It is clear that, for example, the vertex $\rho \to q\bar{q}$ will be non-zero, but the question is whether in any sense this channel dominates the ρ. Most previous calculations in strong interaction physics have implicitly assumed the opposite, i.e. that this channel is negligible, since they have been based on "nearest singularity" type arguments. Thus, for example, one has assumed that in calculating the ρ the $\pi\pi$ channel is dominant, with higher thresholds, e.g. $K\bar{K}$, $\rho\rho$ etc. playing only a small part, and channels which have thresholds as high as 10 BeV being negligible. Provided the couplings to the various channels are all of the same order of magnitude this approximation is certainly reasonable. In a quark model one is making a contrary assumption—namely that the existence of the $\pi\pi$, $K\bar{K}$, $\pi\pi\pi\pi$, etc. channels are *all small perturbations* on the dominant $q\bar{q}$ channels.

In order to make the distinction between these two assumptions clear, let us consider a model[26] of the ρ in which only the two channels $\pi\pi$ and $q\bar{q}$ are included. We consider solving this problem by the matrix N/D method, i.e. we assume that the left-hand cuts of $\boldsymbol{a}$ (a 2×2 matrix) are known (these are the forces) and then solve in the usual way. The solution, we assume, produces a bound state pole at the position of the ρ (actually a resonance of course since it lies above the $\pi\pi$ threshold). This bound state occurs of course in all channels. Now we compare this calculation with two others, namely

A. We solve the $\pi\pi$ problem with the same $\pi\pi$ left-hand cut, first with purely elastic unitarity, and then with a corrected form of unitarity taking into account the inelasticity obtained in the two channel problem.
B. Here we solve as in A, but in the $q\bar{q}$ channel rather than the $\pi\pi$ channel.

Under the first assumption (i.e. $\pi\pi$ channel domination) we would obtain the ρ in calculation A and its position would not be altered very much by the inclusion of the inelasticity factor. However, we would not obtain the ρ in calculation B, even if the correct inelasticity factor is involved; in fact to obtain the correct amplitude in this calculation we would need to insert the ρ through a CDD pole. Similarly under the second assumption (we refer to this as the "extreme" quark model), *calculation* A *would not produce the* ρ— it would have to arise through a CDD pole—but it would be produced in calculation B and the inclusion of the inelasticity factor would make little difference to its position.

In the above discussion we have used the fact—discussed by many authors[27-31]—that a one-channel problem, solved without CDD poles, does not necessarily produce the same result as a many channel problem, also solved without CDD poles, even when the correct analyticity factor is included. Halpern and Atkinson[32] have discussed similar problems in a slightly different context.

It is of course possible to have intermediate situations to the above, e.g. where the ρ appears in calculation B but its position is considerably affected by the presence of the channels with lower threshold—as we shall see below this "modified quark model" is almost certainly the only one that can be regarded as reasonable. Note, however, that it is *impossible* for the ρ to appear as a bound state in both calculation A and calculation B. (These considerations can readily be extended to more than two channels, a given bound state can appear in only one of the corresponding "one-channel" problems.) Thus one can say that in any form of quark model for the ρ, the ρ cannot appear as a consequence of forces in the $\pi\pi$ channel, but must be a CDD *pole on any calculation which does not explicitly include quark channels*. This means that quark models are incompatible with most of the dynamical calculations of strongly interacting particles that have been made.

Faced with two such contradicting pictures we would like to decide between them. Of course if we knew the forces we could do the calculations and readily decide between the two alternatives. However we have only scanty information on the left-hand cuts so this is not possible. We can however see that with our present knowledge the first assumption is clearly *possible*, even if quarks exist, i.e. the couplings of quarks could be sufficiently small (they could still be "strong" couplings) so that they play only a small role at low energies (just as, for example, nobody would doubt that the $\pi\pi \to L_i^5 \overline{L_i^5}$ channel, for example, plays only a small role in the dynamics of the ρ even though it is a "strong" coupling). We now want to see whether the *other* assumption, i.e. the quark model, is in the same sense possible. The situation is rather different here since we *know* the $\rho \to \pi\pi$ coupling, for example, and cannot make it arbitrarily weak. This problem we discuss in the next two sections.

6

MASS SHIFTS DUE TO LOWER THRESHOLD CHANNELS

We consider a model of the ρ meson, involving just the $\pi\pi$ and $q\bar{q}$ channels. For simplicity we treat all particles as having zero spin and isospin. We suppose that the direct $q\bar{q}$ force can be represented by the exchange of a hypothetical X meson, i.e. by the diagram in Fig. 1. The mass, M_X, of the X meson and the residue of the pole $G^2_{q\bar{q}X}$, are two free parameters. We solve the $q\bar{q}$ scattering problem by the N over D method—using the contribution

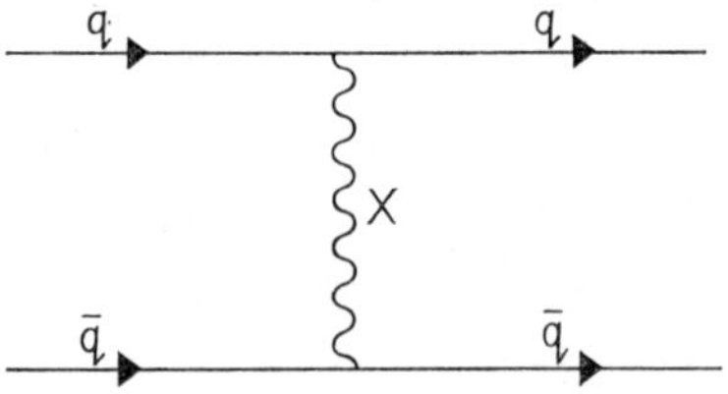

FIG. 1

The $q\bar{q}$ force.

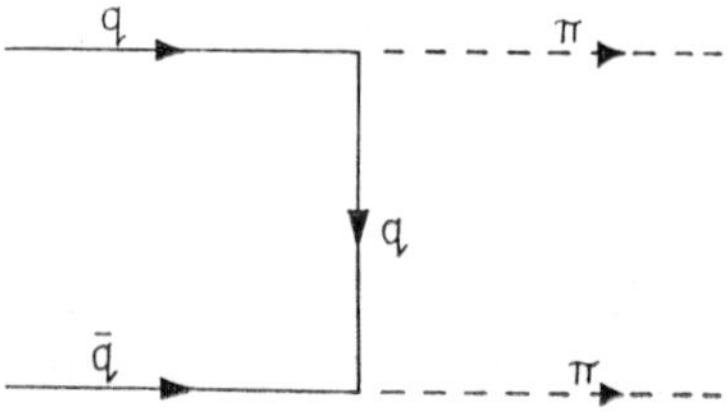

FIG. 2

The $q\bar{q} \to \pi\pi$ force

of Fig. 1 for the left-hand cut, and choose $G^2_{q\bar{q}X}$ so that, for a given M_X, there is a bound state in the neighbourhood of the ρ mass. This gives the one-channel quark-model ρ meson.

To calculate the correction due to the $\pi\pi$ channel we introduce the coupling between the channels due to the lightest particle which can be exchanged, which is the quark, as in Fig. 2. The contribution of this diagram depends on one free parameter, i.e. the residue $G^2_{q\bar{q}\pi}$. We determine this parameter by solving the two channel problem by the matrix N over D method, and fitting the resulting residue of the ρ pole in the $q\bar{q} \to \pi\pi$ amplitude to $G_{q\bar{q}\rho}G_{\pi\pi\rho}$, where $G_{\pi\pi\rho}$ is known from the experimental value of the

ρ lifetime. A comparison of the new ρ mass with its value in the one channel calculation shows how far the $\pi\pi$ channel has affected the mass.

The only free parameter in this calculation is M_X, i.e. essentially the range of the $q\bar{q}$ force, and one of the purposes of this investigation is to see whether the quark model is more likely to be successful in a long range or short range force quark model.

It is instructive to carry out this calculation first in perturbation theory. If it is reasonable to regard the effect of the $\pi\pi$ channel as a small perturbation—as would be expected in a realistic quark model—then a reasonable estimate of its effect on the ρ mass should be obtained from a lowest order calculation. The effect is second order in the coupling force (Fig. 2), so we need an extension of the Dashen-Frautschi method[33] which we now describe.

We denote the s wave partial wave amplitude for $q\bar{q}$ scattering by a_{11}, for $\pi\pi$ scattering by a_{22} and for $q\bar{q} \to \pi\pi$ by a_{12}. We normalize these amplitudes so that the unitarity equation for the 2×2 matrix $\boldsymbol{a}$ reads

$$\text{Im}\, \boldsymbol{a} = \boldsymbol{a}^* \boldsymbol{P} \boldsymbol{a} \qquad ...[6.1]$$

where the diagonal matrix $\boldsymbol{P}$ has elements given by

$$P_{ij} = \delta_{ij}\theta(s-s_i)(4k_i^2/s)^{\frac{1}{2}} \qquad ...[6.2]$$

with

$$4k_1^2 = s-4M_q^2, \qquad ...[6.3]$$

$$4k_2^2 = s-4m_\pi^2, \qquad ...[6.4]$$

$$s_1 = 4M_q^2, \qquad ...[6.5]$$

and

$$s_2 = 4m_\pi^2. \qquad ...[6.6]$$

From equation [6.1] we have

$$\text{Im}\,[\boldsymbol{a}^{-1}] = -\boldsymbol{P} \qquad ...[6.7]$$

on the right-hand cut.

For the unperturbed problem we have no coupling between the two channels. Let $\Delta\boldsymbol{a}$ be the first order change in $\boldsymbol{a}$ due to the diagram in Fig. 2. Since the unitarity equation is unaffected we have, from equation [6.7]

$$\text{Im}\,[(\boldsymbol{a}+\Delta\boldsymbol{a})^{-1}]-\text{Im}\,[\boldsymbol{a}^{-1}] = 0 \qquad ...[6.8]$$

or, to lowest order,

$$\text{Im}\,[\boldsymbol{a}^{-1}\Delta\boldsymbol{a}\,\boldsymbol{a}^{-1}] = 0. \qquad ...[6.9]$$

We put

$$\boldsymbol{a} = \boldsymbol{N}\boldsymbol{D}^{-1} \equiv (\boldsymbol{D}^T)^{-1}\boldsymbol{N}^T \qquad ...[6.10]$$

where $\boldsymbol{N}$ has a left-hand cut and no other singularities and $\boldsymbol{D}$ has a right-hand cut and no other singularities. Then [6.9] becomes

$$\text{Im}\,[(\boldsymbol{N}^T)^{-1}\boldsymbol{D}^T\Delta\boldsymbol{a}\boldsymbol{D}\boldsymbol{N}^{-1}] = 0 \qquad ...[6.11]$$

or, since N is real on the right-hand cut,

$$\text{Im}\,[\boldsymbol{D}^T \Delta \boldsymbol{a} \boldsymbol{D}] = 0 \qquad \ldots[6.12]$$

on the right-hand cut. This means that $\boldsymbol{D}^T \Delta \boldsymbol{a} \boldsymbol{D}$ has only a left-hand cut, so we can write a dispersion relation

$$\boldsymbol{D}^T \Delta \boldsymbol{a} \boldsymbol{D}(s) = \int_{\text{L.H.}} \frac{ds'}{\pi} \frac{\boldsymbol{D}^T\, \text{Im}\,[\Delta \boldsymbol{a}]\, \boldsymbol{D}(s')}{(s'-s)}. \qquad \ldots[6.13]$$

Now $\boldsymbol{D}$ is diagonal and Im $[\Delta \boldsymbol{a}]$ is zero except for the off diagonal (12) element, so [*6.13*] shows, as we expect, that the first order perturbation occurs only in a_{12}, for which we have

$$a_{12}(s) \equiv \Delta a_{12}(s) = D_{11}^{-1} D_{22}^{-1} \int_{\text{L.H.}} \frac{ds'}{\pi} \frac{D_{11}(s')\,\text{Im}\,[a_{12}(s')]\,D_{22}(s')}{(s'-s)}. \qquad \ldots[6.14]$$

We now use unitarity again to obtain the effect of $a_{12}(s)$ upon a_{11}. From [*6.1*] and [*6.2*] we have

$$\text{Im}\,[a_{11}+\Delta a_{11}] = |\,a_{12}\,|^2 \sqrt{\left(\frac{s-4m_\pi^2}{s}\right)} + \theta(s-4M_q^2)|\,a_{11}\,|^2 \sqrt{\left(\frac{s-4M_q^2}{s}\right)} \qquad \ldots[6.15]$$

for $s > 4m_\pi^2$. Working as before we derive

$$\text{Im}\,[D_{11}^2\,\Delta a_{11}] = D_{11}^2\,|\,a_{12}\,|^2 \sqrt{\left(\frac{s-4m_\pi^2}{s}\right)} \qquad \ldots[6.16]$$

for $4m_\pi^2 < s < 4M_q^2$. A more complicated result holds for $s > 4M_q^2$, but since the principal effect of the $\pi\pi$ physical cut comes from the region below $4M_q^2$ we use [*6.16*] for the whole right-hand cut.

Now $D_{11}^2 \Delta a_{11}$ has no other singularities apart from the right-hand cut given by [*6.16*], so we can write a dispersion relation

$$D_{11}^2\,\Delta a_{11}(s) = \int_{4m_\pi^2}^{\infty} \frac{ds'}{\pi} \frac{|\,D_{11}(s')\,|^2\,|\,a_{12}(s')\,|^2}{(s'-s)} \sqrt{\left(\frac{s'-4m_\pi^2}{s'}\right)}. \qquad \ldots[6.17]$$

Equations [*6.14*] and [*6.17*] together give the effect of the perturbation on $a_{11}(s)$.

To calculate the change in the position of the ρ meson we write, near the ρ pole,

$$a_{11} \simeq \frac{G_{\rho q\bar{q}}^2}{s-m_\rho^2} \qquad \ldots[6.18]$$

and

$$\Delta a_{11} \simeq \frac{G_{\rho q\bar{q}}^2}{(s-m_\rho^2)^2}\,\Delta m_\rho^2 + \frac{\Delta G_{\rho q\bar{q}}^2}{s-m_\rho^2}. \qquad \ldots[6.19]$$

Comparing [*6.19*] and [*6.18*] with [*6.17*] we find

$$\Delta m_\rho^2 = \frac{1}{G_{\rho q\bar{q}}^2 [D'_{11}(m_\rho^2)]^2} \int_{4m_\pi^2}^{\infty} \frac{ds'}{\pi} \frac{|D_{11}(s')|^2 \, |a_{12}(s')|^2}{(s'-m_\rho^2)} \sqrt{\left(\frac{s'-4m_\pi^2}{s'}\right)} \qquad \text{...[6.20]}$$

for the shift in the ρ mass.

The contribution of Fig. 2 to $\text{Im}\, a_{12}(s)$ is non-zero in the region $-\infty < s < (4m_\pi^2 - m_\pi^4/M_q^2)$. It is interesting to replace this contribution by a single pole at $s = -\varepsilon$, say, where presumably the best value of ε is small. Thus we put

$$\text{Im}\, a_{12}(s) = \lambda\delta(s+\varepsilon) \qquad \text{...[6.21]}$$

where λ replaces $G_{q\bar{q}\pi}^2$ as the free parameter which is a measure of the $q\bar{q}\pi$ coupling. Then [*6.14*] reduces to

$$a_{12}(s) = -\frac{\lambda}{\pi} \frac{D_{11}(-\varepsilon) D_{22}(-\varepsilon)}{D_{11}(s) D_{22}(s)} \frac{1}{\varepsilon+s}. \qquad \text{...[6.22]}$$

This has a pole at $s = m_\rho^2$, the residue of which is $G_{q\bar{q}\rho} G_{\rho\pi\pi}$, so that

$$G_{q\bar{q}\rho} G_{\rho\pi\pi} = -\frac{\lambda}{\pi} \frac{D_{11}(-\varepsilon) D_{22}(-\varepsilon)}{D'_{11}(m_\rho^2) D_{22}(m_\rho^2)} \frac{1}{\varepsilon+m_\rho^2}. \qquad \text{...[6.23]}$$

We can eliminate λ between [*6.23*] and [*6.22*] to obtain

$$a_{12}(s) = G_{q\bar{q}\rho} G_{\rho\pi\pi} \frac{D'_{11}(m_\rho^2) D_{22}(m_\rho^2)}{D_{11}(s) D_{22}(s)} \frac{\varepsilon+m_\rho^2}{\varepsilon+s} \qquad \text{...[6.24]}$$

which we then insert into [*6.20*] for our final answer

$$\Delta m_\rho^2 = G_{\rho\pi\pi}^2 \int_{4m_\pi^2}^{\infty} \frac{ds'}{\pi} \sqrt{\left(\frac{s'-4m_\pi^2}{s'}\right)} \frac{|D_{22}(m_\rho^2)|^2}{|D_{22}(s')|^2} \frac{(\varepsilon+m_\rho^2)^2}{(\varepsilon+s'^2)^2} \frac{1}{s'-m_\rho^2}. \qquad \text{...[6.25]}$$

We see that this final result, which expresses the required shift, Δm_ρ^2, in terms of the experimentally known coupling, $G_{\rho\pi\pi}^2$, does not depend on the $q\bar{q}$ force since there is no dependence upon D_{11}. Thus, in lowest order, and with a pole approximation to the left-hand cut of the $q\bar{q} \to \pi\pi$ amplitude, the model does not distinguish between the long range force and short range force quark models.

Equation [*6.25*] also shows the dependence of the mass-shift upon $D_{22}(s)$, i.e. upon the $\pi\pi$ force, which we have not so far introduced. However, it appears from [*6.25*] that this dependence is slight unless $D_{22}(m_\rho^2)$ is small compared to unity. This would only occur, however, if there was a resonance in the uncoupled $\pi\pi$ problem near the ρ mass, which in our model would imply two poles in the coupled channel problem and is therefore not realistic. For this reason we have not taken account of any $\pi\pi$ force in our calculations, i.e. we have taken $D_{22} \equiv 1$. Deviations from this will make only a small difference to the result, which would be of no importance since we are only seeking an order of magnitude estimate.

Since the ρ lies above the $\pi\pi$ threshold Δm_ρ^2 has an imaginary part. This is given from [*6.20*] and [*6.14*], or directly from [*6.25*], as

$$\text{Im}\, m_\rho^2 = -\,|\,G_{\rho\pi\pi}\,|^2 \sqrt{\left(\frac{m_\rho^2 - 4m_\pi^2}{m_\rho^2}\right)}, \quad \text{for} \quad m_\rho^2 > 4m_\pi^2 \qquad \ldots[6.26]$$

which is the expected relation. It is clear from [*6.25*] that, in general, the shift in the real part of the mass is of the same order of magnitude as the shift in the imaginary part.

In Squires and Watson[26] numerical results obtained by solving the complete 2×2 matrix $\boldsymbol{N}$ over $\boldsymbol{D}$ method are reported. We reproduce these results in Figs. 3, 4 and 5. For these results M_q was taken to be 5 BeV and the value of $G_{q\bar{q}X}^2$ was chosen so that $m_\rho = 1$ BeV.

FIG. 3

Showing the variation of the $\rho \to \pi\pi$ coupling with the shift in the ρ mass squared, for two values of M_X. The line marked "experimental width" is actually that for which the dimensionless $\rho\pi\pi$ coupling in our model problem agrees with the experimental value.

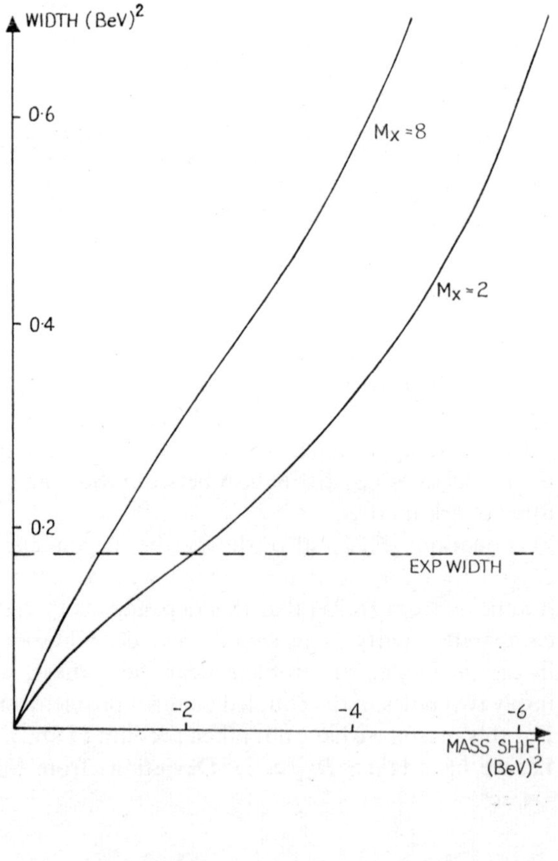

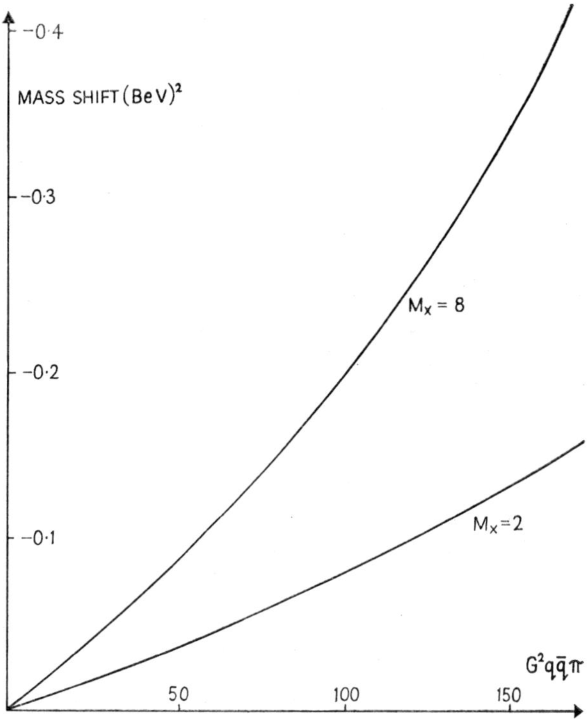

FIG. 4

Showing the variation of the shift of the ρ mass with $G^2_{q\bar{q}\pi}$.

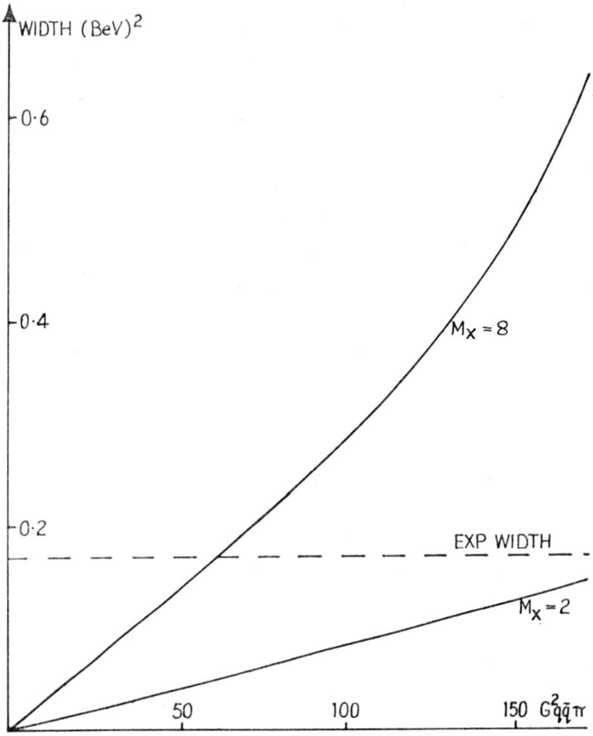

FIG. 5

Showing the variation of the $\rho \to \pi\pi$ width with $G^2_{q\bar{q}a}$.

We see from Fig. 3 that both the effect of higher order terms in Im a_{12}, and the effect of improving the pole approximation to Im a_{12}, result in larger mass shifts for the long range force model than for the short range force model. The approximation of neglecting the lower thresholds is thus better for the short range model, although the difference between the two does not seem to be large.

The actual numerical values for the mass shifts for the experimental $\rho\pi\pi$ coupling can be read off from Fig. 1 and are given by $-\Delta m_\rho^2 = 0{\cdot}11$ and $0{\cdot}20(\text{BeV})^2$ for values of M_X equal to 8 and 0·5 BeV respectively. These figures should not be taken too seriously because of the neglect of spin, but it is reassuring to find that they are quite small. Of course in the true situation there are many other channels which should also be included and so the total mass shift will be much greater than that given by the $\pi\pi$ channel. Note that for closed channels the mass shift is negative definite in first order (this follows from equation [*6.20*] when we recall that $G^2 < 0$) so that all the contributions add together. It seems clear then that the combined effect of the lower threshold channels is to make a very significant contribution to the mass of the ρ, and that calculations in terms of the single $q\bar{q}$ channels are misleading.

With regard to the mass differences between various multiplets in quark models the situation is not quite so clear. There will be appreciable shifts in each mass due to the lower threshold channels, and the differential effect is likely to be of the order of the effect of a single channel. This is clearly a significant contribution to the mass difference between multiplets, but it may not completely invalidate attempts to fit the hadron spectra[34, 35] in terms of simple $q\bar{q}$ and qq forces.

In the model discussed here the ratio $G_{q\bar{q}\rho}/G_{q\bar{q}\pi}$ comes out as 15 and 25 for $M_X = 8$ and 0·5 BeV respectively. It is clear that in a realistic model, taking account of spin and isospin, we would expect these to be of the same order of magnitude because typical ρ and π couplings are similar in magnitude. The deviation of the above ratio from unity is therefore an unsatisfactory feature of the results. Further investigations of this point are in progress.

7

Effect of Lower Threshold Channels on Couplings

Many of the predictions of quark models require the assumption that the interaction of a given particle with a hadron is the sum of the interactions of the particle with the quarks and antiquarks composing the hadron. Here we want to investigate whether the effect of the lower threshold channels makes this assumption invalid.

In order to avoid additional complications like "multiple scattering", etc. we consider the coupling of a weakly interacting particle which we denote by a. Thus our calculation is most directly related to the couplings of a photon, i.e. to the magnetic moment and photon transition calculations referred to in the last section.

We calculate the coupling of a to the ρ. Clearly in a simple quark model we would expect $G_{\rho\rho a}$ to be just the sum of $G_{q\bar{q}a}$ and $G_{\bar{q}qa}$, and we want to study the effect of the $\pi\pi$ channel on this result.

To do this we first ignore the $\pi\pi$ channel and consider the $q\bar{q}$ and $a\rho$ coupled channel problem. From the residue of the ρ pole in the $q\bar{q} \to a\rho$ amplitude we can calculate the $\rho\rho a$ coupling (see Fig. 6). Since the a coupling is weak we can treat it in first order and use the Dashen-Frautschi perturbation theory.[33] We denote the $q\bar{q}$ channel by 1 and the ρa channel by 2, so that, in lowest order, $D_{22} = 1$ and equation [*6.14*] yields

$$\Delta a_{12}(s) = [D_{11}(s)]^{-1} \int_{-\infty}^{0} \frac{ds'}{\pi} \frac{D_{11}(s') \operatorname{Im} a_{12}(s')}{(s'-s)}. \qquad \text{...[7.1]}$$

Near the ρ pole (Fig. 6),

$$\Delta a_{12}(s) \simeq \frac{G_{q\bar{q}\rho} G_{\rho\rho a}}{s-m_\rho^2} \qquad \text{...[7.2]}$$

so, comparing equations [*7.1*] and [*7.2*], we have

$$G_{\rho\rho a} = \frac{1}{G_{q\bar{q}\rho}} \frac{1}{D'_{11}(m_\rho^2)} \int_{-\infty}^{0} \frac{ds'}{\pi} \frac{D_{11}(s') \operatorname{Im} a_{12}(s')}{s'-m_\rho^2}. \qquad \text{...[7.3]}$$

The largest range contributions to $\operatorname{Im} a_{12}(s')$ are those shown in Fig. 7. There are of course many other contributions which are also of lowest order in the a coupling, e.g. Fig. 8. These contributions are of shorter range, so we hope that we will not obtain misleading results by neglecting them.

With some loss of generality, but hopefully without losing any significant features, we shall choose $m_a = m_\rho$. This simplifies the algebra involved and allows us to put

$$[\operatorname{Im} a(s)]_{\text{pole}} = \frac{2\pi G_{q\bar{q}\rho} G_{q\bar{q}a}}{(s-4M_q^2)^{\frac{1}{2}}(s-4m_\rho^2)^{\frac{1}{2}}}, \quad -\infty < s < (4m_\rho^2 - m_\rho^4/M_q^2) \qquad \text{...[7.4]}$$

for the contribution of the two diagrams in Fig. 7. Here we have taken the quark and antiquark couplings to a to be equal.

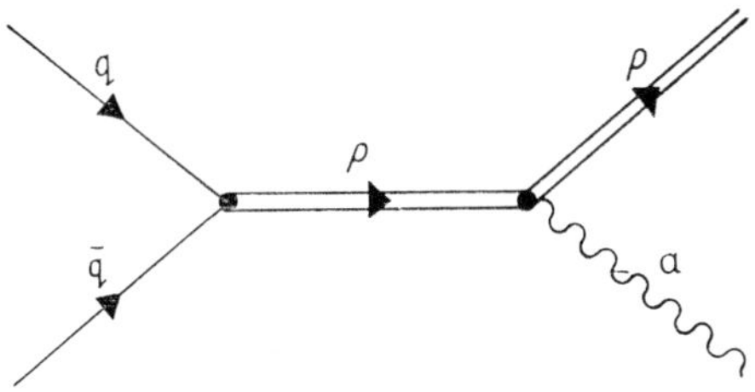

FIG. 6. The ρ-pole in the $q\bar{q} \to a\rho$ amplitude

Inserting [7.4] into [7.3] we obtain

$$G_{\rho\rho a} = 2G_{q\bar{q}a}A_0 \qquad \text{...[7.5]}$$

with

$$A_0 = [D'_{11}(m_\rho^2)]^{-1} \int_{-\infty}^{(4m_\rho^2 - m_\rho^4/M_q^2)} \frac{ds'}{s'-m_\rho^2} \frac{D_{11}(s')}{(s'-4M_q^2)^{\frac{1}{2}}(s'-4m_\rho^2)^{\frac{1}{2}}}. \qquad \text{...[7.6]}$$

This has the expected form; the coupling of a to ρ being the sum of the couplings of a to a quark and an antiquark multiplied by a factor which is of

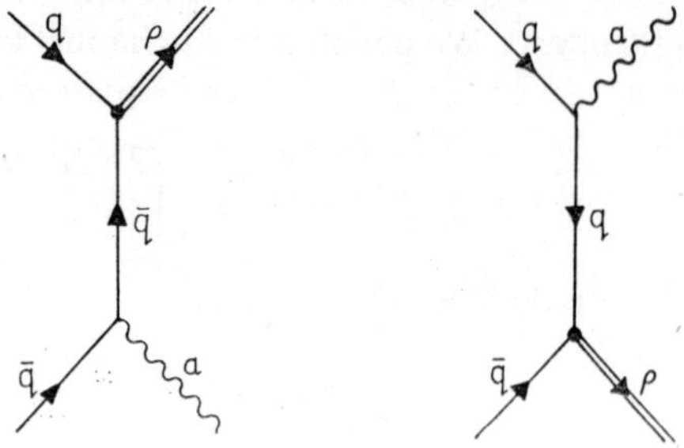

FIG. 7

Diagrams which contribute to $G_{\rho\rho a}^2$.

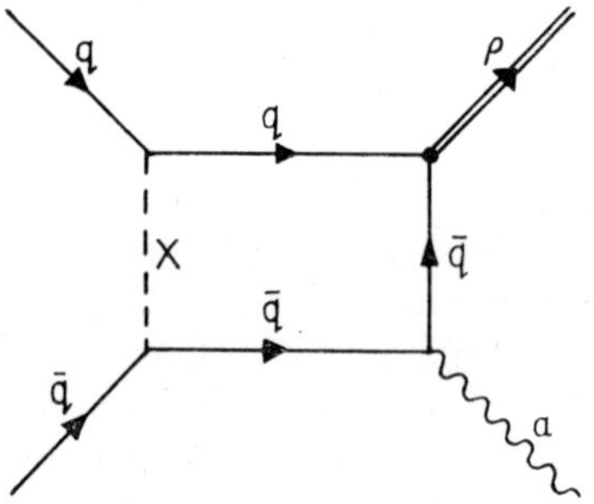

FIG. 8

A neglected diagram which contributes to $G_{\rho\rho a}^2$.

order unity. In fact calculations[26] yield $A_0 = 0{\cdot}91$ and $1{\cdot}33$ for $M_X = 8$ BeV and 0·5 BeV respectively. The main reason why A_0 differs from unity is probably the neglect of diagrams like that of Fig. 8.

We now repeat the calculation with the inclusion of the $\pi\pi$ channel as in the previous section. The result now comes out in the form

$$G_{\rho\rho a} = 2G_{q\bar{q}a}A + 2G_{\pi\pi a}B \qquad \text{...[7.7]}$$

as would be expected, where the second term reflects the presence of the $\pi\pi$ channel. Numerical results[26] are given in Table II.

TABLE II

	$M_X = 8$ BeV	$M_X = 0{\cdot}5$ BeV
A_0	0·91	1·33
A	1·00	1·50
B	0·02	0·57

We see that the contribution of the $\pi\pi$ channel appears to be quite small for the short range case, but is *large* for the long range case. Note that, on the simple quark model, the value of $G_{\pi\pi a}$ would be twice that of $G_{q\bar{q}a}$, so that the effect of the second term of [*7.7*] is enhanced by a factor of two. It would appear then from this calculation to be unlikely that lower threshold channels can be ignored in the calculation of interactions of hadrons, if the long range force model is realistic.

Finally in this section we note that although we have calculated in the on-shell N over D method, there is no reason why our results, at least qualitatively, should not apply to other methods of calculation, e.g. by the Bethe-Salpeter equation. In particular the assumption in a field theory that quarks are the only fundamental fields does not exclude the lower threshold channels; the $\pi\pi$ channel, for example, would arise as a $q\bar{q}\ q\bar{q}$ channel.

8

The "Long Range" Quark Model

Here we shall consider quark models in which the size of the bound states is assumed to be much bigger than the Compton wavelength of the quark, i.e. the important quark-quark (and quark-antiquark) force is mediated by a particle of mass M_X,

$$M_X \ll M_q. \qquad ...[8.1]$$

This assumption has been made in most quark models and its consequences have been extensively reviewed;[36] here we shall only outline its principle features.

Note first a difficulty which seems to arise if we regard the forces as arising from the known mesons. Then the large binding would appear to require very large meson-quark couplings. This is somewhat strange in view of the results of section 10 in which we require quark-quark interaction at high energy to be smaller than nucleon-nucleon cross-sections. This difficulty is probably not so significant for the short range model since very short-range forces would contribute little to the scattering amplitude.

An essential characteristic of such models is that the kinetic energies of the quarks in the bound states are small compared to the quark mass and to the binding energies. To see this we note that the range of the potential is approximately

$$R \sim \hbar/M_X c \qquad ...[8.2]$$

and, from the uncertainty principle, the quark momentum in the lowest state is given by

$$p \sim \hbar/R \sim M_X c. \qquad ...[8.3]$$

Thus, using the non-relativistic approximation, which is immediately justified,

$$T \sim p^2/2M_q \sim \frac{M_X}{M_q}(M_X c^2). \qquad ...[8.4]$$

With $M_q \sim 5$ BeV and $M_X \sim 1$ BeV this gives a kinetic energy of 200 MeV.

Because of this small value for the mean quark kinetic energies, compared to the quark rest mass, it has been suggested that non-relativistic potential models might provide realistic descriptions of quark bound states. This is not necessarily so, however, and depends upon the shape of the potential. Greenberg[37] has shown that for purely attractive Yukawa-type potentials the non-relativistic approximation is unlikely to be adequate.

A more significant consequence of low kinetic energies is that the energy differences between states with different orbital wave function is likely to be small compared to quark masses, i.e. also of the order of a few hundred MeV. This means that it is natural (but not essential) to regard all or most of the known meson and baryon states as being excited states of the $q\bar{q}$ and qqq systems (contrast the situation discussed in section 9). We shall now see what states would be expected on this assumption.

For the mesons we would be limited to $SU(3)$ octets and singlets (a nonet of $U(3)$). The $l = 0$ and $l = 1$ $q\bar{q}$ states are listed in Table III, together with possible assignments.

TABLE III

l	s	JPG	Possible Particles
0	0	0^{--}	π, K, η, η'
0	1	1^{-+}	ρ, K^*, ω, ϕ
1	1	0^{+-}	?
1	1	2^{+-}	f, f', A_2
1	1	1^{+-}	?
1	0	1^{++}	?

On the whole the agreement is encouraging, although there seems to be a shortage of $J^P = 1^+$ mesons. We refer to Dalitz[36] for further details.

For the baryons the situation is more complicated. According to the $SU(6)$ classification the lowest mass baryons should belong to the **56** representation: this requires the 3-quark wavefunction to be completely symmetrical in spin and unitary indices, and hence, assuming that quarks obey Fermi-Dirac statistics, the spatial wavefunction must be antisymmetrical (we discuss other possibilities in section 9). This is an unusual situation since the kinetic energies are smallest in the symmetrical state; however in view of the fact that the kinetic energy is in any case small compared to the potential energy, it is not impossible. There have been various attempts to fit the higher baryon resonances into excited 3-quark systems (e.g. Dalitz[36]) but the lack of experimental data makes this rather speculative.

Given the wavefunction of the meson and baryon states it is possible to construct potential models and to try to fit the masses of the various multiplets to parameters in the potential. Hendry[35] has carried out calculations of this type.

Morpurgo[38] has studied possible $L = 1^+$ three-quark states. There are reasons to believe that these should lie near to the **56**, and they should give rise to further **56** representations with various values of J. However, according to the predictions of a simple quark model, they would not be readily observed since they would not decay into, for example, a nucleon plus a pion.

An alternative treatment is due to Delbourgo, Salam and Strathdee,[39] who use the Bethe-Salpeter equation to compute meson bound states in a quark model. With suitable assumptions regarding the force, mediated by meson exchange, they are able to produce various higher symmetry schemes, e.g. with a scalar meson force they obtain $U(6)_w$ for the bound states, even though they start with just $SU(3)$ symmetry. With a four quark interaction they show that the only possible sets of values for the interaction strength and the cut-off, consistent with there being no "ghosts", gives binding in the $(\mathbf{6}, \bar{\mathbf{6}})$ state of $U(6) \otimes U(6)$ as required. However, this same four-point interaction then gives "ghosts" in qq scattering.

Finally, in this section we emphasize that if this model is correct and all (or at least most) of the known hadrons (including resonances) are bound states of 3-quarks or a quark and an antiquark pair, then in dynamical calculations which do not explicitly indicate quark channels, these hadrons are CDD poles. Thus, for example, the N^* resonance is a CDD pole in πN scattering and the ρ in $\pi\pi$ even if the correct inelasticity is included. This means that essentially all strong interaction dynamical calculations that have so far been made are wrong!

9

The "Short Range" Quark Model

The dynamical calculations in sections 6 and 7 suggest that quark models are more likely to be successful if the quark-quark forces are of short range, i.e. mediated by particles with masses of the order of the quark mass. An immediate consequence of such a model is that the kinetic energies will be large, i.e. of the order of M_q and so will the difference in energy between s and p states. Thus it is likely that present energy elementary particle physics is "low energy" on the scale of quark dynamics, so that only quark (and antiquark) systems in relative s states have been observed.

For the mesons the s states are the $SU(3)$ singlets and octets of spins 0^- and 1^- as shown in Table III. On this model all other meson resonances are made up of two or more mesons with their quark structures unexcited. Thus for example the f_0 would be a $\pi\pi$ (plus $\rho\rho$, etc.) resonance in which the two pions are approximately in their ground state as regards their quark structure. This of course is only an approximation; there will be some contribution from the excited states of the pions, due to the fact that the forces will distort the pion wavefunctions.

It is worth noting that even in the long-range quark model there will exist bound states containing several hadrons approximately in their ground states, but this will happen at a lower energy. Thus, for example, the quark structure of baryons should not play any role in nuclei, which will contain, to a good approximation, nucleons in their ground states.

For the baryons, there are two possibilities. If we assume they contain three quarks in relative *s* states, then the orbital wave function is fully symmetrical (i.e. the **20** of $SU(6)$) or the quarks obey parastatistics.[40] We briefly discuss these two cases.

The first possibility leads to an $SU(3)$ octet of spin $\frac{1}{2}^+$ and an $SU(3)$ singlet of spin $\frac{3}{2}^+$, the former being the baryon octet and the latter being so far unobserved—which need not be surprising since it is strange. Thus on this model the N^* for example is regarded as a πN bound state—rather than as a CDD pole in πN scattering as would be the case in the long range model. The model, however, has the disadvantage that it fails to explain the successes of $SU(6)$ theory, in particular the prediction of the ratio of the proton and neutron magnetic moments (see section 10). This is not necessarily an obstacle to this model since there is no reason why the representation (or even the symmetry) associated with phenomena at the level of quark dynamics ($\gtrsim$ 10 BeV) should be the same as that at the level of present day energies. In fact, there already exists a very familiar illustration of this: in Wigner's original $SU(4)$ classification, which was appropriate to nuclear energy levels (a few MeV), the nucleons were in the lowest (spinor) representation of $SU(4)$, whereas in elementary particle physics they are in the (three index) **20** representation. It would clearly be incorrect, at the level of nuclear physics, to employ the latter representation, since it would not be reasonable to start with an approximation in which the N^* and N have the same mass.

If we abandon Fermi statistics for the quarks, or if we assume that there are three types of quarks distinguished by some as yet unobserved quantum number, then we can again take the **56** representation of $SU(6)$ as the lowest three quark state. This restores the usual $SU(6)$ scheme, and means that the baryon octet and decuplet are CDD poles in non-quark physics. The higher baryon resonances are meson-baryon states in the usual way. Greenberg[40] has discussed the use of parastatistics in quark models.

If the short range quark model has anything to do with physics then one might expect to see the higher *l* states in sufficiently high energy experiments (e.g. energies $\gtrsim$ 10 BeV). These states might show up as structure in, for example, πN scattering in energy regions where on present ideas structure would be absent. One could conceivably find evidence for such states in the failure of forward dispersion relations to be compatible with experiment on the assumption of smooth extrapolations—a possible discrepancy has already been suggested.[41] Alternatively one might see the effects directly in total cross-section curves (see Lautrup, Nielson and Olesen,[42] for example).

It would be interesting in the above context to know the couplings of the quark states with higher *l* values (if these exist). On the basis of the calculation in section 6 the effect of the *mass* of the state would be to make

$$\frac{G_{\rho'\pi\pi}}{G_{\rho\pi\pi}} \sim \frac{1}{100}$$

where ρ' is a supposed quark-antiquark state of mass around $2M_q$. This suggests that the higher mass states might be coupled much more weakly than those of lower mass.

An interesting prediction of the short range model is that the slopes of the Regge trajectories associated with the π and ρ nonets and the baryon octet (and possibly the baryon decuplet) should be very much smaller than the slopes of other trajectories; in fact we would expect

$$\frac{d\alpha}{ds} \lesssim \frac{1}{100} (\text{BeV})^{-2}.$$

There is already some experimental evidence which seems to contradict this, but the general state of Regge-pole fits to high energy scattering is such that this cannot be taken as conclusive.

10

Further Possible Predictions of Quark Models

If the objections to quark models are ignored, and we treat them in the simplest possible way, then they permit a big variety of predictions. It is not too easy to test the validity of many of these predictions since most of the required data is not available (or likely to be for some time in many cases). A further problem is that it is often not clear whether one is really testing the quark model since some of the predictions can be obtained from a symmetry assumption.

First we consider the magnetic moments of protons and neutrons.[43] In addition to the assumption that the hadron wavefunctions are purely the quarks, we need three assumptions:

1. The magnetic moments of nucleons are sums of the magnetic moments of the constituent quarks.
2. The spin-unitary spin part of the wavefunction is symmetrical, and the orbital part has total angular momentum zero.
3. The quarks have Bohr magnetic moments proportional to their charges.

The first of these assumptions requires the neglect of exchange moments and assumes that the quarks inside the nucleon behave as free quarks. It allows us to write

$$\mu_p = \langle p \mid \mu_z^{op} \mid p \rangle \qquad \text{...[10.1]}$$

$$\mu_n = \langle n \mid \mu_z^{op} \mid n \rangle \qquad \text{...[10.2]}$$

where μ_z^{op} is the z-component of the quark magnetic moment operator, and the states $| p\rangle$, $| n\rangle$ are proton and neutron states, in the z-direction, expressed in terms of their constituent quarks.

Assumption 2 would follow from the $SU(6)$ classification scheme for baryons. Since the converse is not true it is a *weaker* assumption—though any failure of $SU(6)$ would almost certainly spoil the purity of the wave-function. The proton and neutron states can then be written as

$$| p\rangle = NA\alpha_1(\alpha_2\beta_3-\alpha_3\beta_2)p'_1(p'_2n'_3-p'_3n'_2) \qquad ...[10.3]$$

$$| n\rangle = NA\alpha_1(\alpha_2\beta_3-\alpha_3\beta_2)n'_1(n'_2p'_3-n'_3p'_2) \qquad ...[10.4]$$

where N is a normalization constant, A antisymmetrizes between the three quarks 1, 2 and 3, and α, β refer to the spin state in the z–direction.

The third of the above assumptions would follow from $SU(3)$ if the charge and magnetic moment operators were assumed to transform in the same way under $SU(3)$. Note that we could modify assumptions 1 and 3 by allowing the quarks inside the nucleon to behave differently from free quarks but still independently, and then making assumption 3 apply to these "quasi-free" quarks. With this assumption we can put

$$\mu_z^{op} = \frac{e\hbar}{2M_q c} g(\tfrac{2}{3}\sigma_{p'z}-\tfrac{1}{3}\sigma_{n'z})$$

where $\sigma_{p'z}$ and $\sigma_{n'z}$ are the quark spin operators in the z direction.

Combining assumptions 1, 2 and 3 we readily obtain

$$\mu_p = \frac{e\hbar}{2M_q c} g \qquad ...[10.5]$$

$$\mu_n = -\tfrac{2}{3}\frac{e\hbar}{2M_q c} g, \qquad ...[10.6]$$

from which the ratio $\mu_p/\mu_n = -\tfrac{3}{2}$ follows. The exact value of this ratio ($-2{\cdot}79/1{\cdot}91$) is remarkably close to this prediction. The theoretical prediction was first obtained by Bég, Lee and Pais,[44] from the assumptions of $SU(6)$ invariance and a photon transforming like a member of the **35**.

Note that the absolute value of the magnetic moments requires the g-factor for the quarks to be large—this is because of the large mass of the quarks. Barton (private communication) has pointed out that since the origin of anomalous magnetic moments lies in the meson field surrounding the baryon or quark there is no reason why the baryon or quark mass should provide the scale, and that anomalous moments of the same *absolute value* would be expected for nucleons and quarks. In any case the fact that the moments are so anomalous (i.e. so far from the Dirac moments) means that the apparent success of the model, in view of assumptions 1 and 3, is surprising.

Becchi and Morpurgo[45] have used the above values of the quark magnetic moments to predict the transition rates for the pure $M1$ transitions,

$$V \rightarrow P+\gamma$$

where V is a vector meson and P a pseudoscalar meson. There are some uncertainties due to the relativistic effects which cannot be included unambiguously in the model employed,‡ and there is not much experimental data. However, the agreement is not discouraging and should certainly stimulate further measurements of these decays. Similar remarks hold for the calculations on $NN^*\gamma$ of Moorhouse.[46]

Note that the above prediction would not hold if we adopted the short range quark model without parastatistics (or an unobserved extra quantum number), since the quarks would then be in the **20** state of $SU(6)$ and a completely false result would emerge. One would then be forced to abandon assumption 1 above.

Rubinstein[44] has calculated electromagnetic mass differences on the assumption that these have the form:

$$\delta M = M - M_0 = \sum_i \delta m_i + \sum_{i \neq j} D_{ij}$$

where M is the mass of a hadron, M_0 its mass when electromagnetic effects are turned off, δm_i is the mass shift of the i-th quark or antiquark due to electromagnetic effects. The baryon octet and decouplet mass-shifts then depend upon 9 parameters (3 δm's and 6 D's), and a similar set apply to the mesons. Note that one does not assume any relation between the D's for the mesons and those for the baryons, and therefore one does not relate meson and baryon mass splittings. Several well known results follow immediately, e.g.

$$\Sigma^+ - \Sigma^- + \Xi^- - \Xi^0 = p - n \qquad \ldots[10.7]$$

$$\Xi^{*-} - \Xi^{*0} = \Xi^- - \Xi^0 \qquad \ldots[10.8]$$

$$\pi^+ - \pi^0 = \rho^+ - \rho^0 \qquad \ldots[10.9]$$

$$K^+ - K^0 = K^{*+} - K^{*0}. \qquad \ldots[10.10]$$

The first of these relations is the Coleman-Glashow relation, and is well satisfied, as is the second. There is inadequate data to test the meson relations which, in fact, would not hold if the D's were allowed to depend on the spin orientations of the quarks as would be the case for magnetic effects.

Federman, Rubinstein and Tulmi[48] and Rubinstein[49] have studied mass splittings due to $SU(3)$ violating effects in a similar spirit; we refer to the original papers for details.

Although these attempts at fitting mass splittings by quark models appear to be successful, they are hard to justify in view of the results mentioned in section 6.

A further class of predictions of quark models concerns scattering cross-sections. At high energies it may be reasonable to assume that a hadron-hadron scattering amplitude is essentially the sum of the individual quark-

‡ The method used in the first part of section 7 should overcome these ambiguities.

quark and quark-antiquark amplitudes.[50-52] This leads to many predictions. For example, in the quark model the proton state consists of the three non-strange quarks $(p'p'n')$ and the π^+ state of the quark-antiquark pair $(p'\bar{n})$—where p' and n' are the quarks with $I_3 = \pm\frac{1}{2}$ respectively. Thus the forward π^+p elastic scattering amplitude is given by

$$\langle \pi^+p \mid \pi^+p\rangle = 2\langle p'p' \mid p'p'\rangle + \langle p'n' \mid p'n'\rangle + 2\langle \bar{n}'p' \mid \bar{n}'p'\rangle + \langle \bar{n}'n' \mid \bar{n}'n'\rangle \qquad \ldots[10.11]$$

where the terms on the right represent forward scattering amplitudes for the quarks. Similarly the forward pp elastic scattering amplitude is given by

$$\langle pp \mid pp\rangle = 4\langle p'p' \mid p'p'\rangle + 2\langle p'n' \mid p'n'\rangle + 2\langle n'p' \mid n'p'\rangle + \langle n'n' \mid n'n'\rangle \qquad \ldots[10.12]$$

If we use the Pomeranchuk theorem to relate $q\bar{q}$ amplitudes to qq amplitudes. use isotopic spin invariance for the quarks, and assume that at high energy charge exchange scattering becomes negligible, then all the amplitudes on the right of the above equations become equal so we obtain

$$\frac{\langle pp \mid pp\rangle}{\langle \pi^+p \mid \pi^+p\rangle} = \frac{3}{2}. \qquad \ldots[10.13]$$

We assume that the forward amplitudes are purely imaginary and use the optical theorem so that this result implies

$$\frac{\sigma^{\text{Total}}(pp)}{\sigma^{\text{Total}}(\pi^+p)} = \frac{3}{2} \qquad \ldots[10.14]$$

a result which agrees remarkably well with the experimental data.

Several other results of a similar nature can be obtained[50, 52] but these require $SU(3)$ symmetry for the quarks, and can also be obtained directly from higher symmetries, e.g. the Johnson-Treiman relations.[53] Lipkin[54] gives a good discussion of the relation between the assumptions made and the results obtained.

Kokkedee and Van Hove[52] consider non-forward amplitudes. For the elastic scattering amplitude of particles A and B they write

$$T_{AB}(s, t) = \sum_{ij} f_i^A(t) f_j^B(t)\, T_{ij}(s_{ij}, t) \qquad \ldots[10.15]$$

where s, t are the usual invariant energy and momentum transfer variables, s_{ij} is the corresponding energy for the collision between the i-th quark and j-th quark, T_{ij} being the appropriate scattering amplitude, the sums over i and j summing over the quarks contained in particles A and B respectively, and $f_i^A(t)$ being the probability amplitude that the i-th quark of A receives a momentum transfer $(-t)$ and A particle remains in its ground state (in non-relativistic theory this is essentially the Fourier transform of the quark wave-function). At very high energies it is probably reasonable to replace $T_{ij}(s_{ij}, t)$

by a single function $ig(t)$, independent of s, where $g(t)$ is likely to be purely real. Then

$$T_{AB}(s, t) = ig(t)\left[\sum_i f_i^A(t)\right]\left[\sum_j f_j^B(t)\right]. \qquad ...[10.16]$$

It is interesting that this immediately gives the factorization property characteristic of Regge-pole theory.[55, 56]

If we assume that the variation of $f_i^A(t)$ and $f_j^B(t)$ with t is negligible—as would certainly be the case in the short range model—then the t dependence of the diffraction peak is given by the factor $g(t)$, which is common to all particles. This would explain the observed agreement in shape between the πN and NN diffraction peaks at high energy. The form factor of the quark on this model, $g(t)$, is the same as that of the nucleon so the quark is effectively as large as a nucleon but about 9 times more transparent (see [*10.12*])—this appears to justify the neglect of multiple scattering which is an essential feature of this model.

Because the complete wavefunction of all the 6 quarks in, for example pp scattering should be antisymmetrical, some structure in the scattering amplitude should occur because of exchanges between the quarks between the two protons. Such effects certainly occur in the scattering of light nuclei; see, for example, Squires, Hodgson and Forest[57] where results for He^3-He^4 scattering are reported and compared with experiment. On the basis of the long range quark model one would expect to see these structure effects at scattering energies of a few hundred MeV. They do not seem to occur experimentally.

Alexander, Lipkin and Scheck[58] have extended these ideas to certain high energy reactions. Their predictions assume the quark model for mesons but make no assumption about baryons, and they write for a transition matrix element $\langle MB \mid M'B'\rangle$

$$\begin{aligned}\langle MB \mid M'B'\rangle &= \langle (q\bar{q})_M B \mid (q\bar{q})_{M'} B'\rangle \\ &= \langle q_M B \mid q_{M'} B'\rangle \delta_{\bar{q}_M \bar{q}_{M'}} + \langle \bar{q}_M B \mid \bar{q}_{M'} B'\rangle \delta_{q_M q_{M'}}.\end{aligned} \qquad ...[10.17]$$

Clearly on this model only one quark can change its state, so that double charge exchange processes and processes in which a meson containing no strange quarks goes into one containing two strange quarks are forbidden. From the latter restriction one obtains equations of the form

$$\frac{\langle \pi^- p \mid \phi n\rangle}{\langle \pi^- p \mid \omega n\rangle} = \frac{\langle \pi^+ p \mid \phi N^{*++}\rangle}{\langle \pi^+ p \mid \omega N^{*++}\rangle} = \frac{\cos\alpha - \sqrt{2}\sin\alpha}{\sin\alpha + \sqrt{2}\cos\alpha} \qquad ...[10.18]$$

and

$$\frac{\langle \pi^- p \mid \eta n\rangle}{\langle \pi^- p \mid X^0 n\rangle} = \frac{\langle \pi^+ p \mid \eta N^{*++}\rangle}{\langle \pi^+ p \mid X^0 N^{*++}\rangle} = \frac{\cos\beta - \sqrt{2}\sin\beta}{\sin\beta + \sqrt{2}\cos\beta} \qquad ...[10.19]$$

where α and β are the mixing angles for the 1^- and 0^- nonets, i.e.

$$\phi = \omega_8 \cos\alpha - \omega_1 \sin\alpha \qquad \text{...[10.20]}$$

$$\omega = \omega_8 \sin\alpha + \omega_1 \cos\alpha \qquad \text{...[10.21]}$$

$$\eta = \eta_8 \cos\beta - \eta_1 \sin\beta \qquad \text{...[10.22]}$$

$$X^0 = \eta_8 \sin\beta + \eta_1 \cos\beta \qquad \text{...[10.23]}$$

The relations in [*10.18*] appear to be in excellent agreement with experiment, i.e. they agree with the known result that ϕ production is very small. The relations for the pseudoscalar mesons cannot at this stage be usefully compared with experiment, owing to lack of data.

In general the predictions of [*10.17*] are difficult to compare with experiment because of the mass deviations from $SU(3)$. Corrections have to be applied for phase space factors and also for the effects of the form factors ($f(t)$ functions in equation [*10.15*]) since the comparisons are not made at the same value of t.

Even with these corrections certain predictions turn out to be badly in violation of experiment. For example the result

$$\langle K^- p \mid \omega\Lambda \rangle = \langle K^- p \mid \rho^0 \Lambda \rangle \qquad \text{...[10.24]}$$

which is obtained from [*10.17*] is badly violated, since ω production is much greater than ρ production. The authors suggest that the explanation for this lies in the fact that ω production is here largely isotropic and that there is some mechanism for this which overwhelms the peripheral (forward-peaked) contribution which might be correctly explained by this model. The precise reason for this large isotropic component is unclear (on any model!).

Rubinstein and Stern[59] have considered $N\bar{N}$ annihilation on the quark model. These assume that $q\bar{q}$ pairs are not likely to annihilate or be produced, so they immediately obtain the experimentally observed small probability of $N\bar{N} \rightarrow 2$ mesons. Other results seem in reasonable agreement with experiment.

Ahmadzadeh[60] has used a Regge pole model of qq and $q\bar{q}$ scattering to reduce the number of parameters. He obtains good results on elastic hadron-hadron scattering.

Cabibbo, Horowitz and Ne'eman[61] have suggested that the various high energy results which have been obtained by quark models could be obtained equally by using a Regge pole model and postulating that the Regge residues γ_C^{AB} obey a $U(3) \otimes U(3)$ algebra. Since we originally thought of quarks as a means of "explaining" algebraic results we appear to have returned to our starting point—which makes this a good point to close.

To summarize we can say that there are many predictions which can simply be made on the basis of quark models. A large proportion of these predictions are not susceptible to experimental test, but those that are appear to be quite good—sometimes very good. Whether this is because quark models are physically "true", whether the successes are just accidents, or whether they merely reflect unknown symmetries, are questions we leave to the reader.

REFERENCES

1. M. GELL-MANN. 1962. *Phys. Rev.* **125**, 1067.
2. Y. NE'EMAN. 1961. *Nucl. Phys.* **26**, 222.
3. S. SAKATA. 1956. *Prog. Theor. Phys.* **16**, 686.
4. M. GELL-MANN. 1964. *Phys. Lett.* **8**, 214.
5. G. ZWEIG. 1964. Unpublished CERN preprint.
6. D. AMATI, H. BACRY, J. NUYTS and J. PRENTKI. 1964. *Nuovo Cim.* **34**, 1732.
7. F. GÜRSEY and L. A. RADICATI. 1964. *Phys. Rev. Lett.* **13**, 173.
8. R. K. ADAIR. *Proceedings of the 1964 Coral Gables Conference.*
9. L. P. LEIPUNER, W. T. CHU, R. S. LARSEN and R. K. ADAIR. 1964. *Phys. Rev. Lett.* **12**, 423.
10. D. R. O. MORRISON. 1964. *Phys. Lett.* **9**, 199.
11. V. HAGOPIAN, W. SELOVE, R. GHRLISH, E. LEBON, R. LANZA, D. RAHM and M. WEBSTER. 1964. *Phys. Rev. Lett.* **13**, 280.
12. W. BLUM, S. BRANDT, V. T. COCCANI, D. CZYZEWSKY, J. DANSYZ, M. JOBES, G. KELLNER, D. MILLER, D. R. O. MORRISON, W. NEALE and J. G. RUSHBROOKE. 1964. *Phys. Rev. Lett.* **13**, 353.
13. G. DOMOKOS and T. FULTON. 1966. *Phys. Lett.* **20**, 546.
14. F. CHILTON, D. HORN and R. J. JABBUR. 1966. *Phys. Lett.* **22**, 91.
15. R. K. ADAIR and N. J. PRICE. 1966. *Phys. Rev.* **142**, 844.
16. A. W. SUNYAR, A. Z. SCHWARZSCHILD and P. I. CONNORS. 1964. *Phys. Rev.* **136**, B1157.
17. T. BOWEN, D. A. DE LISE, R. M. KALBACH and L. B. MORTARA. 1964. *Phys. Rev. Lett.* **13**, 728.
18. YA. B. ZELDOVICH, L. B. OKUN and S. B. PIKEL'NER. 1965. *Sov. Phys. JETP* **8**.
19. V. B. BRAGINSKII. 1966. *Zh. Eksp. Teoret. Fiz. Pisma* **3**, 69.
20. BECCHI, G. GALLINARO and G. MORPURGO. 1965. *Nuovo Cim.* **39**, 409.
21. W. A. CHUPKA, J. P. SCHIFFER and C. M. STEVENS. 1966. *Phys. Rev. Lett.* **17**, 60.
22. G. CHEW. 1962. *Phys. Rev. Lett.* **9**, 233.
23. V. SINGH and B. M. UDGAONKAR. 1963. *Phys. Rev.* **130**, 177.
24. A. WAZED. 1965. *Nuovo Cim.* **39**, 292.
25. M. NIETO. 1966. *Proceedings of the Meeting of the American Physical Society*, Washington, 381, and *Phys. Rev. Lett.* **17**, 43.
26. E. J. SQUIRES and P. J. S. WATSON. 1967. Durham preprint, to be published in *Annals of Physics*.
27. E. J. SQUIRES. 1964. *Nuovo Cim.* **34**, 1751.
28. E. J. SQUIRES and P. J. S. WATSON. 1966. *Nuovo Cim.* **42**, 77.
29. M. BANDER, P. W. COULTER and G. L. SHAW. 1965. *Phys. Rev. Lett.* **14**, 270.
30. D. ATKINSON, K. DIETZ and D. MORGAN. 1966. *Ann. Phys.* (N.Y.) **37**, 77.
31. J. B. HARTLE and C. E. JONES. 1965. *Phys. Rev.* **140**, *B*90.
32. M. B. HALPERN and D. ATKINSON. 1966. *Phys. Rev.* **150**, 1377.
33. R. F. DASHEN and S. C. FRAUTSCHI. 1965. *Phys. Rev.* **135**, 1190.
34. A. N. MITRA. 1966. *Phys. Rev.* **142**, 1119.
35. A. HENDRY, private communication.
36. R. H. DALITZ. 1965. *Proceedings of the Oxford International Conference in Elementary Particles*, 157.
37. O. W. GREENBERG. 1966. *Phys. Rev.* **147**, 1077.
38. G. MORPURGO. 1966. *Phys. Lett.* **22**, 214.
39. R. DELBOURGO, A. SALAM and J. STRATHDEE. 1966. "On the Dynamical Origin of $SU(3)$", Trieste preprint IC/66/41.

40. O. W. Greenberg. 1965. *Phys. Rev. Lett.* **13**, 598.
41. P. Olesen. 1965. "On the comparison of forward pion-nucleon dispersion relations with experiments", Niels Bohr Institute (Copenhagen) preprint.
42. B. Lautrup, P. M. Nielson and P. Olesen. 1965. *Phys. Lett.* **19**, 422 and *Phys. Rev.* **140**, B984.
43. G. Morpurgo. 1965. *Physics* **2**, 95.
44. M. A. Bég, B. W. Lee and A. Pais. 1964. *Phys. Rev. Lett.* **13**, 514.
45. C. Becchi and G. Morpurgo. 1965. *Phys. Rev.* **140**, 687, and *Phys. Lett.* **17**, 352.
46. R. G. Moorhouse. 1966. *Phys. Rev. Lett.* **16**, 771 and 968.
47. H. R. Rubinstein. 1966. *Phys. Rev. Lett.* **17**, 41.
48. P. Federmann, H. R. Rubinstein and I. Talmi. 1966. *Phys. Lett.* **22**, 208.
49. H. R. Rubinstein. 1966. *Phys. Lett.* **22**, 210.
50. E. M. Levin and L. L. Frankfurt. 1965. *JETP Lett.* **2**, 65.
51. H. J. Lipkin and F. Scheck. 1966. *Phys. Rev. Lett.* **16**, 71.
52. J. J. J. Kokkedee and L. Van Hove. 1966. *Nuovo Cim.* **42**, 1827.
53. K. Johnson and S. B. Treiman. 1965. *Phys. Rev. Lett.* **14**, 189.
54. H. J. Lipkin. 1966. *Phys. Rev. Lett.* **16**, 1015.
55. M. Gell-Mann. 1962. *Phys. Rev. Lett.* **8**, 263.
56. V. N. Gribov. 1962. *Sov. Phys. JETP* **15**, 873.
57. E. J. Squires, A. E. Forest and P. E. Hodgson. 1963. *Nucl. Phys.* **42**, 490.
58. G. Alexander, H. J. Lipkin and F. Scheck. 1966. *Phys. Rev. Lett.* **17**, 412.
59. H. R. Rubinstein and H. Stern. 1966. *Phys. Lett.* **21**, 477.
60. A. Ahmadzadeh. 1966. *Phys. Lett.* **22**, 96.
61. N. Cabibbo, L. Horowitz and Y. Ne'eman. 1966. *Phys. Lett.* **22**, 336.